Landscape Ecology: Concepts, Methods and Applications

Landscape Ecology: Concepts, Methods and Applications

Editor: Alex Vedder

Callisto Reference

www.callistoreference.com

Callisto Reference,
118-35 Queens Blvd., Suite 400,
Forest Hills, NY 11375, USA

Visit us on the World Wide Web at:
www.callistoreference.com

ISBN: 978-1-63239-837-6 (Hardback)

The publisher's policy is to use permanent paper from mills that operate a sustainable forestry policy. Furthermore, the publisher ensures that the text paper and cover boards used have met acceptable environmental accreditation standards.

Printed in China.

Cataloging-in-publication Data

Landscape ecology : concepts, methods and applications / edited by Alex Vedder.
p. cm.
Includes bibliographical references and index.
ISBN 978-1-63239-837-6
1. Landscape ecology. 2. Biodiversity. 3. Ecology. I. Vedder, Alex.
QH541.15.L35 L36 2017
577--dc23

Table of Contents

Preface

Landscape ecology is concerned with the interactions between ecological processes and ecosystems. The important studies included in this area are land use, scaling, landscape mosaics, land cover change, landscape pattern analysis, landscape conservation, etc. This field uses the techniques of spatial ecology, landscape patterns, etc. to understand ecological processes. This book traces the progress of this field and highlights some of its key concepts and applications. It brings forth some of the most innovative theories and elucidates the unexplored aspects of landscape ecology. For all readers who are interested in this subject, the case studies included in this text will serve as excellent guide to develop a comprehensive understanding. It will serve as a valuable source of reference for graduate and post-graduate students. This book includes contributions of experts and scientists, which will provide innovative insights into this field.

This book has been the outcome of endless efforts put in by authors and researchers on various issues and topics within the field. The book is a comprehensive collection of significant researches that are addressed in a variety of chapters. It will surely enhance the knowledge of the field among readers across the globe.

It gives us an immense pleasure to thank our researchers and authors for their efforts to submit their piece of writing before the deadlines. Finally in the end, I would like to thank my family and colleagues who have been a great source of inspiration and support.

Editor

1

Spatial Ecological Processes and Local Factors Predict the Distribution and Abundance of Spawning by Steelhead (*Oncorhynchus mykiss*) across a Complex Riverscape

Jeffrey A. Falke[1*¤], Jason B. Dunham[2], Christopher E. Jordan[3], Kristina M. McNyset[1], Gordon H. Reeves[4]

1 Department of Fisheries and Wildlife, Oregon State University, Corvallis, Oregon, United States of America, **2** Forest and Rangeland Ecosystems Science Center, U.S. Geological Survey, Corvallis, Oregon, United States of America, **3** Northwest Fisheries Science Center, National Oceanographic and Atmospheric Agency, Seattle, Washington, United States of America, **4** Pacific Northwest Research Station, United States Forest Service, Corvallis, Oregon, United States of America

Abstract

Processes that influence habitat selection in landscapes involve the interaction of habitat composition and configuration and are particularly important for species with complex life cycles. We assessed the relative influence of landscape spatial processes and local habitat characteristics on patterns in the distribution and abundance of spawning steelhead (*Oncorhynchus mykiss*), a threatened salmonid fish, across ~15,000 stream km in the John Day River basin, Oregon, USA. We used hurdle regression and a multi-model information theoretic approach to identify the relative importance of covariates representing key aspects of the steelhead life cycle (e.g., site access, spawning habitat quality, juvenile survival) at two spatial scales: within 2-km long survey reaches (local sites) and ecological neighborhoods (5 km) surrounding the local sites. Based on Akaike's Information Criterion, models that included covariates describing ecological neighborhoods provided the best description of the distribution and abundance of steelhead spawning given the data. Among these covariates, our representation of offspring survival (growing-season-degree-days, °C) had the strongest effect size (7x) relative to other predictors. Predictive performances of model-averaged composite and neighborhood-only models were better than a site-only model based on both occurrence (percentage of sites correctly classified = 0.80±0.03 SD, 0.78±0.02 vs. 0.62±0.05, respectively) and counts (root mean square error = 3.37, 3.93 vs. 5.57, respectively). The importance of both temperature and stream flow for steelhead spawning suggest this species may be highly sensitive to impacts of land and water uses, and to projected climate impacts in the region and that landscape context, complementation, and connectivity will drive how this species responds to future environments.

Editor: Sofia Consuegra, Aberystwyth University, United Kingdom

Funding: J. A. Falke and K. M. McNyset were supported by funds from the National Research Council and the American Recovery and Reinvestment Act through the USDA Forest Service Pacific Northwest Research Laboratory. J. B. Dunham and G. H. Reeves were supported in part by a collaborative research agreement between U.S. Geological Survey, National Climate Change and Wildlife Science Center, and USDA Forest Service. The funders had no role in study design, data collection and analysis, decision to publish, or preparation of the manuscript.

Competing Interests: The authors have declared that no competing interests exist.

* E-mail: Jeffrey.Falke@alaska.edu

¤ Current address: U.S. Geological Survey, Alaska Cooperative Fish and Wildlife Research Unit, University of Alaska Fairbanks, Fairbanks, Alaska, United States of America

Introduction

Understanding habitat selection by an organism across landscapes represents one of the major challenges in contemporary behavioral ecology [1]. Landscape structure influences the behavior of organisms, which in turn generates patterns we observe in the distribution and abundance of individuals in local habitats [2]. Processes that influence habitat selection involve the interaction of landscape composition and configuration (e.g., [3]) across spatial extents encompassed by landscapes, particularly for species with specific habitat requirements and complex life cycles [4]. Common examples include amphibians which require complementary aquatic and terrestrial habitats to complete their life cycles [5], and birds which require distinctive and spatially discrete locations for feeding and breeding [6].

Although stream fish habitats are less commonly framed in the context of spatial ecological processes, many species have complex life histories and complementary habitat requirements [7], [8]. Spatial variability in habitat conditions across stream networks can produce spatially-structured populations [9], [10]. For example, recent work on Pacific salmon (*Oncorhynchus spp.*) in stream networks suggests that spatial patterns in the composition and configuration of habitat, not local habitat quality, drive patterns of spawning habitat use by Chinook salmon (*O. tshawytscha*; [11]). The challenge in considering spatial processes and their influence on fishes in stream networks lies in: 1) identifying the ability of individuals to move across the network relative to the distribution of available habitats, 2) constructing testable hypotheses about specific spatial processes at play based on the species' dispersal ability, habitat needs, and spatial structuring of the stream network, and 3) quantifying local and spatial network processes in a way that allows for testing predictions derived from their hypothesized influences.

The combination of unique habitat requirements for different life stages, localized movement, and habitat selection suggests that salmonids should be strongly influenced by two key spatial processes: landscape complementation and neighborhood effects [3]. Landscape complementation refers to cases where an organism requires unique, non-substitutable, and spatially discrete habitats to complete its life cycle. Neighborhood effects refer to the influence of adjacent habitat conditions on an organism within a given focal location [12]. Because local and landscape features are inextricably linked, examining species distribution and abundance with reference to a spatially continuous representation of the composition and configuration of habitats within the stream network should contribute to a better understanding of the importance of spatial processes.

The well-known habitat requirements of salmon and trout [13] provide an ideal setting for testing predictions about the influence of spatial processes on species' distributions and abundances. For salmonids, landscape complementation is a key spatial process involving use of a sequence of habitats through the life cycle (e.g., from egg to adult). Early life stages often disperse from natal habitats to neighboring locations that may be more suitable for growth and survival during rearing [14]. These localized movements are likely moderated by the dispersal ability of juvenile fishes, whose smaller body size and poor swimming ability, relative to larger adults, may limit their range of movement (i.e., extents ranging from 10^0–10^1 km; [15]). Intraspecific interactions (e.g., territoriality, dominance hierarchies) are strong during this period and may also influence movement [16]. Later in life, (one to several years, depending on species), juveniles can migrate long distances (10^1–$>10^3$ km) to reach complementary feeding habitats (e.g., large rivers, lakes, estuaries, or to the sea). Finally, adults migrate to reproduce, where natal homing and localized habitat selection ultimately determine spawning locations [17].

In this study, we developed continuous representations (i.e., covariates) describing processes hypothesized to influence the life cycle of steelhead (*Oncorhynchus mykiss*), an anadromous (marine-migratory) salmonid in a large river network. We tested whether predictors measured across spatial scales could explain the occurrence and abundance of steelhead redds (nests). The variables we considered were based on three simple requirements for individuals to successfully reproduce at a given location: 1) spawning grounds must be accessible; 2) reproduction needs to be successful; and 3) offspring must survive to continue the cycle [18]. We used geographic information systems (GIS) and count-based regression, combined with an information theoretic approach to: 1) develop spatially continuous representations of local, neighborhood, and landscape spatial ecological processes that may be important to the entire freshwater life cycle of steelhead; 2) evaluate the relative importance of these covariates in simultaneously predicting the distribution (i.e., occurrence) and abundance of redds; and 3) assess the predictive ability of models representing spatial versus local ecological processes. We evaluated our findings in light of the importance of interactions between complex life cycles, spatial ecological processes, and the distribution and abundance of organisms across heterogeneous landscapes.

Materials and Methods

Ethics Statement

Steelhead redd surveys were conducted by the Oregon Department of Fish and Wildlife (ODFW). No specific permits were required as surveys were strictly observational. No interactions with, or handling of, animals were conducted during sampling. Surveys were conducted on public and private lands. Landowner permissions were obtained prior to accessing privately owned stream reaches.

Study System

Our study area was the John Day River basin, located in eastern Oregon, USA (Figure 1). The John Day River is the second longest free flowing (i.e., undammed) river in the conterminous United States, draining approximately 20,000 km^2, with a stream network comprised of over 15,000 stream km. Flow regimes vary across the basin with snow melt in most areas generating peak flows in spring, with base flows strongly augmented by groundwater in some locations [19].

Natural populations of steelhead in the John Day River are currently listed as threatened under the U.S. Endangered Species Act (ESA) owing to population declines caused by habitat alteration, hydropower, interactions with hatchery fish, and overharvest [20]. Steelhead in the John Day River basin consist of five naturally reproducing, genetically distinct spawning populations (hereafter referred to as TRT populations in reference to their designation by Technical Recovery Teams; [21]), which conform approximately to 5th field hydrologic units (Figure 1; [22]).

Field Sampling

Steelhead spawning surveys were conducted as part of ongoing status and trend monitoring by ODFW from 2004–2010. Sites were randomly selected using a generalized random tessellation survey (GRTS) design [23] from a sampling frame comprised of 4,269 stream km thought to support steelhead spawning (ODFW, *unpublished data*; Figure 1). The sampling frame was limited to exclude reaches above known natural and anthropogenic barriers to steelhead access (e.g., gradient barriers, waterfalls, culverts). Management goals were to sample approximately 50 2-km long sites per year. Sites were visited multiple times within the steelhead spawning season (February-June). During each visit, redds were visually identified from alongside the stream based on standardized protocols [24], their locations flagged, and new redd observations were recorded. Therefore, the total number of redds observed during a season at a survey site was the sum of the number of new redds across site visits. We assumed that sites containing more redds were more suitable spawning locations than those with few or no redds based upon the hypothesis that females should select sites with environmental conditions that maximize offspring survival. We took advantage of detailed temporal information within the redd survey data to calculate our response variable: the maximum number of individual steelhead redds observed at a site across years. We felt justified in using this approach for several reasons. First, the occurrence state of redds (i.e., present or absent) was temporally consistent. State changes occurred at only 4 of the 72 sites (6%) for which multi-year data were available. Second, counts were also temporally consistent among years at individual sites, as the average difference between maximum and minimum counts at a site was 2.34 redds (SD = 0.79). Last, we considered the effects of site length (km) and the number of years a site was sampled on redd occurrence and abundance. We found no relationship between site length and maximum redd counts via linear regression (see Results), or between presence and absence of redds and the number of years a site was sampled via logistic regression (see Results).

Covariates

We used the FLoWS version 9.3 toolbox [25] for AcrGIS (ver. 9.3.1, ESRI, Redlands, CA) to create a digital hydrologic network

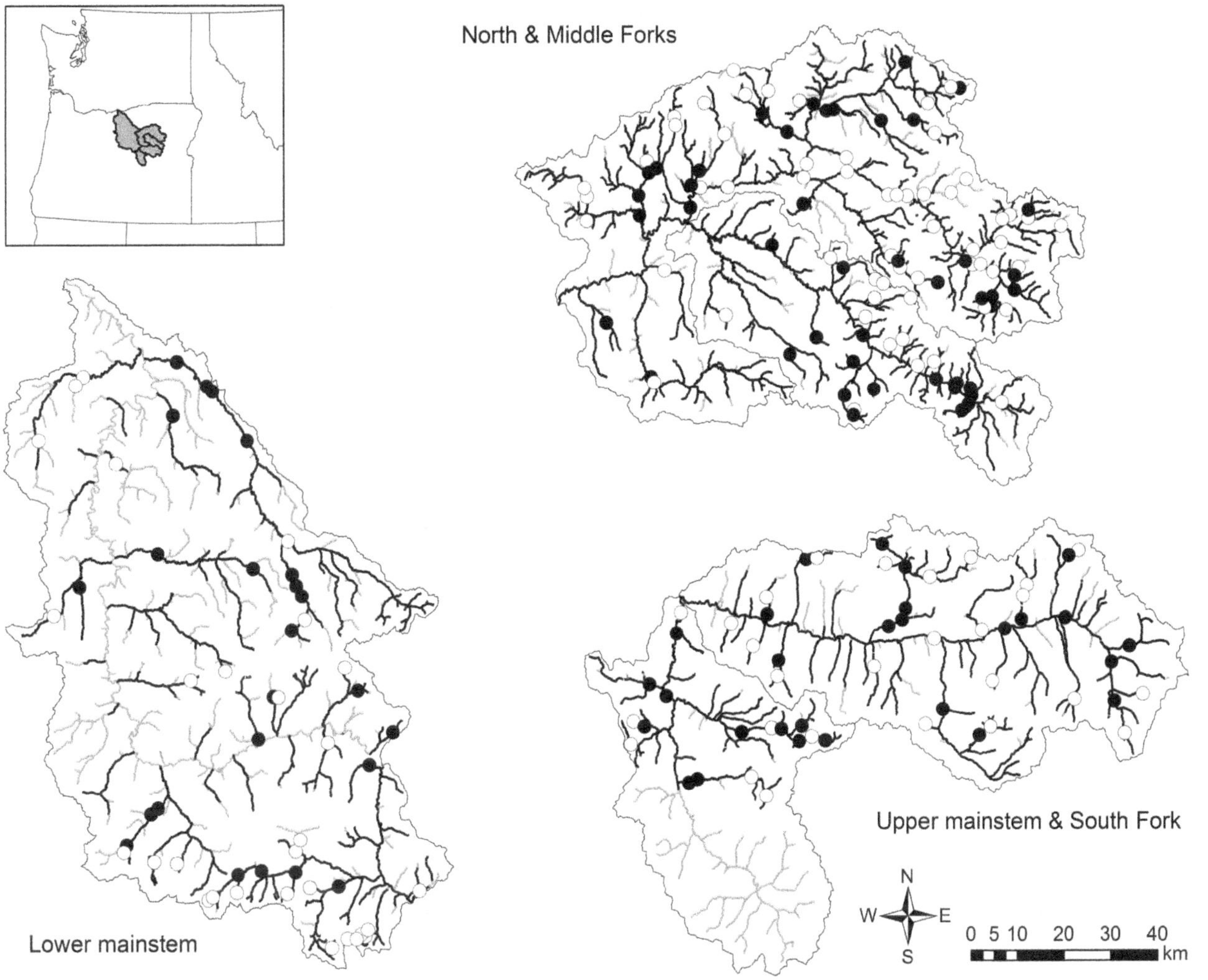

Figure 1. Map of the study area in the John Day River basin, Oregon, USA. Locations of 209 steelhead redd surveys conducted within five catchments (inset) from 2004–2010 are shown (circles). Filled circles are sites where redds were observed; redds were not observed at sites represented by an open circle. Bold stream lines indicate stream reaches that potentially support spawning and early rearing of steelhead (Oregon Department of Fish and Wildlife, *unpublished data*).

for the John Day River basin based on the 1:100000 scale National Hydrography Dataset (http://nhd.usgs.gov/). Values of covariates (described below) were attributed to stream reaches within the network at one of two spatial grains: 1) confluence-to-confluence stream reaches, hereafter referred to as valley segments (VS), or 2) 200-m reaches (200 M) nested within the valley segments. The 200 M reaches were pre-attributed with basic hydrologic and geomorphic information, including elevation (m), floodplain width (m), and drainage area (km^2), which were used to calculate covariates [26].

Covariates were applied to stream reaches at two ecologically-meaningful spatial scales: within each 2-km long redd survey site, and for the "neighborhood" of stream reaches surrounding each site (hereafter referred to as local- and neighborhood-scales, respectively; Figure 2). Neighborhoods were delineated along the stream network using a recursive algorithm that measured the instream distance (km) from the midpoint of each site to every other 200 M reach within a pre-defined distance (A. Brookes, Western Ecology Division, U.S. Environmental Protection Agency, *unpublished data*). These neighborhoods encompassed all surrounding reaches including adjacent reaches and tributaries (Figure 2). We used a 5-km neighborhood size as a conservative estimate of juvenile steelhead dispersal distance [27]. We initially considered neighborhood sizes that encompassed 5 to 30 km, but found little difference in estimates of neighborhood characteristics summarized at the different distances.

We calculated seven covariates to represent key processes in the freshwater life cycle of steelhead that included: adult survival and spawning site accessibility, characteristics of the depositional environment, and juvenile survival. The first covariate we developed was a representation of stream size. Steelhead typically select spawning reaches across a broad range of mid-sized streams [28]. We characterized stream size as mean annual flow (MA; m^3/sec) which was calculated using a variable infiltration capacity (VIC) macroscale hydrologic model developed for the Interior Columbia River basin [29].

Adult salmon and steelhead expend considerable energy while migrating from the marine environment to freshwater spawning locations and this influences adult pre-spawning survival [30]. We estimated the amount of energetic work (WORK) required by spawners to reach spawning sites as the distance from the site (midpoint) to the John Day River basin outlet (km) multiplied by

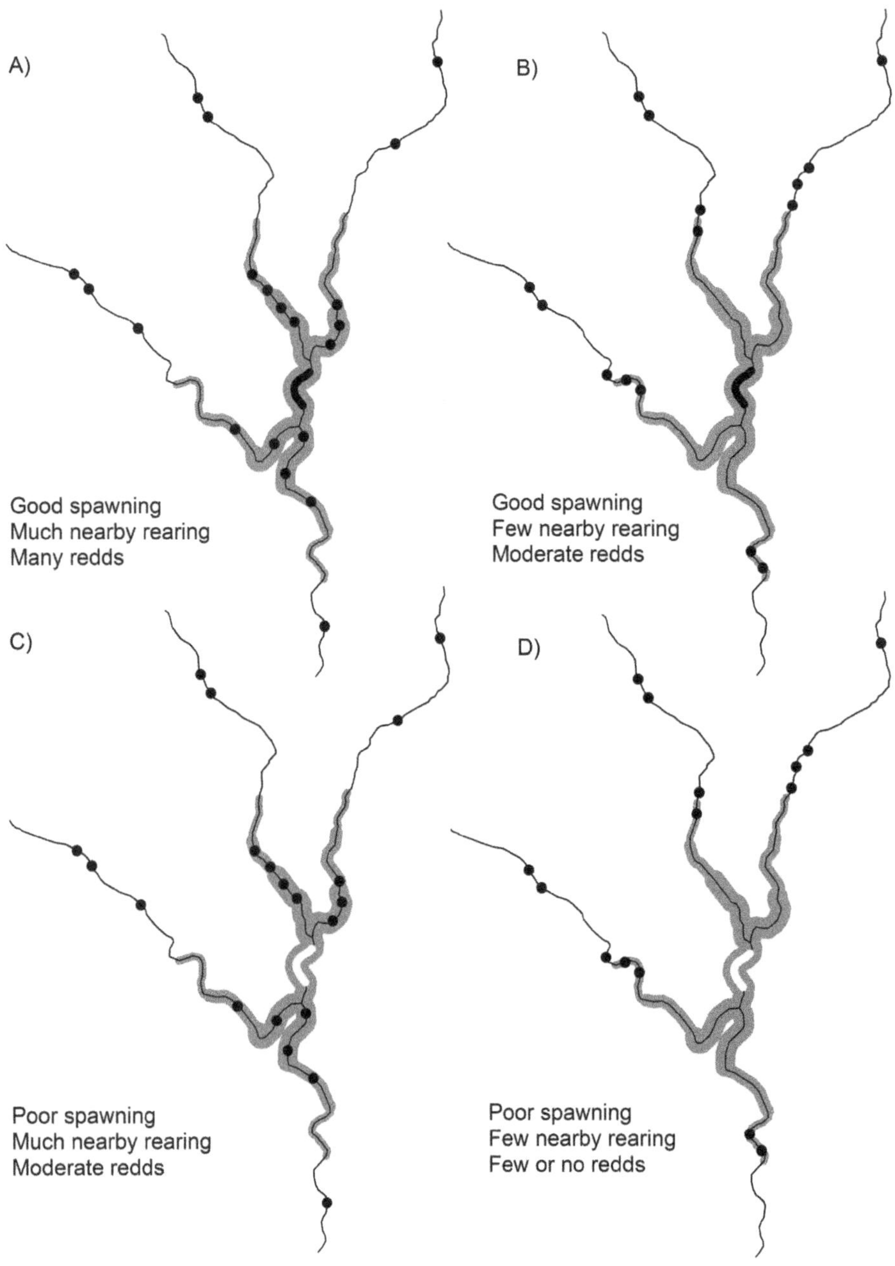

Figure 2. Conceptual model of hypothesized effects of landscape complementation on the occurrence of steelhead redds in stream networks. Likelihood of steelhead redd occurrence is indicated by black (= high likelihood) or white (= low likelihood) fill inside of the focal reach located within the center of each network. The likelihood of steelhead redd occurrence is hypothesized to increase when habitats utilized by juveniles (points) are abundant and located closer (along the stream network) to spawning reaches. Grey shading shows that the probability of juvenile movement exponentially declines with increasing stream distance from their natal reach (i.e., dispersal kernel).

the elevation (m) at the site. In the context of migration, distance represents travel time, and elevation is a surrogate for the average flow velocities encountered en route [31].

Locally, steelhead require suitably sized gravels for successful spawning [32]. However, at the neighborhood-scale the presence of suitable substrate may be correlated with habitat features (e.g., cover, pool depth, food resources) that influence the abundance and survival of offspring. As a result, females may choose spawning sites based not only on gravel size in the immediate area, but also on habitat potential for juvenile rearing within close proximity (Figure 2). In other words, if habitat in the surrounding neighborhood is of high quality a site may support more spawning than expected based only on local conditions. To address this hypothesis, we calculated median grain size (D50) based on methods developed by [33] (see Methods S1 for detailed methods of the D50 calculations). For our local-scale substrate metric ($D50_{SITE}$), we calculated the proportion of reaches within a site that had an estimated D50 within the 25th and 75th percentile of substrate sizes used by steelhead (range 10–48 mm; [32]).

We also hypothesized that substrate characteristics within neighborhoods could influence selection of spawning locations (Figure 2). For our neighborhood metric, we calculated an index that weighted reaches based on their distance from the survey reach and their D50, using the sum of an exponential decay function:

$$D50_{NEB} = \sum_{j \neq 1} (\frac{1}{d_{ij}}) D50_j \times 100 \quad (1)$$

where $D50_j$ is 1 if D50 in reach j is between 10 and 48 mm, and 0 if not, and d is the distance between the midpoint of the survey reach i and the downstream end of reach j. In all cases D50 for reaches where redds were surveyed was excluded from neighborhood estimates.

Scouring flows are a significant source of mortality of eggs and fry through redd scour [34], and downstream displacement [35], respectively. As a result, females should avoid spawning at sites with a greater frequency of spring high flows that lead to a higher likelihood of scouring. We calculated a metric representing the frequency of high flows from Feb 1 to June 30 (S95) based on VIC model estimates [29]. Based on the results of the redd survey data, this period is arguably when redds are most susceptible to scour and fry are most likely to be displaced by high flows. The S95 was a count of the number of days that flows were in the top 5% within the Feb 1 to Jun 30 period.

Water temperature has strong influences on growth and survival of juvenile steelhead [36], and water temperature variability is an important component of aquatic habitat in the John Day River basin [37], [38]. We characterized the thermal regime for all valley segments in the John Day River basin by cumulative growing season degree days (sum of mean daily water temperatures (°C) from Jun 1 to Sep 31; GSDD). Mean daily water temperatures were estimated continuously across the John Day River basin using a riverscape temperature model (K. McNyset *unpublished manuscript*; see Methods S2 for detailed methods of the GSDD calculations). Preliminary analyses suggested that various temperature metrics at local- and neighborhood-scales (e.g., GSDD, mean summer temperature, etc.) were highly correlated (Pearson's r>0.9), so to avoid issues with multicollinearity we included only GSDD in our models. The GSDD were estimated for each VS (i.e., 1–10 km scale), thus we considered it to be a neighborhood-scale predictor.

Finally, we included a categorical predictor (TRT) to identify the sub-population within which each site was located to explore among-population differences in patterns of redd abundance, and to account for residual spatial variation at larger (e.g.,>neighborhood) scales.

Data Analysis

We used the "pscl" library in Program R to develop count-based regression models to predict the occurrence and the maximum number of redds at a site as a function of covariates measured at local and neighborhood scales. Specifically, we used a hurdle count regression model, which is a two-component model with a truncated count component for positive counts and a hurdle component that models the zero counts [39], [40]. The concept underlying the hurdle model is that a binomial probability model governs the binary outcome of whether a count variable has a zero or a positive value. If the value is positive, the "hurdle is crossed," and the conditional distribution of the positive values is governed by a zero-truncated count model. This model structure allowed us to simultaneously model the probability of occurrence and the abundance of redds, and thereby investigate processes that lead to presence and absence of redds, and the number of redds observed at a spawning survey site. The binomial (occurrence) component was modeled using the binomial distribution with a logit link and the count (abundance) component with a negative binomial distribution with logit link.

We assessed the potential for multicollinearity among covariates using the variance inflation factor (VIF) statistic. Covariates with a VIF>10 were removed from analyses [41]. Covariates were log or arcsine-square root transformed as needed to meet normality assumptions of linear models. We also checked for problems associated with influential outliers by visually examining standardized residuals and for spatial dependency in model residuals (i.e., autocorrelation) using an empirical variogram (Figure S3; [42]).

We constructed a set of candidate models to represent biologically relevant combinations of processes measured at the local and neighborhood scales. Our candidate model set was formulated to represent alternative hypothesized effects of these processes on redd abundance, occurrence, or both (Table 1). We also considered a global model that included all terms. We then used an information-theoretic approach to find the most parsimonious set of independent variables to estimate the probability of occurrence and the maximum count of redds at a survey site [43]. We used Akaike's information-criterion (AIC) to select the best approximating models by comparing candidate models. The AIC values were adjusted for small sample size (AIC_c), and Akaike weights (w_i) were calculated. The model with the lowest AIC_c and the highest w_i was considered the best model. To account for model uncertainty, we used model averaging to calculate parameter estimates and variances from models in the confidence model set (w_i>0.05), and made inferences using this composite model.

Model Diagnostics

We evaluated the predictive ability of the occurrence and abundance components of our hurdle model separately to evaluate their suitability for management applications (e.g., survey design optimization, prioritizing habitat restoration). For the occurrence component, we used the "PresenceAbsence" library in Program R to calculate the percentage of sites correctly classified (PCC; cutoff = 0.5) and area under the curve (AUC) statistics. Standard deviations were computed for each metric. As insufficient data were available for an independent evaluation dataset and subsequent cross-validation analysis, we used the "0.632+" bootstrap evaluation method [44], [45] to assess the predictive accuracy of our abundance model. This bootstrapping method uses an optimism estimate to adjust model evaluation statistics and provides a nearly unbiased estimate of the external predictive performance of the model. We used a range of evaluation statistics to assess different aspects of the abundance model predictive performance. Bootstrap estimates were calculated on observed vs. predicted redd counts using Pearson's (r) and Spearman's rank (ρ) correlations, average error, and the root mean square error (RMSE). The average optimism was calculated for each statistic across 200 replicate bootstraps.

Results

Two-hundred and nine sites (Figure 1) were surveyed for steelhead redds across the seven year period (2004–2010). Of the 4,269 stream km within the sampling frame, 411 km (10%) were sampled at least once. Maximum redd counts at a site ranged from 0 to 17 redds (mean = 2.09, SE = 0.25; Figure S4), with redds being observed at 96 of the 209 sites (46%). On average, redd densities along these 2-km stream reaches were low. At the majority of sites where redds were observed, maximum counts were less than 10 redds (87%), and about half of occupied sites contained only 1 or 2 redds (47%). We found no relationship between sampling effort (site length and number of years a site was sampled) on either presence and absence of redds using multiple logistic regression (F = 1.54, P = 0.479) or maximum counts (>0) based on negative binomial count regression (F = 2.18, P = 0.827).

Table 1. Candidate hurdle count regression models used to estimate occurrence and abundance of steelhead redds in the John Day River basin, Oregon.

Model[1]	Scale[2]	Hypothesis
MA^a+$WORK^a$+$S95^b$+$GSDD^{a,b}$+ $D50_{SITE}{}^a$+ $D50_{NEB}{}^a$+TRT^b	Mixture	Global model
MA^a+$WORK^a$+$S95^b$+$GSDD^{a,b}$+ $D50_{SITE}{}^a$+ $D50_{NEB}{}^a$	Mixture	Global model without TRT
$WORK^a$+$GSDD^{a,b}$+ $D50_{NEB}{}^a$+TRT^b	Neighborhood	Neighborhood+TRT
MA^a+ $S95^b$+$D50_{SITE}{}^a$+TRT^b	Site	Site+TRT
$WORK^a$+$GSDD^{a,b}$+ $D50_{NEB}{}^a$	Neighborhood	Neighborhood-only
MA^a+ $S95^b$+$D50_{SITE}{}^a$	Site	Site-only
MA^a+$WORK^a$+TRT^b	Mixture	Adult Survival/Access+TRT
$S95^b$+ $D50_{SITE}{}^a$+ $D50_{NEB}{}^a$+TRT^b	Mixture	Depositional Environment+TRT
$GSDD^{a,b}$+TRT^b	Mixture	Juvenile Survival+TRT
MA^a+$WORK^a$	Mixture	Adult Survival/Access only
$S95^b$+ $D50_{SITE}{}^a$+ $D50_{NEB}{}^a$	Site	Depositional Environment only
$GSDD^{a,b}$	Neighborhood	Juvenile Survival only

[1]Covariates were applied to the occurrence (a) and/or abundance (b) model components.
[2]Models were formulated to address hypotheses at two scales (Site and Neighborhood) and in combinations (Mixture).

As such, we did not account for these terms in our hurdle models. The model that included only neighborhood-scale covariates had the lowest AIC_c and the highest w_i. Considerable model uncertainty existed as the site-only model and models that included the TRT covariate had w_i values>0.05 (Table 2). However, models representing only a single aspect of the steelhead life-cycle, and the global model that included all terms, had essentially no support (w_i <<0.05).

Occurrence of Redds

Results of the composite (i.e., model-averaged) model indicated that the probability of steelhead redd occurrence increased with: 1) stream size; 2) the proportion of suitable D50 within a site; 3) the amount of suitable D50 in close proximity to a site; and 4) growing season degree-days (Figure 3,Table 3). For a given proportion of suitable D50 within a site, the probability of redd occurrence more than doubled when availability of suitable D50 was greater in the surrounding neighborhood (Figure 4). Following model-averaging WORK was not significantly different than zero (i.e., confidence limits overlapped zero).

Table 2. Model selection metrics for hurdle count regression models fit to occurrence and abundance data for steelhead redds at 209 sites in the John Day River basin, Oregon.

Model[1]	K[2]	L-L	AIC_c	ΔAIC_c	w_i
Neighborhood-only	7	−538.70	1091.5	0	0.413
Site-only	6	−539.90	1091.9	0.4	0.347
Neighborhood+TRT	11	−536.01	1094.2	2.7	0.110
Site+Hatchery	10	−537.87	1095.9	4.4	0.047
Global model without TRT	10	−537.87	1095.9	4.4	0.047
Global model	12	−540.76	1105.7	14.2	<0.001

[1]Model results are ranked by AIC_c from best to worst, and Akaike weights ($w_{i,}$)>0.05 are also shown.
[2]*K* is the number of estimated parameters, L-L is the log-likelihood, and ΔAIC_c is the difference in AIC_c relative to the best model (see [41] for details).

Abundance of Redds

The maximum number of redds at a site was predicted to be higher at sites with warmer thermal regimes (e.g., more GSDD), and lower at sites with more days of scouring flows (S95). At a given value of S95, expected redd counts varied across sub-basins even with GSDD held constant. For example, sites in the Lower Mainstem sub-basin close to the Columbia River were predicted to contain twice as many redds relative to sites in the North Fork sub-basin. Visual examination of the empirical semi-variogram (see Supplementary Materials) based on composite model residuals versus in-stream distance (km) indicated that the spatial pattern of the composite model residuals was not different from a random spatial pattern.

Model Performance

Predictive performance of both the occurrence and abundance components of the composite model were generally better than the performance of either the non-averaged neighborhood- or site-only models (Table 4). The composite model PCC was 18% higher than the site-only model, but just 2% higher than the neighborhood-only model. The AUC statistic also was highest for the composite model, but differences among models were low. Redd abundance component evaluation statistics derived from bootstrap analysis indicated that the composite and neighborhood-only models performed best. Correlations between observed and predicted values for both Pearson's and Spearman's rank statistics were highest, and average model error was lowest, for the neighborhood-only model. Root mean square error was lowest for the neighborhood model (Table 4). Overall, the site-only model performed poorly in comparison to models that included neighborhood scale predictors.

Discussion

Our results emphasize the importance of spatial ecological processes that can drive habitat selection and ultimately patterns of presence and abundance at broad extents. More specifically, we found that selection of spawning locations by steelhead across a complex riverscape was strongly associated with habitats thought to provide complementary support to different life stages.

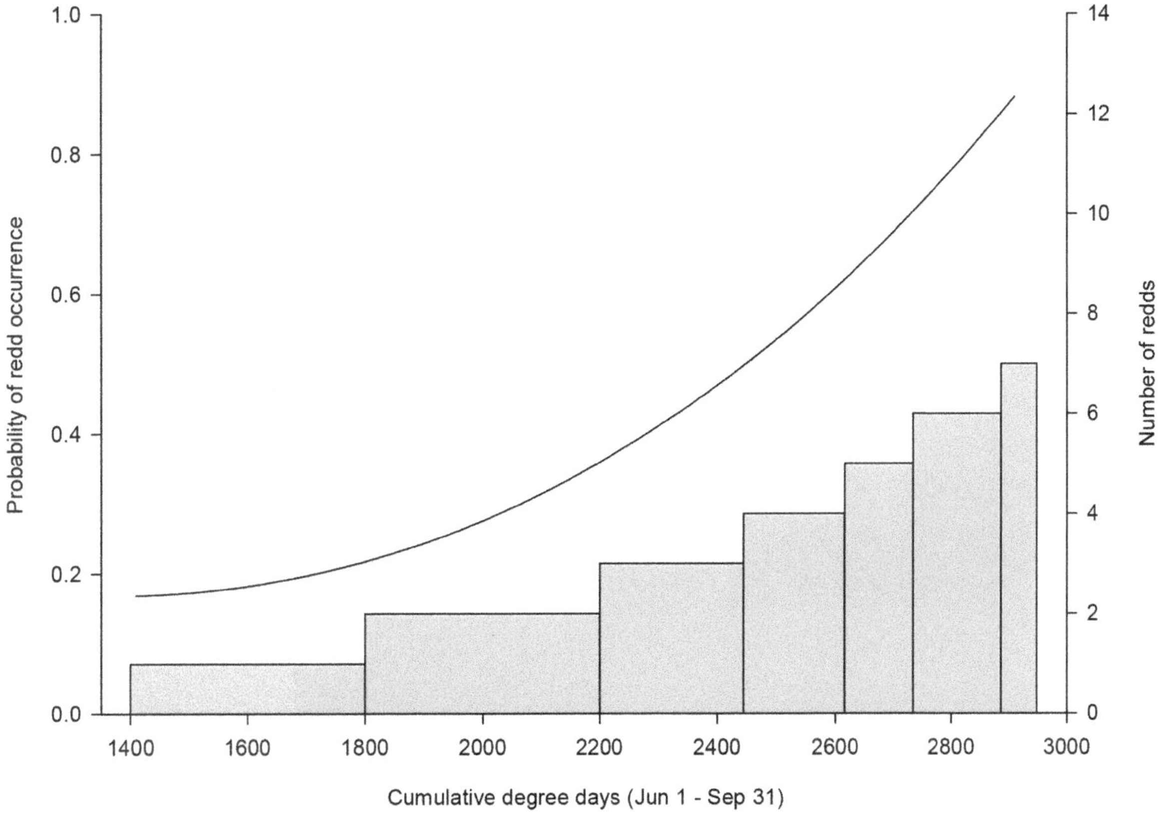

Figure 3. Probability of occurrence (line, left y-axis) and abundance of steelhead redds (bars, right y-axis) as a function of cumulative degree days (GSDD; °C) estimated from a hurdle count regression model. Horizontal thickness of bars indicates the approximate range of degree days predicted for a given count estimate.

Table 3. Standardized model-averaged parameter estimates, unconditional SE values, and 95% confidence limits (CLs) for covariates predicting the occurrence (binomial model) and abundance (count model) of steelhead redds in the John Day River basin, Oregon.

	Covariate	Parameter estimate	SE	Lower 95% CL	Upper 95% CL
Occurrence model[1]	Intercept	−2.848	1.100	−4.993	−0.703
	MA	0.216	0.014	0.189	0.243
	$D50_{SITE}$	0.243	0.019	0.207	0.280
	$D50_{NEB}$	0.279	0.016	0.248	0.309
	GSDD	0.694	0.048	0.600	0.787
	WORK	−0.004	0.007	−0.018	0.010
Abundance model[1,2]	Intercept	0.240	0.021	0.198	0.282
	GSDD	0.683	0.063	0.560	0.806
	S95	−0.047	0.006	−0.058	−0.036
	TRTMF	0.014	0.086	−0.154	0.183
	TRTNF	−0.130	0.023	−0.174	−0.085
	TRTSF	0.000	0.061	−0.118	0.118
	TRTUM	0.029	0.094	−0.154	0.212
	Log(θ)[3]	−0.276	0.356	−0.970	0.419

[1]Results are based on the top five hurdle count regression models, which were responsible for 96% of the collective model weight (see Table 2).
[2]Levels for the TRT covariate are Lower Mainstem = Intercept, Middle Fork = TRTMF, North Fork = TRTNF, South Fork = TRTSF, and Upper Mainstem = TRTUM.
[3]Log(θ) is the dispersion parameter.

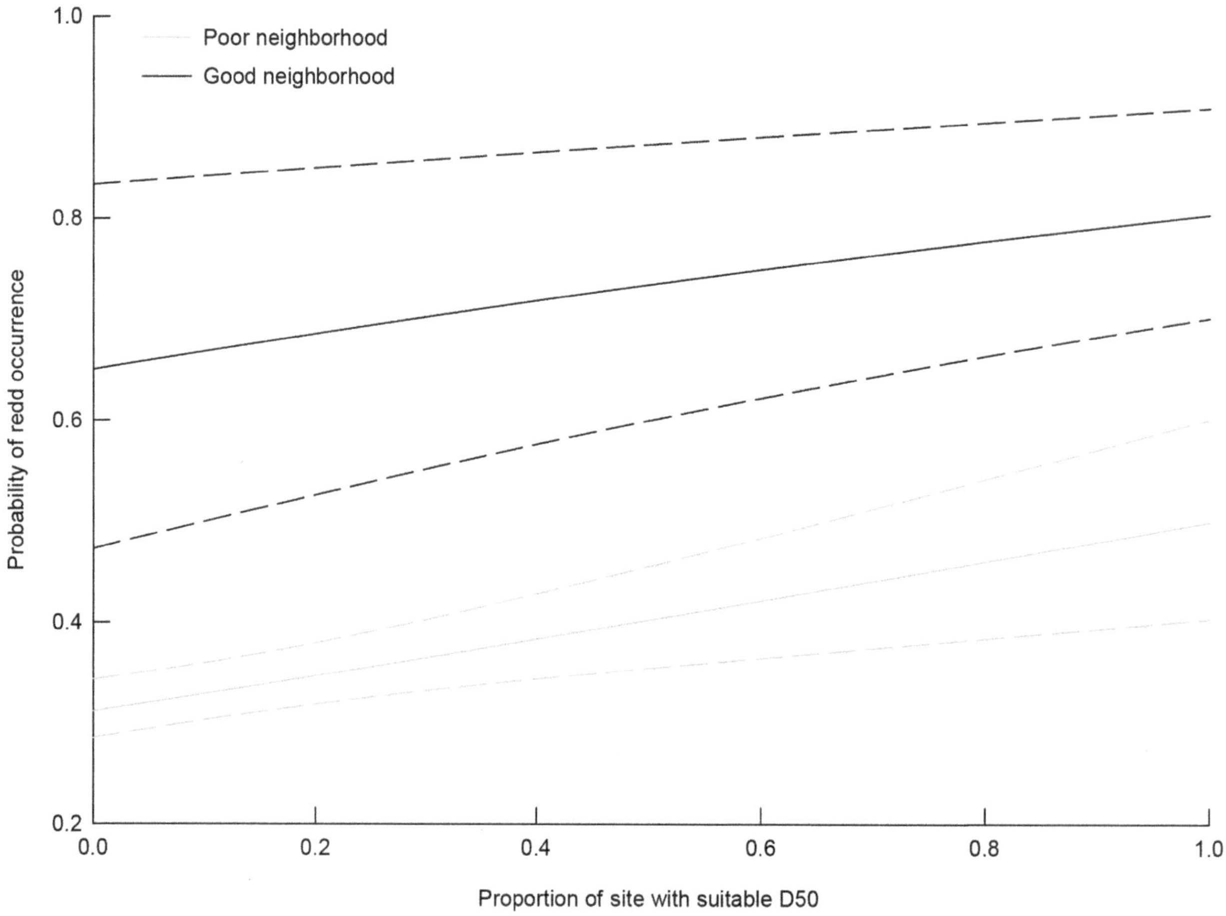

Figure 4. Probability of steelhead redd occurrence as a function of the proportion of a spawning survey site with suitable spawning substrates ($D50_{SITE}$), in two neighborhood types : 1) Good has high amounts of suitable substrate in nearby reaches; and 2) Poor has low amounts. Estimates are from a hurdle count regression model. Dashed lines are 95% confidence intervals.

Furthermore, habitat characteristics at both local and neighborhood extents were important in predicting the distribution and abundance of steelhead redds across a diverse riverscape. Landscape complementation is not often explicitly treated when modeling presence or abundance of a species [46]. We found two benefits to a more explicit treatment of complementation: a better understanding of spatial processes that influence a species, and improved prediction of species presence or abundance. In the case of steelhead, patterns we observed may be explained by both complementation and connectivity. Complementation was addressed specifically in our model by including metrics describing the quality of rearing and spawning locations near focal sites. Connectivity was implicitly factored into our predictive models by including a variable to account for the cost of travel to spawning locations as well as connectivity to reaches within a neighborhood. In other words, the synergistic effect of complementation and connectivity proved to be a powerful predictor of spawning locations and abundance of redds constructed by steelhead.

Table 4. Model prediction diagnostics for three hurdle regression models predicting the occurrence (binomial) and abundance (count) of steelhead redds in the John Day River basin, Oregon.

	Occurrence[1]		Abundance[2]			
Model	**PCC**	**AUC**	***r***	***ρ***	**AVE_{error}**	**RMSE**
Composite	0.80±0.03	0.90±0.02	0.79	0.74	1.17	3.37
Neighborhood-only	0.78±0.02	0.87±0.03	0.82	0.80	0.94	3.93
Site-only	0.62±0.05	0.81±0.04	0.66	0.62	2.44	5.57

[1]For the occurrence component, percent correctly classified (PCC; cutoff = 0.5) and area under the curve (AUC) statistics with standard deviations are presented.
[2]For the abundance component, the results of a "0.632+" bootstrap evaluation of Pearson's *r*, Spearman's *ρ*, average error (AVE_{error}), and root mean square error (RMSE) of observed versus predicted redd counts are shown.

Landscape supplementation was not explicitly addressed in this study though it could also be important in predicting steelhead spawning locations and redd counts. Landscape supplementation is a mechanism by which organisms supplement resource acquisition by making use of identical resources located in nearby patches [3]. For example, a recent analysis of passerine birds showed that abundance was higher in natural locations where agriculturally-modified habitats provided supplementary feeding opportunities [6]. We suspect that landscape supplementation is likely to be important for steelhead as well, for analogous reasons. Landscape supplementation may provide redundancy in availability of key habitat requirements, thereby promoting persistence of steelhead in the face of environmental changes such as seasonal variability in flows and temperatures or episodes of major habitat rearrangements from catastrophic disturbances [47]. In practice it can be challenging to precisely describe and model a specific landscape process (e.g., complementation vs. neighborhood effects) that is clear and distinctive in concept ([3], [48]; Figure 2). Regardless, a more explicit consideration of spatial ecological processes clearly improves our understanding of how species presence and abundance is expressed across broad extents.

The importance of spatial landscape processes to stream fishes has rarely been considered explicitly (but see [49–51]) despite being widely acknowledged [7], [10], [52]. Two explanations may account for this apparent discord. Spatial ecological processes are: 1) well-known, but not expressed within the rubrics of aquatic landscape ecology and 2) viewed inconsistently with respect to their relation to ecological scale (e.g., neighborhoods vs. metapopulations). However, there is little question that spatial landscape processes are implicitly acknowledged for stream fishes. For example, [53] found that the distance among spawning and rearing habitats was a fundamental factor explaining recruitment and population size of fishes in the Danube River. Moreover, the literature includes numerous references to distinctive life stages of stream fishes and their ecological requirements (e.g., [8]). However, more often than not these life stages and their requirements are studied in isolation, and thus connections among them (i.e., landscape complementation; see above discussion) are not clear (but see [10]). On the other hand, the importance of habitat connectivity is well-acknowledged and is the focus of intensive study (see review by [54]). Arguably, combining approaches to develop a more explicit treatment of connectivity should improve our understanding of stream fish ecology as evidenced by the results herein.

Predicting the distribution and abundance of organisms across complex landscapes is challenging as it requires accurately and precisely capturing habitat heterogeneity at scales that are relevant to the life history of the organism in question [55–57]. Past approaches to modeling effects of landscape configuration and composition on salmon and trout have relied on a patch-based approach [9], [11], [58], but relative to species in these studies, *O. mykiss* (i.e., steelhead and rainbow trout) is a habitat generalist that can occupy a wide range of habitat conditions [59]. The broad extent we addressed certainly encompassed multiple local populations of steelhead, but the boundaries delineating these populations were far from clear. Thus, we adopted an approach based on the concept of ecological neighborhoods that were delineated by the presumed movement behavior of steelhead. Unlike a patch-based approach, which is often focused on the importance of habitat size and connectivity, our neighborhoods were constant in size, but variable in their composition and connectivity. Whereas we were unable to evaluate the importance of neighborhood size or extent, we found that simply accounting for neighborhood effects dramatically improved our ability to predict presence and abundance of a generalist species. Given that many stream fishes are generalists [8] and that stream networks represent continuous gradients rather than discrete boundaries [60], [61], the concept of ecological neighborhoods seems widely applicable.

Implications

Based on our rationale, what are some general guidelines for predicting the distribution and abundance of organisms with complex life cycles across large, heterogeneous landscapes? First, when spatial processes such as landscape complementation and neighborhood effects may be influential, an attempt should be made to measure covariates continuously, thus allowing the ecology of the organism to dictate the scale of the response. In an analysis such as the one we have proposed, failure to account for the habitat matrix may lead to biased estimates of population spatial structure [5], [62], [63]. Second, predictors should represent (spatial and behavioral) processes that are of primary importance in determining habitat suitability, not only at specific life stages but across the life cycle of an organism. These measures should incorporate knowledge of how individuals disperse across the landscape, select habitat in which to reproduce, and successfully produce offspring that survive to promote population persistence [64]. For our steelhead example we focused on access to spawning habitats, suitability of spawning gravels and the likelihood of disturbance of redds, and availability of conditions that subsequently support growth of offspring (i.e., thermal regime). We considered these factors to be of primary importance to persistence of steelhead in freshwater. Third, our results suggest that the effects of local habitat conditions should not be ignored. Although models that included local effects did not perform as well in a predictive sense as those that included neighborhood-scale measurements, we could not rule out their importance based upon model selection. We suggest that accounting for both neighborhood and local conditions are important, as based on the VIF analysis local habitat and neighborhood conditions were not good predictors of one another. Future work predicting steelhead spawning distributions could better characterize local-scale conditions by linking specific locations of redds (e.g., GPS coordinates) to individual reaches, thus providing information regarding the spatial arrangement (e.g., clumped vs. dispersed) of suitable reaches within a site. Estimates of substrate suitability at finer scales (e.g., channel units: pools, riffles, runs) might also improve predictive models that incorporate local-scale processes. Unfortunately, continuous mapping of geomorphic controls at very fine spatial grains remains difficult, although new technologies hold promise (e.g., LiDAR; [65]). Last, use of empirical data collected under a spatially balanced survey design, and statistical methodology that allows for the evaluation of alternative hypotheses are critical for an unbiased view of the influence of spatial ecological processes on the distribution and abundance of organisms in complex landscapes.

This study has numerous implications for evaluating and managing future climate impacts on steelhead. Changes in thermal and flow regimes are anticipated with regional climate warming [66]. We found that water temperature and stream flows most strongly influenced the distribution and abundance of steelhead redds. Climate change effects on the distribution of *O. mykiss* have only considered the resident form (i.e., non-anadromous rainbow trout) which appear to be the most resilient trout species with respect to climate change [19]. Whereas *O. mykiss* as a species may be resilient in the face of climate change, implications of climate change for life histories is another question. Expression of a resident (rainbow trout) or migratory (steelhead) pathway has been linked to thermal regime [67] and stream size [28], both of which were important in our models predicting the distribution and abundance of steelhead redds. More work is needed to better understand controls on migratory life history expression in this species, but climate-linked changes in life history expression seem quite plausible (see also [68]). Accordingly, the response of this species to changing climates will require an understanding of not only how climate affects stream flows and temperatures in both local and landscape contexts, but also how these factors may change the species itself.

In an applied context, the results of this work are generally relevant for species with complementary habitat needs, and have specific implications for conservation of steelhead, which is threatened across much of its range [20]. It is clear that presence and abundance of spawning locations were strongly tied to spatial landscape processes, namely connectivity, landscape complementation, and neighborhood effects. Quantifying these processes was possible using a combination of previously derived measures (e.g., VIC model output: MA, S95) and those generated by us (e.g., D50, WORK). The importance of these processes provides a fundamentally different view of why steelhead use specific locations for spawning. In essence, suitable characteristics of sites are a necessary but often insufficient condition for the persistence of steelhead populations. Accordingly, restoration of stream habitats can benefit from this broader perspective [69], [70]. For example, restoration of a specific site for a specific life stage may fail if steelhead are less likely to use the location due to limited connectivity, lack of complementary habitats for other life stages, or unsuitable neighborhood characteristics. In other words, the site itself is only one of several factors that will influence the outcome of a given restoration effort. Currently, there are many active efforts aimed at site-based restoration of salmon and trout habitat, but whether or not these efforts have resulted in positive biological responses is unclear. A major contributing factor is lack of suitable monitoring of these local efforts [71], but failure to consider spatial landscape processes is also a strong possibility [72]. Whereas active restoration at sites can produce short-term results, passive attempts to manage land and water uses at broader extents may be more successful in restoring natural landscape processes and fish populations over the longer term. Such an approach may be particularly important in the John Day River basin where recovery from widespread impacts of historical human activities will require decades [73].

Supporting Information

Figure S1 Empirical variogram [1] of residuals from a hurdle count regression model predicting the distribution and abundance of steelhead redds in the John Day River basin, Oregon, USA. The variogram depicts the semivariance (y-axis) as a function of separation distance (x-axis) for 209 sites with 2.5^{th} and 97.5^{th} percentiles from 5000 permutations (dashed lines).

Figure S2 Frequency distribution of maximum steelhead redd counts in the John Day River basin, Oregon, USA collected from 2004–2010.

Methods S1 Description of D50 and channel type classification methods.

Methods S2 Description of methods for the water temperature model.

Acknowledgments

We gratefully acknowledge assistance with data gathering and analyses provided by A. Brookes, P. Kaufmann, T. Kincaid, A. Przeszlowska, J. Van Sickle, C. Volk, M. Weber, and S. Wenger. Redd survey data were provided by the Oregon Department of Fish and Wildlife, collected under the supervision of J. Ruzycki. Earlier versions of the manuscript benefitted from comments by K. Gido and anonymous reviewers. Use of trade or firm names in this manuscript is for reader information only and does not constitute endorsement of any product or service by the U.S. Government.

Author Contributions

Analyzed the data: JF KM. Contributed reagents/materials/analysis tools: JF JD KM. Wrote the paper: JF JD CJ KM GR.

References

1. Lima SL, Zollner PA (1996) Towards a behavioral ecology of ecological landscapes. Trends Ecol Evol 11: 131–135.
2. Belisle M (2005) Measuring landscape connectivity: The challenge of behavioral landscape ecology. Ecology 86: 1988–1995.
3. Dunning JB, Danielson BJ, Pulliam HR (1992) Ecological processes that affect populations in complex landscapes. Oikos 65: 169–175.
4. Andren H, Delin A, Seiler A (1997) Population response to landscape changes depends on specialization to different landscape elements. Oikos 80: 193–196.
5. Pope SE, Fahrig L, Merriam NG (2000) Landscape complementation and metapopulation effects on leopard frog populations. Ecology 81: 2498–2508.
6. Brotons L, Wolff A, Paulus G, Martin JL (2005) Effect of adjacent agricultural habitat on the distribution of passerines in natural grasslands. Biol Conserv 124: 407–414.
7. Schlosser IJ (1995) Critical landscape attributes that influence fish population-dynamics in headwater streams. Hydrobiologia 303: 71–81.
8. Matthews WJ (1998) Patterns in freshwater fish ecology: Kluwer Academic Pub.
9. Dunham JB, Rieman BE, Peterson JT (2002) Patch-based models to predict species occurrence: Lessons from salmonid fishes in streams. In: Scott JM, Heglund PJ, Morrison ML, Haufler JB, Raphael MG et al., editors. Predicting species occurrences. Issues of accuracy and scale: Island Press. 327–334.
10. Falke JA, Fausch KD (2010) From metapopulations to metacommunities: Linking theory with empirical observations of the spatial population dynamics of stream fishes. In: Gido KB, Jackson DA, editors. Community ecology of stream fishes: Concepts, approaches and techniques. Bethesda, MD: American Fisheries Society. 207–233.
11. Isaak DJ, Thurow RF, Rieman BE, Dunham JB (2007) Chinook salmon use of spawning patches: Relative roles of habitat quality, size, and connectivity. Ecol Appl 17: 352–364.
12. Addicott JF, Aho JM, Antolin MF, Padilla DK, Richardson JS, et al. (1987) Ecological neighborhoods - scaling environmental patterns. Oikos 49: 340–346.
13. Quinn TP (2005) The behavior and ecology of Pacific salmon and trout: University of Washington Press.
14. Kahler TH, Roni P, Quinn TP (2001) Summer movement and growth of juvenile anadromous salmonids in small western Washington streams. Can J Fish Aquat Sci 58: 1947–1956.
15. Rodriguez MA (2002) Restricted movement in stream fish: The paradigm is incomplete, not lost. Ecology 83: 1–13.
16. Reeves GH, Sleeper JD, Lang DW (2011) Seasonal changes in habitat availability and the distribution and abundance of salmonids along a stream gradient from headwaters to mouth in coastal Oregon. Trans Am Fish Soc 140: 537–548.
17. Cury P (1994) Obstinate nature - an ecology of individuals - thoughts on reproductive-behavior and biodiversity. Can J Fish Aquat Sci 51: 1664–1673.
18. Beechie T, Moir H, Pess G (2008) Hierarchical physical controls on salmonid spawning location and timing. In: Sear DA, DeVries P, editors. Salmon spawning habitat in rivers: Physical controls, biological responses, and approaches to remediation. Bethesda, MD: American Fisheries Society. 83–102.
19. Wenger SJ, Isaak DJ, Luce CH, Neville HM, Fausch KD, et al. (2011) Flow regime, temperature, and biotic interactions drive differential declines of trout species under climate change. Proc Natl Acad Sci U S A 108: 14175–14180.
20. Good TP, Waples RD, Adams P (2005) Updated status of federally listed ESUs of West Coast salmon and steelhead. U.S. Department of Commerce, NOAA Technical Memorandum NMFS-NWFSC-66.
21. McClure MM, Holmes EE, Sanderson BL, Jordan CE (2003) A large-scale, multispecies status, assessment: Anadromous salmonids in the Columbia River basin. Ecol Appl 13: 964–989.
22. Seaber PR, Kapinos FP, Knapp GL (1994) Hydrologic unit maps. U.S. Geological Survey Water Supply Paper 2294. United States Government Printing Office, Washington, D.C.
23. Stevens DL, Olsen AR (1999) Spatially restricted surveys over time for aquatic resources. J Agric Biol Envir S 4: 415–428.
24. Susac GL, Jacobs SE (1999) Evaluation of spawning ground surveys for indexing the abundance of adult winter steelhead in Oregon coastal basins. Oregon Department of Fish and Wildlife. Annual Progress Report F145-R-08. Portland, Oregon.

25. Theobald DM, Norman JB, Peterson EE, Ferraz S, Wade A, et al. (2006) Functional linkage of watersheds and streams (FLoWs) user's manual: network-based ArcGIS tools to analyze freshwater ecosystems. Natural Resource Ecology Laboratory, Colorado State University, Fort Collins, Colorado. Available: http://www.nrel.colostate.edu/projects/starmap Accessed 2013 Oct 14.
26. Interior Columbia Technical Recovery Team [ICTRT] (2007) Viability criteria for application to Interior Columbia Basin salmonid ESUs. U.S. Department of Commerce, NOAA Fisheries Service. Available: http://www.nwfsc.noaa.gov/trt/trt_documents/ictrt_viability_criteria_reviewdraft_2007_body.pdf Accessed 2012 25 Aug.
27. Tattam IA (2007) Seasonal life history of Oncorhynchus mykiss in the South Fork John Day River basin, Oregon. CorvallisOR: Oregon State University. 139 p.
28. Mills JS, Dunham JB, Reeves GH, McMillan JR, Zimmerman CE, et al. (2012) Variability in expression of anadromy by female Oncorhynchus mykiss within a river network. Environ Biol Fishes 93: 505–517.
29. Wenger SJ, Luce CH, Hamlet AF, Isaak DJ, Neville HM (2010) Macroscale hydrologic modeling of ecologically relevant flow metrics. Water Resources Research 46.
30. Hinch SG, Diewert RE, Lissimore TJ, Prince AMJ, Healey MC, et al. (1996) Use of electromyogram telemetry to assess difficult passage areas for river-migrating adult sockeye salmon. Trans Am Fish Soc 125: 253–260.
31. Crossin GT, Hinch SG, Farrell AP, Higgs DA, Lotto AG, et al. (2004) Energetics and morphology of sockeye salmon: Effects of upriver migratory distance and elevation. J Fish Biol 65: 788–810.
32. Kondolf GM, Wolman MG (1993) The sizes of salmonid spawning gravels. Water Resources Research 29: 2275–2285.
33. Buffington JM, Montgomery DR, Greenberg HM (2004) Basin-scale availability of salmonid spawning gravel as influenced by channel type and hydraulic roughness in mountain catchments. Can J Fish Aquat Sci 61: 2085–2096.
34. Montgomery DR, Buffington JM, Peterson NP, SchuettHames D, Quinn TP (1996) Stream-bed scour, egg burial depths, and the influence of salmonid spawning on bed surface mobility and embryo survival. Can J Fish Aquat Sci 53: 1061–1070.
35. Fausch KD, Taniguchi Y, Nakano S, Grossman GD, Townsend CR (2001) Flood disturbance regimes influence rainbow trout invasion success among five holarctic regions. Ecol Appl 11: 1438–1455.
36. McCarthy SG, Duda JJ, Emlen JM, Hodgson GR, Beauchamp DA (2009) Linking habitat quality with trophic performance of steelhead along forest gradients in the South Fork Trinity River watershed, California. Trans Am Fish Soc 138: 506–521.
37. Torgersen CE, Price DM, Li HW, McIntosh BA (1999) Multiscale thermal refugia and stream habitat associations of chinook salmon in northeastern Oregon. Ecol Appl 9: 301–319.
38. Torgersen CE, Faux RN, McIntosh BA, Poage NJ, Norton DJ (2001) Airborne thermal remote sensing for water temperature assessment in rivers and streams. Remote Sensing of Environment 76: 386–398.
39. Cameron AC, Trivedi PK (1998) Regression analysis of count data: Cambridge University Press.
40. Zeileis A, Kleiber C, Jackman S (2008) Regression models for count data in R. J Stat Soft 27: 1–25.
41. ter Braak CJF, Smilauer P (2002) CANOCO reference manual and Canodraw for Windows user's guide: Software for canonical community ordination (version 4.5).
42. Ganio LM, Torgersen CE, Gresswell RE (2005) A geostatistical approach for describing spatial pattern in stream networks. Front Ecol Environ 3: 138–144.
43. Burnham KP, Anderson DR (2002) Model selection and multimodel inference: A practical information-theoretic approach. New York, USA: Springer-Verlag.
44. Steyerberg EW, Harrell FE, Borsboom GJJM, Eijkemans MJC, Vergouwe Y, et al. (2001) Internal validation of predictive models: Efficiency of some procedures for logistic regression analysis. J Clin Epidemiol 54: 774–781.
45. Potts JM, Elith J (2006) Comparing species abundance models. Ecol Model 199: 153–163.
46. Fahrig L (2003) Effects of habitat fragmentation on biodiversity. Annual Review of Ecology Evolution and Systematics 34: 487–515.
47. Reeves GH, Benda LE, Burnett KM, Bisson PA, Sedell JR (1995) A disturbance-based ecosystem approach to maintaining and restoring freshwater habitats of evolutionary significant units of anadromous salmonids in the Pacific Northwest. In: Nielsen JL, Powers DA, editors. Evolution and the aquatic ecosystem: Defining unique units in population conservation. Bethesda, MD: American Fisheries Society Press. 334–349.
48. Taylor PD, Fahrig L, Henein K, Merriam G (1993) Connectivity is a vital element of landscape structure. Oikos 68: 571–573.
49. White SM, Rahel FJ (2008) Complementation of habitats for Bonneville cutthroat trout in watersheds influenced by beavers, livestock, and drought. Trans Am Fish Soc 137: 881–894.
50. Le Pichon C, Gorges G, Boet P, Baudry J, Goreaud F, et al. (2006) A spatially explicit resource-based approach for managing stream fishes in riverscapes. Environ Manage 37: 322–335.
51. Kim M, Lapointe M (2011) Regional variability in Atlantic salmon (Salmo salar) riverscapes: A simple landscape ecology model explaining the large variability in size of salmon runs across Gaspe watersheds, Canada. Ecol Freshwat Fish 20: 144–156.
52. Schlosser IJ, Angermeier PL (1995) Spatial variation in demographic processes of lotic fishes: Conceptual models, emperical evidence, and implications for conservation. In: Nielsen JL, Powers DA, editors. Evolution and the aquatic ecosystem: Defining unique units in population conservation. Bethesda, MD: American Fisheries Society Press. 392–401.
53. Schiemer F, Spindler T, Wintersberger H, Schneider A, Chovanec A (1991) Fish fry associations: Important indicators for the ecological status of large rivers. Internationale Vereinigung fuer Theoretische und Angewandte Limnologie Verhandlungen IVTLAP 24: 2497–2500.
54. Fullerton AH, Burnett KM, Steel EA, Flitcroft RL, Pess GR (2010) Hydrological connectivity for riverine fish: Measurement challenges and research opportunities. Freshwat Biol 55: 2215–2237.
55. Cooper SD, Diehl S, Kratz K, Sarnelle O (1998) Implications of scale for patterns and processes in stream ecology. Aust J Ecol 23: 27–40.
56. Fausch KD, Torgersen CE, Baxter CV, Li HW (2002) Landscapes to riverscapes: Bridging the gap between research and conservation of stream fishes. Bioscience 52: 483–498.
57. Peterson JT, Dunham JB (2010) Scale and fisheries management. In: Hubert WA, Quist MC, editors. Inland fisheries management in North America. Bethesda, MD: American Fisheries Society Press. 81–105.
58. Koizumi I, Maekawa K (2004) Metapopulation structure of stream-dwelling Dolly Varden charr inferred from patterns of occurrence in the Sorachi River basin, Hokkaido, Japan. Freshwat Biol 49: 973–981.
59. Crawford SS, Muir AM (2008) Global introductions of salmon and trout in the genus Oncorhynchus: 1870–2007. Rev Fish Biol Fish 18: 313–344.
60. Vannote RL, Minshall GW, Cummins KW, Sedell JR, Cushing CE (1980) River continuum concept. Can J Fish Aquat Sci 37: 130–137.
61. Montgomery DR, Buffington JM (1997) Channel-reach morphology in mountain drainage basins. Geological Society of America Bulletin 109: 596–611.
62. Wiens JA (1995) Habitat fragmentation: Island v landscape perspectives on bird conservation. Ibis 137: S97–S104.
63. Ricketts TH (2001) The matrix matters: Effective isolation in fragmented landscapes. Am Nat 158: 87–99.
64. Knowlton JL, Graham CH (2010) Using behavioral landscape ecology to predict species' responses to land-use and climate change. Biol Conserv 143: 1342–1354.
65. McKean J, Nagel D, Tonina D, Bailey P, Wright CW, et al. (2009) Remote sensing of channels and riparian zones with a narrow-beam aquatic-terrestrial LiDAR. Remote Sensing 1: 1065–1096.
66. Chang H, Jones J (2010) Climate change and freshwater resources in Oregon. Oregon Climate Assessment Report (OCAR): 69–150.
67. McMillan JR, Dunham JB, Reeves GH, Mills JS, Jordan CE (2012) Individual condition and stream temperature influence early maturation of rainbow and steelhead trout, Oncorhynchus mykiss. Environ Biol Fishes 93: 343–355.
68. Jonsson B, Jonsson N (2009) A review of the likely effects of climate change on anadromous Atlantic salmon Salmo salar and brown trout Salmo trutta, with particular reference to water temperature and flow. J Fish Biol 75: 2381–2447.
69. Beechie TJ, Bolton S (1999) An approach to restoring salmonid habitat-forming processes in Pacific Northwest watersheds. Fisheries 24: 6–15.
70. Beechie TJ, Sear DA, Olden JD, Pess GR, Buffington JM, et al. (2010) Process-based principles for restoring river ecosystems. Bioscience 60: 209–222.
71. Katz SL, Barnas K, Hicks R, Cowen J, Jenkinson R (2007) Freshwater habitat restoration actions in the Pacific Northwest: A decade's investment in habitat improvement. Restor Ecol 15: 494–505.
72. Bisson PA, Dunham JB, Reeves GH (2009) Freshwater ecosystems and resilience of Pacific salmon: Habitat management based on natural variability. Ecol Soc 14: 45 [online].
73. McIntosh BA, Sedell JR, Thurow RF, Clarke SE, Chandler GL (2000) Historical changes in pool habitats in the Columbia River basin. Ecol Appl 10: 1478–1496.

2

Accounting for Imperfect Detection in Ecology

Kenneth F. Kellner*, Robert K. Swihart

Department of Forestry and Natural Resources, Purdue University, West Lafayette, Indiana, United States of America

Abstract

Detection in studies of species abundance and distribution is often imperfect. Assuming perfect detection introduces bias into estimation that can weaken inference upon which understanding and policy are based. Despite availability of numerous methods designed to address this assumption, many refereed papers in ecology fail to account for non-detection error. We conducted a quantitative literature review of 537 ecological articles to measure the degree to which studies of different taxa, at various scales, and over time have accounted for imperfect detection. Overall, just 23% of articles accounted for imperfect detection. The probability that an article incorporated imperfect detection increased with time and varied among taxa studied; studies of vertebrates were more likely to incorporate imperfect detection. Among articles that reported detection probability, 70% contained per-survey estimates of detection that were less than 0.5. For articles in which constancy of detection was tested, 86% reported significant variation. We hope that our findings prompt more ecologists to consider carefully the detection process when designing studies and analyzing results, especially for sub-disciplines where incorporation of imperfect detection in study design and analysis so far has been lacking.

Editor: Jeffrey Buckel, North Carolina State University, United States of America

Funding: Funding was provided by the Purdue University Graduate School and the Purdue Department of Forestry and Natural Resources. The funders had no role in study design, data collection and analysis, decision to publish, or preparation of the manuscript.

Competing Interests: The authors have declared that no competing interests exist.

* Email: kkellner@purdue.edu

Introduction

Measuring the abundance and distribution of organisms is a primary goal of ecology, conservation, and management [1]. These and related parameters (e.g. vital rates, species diversity) are essential to understanding population and metapopulation dynamics, community assembly, trophic interactions, conservation of threatened and endangered species, and the effects of management. Numerous advances in measuring these parameters have occurred over the past several decades, nearly all of which rely on counts and/or observations of organisms [2]. Unfortunately, in most cases complete counts (i.e., censuses) are impossible due to logistical constraints and the cryptic nature of many species [3]. Therefore, inference is based on a sample from the broader population of interest. Perfect or invariant detection is frequently assumed in count-based plant and animal studies [2]. Unfortunately, detection is rarely either perfect or constant due to observer error [4], species rarity [5] or because detection varies with confounding variables such as environmental conditions [6]. Regardless of its cause, we will refer to this condition henceforth as imperfect detection.

When detection is imperfect, additional steps are needed to improve inference. Failure to do so can result in biased estimation and erroneous conclusions. Numerous studies have demonstrated that detection varies among species, over time, and among habitats, and there may be serious consequences when this variability is ignored. For example, failure to correct for imperfect detection may result in bias in estimated relationships with ecological covariates [6], [7], estimates of species distribution or abundance that are inaccurate or mask trends [8–11], improper selection of indicator species [12], and misinterpreted components of fitness such as size-dependent survival and senescence [13]. These errors can misinform management and policy and erode trust in ecologists.

Authors of several landmark papers in the past century have (1) alerted the scientific community to the harmful effects of imperfect detection and (2) designed experimental and statistical approaches that explicitly incorporate detection probability. Petersen [14] and Lincoln [15] recognized the limitations of simple counts and proposed a basic method to account for imperfect detection in abundance estimation using capture histories of marked organisms. Their approach laid the foundation for future methods of estimating abundance and survival based on marked animals including the Cormack-Jolly-Seber model [16–18], the robust design model [19], [20], and numerous others with increasing complexity and ability to account for variation in detection [21–24]. Recent advances have merged analysis of animal movement with capture-recapture, resulting in spatial capture-recapture models to estimate density and other parameters of interest [25], [26]. When identifying individual organisms is infeasible, repeated counts [27] [4], [28] and distance sampling [29] may be useful to account for detection when estimating abundance, as long as within-sample double-counting of individuals is avoided.

Other population and community metrics (occurrence/occupancy, local colonization and extinction, richness, diversity, and turnover) can also be biased when imperfect detection is ignored [6], [30–32]. Methods exist to estimate these parameters while accounting for detectability, if data are collected in a way that allows the detection process to be modeled [1], [11], [31], [33–35]. This is often achieved through replicate surveys of sampling sites [31], [33], [34], [36] although other methods such as the collection of times to detection are also possible [37], [38]. Software packages have been developed to make the modeling advances described above more accessible to ecologists, including programs CAPTURE [39], MARK [40], DISTANCE [41], PRESENCE [42], COMDYN [43], and E-Surge [44], as well as numerous R packages including 'marked' [45] 'unmarked' [46] and 'secr' [47]. Recently, Bayesian modeling approaches have been introduced to ecologists as an alternative means of fitting complex population and community models [1], [26], [48–51].

Despite numerous articles alerting ecologists to the consequences of imperfect detection, and despite available models and software capable of addressing imperfect detection, we commonly encounter refereed publications that fail to acknowledge or account for the presence of non-detection error. These casual observations beg the question: how prevalent are methods that incorporate imperfect detection in the ecological literature, and how does this differ among various types of studies and over time? Identifying these patterns is crucial to understand barriers to incorporation of imperfect detection and to target areas in which ecological inference might be improved.

We posited that the expanding availability of analysis methods, software, and computing power has increased the probability that, over time, researchers account for imperfect detection. We suspected that adoption has varied among areas of ecological inquiry (e.g. [10], [52]), so we asked whether the probability of accounting for imperfect detection has varied with the type of organism studied, the level of biological organization studied, or the spatial scale studied. Because of the taxonomic focus of many of the researchers who have developed methods for addressing imperfect detection, we predicted that a greater proportion of studies focused on fish, mammals, and birds would incorporate imperfect detection than other groups (e.g. plants, invertebrates); however we expected this difference to have declined over time. We also predicted that studies conducted at higher levels of organization (e.g. communities vs. individual animals) and with greater spatial extents (e.g. regional or landscape vs. local) would be less likely to incorporate imperfect detection due to the difficulties of implementing more complex study designs with limited resources. To test our predictions, we conducted a quantitative review of ecological literature spanning 40 years, 6 taxonomic groups, 5 commonly reported parameters, 2 levels of biological organization, and 3 spatial scales to determine how the use of statistical methods that incorporate imperfect detection has varied among these variables and over time.

Materials and Methods

Article Selection

A census of the literature was impractical, so we adopted a stratified sampling approach (Figure 1). We selected a subset of 10 journals to include in the study, chosen for their impact factors, long publication history, and coverage of a range of taxonomic groups (birds, fish, mammals, herpetofauna, invertebrates, and plants; Table 1). For each of the journals, we selected 5 years from which to sample papers: 1971, 1981, 1991, 2001, and 2011.

Within each journal/year combination, we examined all articles (starting in the first issue), identifying those that fit our criteria for inclusion through careful reading of the abstract, methods, and results. Three criteria were defined: (1) the study had to be focused on one or more of the target taxonomic groups; (2) the study had to measure one or more of a set of parameters of interest, composed of abundance, occurrence, survival, richness or diversity, and extent or size of species range (selected because they are among those most often studied by ecologists and susceptible to the biases of imperfect detection); (3) the study had to be implemented in a way in which detection was likely imperfect, i.e., there was no evidence to support the feasibility of a complete census or to ensure constancy of detection. If fewer than 15 articles were selected in a given journal/year combination, we continued our search into the following year(s) until we had a minimum of 15 articles that fit the criteria.

Data Collection

We recognized that our ability to correctly judge whether a study incorporated imperfect detection was likely less than perfect itself. We therefore implemented a replicated sampling approach commonly used as a means of parameter estimation when detection is imperfect. The objectives, methods, results, and discussion of each article were carefully read by the first author (KFK). The journal type (single-taxon focus vs. broader focus) year, taxa and parameters measured, experimental scale (population or community level), and spatial extent (local or landscape) were recorded for each article, as well as the presence/absence of an approach to account for imperfect detection. For the binary response variable, the presence of an approach included a statistical method of estimating detectability or an acknowledgment by the authors of imperfect detection as a potential issue. Articles that accounted for imperfect detection were further examined to determine (1) if they explicitly reported detection probabilities and (2) if and how detection varied (e.g. over time or among species). The second author (RKS) independently read and scored a subset of all articles in the same manner. Multiple "sampling occasions" for this subset of articles allowed for estimation of a detection parameter, analogous to a repeated-sample design for occupancy estimation [1], [31]. The approach we used assumes that there are no "false positives", i.e., studies identified as accounting for imperfect detection when in truth they did not. We took a conservative approach to identifying studies that accounted for imperfect detection and believe false positives were negligible. Information about all articles included in the quantitative review is contained in the Dataset S1.

Analysis

Using this repeated-sample data, we simultaneously modeled the probability that an article incorporated imperfect detection (hereafter pIID) and the probability that we were able to detect this when reading the articles (hereafter pR) using a hierarchical logistic regression model [1]. Year of publication, journal type, taxon and parameter(s) measured, scale, and spatial extent were considered as covariates on pIID. Taxon and reader (KFK or RKS) were considered as covariates on pR. In addition to the full model, we fit separate models for each taxon; these models did not have taxa as predictor variables but were otherwise identical to the full model. We explored models with interacting effects; however, finding no significant interactions, our final models included only additive effects. Models were fit in a Bayesian framework using JAGS [53], called from R [54] using the package R2jags [55]. We used two methods for assessing the statistical importance of each covariate in the models. First, we calculated a 95% credible interval for each parameter estimate based on the posterior distribution. Second, we calculated a parameter f representing the fraction of the sampled posterior distribution with the same sign as

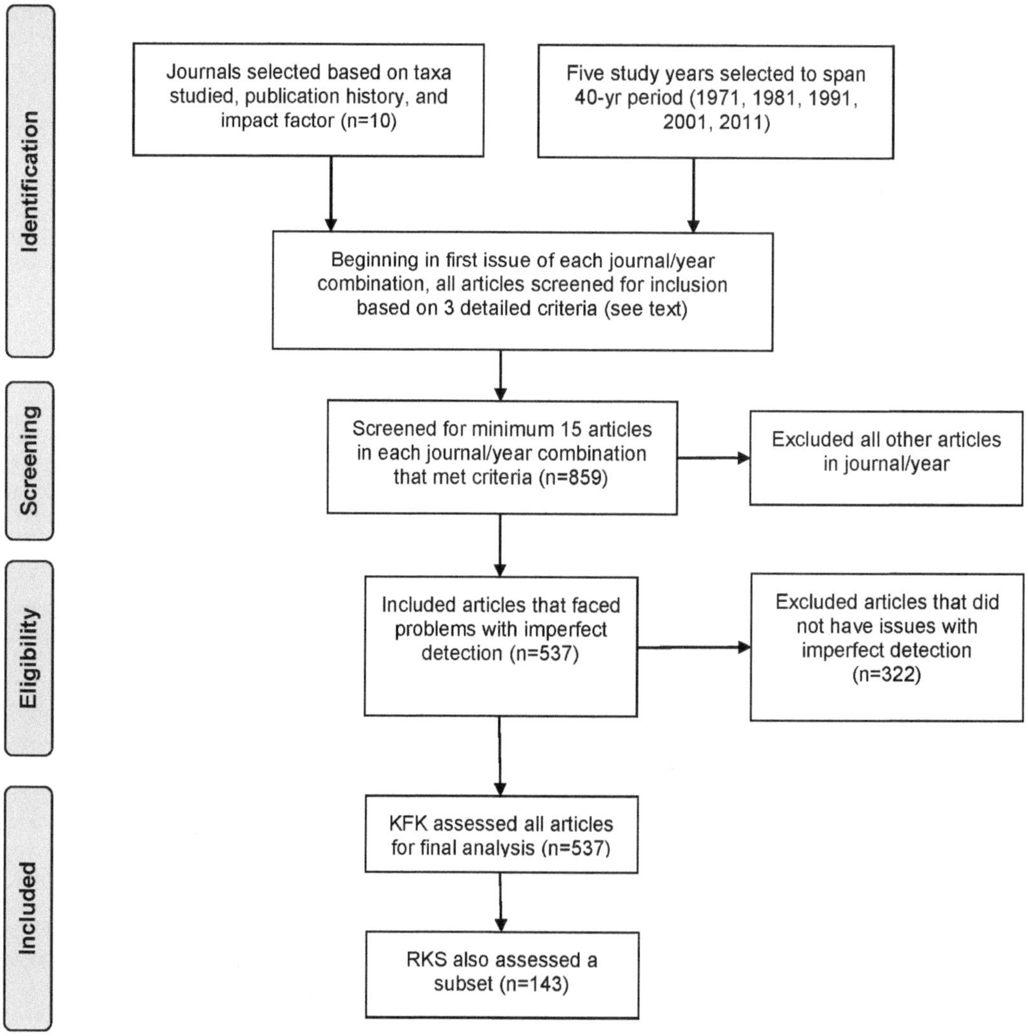

Figure 1. PRISMA flow diagram detailing how articles were selected for inclusion in the quantitative literature review.

Table 1. Selected characteristics of the 10 journals included in a quantitative review examining use of methods that incorporate imperfect detection in wildlife and plant studies.

Journal Name	Taxa	Years Published	Impact Factor
Journal of Ecology	Plants	1913-	5.04
Journal of Animal Ecology	Multiple	1932-	4.94
Ecology	Multiple	1920-	4.85
Journal of Biogeography	Multiple	1974	4.54
Ecological Entomology	Invertebrates	1836-	1.95
The Auk	Birds	1884-	1.81
Journal of Mammalogy	Mammals	1919-	1.74
Transactions of the American Fisheries Society	Fish	1872-	1.55
Journal of Wildlife Management	Multiple	1937-	1.36
Herpetologica	Reptiles, Amphibians	1936-	1.08

Journals were chosen for inclusion in the review on the basis of taxonomic groups studied, years published, and impact factor (from Journal Citation Reports 2011–2012).

the mean; this value reflects our level of certainty that the parameter estimate is positive or negative.

Results

A total of 537 articles surveyed by KFK fit the criteria and were included in the review. A subset (n = 143, 27%) was also examined by RKS. The most common parameters estimated were abundance (n = 377 articles), occupancy/occurrence (n = 121) and richness (n = 103), whereas survival (n = 66) and species range/distribution (n = 31) were less common. Birds (n = 154), invertebrates (n = 150), and mammals (n = 131) were the most commonly studied taxa.

Overall, 23%±1.8 (123/537, mean±standard error) of articles addressed imperfect detection. There was a positive effect of year on *p*IID (*f* = 0.99, Table 2): the yearly mean percent of articles that addressed imperfect detection generally increased from 25%±5.9, 14%±4.0, and 23%±4.6 in 1971, 1981, and 1991, respectively, to 29%±5.1 and 35%±5.1 in 2001 and 2011. The same increasing trend over time appeared when taxa were modeled separately, with the exception of fish and plant articles (Figure 2, Table 3). Taxonomic group generally was an important covariate on *p*IID. Specifically, studies of fish were more likely (*f* = 0.97) to account for imperfect detection than the average (43%±5.9 of articles for fish vs. 23%±1.8 overall). In contrast, articles focused on plants (*f* = 0.99) and invertebrates (*f* = 0.95) were less likely to do so (1.4%±1.3 and 9.0%±2.7, respectively; Table 2). The remaining taxonomic groups (mammals, herpetofauna, and birds) were positively related to *p*IID, but with a lesser degree of certainty (Table 2). Articles in journals that focused on a single taxon were less likely (*f* = 0.98) to incorporate imperfect detection than multi-taxa journals.

The parameter(s) estimated in each article also correlated with *p*IID. Articles that measured survival were more likely (*f* = 1.00) to incorporate imperfect detection than the mean (50%±6.1 vs. 23%±1.8); abundance was also positively related to *p*IID (Table 2). In contrast, papers that measured richness, occurrence/occupancy, and range/distribution were less likely to incorporate imperfect detection (6.0%±2.3, 10%±2.8, and 3.2%±3.2; Table 2). Articles examining entire communities were less likely to incorporate imperfect detection than studies focused on a single species (*f* = 0.95, Table 2). Spatial scale, in contrast, did not affect *p*IID (*f* = 0.67, Table 2). The effects of parameters and scale on *p*IID were similar when taxonomic groups were modeled separately, with the exception of a positive effect of community studies on *p*IID for bird studies (Table 3).

The parameter *p*R did not differ between taxonomic groups. There was an observer effect, but it was small (Table 2). Just 10 of 143 papers (7%±2.1) examined by both readers had different assessments for the inclusion of imperfect detection.

Among articles that accounted for imperfect detection, 62 (50%) also reported information about detection probabilities in some form. Of these, 48%, 33% and 19% reported *maximum* estimated single-survey detection probabilities (i.e., probability of detection on a single sampling occasion) less than 0.7, 0.5, and 0.3, respectively. Most papers reported at least one detection probability significantly less than 1; 70% had at least one estimated detection probability less than 0.5, and 50% had at least one less than 0.3. Reported detection probabilities varied with at least one covariate in 86% of papers that tested for heterogeneity; the most common covariates were time (44%), species (25%), site or population (19%), and methodology (e.g. effort, observer effects; 19%).

Discussion

Taxonomic Groups

Our examination of ecology papers across 5 decades, 10 journals, 5 taxa, and other parameters of interest confirmed our initial observations: the majority (77%) of ecological studies failed to acknowledge or correct for imperfect detection when doing so likely would have been appropriate. As we predicted, there was considerable contrast among taxa: vertebrates (fish, mammals, birds, reptiles and amphibians) were more likely to account for

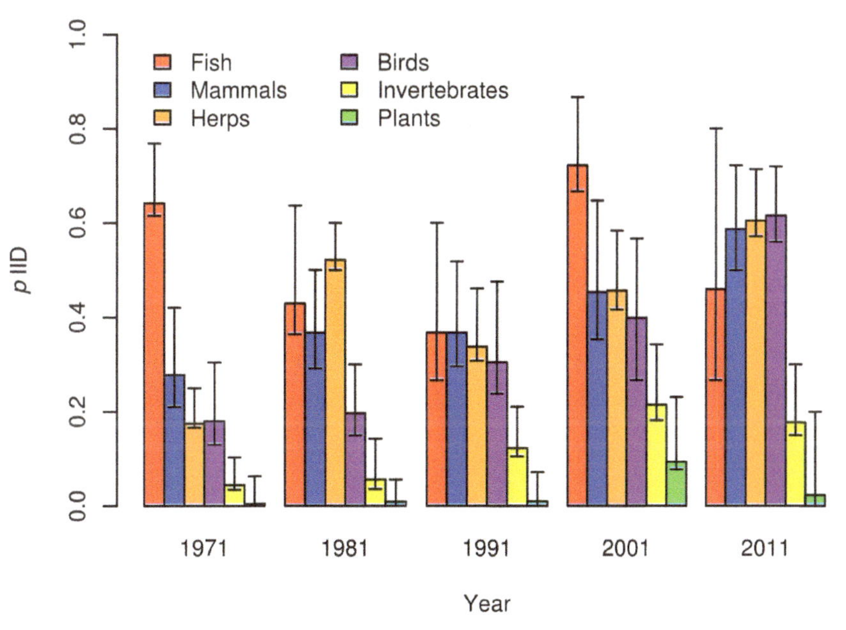

Figure 2. Mean estimated probability (p*IID*) that an article of a given taxon and year incorporated statistical methods to account for imperfect detection, based on output from the hierarchical logistic regression model. Error bars represent 95% credible intervals around the mean.

Table 2. Estimated parameter values from a hierarchical logistic regression relating covariates (taxa, parameter estimated, year, journal type, and experimental scale) on the probability a given study accounted for imperfect detection.

Parameter	Estimate	95% Credible Interval	f[b]
Year	**0.45**[a]	(0.23, 0.66)	0.99
Fish	**1.01**	(0.01, 2.13)	0.97
Mammals	0.38	(−0.51, 1.49)	0.75
Herps	0.53	(−0.38, 1.50)	0.85
Birds	0.33	(−0.70, 1.22)	0.76
Invertebrates	-0.83	(−1.83, 0.16)	0.95
Plants	**-1.70**	(−2.95, −0.35)	0.99
Abundance	0.49	(−0.26, 1.18)	0.90
Occurrence	-0.60	(−1.42, 0.14)	0.95
Survival	**0.91**	(0.23, 1.65)	1.00
Richness	**-1.10**	(−1.81, −0.25)	0.99
Range	-0.92	(−2.17, 0.22)	0.94
Scale (1 = community)	-0.51	(−1.10, 0.08)	0.95
Spatial extent (1 = landscape)	-0.12	(−0.69, 0.45)	0.67
Journal type (1 = single taxon)	**-0.60**	(−1.12, −0.03)	0.98
Detection Covariates			
Observer (1 = RKS)	**1.33**	(0.16, 2.51)	0.98
Fish	-0.02	(−1.37, 1.51)	0.52
Mammals	-0.68	(−2.02, 0.52)	0.85
Herps	0.83	(−0.52, 2.36)	0.85
Birds	-0.62	(−1.80, 0.64)	0.84
Invertebrates	-0.14	(−1.70, 1.48)	0.57
Plants	-0.21	(−2.35, 1.77)	0.55

[a]Bolded parameters had 95% credible intervals that did not overlap 0 and are considered statistically important.
[b]Proportion of the parameter's posterior distribution with same sign as the mean value; values approaching 1 reflect increasing certainty that the parameter is different from 0.

imperfect detection than other taxa (plants and invertebrates). The earliest proponents of accounting for imperfect detection (e.g. [14], [15]) focused on vertebrates. The generally lower overall abundance, greater movement capability, and cryptic nature of vertebrates, relative to plants and invertebrates, may have encouraged earlier adoption by vertebrate ecologists of study designs and statistical methods to correct for limitations of sampling and detection. Many of these methods were published in taxon-specific journals or tailored to specific taxa (e.g. removal sampling of fish or mark-release-recapture of mammals and birds), so their adaptation to studies of other taxa was likely limited. Interestingly, single-taxon journals actually had a lower proportion of studies incorporate imperfect detection, likely reflecting the small number of studies in plant- and invertebrate-focused journals that did so.

The proportion of studies that accounted for imperfect detection increased over the 40-year period of our review for all vertebrate taxa except fish, which had a comparatively high proportion throughout (Figure 2). This positive trend presumably reflects greater awareness among vertebrate ecologists of the risks associated with failing to account for imperfect detection and a correspondingly greater availability of sampling and statistical methods designed specifically to account for errors of omission. While we did not collected detailed data on the specific statistical method(s) used to account for imperfect detection, we observed a corresponding generally positive trend in the diversity of statistical approaches used over time. Early studies (prior to 1981) of fish and mammals primarily used simple mark-release-recapture methods like the Lincoln-Peterson index [14], [15]. From 1981–2001 more complex mark-release-recapture methods [16–18] were most commonly used, but methods were still focused on specific taxa (primarily mammals and fish). From 2001 onward, the diversity of approaches greatly increased for most taxa thanks to the widespread introduction of occupancy modeling and related hierarchical modeling approaches applicable to wide variety of species [1], [31], [33], [36].

While the trend over time in adoption of methods that incorporate imperfect detection is less pronounced for plants and invertebrates (Figure 2), it is nevertheless positive and likely will continue as publications on these taxa draw attention to the issue [3], [56–59]. For example, [10] estimated detection probability for plants surveyed in the Swiss Biodiversity Monitoring program. Based on a random sample of 100 species (of 1700 detected), median single-survey detection probability was 0.74 (range 0.03–0.99) for the spring survey and 0.82 (range 0.03–0.99) for the late summer survey. Thus, distribution maps based on a single survey risk modeling the joint patterns of occurrence and detection. These two parameters potentially could be disentangled if they rely on covariates that are not identical and sample sizes are reasonably large [60], but these constraints suggest that researchers should use caution in making inferences from a single survey.

Table 3. Estimated parameter values (with 95% credible intervals in parentheses) from a series of hierarchical logistic regressions (separated by taxa) relating covariates (parameter estimated, year, journal type, and experimental scale) on the probability a given study accounted for imperfect detection.

	Taxon					
Parameter	**Fish**	**Mammals**	**Herps**	**Birds**	**Inverts**	**Plants**
Year	−0.19 (−0.66, 0.22)	**0.60**[a] (0.12, 1.16)	**0.44** (0.02, 1.01)	**0.79** (0.32, 1.30)	**0.54** (0.08, 1.07)	0.18 (−1.00, 1.68)
Abundance	0.51 (-0.83, 1.95)	**1.31** (0.09, 2.64)	0.38 ((-0.82, 1.69)	0.55 (-0.76, 2.00)	-0.67 (-1.99, 0.62)	0.08 (-1.84, 1.95)
Occurrence	-0.34 (-1.83, 1.20)	-0.20 (-1.62, 1.22)	-0.11 (-1.44, 1.21)	-1.05 (-2.38, 0.18)	-0.99 (-2.43, 0.27)	-0.15 (-1.89, 1.68)
Survival	1.17 (-0.34, 2.63)	0.09 (-1.23, 1.37)	0.55 (-0.84, 1.80)	**1.33** (0.09, 2.74)	1.09 (-0.38, 2.49)	-0.06 (-1.92, 1.81)
Richness	-0.58 (-2.08, 0.92)	-0.42 (-1.87, 1.22)	-0.98 (-2.72, 0.49)	-1.05 (-2.33, 0.11)	-0.16 (-1.31, 1.13)	-0.18 (-2.20, 1.69)
Range	-0.53 (-2.36, 1.17)	-0.53 (-2.35, 1.17)	-0.77 (-2.41, 0.85)	0.23 (-1.93, 0.15)	-0.45 (-2.17, 1.38)	-0.07 (-1.90, 1.69)
Scale (1 = community)	0.22 (-0.79, 1.27)	-0.55 (-1.60, 0.45)	-0.25 (-1.51, 1.08)	**1.27** (-2.46, −0.16)	-0.68 (-1.95, 0.61)	-0.52 (-2.51, 1.44)
Spatial extent (1 = landscape)	0.44 (-0.69, 1.65)	-0.77 (-1.98, 0.43)	-0.62 (-1.89, 0.46)	0.05 (-1.16, 1.30)	-0.10 (-1.36, 1.13)	-0.40 (-2.29, 1.65)
Journal type (1 = single taxon)	0.08 (-1.08, 1.14)	-0.78 (-1.79, 0.34)	0.05 (-1.20, 1.27)	-0.79 (-1.93, 0.15)	-0.03 (-1.18, 1.27)	0.12 (-1.75, 2.02)
Detection covariates						
Observer (1 = RKS)	0.64 (-1.15, 1.22)	0.01 (-2.22, 1.79)	0.23 (-1.43, 1.89)	0.01 (-2.13, 2.16)	0.08 (-1.83, 1.99)	-0.12 (-1.99, 1.62)

[a]Bolded parameters had 95% credible intervals that did not overlap 0 and are considered statistically important.

Parameters Estimated

Each parameter measured in ecological studies presents unique logistical and statistical challenges. For example, estimating survival inherently requires marking individuals and following them through time, with collected data likely including multiple sampling events. Correcting for imperfect detection is a logical extension when these data are available, and the majority of survival studies did so. Important exceptions include survival studies that do not correct for conflation of non-detection with mortality, in which case an individual that is not detected at a given sampling occasion is assumed to be dead. The relatively high proportion of abundance studies that accounted for imperfect detection likely reflects a long history and proliferation of approaches to account for undetected individuals in abundance estimates [2], [61]. A notable exception to this pattern is the widespread use of "catch per unit effort" (CPUE) and similar metrics as indices of abundance. CPUE does not require marked organisms but implicitly assumes that detectability does not vary across time or experimental site, which has been shown to introduce bias into estimation [22], [62], [63].

Richness, occurrence, and range are closely related parameters. Richness can be represented as the sum of all species' occurrences at a site, and range loosely corresponds to species occurrence across a large geographic area. They shared an additional characteristic in our study – a low probability of incorporating imperfect detection. Estimating these parameters (especially at the community scale) can require sampling for many species at numerous sites, requiring potentially difficult or costly repeated samples and/or careful allocation of sampling effort to obtain the data necessary to estimate detection. In addition, sampling designs and statistical methods that incorporate imperfect detection into estimation of occupancy and richness have appeared in the ecological literature only recently [31], [33], [34], [36]. For estimates of species range and distribution, a further issue is that researchers often have relied on historical presence-only data (e.g. from museum collections) which makes estimation of occupancy probability and subsequent inferences difficult [64] but not impossible [65–67], at least in a relative sense [68].

Implications

We concede that it is impossible to know the extent of bias in past studies that have ignored imperfect detection. It certainly is possible that bias may have been small in some of the studies that failed to incorporate detection error, because detection probability was either high or invariant. However, for the subset of papers in our study that estimated detection probability, the median values of minimum and maximum single-survey detectability were 0.29 and 0.71, respectively, and 70% had at least one estimated detection probability less than 0.5, an indication that detection can commonly be much less than perfect. Admittedly, these studies are not a random subset of the studies we considered, as their authors presumably considered imperfect detection to be problematic in their study system. Still, the magnitude and frequency of bias seen in these and other studies (e.g. [6–13]) suggests strongly that imperfect detection is common and can weaken inference for many types of ecological processes. For example, simulations have demonstrated that ignoring imperfect detection, as is done with presence-absence and presence-only data, can dramatically diminish a model's capacity to identify environmental correlates of species distributions [32].

Even if detection is imperfect, bias may not affect inference adversely for some purposes if detection is invariant. For instance, population indices are commonly used to monitor trends and assume constant detectability relative to abundance [69], [70], even though evidence of variation is common [71] and led [72] to recommend that the burden of proof should be shifted to demonstrate that detection probability is invariant. Our results provide support and warrant expansion of their recommendation; in 86% of papers from our study that examined the constancy assumption, detectability varied significantly. Rather than assume that detection is perfect, it seems prudent, then, for field ecologists to assume that detection probabilities differ and to require evidence of their equivalence before using indices or other measures that conflate this nuisance parameter with the parameter(s) of ecological interest.

Collectively, our findings should concern ecologists in search of stronger inference, and particularly managers and policy makers whose decisions often depend on accurate knowledge of species presence or abundance. Accounting for imperfect detection has been the exception and not the rule in ecology across most subdisciplines and study types. We hope the patterns we have identified will encourage ecologists to consider carefully the detection process when designing studies and analyzing results. This is particularly important with taxonomic groups and parameters we highlighted that, in the past, have generally not accounted for imperfect detection.

We recognize that the methods developed thus far to deal with imperfect detection are not a panacea for estimation. For instance, methods to address imperfect detection, as with all estimation approaches, rely on assumptions that need to be carefully evaluated and may not be suitable for certain species or systems (e.g., [73], [74]). Further, addressing imperfect detection will not solve other forms of hidden bias that can afflict observational studies [75]. But we disagree with the suggestion that ignoring imperfect detection may be preferable to accounting and adjusting for it [76], [77]. We believe that a more suitable approach to improved estimation is to minimize bias associated with model assumptions generally, including the issue of detection, by careful consideration of study design, data collection, and statistical analysis [78]. Using hierarchical occupancy models as an example, [78] showed that, in virtually all realistic cases, accounting for detectability reduces the bias in the estimation of occupancy relative to naïve models even when detectability is heterogeneous. Of course reliable inference depends on sampling methods that produce reasonable odds of detection given presence; no estimator will be particularly helpful when applied to data on populations or species that are "invisible" to sampling [79].

We restricted our quantitative review to a subset of ecological parameters in which imperfect detection may play an important role. But there certainly are other parameters of ecological interest for which inference may be improved by accounting for imperfect detection. For example, most studies of seed dispersal by animals have ignored the implications of seeds that were undetected (but see [80]). Recently, though, methods of varying complexity have been developed to account for the effects of imperfect detection on estimates of seed dispersal and survival [81–83]. In general, the current proliferation of detection methods [84–86] and statistical techniques for modelling detectability [1], [26], [49] are applicable to a wide range of taxonomic groups, parameters, and experimental scales and present an unprecedented opportunity for ecologists to provide more robust estimation and inference.

Supporting Information

Dataset S1 Articles included in the literature review. A spreadsheet containing information about papers included in this quantitative literature review, including file names, journals, taxa studied, parameters estimated, and other related values. Detailed metadata are included in the spreadsheet.

Checklist S1 PRISMA Summary details and locations within the text of specific components of the systematic review conducted in this paper, based on the 2009 PRISMA group guidelines.

Acknowledgments

Roxanne Pourshoushtari compiled the papers included in this review. Nate Lichti, Rita Blythe, Mekala Sundaram, Tim Smyser, Byju Govindan, and two anonymous reviewers provided helpful comments on the manuscript.

Author Contributions

Conceived and designed the experiments: KFK RKS. Performed the experiments: KFK RKS. Analyzed the data: KFK. Contributed reagents/materials/analysis tools: KFK RKS. Contributed to the writing of the manuscript: KFK RKS.

References

1. Royle JA, Dorazio RM (2008) Hierarchical modeling and inference in ecology. Burlington, MA: Elsevier Academic Press. 464 p.
2. Williams B, Nichols JD, Conroy M (2002) Analysis and management of animal populations. San Diego, CA: Elsevier Academic Press. 817 p.
3. MacKenzie DI, Nichols JD, Sutton N (2005) Improving inferences in population studies of rare species that are detected imperfectly. Ecology 86: 1101–1113.
4. Nichols JD, Hines JE, Sauer JR, Fallon FW, Fallon JE, et al. (2000). A double-observer approach for estimating detection probability and abundance from point counts. Auk 117: 393–408.
5. Dettmers R, Buehler DA, Bartlett JG, Klaus NA (1999) Influence of point count length and repeated visits on habitat model performance. J Wildl Manage 63: 815–823.
6. Gu W, Swihart RK (2004) Absent or undetected? Effects of non-detection of species occurrence on wildlife-habitat models. Biol Conserv 116: 195–203.
7. Zipkin EF, Royle JA, Dawson DK, Bates S (2010) Multi-species occurrence models to evaluate the effects of conservation and management actions. Biol Conserv 143: 479–484.
8. Kéry M, Schmidt BR (2008) Imperfect detection and its consequences for monitoring for conservation. Community Ecol 9: 207–216.
9. Buckland ST, Studeny AC, Magurran AE, Newson SE (2011) Biodiversity monitoring: the relevance of detectability. In: Magurran AE, McGill BJ, editors. Biological diversity: frontiers in measurement and assessment. Oxford. Oxford University Press. 25–36.
10. Chen G, Kéry M, Plattner M (2013) Imperfect detection is the rule rather than the exception in plant distribution studies. J Ecol 101: 183–191.
11. Tingley MW, Beissinger SR (2013) Cryptic loss of montane avian richness and high community turnover over 100 years. Ecology 94: 598–609.
12. Urban NA, Swihart RK, Malloy MC, Dunning JB Jr (2012) Improving selection of indicator species when detection is imperfect. Ecol Indic 15: 188–197.
13. Gimenez O, Viallefont A, Charmantier A, Pradel R, Cam E, et al. (2008) The risk of flawed inference in evolutionary studies when detectability is less than one. Am Nat 172: 441–448.
14. Petersen CG (1896) The yearly immigration of young plaice into the Limfjord from the German Sea. Report of the Danish Biological Station 6: 1–48.
15. Lincoln FC (1930) Calculating waterfowl abundance on the basis of banding returns. Washington, DC: US Department of Agriculture. Circular No.118.
16. Cormack RM (1964) Estimates of survival from the sighting of marked animals. Biometrika 51: 429–438.
17. Jolly GM (1965) Explicit estimates from capture-recapture data with both death and immigration-stochastic model. Biometrika 52: 225–247.
18. Seber GAF (1965) A note on the multiple-recapture census. Biometrika 52: 249–259.
19. Pollock KH (1982) A capture-recapture design robust to unequal probability of capture. J Wildl Manage 46: 752–757.
20. Pollock KH, Nichols JD, Brownie C, Hines JE (1990) Statistical inference for capture-recapture experiments. Wildl Monogr No.107.
21. Otis DL, Burnham KP, White GC, Anderson DR (1978) Statistical inference from capture data on closed animal populations. Wildl Monogr 62: 1–135.
22. Nichols JD, Pollock KH (1983) Estimation methodology in contemporary small mammal capture-recapture studies. J Mammal 64: 253–260.
23. Royle JA, Young KV (2008) A hierarchical model for spatial capture-recapture data. Ecology 89: 2281–2289.
24. Royle JA, Dorazio RM (2012) Parameter-expanded data augmentation for Bayesian analysis of capture-recapture models. J Ornithol 152: 521–537.
25. Ergon T, Gardner B (2014). Separating mortality and emigration: modelling space use, dispersal, and survival with robust-design spatial capture-recapture data. Methods Ecol Evol, doi:10.1111/2041-210X.12133.
26. Royle JA, Chandler RB, Sollmann R, Gardner B (2014) Spatial capture-recapture. Waltham, MA: Elsevier Academic Press. 612 p.
27. Caughley G, Goddard J (1972) Improving the estimates from inaccurate censuses. J Wildl Manage 36: 135–140.
28. Royle JA (2004) *N*-mixture models for estimating population size from spatially replicated counts. Biometrics 60: 108–115.
29. Buckland ST, Anderson DR, Burnham KP, Laake JL (1993) Distance sampling: estimating abundance of biological populations. London: Chapman and Hall. 446 p.
30. Boulinier T, Nichols JD, Sauer JR, Hines JE, Pollock KH (1998) Estimating species richness: the importance of heterogeneity in species detectability. Ecology 79: 1018–1028.
31. MacKenzie DI, Nichols JD, Hines JE, Knutson MG, Franklin AB (2003) Estimating site occupancy, colonization, and local extinction when a species is detected imperfectly. Ecology 84: 2200–2207.

32. Lahoz-Monfort JJ, Guillera-Arroita G, Wintle BA (2014) Imperfect detection impacts the performance of species distribution models. Glob Ecol Biogeogr 23: 504–515.
33. MacKenzie DI, Nichols JD, Lachman GB, Droege S, Royle JA, et al. (2002) Estimating site occupancy rates when detection probabilities are less than one. Ecology 83: 2248–2255.
34. Dorazio RM, Royle JA, Söderström B, Glimskär A (2006) Estimating species richness and accumulation by modeling species occurrence and detectability. Ecology 87: 842–854.
35. Rota CT, Fletcher RJ, Evans JM, Hutto RL (2011) Does accounting for imperfect detection improve species distribution models? Ecography 34: 659–670.
36. Tyre AJ, Tenhumberg B, Field SA, Niejalke D, Parris K, et al. (2003) Improving precision and reducing bias in biological surveys: estimating false-negative error rates. Ecol Appl 13: 1790–1801.
37. Garrard GE, Bekessy SA, McCarthy MA, Wintle BA (2008) When have we looked hard enough? A novel method for setting minimum survey effort protocols for flora surveys. Austral Ecol 33: 986–998.
38. Guillera-Arroita G, Morgan BJT, Ridout MS, Linkie M (2011) Species occupancy modeling for detection data collected along a transect. J Agr Biol Envir St 16: 301–317.
39. White GC, Burnham KP, Otis DL, Anderson DR (1978) User's manual for Program CAPTURE. Logan, UT: Utah State University Press.
40. White GC, Burnham KP (1999) Program MARK: survival estimation from populations of marked animals. Bird Study 46 Supplement: 120–138.
41. Thomas L, Buckland ST, Rexstad EA, Laake JL, Strindberg S, et al. (2010) Distance software: design and analysis of distance sampling surveys for estimating population size. J Appl Ecol 47: 5–14.
42. Hines JE (2006) PRESENCE2– software to estimate patch occupancy and related parameters. Laurel, MD: USGS Patuxent Wildlife Research Center. Available: http://www.mbr-pwrc.gov/software/presence.html.
43. Hines JE, Boulinier T, Nichols JD, Sauer JR, Pollock KH (1999) COMDYN: Software to study the dynamics of animal communities using a capture-recapture approach. Bird Study 46: 209–217.
44. Choquet R, Rouan L, Pradel R (2009) Program E-Surge: a software application for fitting multilevel models. Environ Ecol Stat 3: 845–865.
45. Laake J, Johnson D, Conn P (2014) Marked: R code for mark-recapture analysis. R package version 1.1.6. Available: http://cran.r-project.org/web/packages/marked/.
46. Fiske I, Chandler R, Miller D, Royle A, Kéry M (2014) Unmarked: models for data from unmarked animals. R package version 0.10–4. Available: http://cran.r-project.org/web/packages/unmarked/.
47. Efford M (2014) Secr: spatially explicit capture-recapture. R package version 2.8.2. Available: http://cran.r-project.org/web/packages/secr/.
48. Clark JS (2007) Models for ecological data: an introduction. Princeton, NJ: Princeton University Press. 617 p.
49. Kéry M (2010) Introduction to WinBUGS for ecologists. Elsevier Academic Press, Burlington, MA. 320 p.
50. Kéry M, Schaub M (2012) Bayesian population analysis using WinBUGS. Waltham, MA: Elsevier Academic Press. 554 p.
51. Iknayan KJ, Tingley MW, Furnas BJ, Beissinger SR (2014) Detecting diversity: emerging methods to estimate species diversity. Trends Ecol. Evol. 29: 97–106.
52. Mazerolle, Bailey LL, Kendall WL, Royle JA, Converse SJ, et al. (2007) Making great leaps forward: accounting for detectability in herpetological field studies. J Herpetol 41: 672–689.
53. Plummer M (2003) JAGS: A program for analysis of Bayesian graphical models using Gibbs sampling.Proceedings of the 3rd International Workshop on Distributed Statistical Computing, 20–22 Mar 2003, Vienna. Vienna, Austria: Technische Universität Wien.
54. R Development Core Team (2013) R: A language and environment for statistical computing.ISBN 3-900051-07-0. R Foundation for Statistical Computing. Vienna, Austria. Available: http://www.R-project.org.
55. Su Y, Yajima M (2014) R2jags: A package for running jags from R. R package version 0.04–03. Available: http://cran.r-project.org/web/packages/R2jags/.
56. Shefferson RP, Sandercock BK, Proper J, Beissinger SR (2001) Estimating dormancy and survival of a rare herbaceous perennial using mark-recapture models. Ecology 82: 145–156.
57. Chen G, Kéry M, Zhang J, Ma K (2009) Factors affecting detection probability in plant distribution studies. J Ecol 97: 1383–1389.
58. Alexander HM, Slade NA, Kettle WD, Pittman GL, Reed AW (2009) Detection, survival rates and dynamics of a cryptic plant, *Asclepias meadii*: applications of mark-recapture models to long-term monitoring studies. J Ecol 97: 267–276.
59. Alexander HM, Reed AW, Kettle WD, Slade NA, Bodbyl Roels SA, et al. (2012) Detection and plant monitoring programs: lessons from an intensive survey of *Asclepias meadii* with five observers. PloS One 7: e52762.
60. Lele SR, Moreno M, Bayne E (2012) Dealing with detection error in site occupancy surveys: what can we do with a single survey? J Plant Ecol 5: 22–31.
61. Nichols JD (1992) Capture-recapture models. BioScience 42: 94–102.
62. Pollock KH, Nichols JD, Simons TR, Farnsworth GL, Bailey LL, et al (2002) Large scale wildlife monitoring studies: statistical methods for design and analysis. Environmetrics 13: 105–119.
63. Kellner KF, Urban NA, Swihart RK (2013) Short-term responses of small mammals to timber harvest in the United States Central Hardwood Forest region. J Wildl Manage 77: 1650–1663.
64. Yackulic CB, Chandler R, Zipkin EF, Royle JA, Nichols JD, et al. (2013) Presence-only modelling using MAXENT: when can we trust the inferences? Method Ecol Evol 4: 236–243.
65. Phillips S, Anderson R, Schapire R (2006) Maximum entropy modeling of species geographic distributions. Ecol Modell 190: 231–259.
66. Dorazio RM (2012) Predicting the geographic distribution of a species from presence-only data subject to detection errors. Biometrics 68: 1303–1312.
67. Royle JA, Chandler RB, Yackulic C, Nichols JD (2012) Likelihood analysis of species occurrence probability from presence-only data for modelling species distributions. Methods Ecol Evol 3: 545–554.
68. Merow C, Silander JA Jr (2014) A comparison of Maxlike and Maxent for modelling species distributions. Methods Ecol Evol 5: 215–225.
69. Burnham KP (1981) Summarizing remarks: environmental influences. Stud Avian Biol 6: 324–325.
70. Thompson WL (2002) Towards reliable bird surveys: accounting for individuals present but not detected. Auk 119: 18–25.
71. Ellingson AR, Lukacs PM (2003) Methods for regional landbird monitoring: a reply to Hutto and Young. Wildl Soc Bull 31: 896–902.
72. MacKenzie DI, Kendall WL (2002) How should detection probabilities be incorporated into estimates of relative abundance? Ecology 83: 2387–2393.
73. Bart J, Droege S, Geissler P, Peterjohn B, Ralph CJ (2004) Density estimation in wildlife surveys. Wildl Soc Bull 32: 1242–1247.
74. Johnson DH (2008) In defense of indices: the case of bird surveys. J Wildl Manage 72: 857–868.
75. Rosenbaum PR (2005) Observational study. In: Everitt BS, Howell DC, editors. Encyclopedia of Statistics in Behavioral Science. Hoboken, NJ: John Wiley and Sons. 1451–1462.
76. Hutto RL, Young JS (2003). On the design of monitoring programs and the use of population indices: a reply to Ellingson and Lukacs. Wildl Soc Bull 31: 903–910.
77. Welsh AH, Lindenmayer DB, Donnelly CF (2013). Fitting and interpreting occupancy models. PloS One 8: e52015.
78. Guillera-Arroita G, Lahoz-Monfort JJ, MacKenzie DI, Wintle BA, McCarthy MA (2014). Ignoring imperfect detection in biological surveys is dangerous: a response to 'fitting and interpreting occupancy models'. PloS One 9: e99571.
79. MacKenzie DI, Nichols JD, Royle JA, Pollock KH, Bailey LL, et al. (2006) Occupancy estimation and modeling: inferring patterns and dynamics of species occurrence. New York: Academic Press. 324 p.
80. Cooper CB, Daniels SJ, Walters JR (2008) Can we improve estimates of juvenile dispersal distance and survival? Ecology 89: 3394–3361.
81. Moore JE, McEuen AB, Swihart RK, Contreras TA, Steele MA (2007) Determinants of seed removal distance by scatter hoarding rodents in deciduous forests. Ecology 88: 2529–2540.
82. Gilroy JJ, Virzi T, Boulton RL, Lockwood JL (2012) A new approach to the "apparent survival" problem: estimating true survival rates from mark-recapture studies. Ecology 93: 1509–1516.
83. Lichti NI, Steele MA, Zhang H, Swihart RK (2014) Mast species composition alters seed fate in North American rodent-dispersed hardwoods. Ecology 95: 1746–1758.
84. Boulanger J, Kendall KC, Stetz JB, Roon DA, Waits LP, et al. (2008) Multiple data sources improve DNA-based mark-recapture population estimates of grizzly bears. Ecol Appl 18: 577–589.
85. O'Connell AF, Nichols JD, Karanth KU (2011) Camera traps in animal ecology: methods and analyses. New York: Springer. 280 p.
86. Rees HC, Maddison B, Middleditch DJ, Patmore JRM, Gough KC (2014) The detection aquatic animal species using environmental DNA – a review of eDNA as a survey tool in ecology. doi:10.1111/1365-2664.12306.

Biodiversity Patterns along Ecological Gradients: Unifying β-Diversity Indices

Robert C. Szava-Kovats*, Meelis Pärtel

Institute of Ecology and Earth Sciences, University of Tartu, Tartu, Estonia

Abstract

Ecologists have developed an abundance of conceptions and mathematical expressions to define β-diversity, the link between local (α) and regional-scale (γ) richness, in order to characterize patterns of biodiversity along ecological (i.e., spatial and environmental) gradients. These patterns are often realized by regression of β-diversity indices against one or more ecological gradients. This practice, however, is subject to two shortcomings that can undermine the validity of the biodiversity patterns. First, many β-diversity indices are constrained to range between fixed lower and upper limits. As such, regression analysis of β-diversity indices against ecological gradients can result in regression curves that extend beyond these mathematical constraints, thus creating an interpretational dilemma. Second, despite being a function of the same measured α- and γ-diversity, the resultant biodiversity pattern depends on the choice of β-diversity index. We propose a simple logistic transformation that rids beta-diversity indices of their mathematical constraints, thus eliminating the possibility of an uninterpretable regression curve. Moreover, this transformation results in identical biodiversity patterns for three commonly used classical beta-diversity indices. As a result, this transformation eliminates the difficulties of both shortcomings, while allowing the researcher to use whichever beta-diversity index deemed most appropriate. We believe this method can help unify the study of biodiversity patterns along ecological gradients.

Editor: Jack Anthony Gilbert, Argonne National Laboratory, United States of America

Funding: This study was supported by Estonian Research Council institutional research funding IUT20-29 and by the European Union through the European Regional Development Fund (Centre of Excellence FIBIR). The funders had no role in study design, data collection and analysis, decision to publish, or preparation of the manuscript.

Competing Interests: The authors have declared that no competing interests exist.

* Email: robszav@ut.ee

Introduction

Ecologists have long expressed interest in spatial and environmental effects on biodiversity, specifically the effects with respect to β-diversity, which quantifies the similarity of species assemblage among sites and represents the link between local (α) diversity and regional (γ) diversity [1]. Consensus on a standard definition of β-diversity, however, has been elusive and has been the subject of considerable debate – both in terms of its fundamental essence [1–5] and its mathematical relationship with α- and γ-diversity [2,4,6–8]. Much mathematical discussion revolves on whether β-diversity is defined better by a multiplicative (i.e., $\alpha\beta_M = \gamma$) [9] or an additive (i.e., $\alpha + \beta_A = \gamma$) [10,11] partition.

Over time, many different expressions for "β-diversity" have been proposed, which, in addition to considerable ambiguous terminology, has led to confusion within the ecological community [2,4,12]. Although each expression addresses the compositional similarity among sites, each is also a different measure of compositional similarity and thus, describes a somewhat different concept. The choice of beta-diversity index is, therefore, dependent on the researcher's ecological application [4]. One set of beta-diversity indices, the "classical metrics" [1], defines α-diversity in terms of the mean diversity in local sites and γ-diversity as the composite of these local sites.

One popular approach to quantify spatial or environmental patterns of biodiversity is regression analysis of a classical β-diversity index against one or more spatial or environmental gradients [13–19]. This approach, however, is subject to two shortcomings that can undermine the validity of the resultant diversity patterns. First, classical β-diversity indices are mathematically constrained, i.e., each index features a lower limit when all local sites are compositionally the same and an upper limit when all local sites are unique. Regression of such β-diversity indices can result in a predicted estimation that crosses these limits, thus violating their mathematical constraints. Such instances cast doubt on the validity of the regression diagnostics, e.g., correlation, residuals, t-tests etc. Second, the regression analysis is dependent on the choice of β-diversity index, thus a biodiversity pattern resultant from one β-diversity index may be radically different from another β-diversity index.

We introduce a simple transformation that rids commonly used classical β-diversity indices of their lower and upper constraints, thereby eliminating the risk of regression analysis producing non-interpretable regression curves. Moreover, regression of these transformed indices against spatial or environmental gradients yields identical relationships for these commonly used classical β-diversity indices. As a result, this transformation eliminates the difficulties of both shortcomings, while allowing the researcher to use whichever β-diversity index deemed most appropriate.

Table 1. Mathematical definitions and expressions of beta-diversity indices.

β-index	Function	Low	High
β_{Md}	γ/α	1	N
$\beta_{Md\text{-}1}$	$(\gamma-\alpha)/\alpha$	0	$N-1$
β_{Pt}	$(\gamma-\alpha)/\gamma$	0	$1-(1/N)$

Low and High represent the lower and upper limits of beta-diversity indices as a function of the number of local sites (N).

Methods: Mathematical Properties of Logistic-Transformed β-Diversity Indices

Classical β-diversity indices are derived from the measured local diversity (α) – usually expressed as the arithmetic mean of species richness in N local sites – and regional diversity (γ), the measured species richness of all N local sites combined. β-diversity is often expressed by one of three β-diversity indices, which are described here using the terminology in Tuomisto [4]. True β-diversity, β_{Md}, expressed as γ/α, quantifies the number of compositional units in the collection. β_{Md} is equivalent to β_M, the basic multiplicative diversity partition. $\beta_{Md\text{-}1}$, expressed as $\gamma/\alpha-1$, provides the number of complete turnovers among the compositional units in the collection. Proportional species turnover, β_{Pt}, expressed as $(\gamma-\alpha)/\gamma$, described the proportion of species found regionally that are not found locally.

All three β-diversity indices, β_{Md}, $\beta_{Md\text{-}1}$, and β_{Pt}, assume a minimum value when all local sites are compositionally identical and a maximum when all local sites are compositionally unique. The lower limit for $\beta_{Md\text{-}1}$ and β_{Pt} is zero and for β_{Md} unity. The respective upper limits are a function of N (Table 1).

A logistic transformation ($y=\ln[x/(1-x)]$ is a standard method by which to treat data that are constrained by upper and lower limits [20]. An analogous transformation applied to these β-diversity indices takes the generalized form, $\beta^*=\ln[(\beta-\beta_{Min})/(\beta_{Max}-\beta)]$, where β_{Max} and β_{Min} are the respective upper and lower limit of a β-diversity index, β. Index β can be recovered from β^* by $\beta=\exp(\beta^*)/[1+\exp(\beta^*)](\beta_{Max}-\beta_{Min})+\beta_{Min}$. This retransformation allows the depiction of beta-diversity in its more familiar constrained units, as a logistic variable may be difficult to visualize. The beta-diversity relationship with a gradient can be plotted simply as the retransformed β^* as a function of the gradient. Any value of β^* will — upon retransformation into β — adhere to the mathematical constraint, $\beta_{Min}<\beta<\beta_{Max}$. Given, for example, a set of $N=10$ local sites with $\alpha=50$ and $\gamma=100$,

Table 2. Values of γ- and α-diversity along a hypothetical ecological gradient for three scenarios.

Gradient	Scenario			
	γ	α(A)	α(B)	α(C)
1	10	9.5	2	9.5
2	20	18.6	6	10
3	40	34	6	38
4	30	18	10.5	9
5	50	37.5	17.5	31.2
6	80	24	12	46.5
7	70	21	8.4	14
8	100	12	12	40
9	80	9.6	12	24
10	90	13.5	9.9	35.8

Number of local sites (N) = 10.

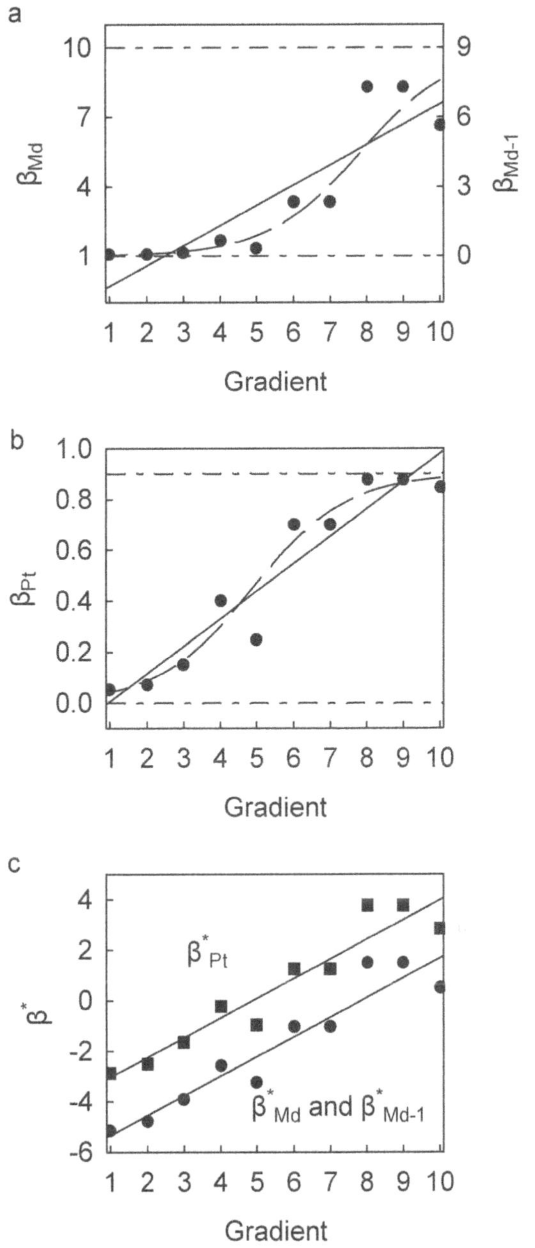

Figure 1. Scatterplots of beta-diversity indices against hypothetical ecological gradient for Scenario A. (a) β_{Md} (left axis) and $\beta_{Md\text{-}1}$ (right axis); linear regression trends, β_{Md} $y=0.87x-1.17$, for $\beta_{Md\text{-}1}$ $y=0.87x-2.17$ ($r=0.88$ for both). (b) β_{Pt}; linear regression trend, $y=0.11x-0.10$ ($r=0.95$). (c) β^*_{Md}, $\beta^*_{Md\text{-}1}$ (circles) and β^*_{Pt} (squares); linear regression trends, for β^*_{Md} $\beta^*_{Md\text{-}1}$ $y=0.78x-6.10$, for β^*_{Pt} $y=0.78x-3.80$ ($r=0.91$ for all). Dashed trends in (a) and (b) depict linear trends of β^*_{Md} (and $\beta^*_{Md\text{-}1}$) and β^*_{Pt} retransformed to β_{Md} (and $\beta_{Md\text{-}1}$) and β_{Pt}, respectively. See Table 1 for description of beta-diversity indices and Table 2 for data for Scenario A.

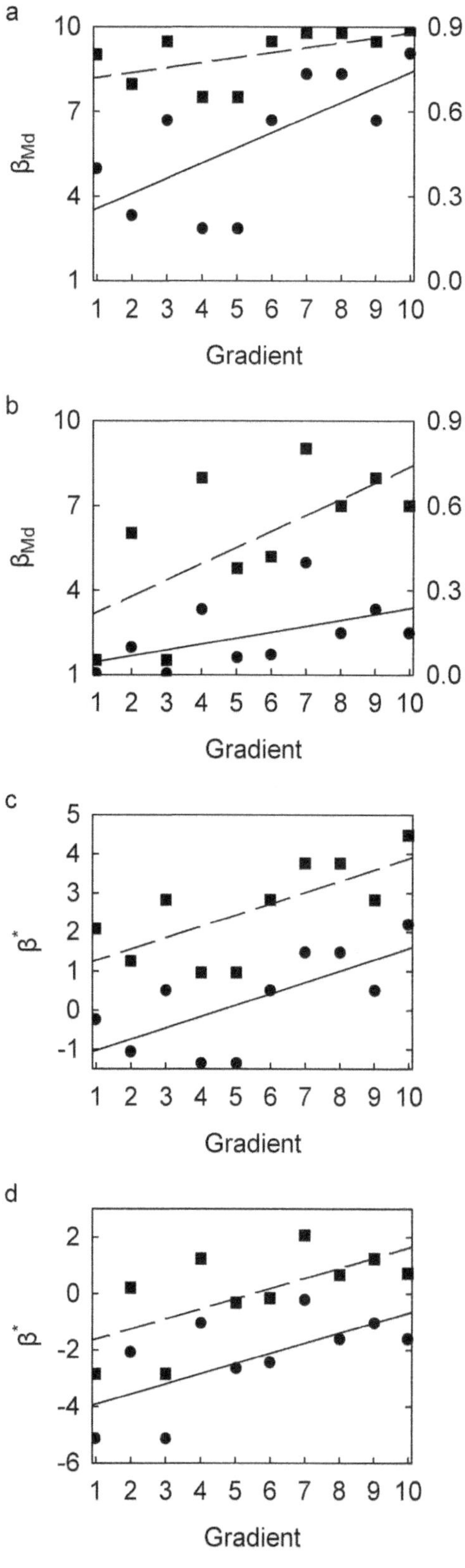

Figure 2. Scatterplots of beta-diversity indices against hypothetical ecological gradient. (a) Scenario B: β_{Md} (circles, left axis) and β_{Pt} (squares, right axis); linear regression trends, for β_{Md} $y=0.54x+3.02$ ($p=0.026$), for β_{Pt} $y=0.02x+0.70$ ($p=0.097$). (b) Scenario C: β_{Md} (circles, left axis) and β_{Pt} (squares, right axis); linear regression trends, for β_{Md} $y=0.21x+1.25$ ($p=0.121$), for β_{Pt} $y=0.06x+1.64$ ($p=0.036$). (c) Scenario B: β^*_{Md} (circles) and β^*_{Pt} (squares); linear regression trends, for β^*_{Md} $y=0.29x-1.32$, for β^*_{Pt} $y=0.29x+0.98$ ($p=0.023$ for both). (d) Scenario C: β^*_{Md} (circles) and β^*_{Pt} (squares); linear regression trends, for β^*_{Md} $y=0.36x-4.27$, for β^*_{Pt} $y=0.36x-1.97$ ($p=0.039$ for both). See Table 1 for description of beta-diversity indices and Table 2 for data for Scenario A.

β_{Md}, $\beta_{Md\text{-}1}$, and β_{Pt} are 2, 1, and 0.5, respectively. Corresponding values of β^*_{Md}, $\beta^*_{Md\text{-}1}$, and β^*_{Pt} are −2.08, −2.08, and 0.22, respectively. Note that because $\beta_{Md}-\beta_{Md\text{-}1}=1$, $\beta^*_{Md}=\beta^*_{Md-1}$, and that $\beta^*_{Pt}=\beta^*_{Md}+\ln(N)=\beta^*_{Md\text{-}1}+\ln(N)$.

Results: Illustrative Example

Table 2 provides a set of illustrative data of α- and γ-diversity along a hypothetical environmental or spatial gradient to compare the performance of these commonly used classical β-diversity indices with their logistic-transformed equivalents. Ten points along the gradient were assigned values of γ and different values of α for three scenarios. In all scenarios, $N=10$, thus the upper limits for β_{Md}, $\beta_{Md\text{-}1}$, and β_{Pt} are 10, 9, and 0.9, respectively.

In Scenario A, β_{Md}, $\beta_{Md\text{-}1}$, and β_{Pt} all exhibit a statistically significant positive relationship along the gradient. The linear regression curves of β_{Md} and $\beta_{Md\text{-}1}$ cross below the lower limits, 1 and 0 respectively, at gradient values <~3 (Figure 1a). The linear regression curve of β_{Pt}, derived from the same set of α and γ, crosses the upper limit at gradient values >~9 (Figure 1b). In other words, Scenario A provides an example of two different β-diversity indices violating their mathematical constraints at opposite ends of the gradient. The logistic-transformed equivalents, β^*_{Md}, $\beta^*_{Md\text{-}1}$, and β^*_{Pt}, when regressed along the gradient yield statistically significant linear relationships with identical values for all regression parameters except for the intercept (Figure 1c). The intercept of β^*_{Pt} is 2.30 greater than that of β^*_{Md} and $\beta^*_{Md\text{-}1}$, which is equivalent to exp(N), i.e., exp(10). Moreover, the resultant regression curves when retransformed into their original β-diversity indices adhere to their respective upper and lower constraints (Figure 1a,b).

The linear regression curves of β_{Md}, $\beta_{Md\text{-}1}$, and β_{Pt} in Scenarios B and C all lie within the upper and lower limits. However, the relationship between β_{Pt} and the gradient in Scenario B is statistically significant at 95% confidence ($p=0.026$), whereas those between β_{Md} and $\beta_{Md\text{-}1}$ are non-significant ($p=0.097$) (Figure 2a). This result is opposite in Scenario C: the relationship between β_{Pt} and the gradient is statistically non-significant ($p=0.121$), whereas those between β_{Md} and $\beta_{Md\text{-}1}$ are significant ($p=0.036$) (Figure 2b). In addition, the magnitude of the residuals in Scenarios B and C differ between β_{Md} (or $\beta_{Md\text{-}1}$) and β_{Pt}. By contrast, linear regression of β^*_{Md}, $\beta^*_{Md\text{-}1}$, and β^*_{Pt} against the gradient results in identical models for both Scenario B and C with the exception of their respective intercepts. The relationship is significant for both Scenario B ($p=0.023$) and Scenario C ($p=0.039$) and both relationships are independent of the choice of beta-diversity index. In addition, this choice has no effect on the resultant residual patterns (Figure 2c,d).

Discussion

Space structure is an often neglected consideration in statistical analysis. All conventional multivariate statistical analysis assume that the data occupy Euclidean space [21], yet this is rarely the case. A constrained space structure can usually be readily transformed into Euclidean space; log-transformation of positive-only data is a well-known example. Likewise, logistic transformation can be used to place data constrained by upper and lower limits into an unbounded line in Real space. Logistic transformation is commonly used for data restricted to values between one and unity (as in odds ratios) [20,22], but can also be performed on constrained classical β-diversity indices. As our illustrative example shows, regression analysis of "raw" classical β-diversity indices can result in regression curves that violate their imposed mathematical constraints, thereby undermining the validity of the regression model and its interpretation. By contrast, regression of logistic-transformed β-diversity indices excludes the possibility of violating these mathematical constraints, thereby eliminating any "impossible" results that would undermine the regression model.

Logistic transformation also eliminates the effects of the choice of these classical β-diversity indices. Although regression of β_{Md} and β_{Md-1} results in identical diversity patterns (the intercept using β_{Md-1} is one less than that using β_{Md}), neither is equivalent to β_{Pt}, which results in a different diversity pattern. This dilemma is especially worrisome; it seems intuitive that a single measured set of α- and γ-diversity should lead to a unique diversity pattern even if β-diversity is expressed by different indices. Our illustrative example shows that a diversity pattern along an ecological gradient can be statistically significant using one index and non-significant using another. Even if the interpretational contrast is not so dramatic, the difference among β-diversity indices can still affect the residuals, which can influence outlier detection, particularly when data points lie close to the limits. By contrast, logistic transformation results in a unique diversity pattern for all three logistic-transformed indices. The difference between β^*_{Md} (and β^*_{Md-1}) and β^*_{Pt} is simply a function of the number of local sites. As a result, all regression parameters (except the intercept) are identical for all β-diversity indices.

Although the illustrative examples shown reflect simple linear regression relationships, the two advantages of the logistic transformation are maintained regardless of the actual relationship, i.e. even if the relationship between logistic-transformed beta-diversity and ecological gradient is markedly non-linear. As a result, the researcher is free to choose any appropriate regression model, including many non-linear models, to describe the beta-diversity pattern. For instance, a polynomial or piecewise (i.e., segmented) regression model could be used to characterize a unimodal relationship. Positive-only regression models, such as logarithmic and exponential models, are excluded, because logistic beta-diversity indices can assume negative values. The lack of equivalency among the "raw" classical beta-diversity indices can lead to inconsistency in describing even the qualitative nature of biodiversity patterns. For example, whereas the relationship of β_{Pt} with the gradient in Figure 2b is linear, the relationship with β_{Md} and β_{Md-1} is arguably non-linear (Figure 2a). As a result, using β_{Md} or β_{Md-1} could lead to the conclusion that beta diversity increases only along the upper part of the gradient, whereas using β_{Pt} suggests that beta-diversity increases along the entire gradient.

The lack of equivalency notwithstanding, attempts to circumvent the constraints on classical β-diversity indices by other methods are fraught with difficulty. For instance, the arc-sine transformation, despite its long tradition of use to mitigate constrained data such as percentage data, has been the subject of criticism and is rapidly running out of favour [22–24]. Moreover, the results of regression on arc-sine transformed indices remains dependent on the choice of index. Regression of logistic-transformed indices typically presumes that the data contain no regions in which all the local sites are either unique or identical; logistic transformation is impossible for a data point located exactly on an upper or lower limit. However, methods are available to perform regression analysis based on a logistic-transformed response variable that includes otherwise non-transformable data [25,26].

Although we have demonstrated the use of logistic transformation on the regression of classical β-diversity indices, the transformation can also be used to circumvent violation of the constraints of pairwise "multivariate measures" [1], such as the Jaccard and Sørensen indices, which are constrained to values between 0 and 1. The Jaccard and Sørensen indices are equivalent to β_{Pt} and β_{Md-1}, respectively, for a pair of sites. As such, the logistic transformation can also result in a unique diversity pattern for the Jaccard and Sørensen indices. Logistic-transformation of β-diversity indices is not limited to simple presence-absence estimates with α-diversity calculated as the arithmetic mean. A similar approach can be followed by expressing α-diversity as a geometric mean, as the maximum richness in a set of local sites [27,28], or by expressing diversity in terms of "effective numbers of species", which incorporates species abundance [29–31]. "Effective numbers of species" can be calculated from, for instance, Shannon or Simpson indices. Moreover, the method can also be applied to phylogenetic and functional richness [32]. We suggest the logical transformation of β-diversity indices as a means of improving and simplifying the interpretation of diversity patterns along ecological gradients.

Author Contributions

Conceived and designed the experiments: RS MP. Performed the experiments: RS. Analyzed the data: RS. Contributed reagents/materials/analysis tools: RS MP. Wrote the paper: RS MP.

References

1. Anderson MJ, Crist TO, Chase JM, Vellend M, Inouye BD, et al. (2011) Navigating the multiple meanings of β diversity: a roadmap for the practicing ecologist. Ecology Letters 14: 19–28.
2. Jurasinski G, Koch M (2011) Commentary: do we have a consistent terminology for species diversity? We are on the way. Oecologia 167: 893–902.
3. Jurasinski G, Retzer V, Beierkuhnlein C (2009) Inventory, differentiation, and proportional diversity: a consistent terminology for quantifying species diversity. Oecologia 159: 15–26.
4. Tuomisto H (2010) A diversity of beta diversities: straightening up a concept gone awry. Part 1. Defingring beta diversity as a function of alpha and gamma diversity. Ecography 33: 2–22.
5. Legendre P, De Cáceras M (2013) Beta diversity as the variance of community data: dissimilarity coefficients and partitioning. Ecology Letters 16: 951–963.
6. Ellison AM (2010) Partitioning diversity. Ecology 91: 1962–1963.
7. Ricotta C (2010) On beta diversity decomposition: Trouble shared is not trouble halved. Ecology 91: 1981–1983.
8. Veech JA, Summerville KS, Crist TO, Gering JC (2002) The additive partitioning of species diversity: recent revival of an old idea. Oikos 99: 3–9.
9. Whittaker RH (1960) Vegetation of the Siskiyou Mountains, Oregon and California. Ecological Monographs 30: 279–338.
10. Lande R (1996) Statistics and partitioning of species diversity, and similarity among multiple communities. Oikos 76: 5–13.
11. MacArthur R, Recher H, Cody M (1966) On the relation between habitat selection and species diversity. American Naturalist 100: 319–332.
12. Tuomisto H (2010) A consistent terminology for quantifying species diversity? Yes, it does exist. Oecologia 164: 853–860.
13. Angeler DG, Drakare S (2013) Tracing alpha, beta, and gamma diversity responces to environmental change in boreal lakes. Oecologia 172: 1191–1202.
14. Janišová M, Michalcová D, Bacaro G, Ghisla A (2014) Landscape effects on diversity of semi-natural grasslands. Agriculture, Ecosystems and Environment 182: 47–58.
15. Lezama F, Santiago B, Altesor A, Cesa A, Chaneton EJ, et al. (2014) Variation of grazing-induced vegetation changes across a large-scale productivity gradient. Journal of Vegetation Science 25: 8–21.

16. Mori AS, Shiono T, Koide D, Kitagawa R, Ota AT, et al. (2013) Community assembly processes shape an altitudinal gradient of forest biodiversity. Global Ecology and Biogeography 22: 878–888.
17. Qian H, Chen S, Mao L, Ouyang Z (2013) Divers of β-diversity along latitudinal gradients revisited. Global Ecology and Biogeography 22: 659–670.
18. Qian H, Jong-Suk S (2013) Latitudinal gradients of associations between beta and gamma diversity of trees in forest communities in the New World. Journal of Plant Ecology 6: 12–18.
19. Rocha-Ortega M, Favila ME (2013) The recovery of ground ant diversity in secondary Lacandon tropical forests. Journal of Insect Conservation 17: 1161–1167.
20. Bolker BM (2008) Ecological Models and Data in R. Princeton, NJ: Princeton University Press. 396 p.
21. Bacon-Shone J (2011) A short history of compositional data analysis: Theory and Applications. In: V Pawlowsky-Glahn and A Buccianti, editors.Compositional Data Analysis.West Sussex: John Wiley & Sons, Ltd. pp.3–11.
22. Wilson K, Hardy CW (2002) Statistical analysis of sex ratios: an introduction. In: I. C. W Hardy, editor editors.Sex Ratios: Concepts and Research Methods. Cambridge: Cambridge University Press. pp.48–92.
23. Milligan GW (1987) The use of the Arc-Sine Transformation in the Analysis of Variance. Educational and Psychological Management 47: 563–573.
24. Warton DI, Hui FKC (2011) The arcsine is asinine: the analysis of proportions in ecology. Ecology 92: 3–10.
25. Papke LE, Wooldridge JM (1996) Econometric methods for fractional response variables with an application to 401(k) plan participation rates. Journal of Applied Economoetrics 11: 619–632.
26. Ramalho EA, Ramalho JJS, Murteira JMR (2011) Alternative estimating and testing empirical strategies for fractional regression models. Journal of Economic Surveys 25: 19–68.
27. Harrison S, Ross SJ, Lawton JH (1992) Beta diversity on geographic gradients in Britain. Journal of Animal Ecology 61: 151–158.
28. Magurran AE (2004) Measuring Biological Diversity.. Oxford: Blackwell Publishing. 256 p.
29. Hill MO (1973) Diversity and evenness: a unifying notation and its consequences. Ecology 54: 427–432.
30. Jost L (2006) Entropy and diversity. Oikos 113: 363–375.
31. Jost L (2007) Partitioning diversity into independent alpha and beta components. Ecology 88: 2427–2439.
32. Scheiner SM (2012) A metric of biodiversity that integrates abundance, phylogeny, and function. Oikos 121: 1191–1202.

4

Development by Design in Colombia: Making Mitigation Decisions Consistent with Conservation Outcomes

Shirley Saenz[1], Tomas Walschburger[1], Juan Carlos González[2], Jorge León[3], Bruce McKenney[4], Joseph Kiesecker[5]*

1 Northern Andes Program, The Nature Conservancy, Bogota, Colombia, **2** Northern Andes Program, The Nature Conservancy, Quito, Ecuador, **3** Latin America Program, The Nature Conservancy, Cartagena, Colombia, **4** Global Conservation Lands Program, The Nature Conservancy, Charlottesville, Virginia, United States of America, **5** Global Conservation Lands Program, The Nature Conservancy, Fort Collins, Colorado, United States of America

Abstract

Mitigation policy and regulatory frameworks are consistent in their strong support for the mitigation hierarchy of: (1) avoiding impacts, (2) minimizing impacts, and then (3) offsetting/compensating for residual impacts. While mitigation frameworks require developers to avoid, minimize and restore biodiversity on-site before considering an offset for residual impacts, there is a lack of quantitative guidance for this decision-making process. What are the criteria for requiring impacts be avoided altogether? Here we examine how conservation planning can guide the application of the mitigation hierarchy to address this issue. In support of the Colombian government's aim to improve siting and mitigation practices for planned development, we examined five pilot projects in landscapes expected to experience significant increases in mining, petroleum and/or infrastructure development. By blending landscape-level conservation planning with application of the mitigation hierarchy, we can proactively identify where proposed development and conservation priorities would be in conflict and where impacts should be avoided. The approach we outline here has been adopted by the Colombian Ministry of Environment and Sustainable Development to guide licensing decisions, avoid piecemeal licensing, and promote mitigation decisions that maintain landscape condition.

Editor: Zoe G. Davies, University of Kent, United Kingdom

Funding: Funding was provided by the Colombian Ministry of Environment, The Nature Conservancy, World Wildlife Fund, and Conservation International. The funders had no role in study design, data collection and analysis, decision to publish, or preparation of the manuscript.

Competing Interests: The authors have declared that no competing interests exist.

* E-mail: jkiesecker@TNC.ORG

Introduction

The world is transforming rapidly, with world population projected to grow to over 9 billion by 2050 and increasing demands for food, water, energy, minerals, and other resources [1]. Over the next two decades, energy and mining companies will invest unprecedented sums – well over 20 trillion dollars – in projects around the world, from China to Colombia [1,2]. Such projects can pose a significant challenge for biodiversity conservation, especially because in most cases the environmental mitigation process for addressing biodiversity impacts is piecemeal, opaque, and inadequate for delivering effective conservation outcomes [3–6].

Mitigation offers the opportunity to address the impacts of development through application of the mitigation hierarchy: avoid, minimize, restore and offset [3,7]. But there are many problems with how mitigation is applied [3–6,8]. Traditionally, mitigation has been carried out on a project-by-project basis; specific measures are implemented to mitigate project impacts at a site, usually on or adjacent to the impact site [9]. Applying the mitigation hierarchy on a project-by-project basis, often at small spatial extents, underestimates the cumulative impacts of multiple current or future development projects and undermines the hierarchy's purpose and utility [10]. Thus existing mitigation measures do not address cumulative impacts associated with development; do not provide a structured decision-making framework to determine when projects can proceed or should be avoided; and do not harness the full potential of offsets (conservation actions applied away from the development site).

Blending systematic conservation planning and mitigation decision making

Landscape-level conservation planning is the process of locating, configuring and managing areas to maintain viability of biodiversity and other natural features [11,12]. A conservation portfolio (= priority sites), the end product of conservation planning, is a set of areas selected to represents the full distribution and diversity of these features [13]. Often plans utilize an optimization approach automated with spatial analysis tools such as Marxan [14], where the portfolio is designed to meet the minimum viability needs of each biological target in a configuration that minimizes the amount of area selected [14,15]. The key feature of a conservation plan is the clear articulation of a biodiversity vision that incorporates the full range of biological features, how they are currently distributed, and the minimum needs of each feature to maintain long-term persistence [16–18].

Landscape-level conservation plans can be used to guide the application of the mitigation hierarchy [10,19]. Where plans have already been completed, proposed developments can be mapped and assessed relative to the conservation portfolio and areas overlapping could trigger the avoidance of development impacts. Alternatively, overlap between the portfolio and proposed development may result in a "re-drawing" of the conservation portfolio to re-capture habitat needed to meet the minimal needs of the biodiversity features impacted by development. If minimum viability needs cannot be met elsewhere within the study-area, the proposed development would need to minimize impacts to the degree that maintains biodiversity values or avoid impacts altogether [10]. If adopted, the latter would provide an opportunity to avoid conflict between potential development and areas that are critical for biodiversity and provide the structure to guide decisions regarding which step in the mitigation hierarchy should be applied in response to proposed development [9,10,20].

Colombia: a case study in mitigation planning

Recognizing the need to improve its existing regulatory framework for mitigation, the Colombian government asked The Nature Conservancy in 2008 to help develop an approach to guide better decision making around siting and mitigation of future development [8]. While the existing regulatory system required development projects to avoid impacts to certain national and regional protected areas and ecosystems such as wetlands, mangroves, mountain grasslands "paramos," the government was seeking a more comprehensive mitigation framework that considers cumulative impacts and provides for the maintenance of minimal needs for biodiversity and ecosystems services [8].

Colombia is home to significant petroleum and mineral deposits, but also harbors some of the most biologically diverse places on Earth, with some of the highest known bird, amphibian and invertebrate species diversity [21–23]. Not surprisingly, some energy resources and mineral deposits intersect areas of significant biological value [24]. Conservation of biodiversity in Colombia is under threat, in part, because the Colombian government has authorized exploration of about 24 million hectares of the estimated 79 million hectares still remaining in natural ecological systems within the country [8,25]. The increase in development forecasted for Colombia may be compatible with biodiversity conservation if properly sited, but will still pose a challenge for conservation because of the large area potentially impacted and the resulting habitat loss and fragmentation. Many impacts can be mitigated or eliminated with appropriate siting and planning for development [26].

There is broad agreement among scholars, scientists, policy-makers, and regulators that the first and most important step in the mitigation hierarchy, avoidance, is ignored more often than it is implemented [4,5]. What are the criteria for requiring impacts be avoided altogether? Here we outline a conservation planning framework for proactively identifying future development that would be incompatible with long-term conservation goals (Figure 1). The conservation portfolios developed through this process can be used to identify areas where future development impacts should be avoided [10,19,20]. We focus on five landscapes where development is projected to increase rapidly over the coming years. This includes the expansion of coal mining in the Cesar River Valley, gold mining in Sur de Bolivar, highway development in Macarena, oil and gas development in Casanare and expansion of a sea port in the Bahia de Tribuga in Choco (Figure 2, Table 1). The approach we outline here has been adopted by the Colombian Ministry of Environment and Sustainable Development to guide licensing decisions, avoid piecemeal licensing, and promote mitigation decisions that maintain good landscape condition.

Methods

Our objective was to design an approach for identifying a conservation portfolio that, taking into account potential future development, maintains a minimum of 10 percent of the occurrence patterns of all terrestrial ecological systems in select pilot landscapes. We adopted a systematic conservation planning approach widely used to develop protected area networks that involves the following: 1) compile a list of important species and habitat types known collectively as 'biodiversity features', 2) collect spatial data on each of the biodiversity features targeted for protection, 3) set representation targets for the minimum amount of each feature intended for protection, 4) use spatial conservation prioritization software 'Marxan' in conjunction with expert based opinion to identify priority areas that meet representation targets, and 5) assess the cumulative impacts of future development and provide a regional context to better guide which step of the mitigation hierarchy should be applied (i.e. avoidance versus offsets). We sought to embed mitigation decisions into a landscape context. Previously, mitigation decisions were focused on the mining or oil and gas concessions and a small buffer around these areas defined often by the companies as part of the Environment Impact Assessments (Saenz & Walschburger unpublished data). Here we created project boundaries based on the main watersheds and administrative boundaries of municipalities within the project areas. A working group that included representatives from environmental authorities, research institutes, academia, indigenous and afro-Colombian groups, local and regional authorities and non-governmental organizations was formed to guide the process for each pilot landscape. This group helped provide the most current spatial data on our biological targets, assessments of the predictive models being developed, as well as insight into the approach being developed. We sought to apply rigorous, objective measures of conservation value whenever possible, recognizing that a quantitative assessment would require expert review and refinement.

Compile list of representative biodiversity features

Biodiversity cannot easily be measured completely and directly. Practitioners address this issue by selecting a set of biodiversity components that adequately represent the range of biological phenomena in the project area and can be measured effectively given existing resources. We selected a set of focal targets with sufficient breadth and depth using the "coarse-filter/fine-filter" approach consistent with The Nature Conservancy's Ecoregional Planning approach [27]. *Coarse filter* generally refers to "ecosystems"; in a more practical sense, it refers to mapped units of vegetation (Figure 3). The basic idea is that conserving a sample of each distinct vegetation type, in sufficient abundance and distribution, is an efficient way to conserve the majority of biological phenomena in the target area [28]. *Fine filter* generally refers to individual species, with specific habitat requirements or environmental relationships that are not adequately captured by the coarse filters (Figure 4). Narrow endemics and extreme habitat specialists, species with restrictive life histories, or those species that have experienced significant loss of habitat and/or are particularly sensitive to human perturbations fall into this category (i.e. IUCN Red List species). For our case studies we generated a list of biodiversity features using the coarse filter, fine filter criteria [29–33]. Details on targets used in each case study can be found in Table 2 and Table S1.

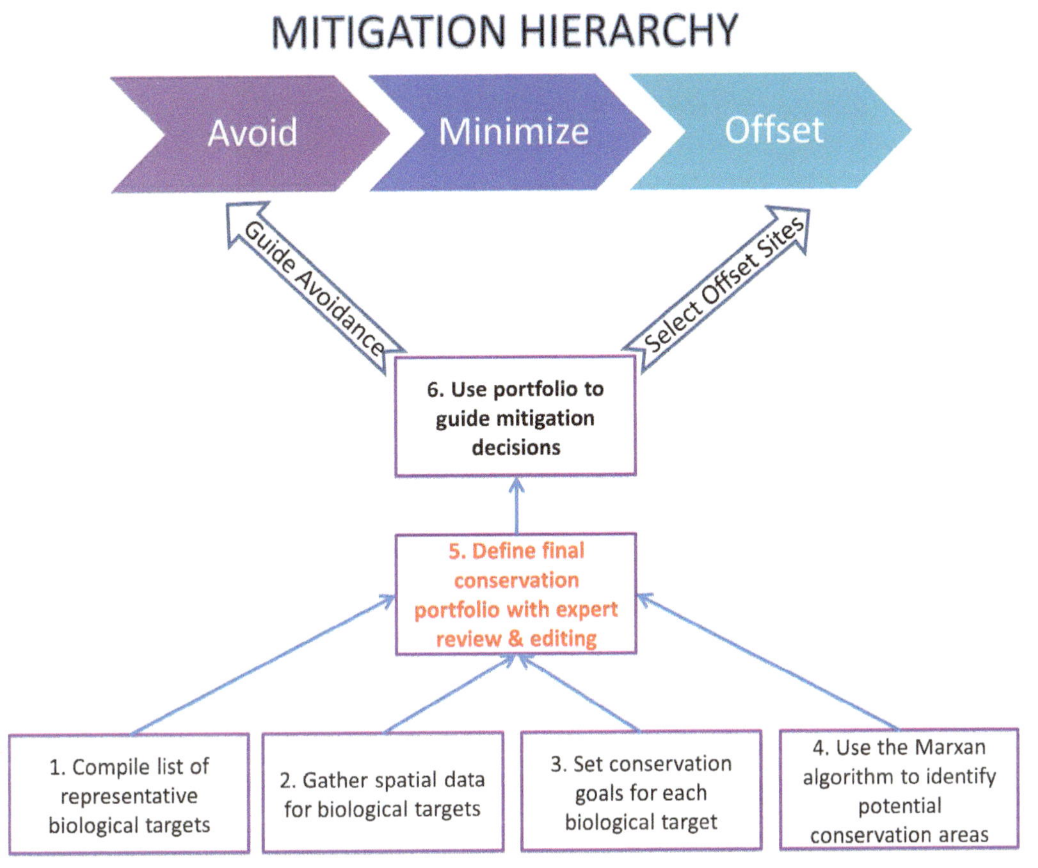

Figure 1. Steps used to blend landscape level conservation planning with the mitigation hierarchy, highlighting the decisions the current analysis is intended to inform: areas to avoid and the selection of offset sites.

Spatial data for biodiversity features

We used a combination of point survey data, vegetation cover estimations and predictive model estimations to represent the spatial distribution of selected targets. For all projects we utilized the national landcover data set with maps produced at the 1:500,000 scale [34]. At the pilot landscape level we refined this product to capture the location and delineation of each ecosystem unit with a remotely sensed exercise focused on each of the five pilot areas (Figure 3). For identification of ecosystems within each pilot landscape, we relied on a 1:100,000 scale land cover map generate with imagery from remote sensing sensors ASTER, and ETM + Landsat, taken between 2000 and 2008. The maps were obtained through the Landsat and TerraLook collections held in the USGS archive by using USGS Global Visualization Viewer - GLOVIS-, and remote sensing sensor Cbers2, obtained from Instituto Nacional de Pesquisas Espaciais (INPE) archive [29–33]. The remotely sensed data along with the national landcover map were used to generate a preliminary map of ecosystems units [29–33]. Where available we also utilized ecoregional assessments conducted by Galindo et al. [35,36].

Con este producto se genero un taller de expertos en el que se discutió la leyenda del mapa de ecosistemas, la delimitación y ubicación geográfica de cada una de las unidades y su pertinencia como objeto de conservación de filtro grueso.To select fine filter species targets we started with base species maps produced by NatureServe [37]. We also reviewed ecoregional assessments that were available for all of the five project areas and used species lists generated from these analyses [29–33,35,36]. Where existing species models were not available we settled on a simple approach of using deductive models by identifying the habitat preferences for each species creating binary models of suitable habitat through a series of GIS overlays based on: slope, aspect, topographic roughness, elevation (DEM), and vegetation type. The resulting species distribution models were subsequently validated by local experts (Figure 4).

Algunos criterios utilizados para la definición de la distribución potencial de las especies están basados en: rangos altitudinales, áreas geográficas descritas (municipios, departamentos, cuencas, toponimia, etc.), cartografía base y los mapas de Nature Serve (2007), donde para algunos casos de anfibios, aves y mamíferos fue de mucha utilidad

Setting conservation goals for biodiversity features

For each project we aimed to develop a portfolio of areas that would maintain Las metas de conservación para ecosistemas o filtro grueso se establecieron tomando como referencia el mínimo de 10% para ambientes terrestres según UICN. viability of ecosystems. Based on goals adopted by Colombian government we sought to maintain a minimum of 10 percent of the occurrence patterns of all terrestrial ecological systems [38].Para el caso de ecosistemas estratégicos (bosques de galería, humedales, paramos, cuerpos de agua) se definió una meta de conservación del 100% dado el régimen de manejo y su importancia por prestación de servicios ecosistémicos.Para el caso de ecosistemas estratégicos (bosques de galería, humedales, paramos, cuerpos de agua) se definió una meta de conservación del 100% dado el régimen de

Figure 2. Location of development by design pilot projects within Colombia.

manejo y su importancia por prestación de servicios ecosistémicos. Given the importance of ecosystem services provided by some ecological systems (riparian forests, wetlands, moors, bodies of water), we sought to conserve 100 percent of these systems. To set goals for ecological systems we utilized an approach developed by Pressey and Taffs [39] in which the amount of the current

Table 1. Pilot project site descriptions.

Pilot Project	Total Area (ha)	Conservation Portfolio Area (ha) and (%) of the Project Area	Future Potential Development (ha)	Percentage of the conservation portfolio that overlaps with future development areas	Area within Conservation Portfolio that Overlaps with Future Potential Development (ha)	Percentage of Future Potential Development Area that Overlaps with Conservation Portfolio
Coal Mining in Cesar	1,285,592	509,725 (39.6%)	55,668	2.95%	15,046	27.03%
Gold Mining in Sur de Bolivar	1,668,565	1,260,600 (75.5%)	117,975	6.77%	85,308	72.31%
Port in Bahia Tribuga Choco	340,265	232,803 (68.4%)	639	0.26%	601	94.02%
Macarena Road in Meta	811,457	330,600 (40.7%)	16,712	0.86%	2,837	16.97%
Oil and Gas in Casanare	1,892,780	717,700 (37.9%)	687,367	30.57%	219,367	31.91%

distribution needed to be conserved is calculated as a function of the goal as well as a function of the current distribution and landscape condition. Thus for each ecological system we calculated a distribution and condition metric.La ecuación propuesta para la estimación de la meta de conservación para cada uno de los objetos de conservación de filtro grueso fue la siguiente: The proposed equation for estimating the conservation goal for each of the terrestrial ecological systems was as follows (Table 3 shows an example of this calculation):

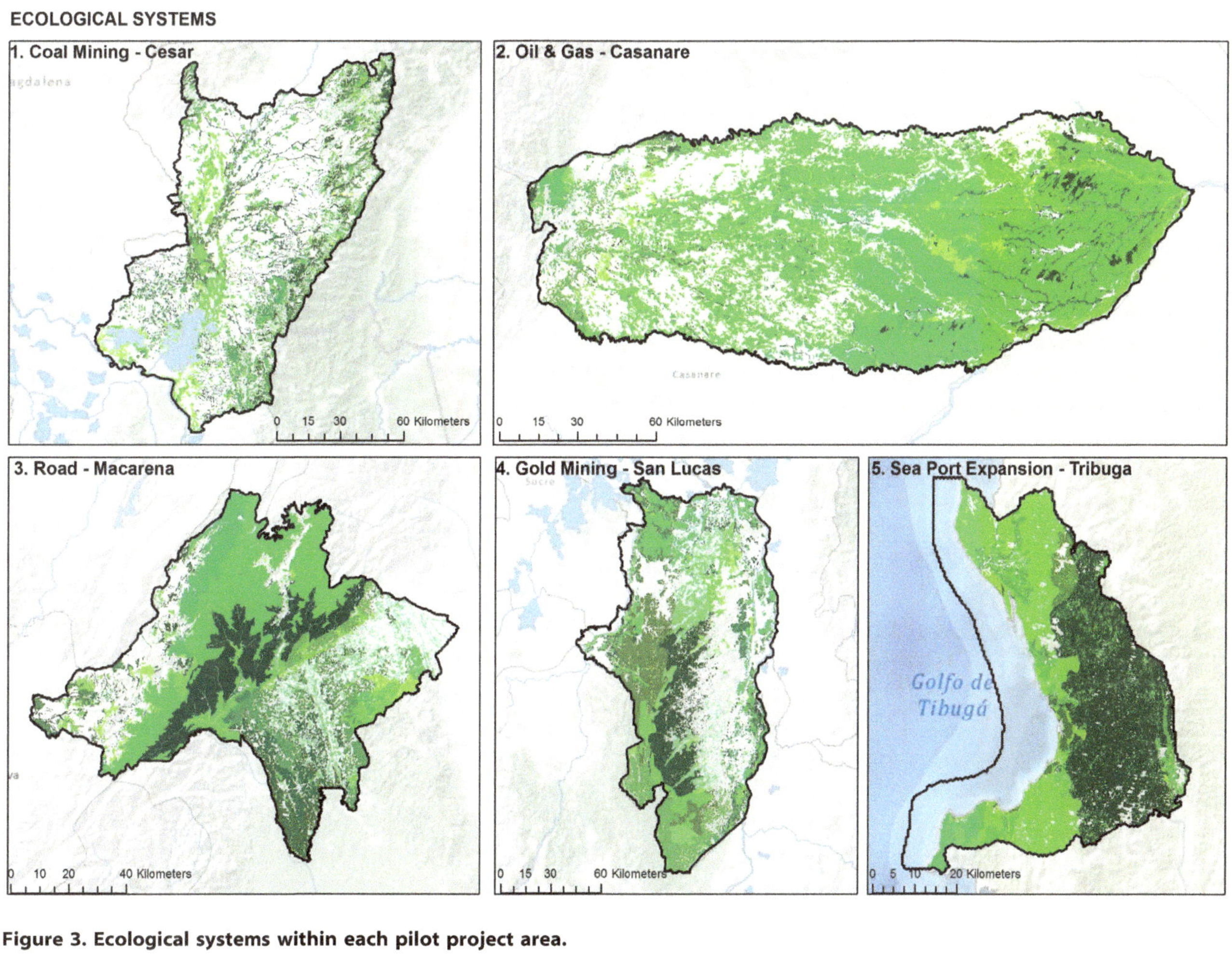

Figure 3. Ecological systems within each pilot project area.

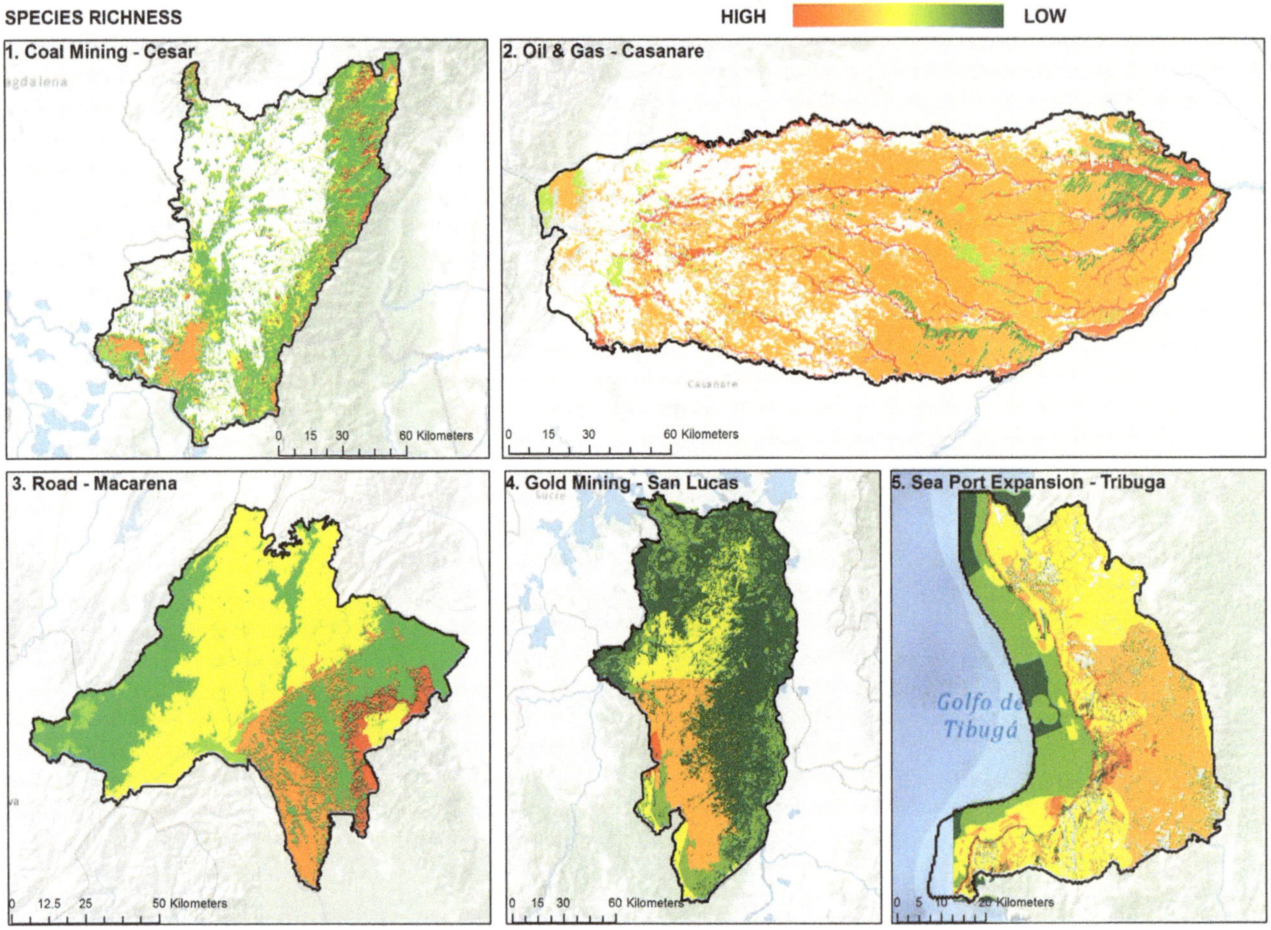

Figure 4. Species richness within each pilot project area.

Conservation Goal = [(10% * Distribution) + (5% * Condition)] * 1.5

Goals were lower for widely distributed ecological systems compared to rarer systems. In addition goals were higher for ecological systems with lower condition relative to those with higher condition. Condition was based on a human disturbance index described below. This approach was used in three projects (coal mining in the Cesar River Valley, gold mining in Sur de Bolivar, and expansion of a sea port in Bahia Tribuga). For the remaining projects a slight modification was added following the approach used by Galindo et al. [35,36] that incorporated future threat when calculating the goal. The proposed equation for estimating the conservation goal for each of the terrestrial ecological systems was:

Goal = (Rarity + Condition + Threat)

As above goals were higher for rare ecological systems and those with higher condition values. Ecological systems that are highly threatened also had a higher conservation goal. For the marine area in Tribuga, we used the conservations goals developed by Alonso et al. [40] in their marine ecoregional planning exercise for the Colombian Pacific. In all cases goals were between 35 and 100 percent and equate to the percent of the current area remaining

Table 2. Details of conservation portfolio design process using Marxan analysis for each of the 5 pilot landscapes.

PILOT PROJECT	PLANNING UNIT (Hexagon ha)	TOTAL NUMBER OF HEXAGONS	BOUNDARY LENGHT MODIFIER	NUMBER OF RUNS
Coal Mining in Cesar	25	490,327	0.2	500
Gold Mining in Sur de Bolivar	300	5,846	1.750	100
Port expansion in Bahia Tribuga Choco	20	16,527	1.0	100
Macarena Road in Meta	100	8,510	10	500
Oil and Gas in Casanare	100	19,424	10	500

Table 3. Example of conservation goal calculation for representative ecological system.

INDICATOR	ATTRIBUTE	WEIGHT (5-1)	SCORE	TOTAL WEIGHT
	Location in biome	5-1	5	
DISTRIBUTION	Rarity	5-1	4	4.3
	Location in watershed	5-1	4	
	Nearest neighbor distance	5-1	3	
CONDITION	Average proximity index	5-1	5	3.3
	Weighted Index Form	5-1	2	

Example from San Lucas Orobiome Gold Mining Pilot for Erosional Mountain Forest Ecological System.
GOAL = [(10% * Distribution) + (5% * Condition)] * 1.5
GOAL = [(10% * 4.3) + (5% * 3.3)] * 1.5
GOAL = 90%

for that ecological system. For example, a goal of 100 percent means that all of the current area remaining should be included in the conservation portfolio.

We choose not to assign specific goals to species, instead using species data to supplement the selection of sites included in the conservation portfolio. Once species models were compiled we summarized the species richness of these select target species across the pilot landscapes [37], (Figure 4). We placed the areas into categories based on Natural Jenks [41]. Details on the number of species for each category can be found in Table S1. Areas with the highest species richness were given priority for selection. All of the pilot landscapes followed this approach with the exception of the Bahia Tribuga port expansion project where we used species goals established by Alonso et al. [40].

Selecting potential conservation areas with Marxan

To design a conservation portfolio we applied the Marxan (version 1.8.2) site-selection algorithm developed by Ball and Possingham [14]. Marxan is a siting tool for landscape conservation analysis that explicitly incorporates spatial design criteria into the site selection process. Marxan utilizes an algorithm called "simulated annealing with iterative improvement" as a heuristic method for efficiently selecting regionally representative sets of areas for biodiversity conservation [42]. Marxan allows inputs of target occurrences represented as points, polygons in a GIS environment, and allows for conservation goals to be stated in a variety of ways, such as percent area, or numbers of point occurrences etc. The program also allows for the integration of many available spatial datasets on land use pattern and conservation status, and enables a rapid evaluation of alternative configurations. The ultimate objective is to identify a portfolio of planning units that meets the goals established for each biodiversity feature at lowest cost (i.e. cost = condition, conservation cost in dollars, size of the reserve etc.). Marxan is not designed to act as a stand-alone reserve design solution. Its effectiveness is dependent upon the involvement of expert opinion for review and revision. Marxan is sensitive to a number of the settings that are selected by the users (i.e. Boundary Length Modifier (BLM), number of runs and iterations). In all cases we assessed the sensitivity of solution sets to different values of these variables. For example to examine how the BLM affected spatial patterns of the solution we set BLM at zero and iteratively increased it by factors of ten. In consultation with the working group we visually inspected the results and examined how variation in the BLM affected the desired degree of clustering and cost. In all cases we assessed the sensitivity of variables in consultation with the working group [43]. Information on the levels for these settings can be found in Table 2 and Table S1. For all analyses we selected hexagons (derived from a uniform grid) as the unit of analysis for running Marxan that were of sufficient spatial resolution to represent biodiversity features and also large enough to permit efficient analyses across broad landscape scales (see Table 2 & Table S1 for specific rules used for each pilot area). The effectiveness of a contiguous set of hexagon units for defining natural variability, especially among spatially heterogeneous data sets, is well documented [44]. Each hexagon was populated by summing the area of suitable habitat for the targeted ecological systems or species richness.

To select conservation areas, we developed a set of decision rules. First, we guided site selection to areas with low levels of human disturbance [45]. Identification and quantification of current impacts is an integral part of ecoregional planning because the success of conservation strategies to protect species, landscapes, and ecosystems are dependent on the condition of the sites chosen [46]. The spatial characterization of the impacts is one of the determining factors in the selection of priority areas for conservation, because it directly or indirectly influences the viability of conservation of biodiversity [12,33]. We modeled six types of impacts for the five study areas: accessibility, livestock, agriculture, mining, energy and illicit crops (for areas where applicable, Figure 5) and combined these factors to calculate an index of cumulative impacts. This impact index is a component of the "cost" function utilized by Marxan [14,43]. As a result it was more costly to select planning units that had a high level of impact as represented by these datasets. Given the difficulty of restoration as well as uncertainty around maintaining conservation values in areas of high human influence, we felt it necessary to select areas in good condition (= areas with the lowest cost) and seek to keep these systems from becoming degraded rather than attempt to focus on restoration as a primary conservation action. Although, given the extensive amount of land conversion in these landscapes, it was necessary to select some degraded sites in order to meet conservation goals for some ecological systems. Next, we locked in areas [14,43] currently listed in the countries protected areas network [29–33,35,36]. We felt that this last rule was critical given the commitment already made to conservation of these areas and the need to complement ongoing efforts to maintain biodiversity of the country.

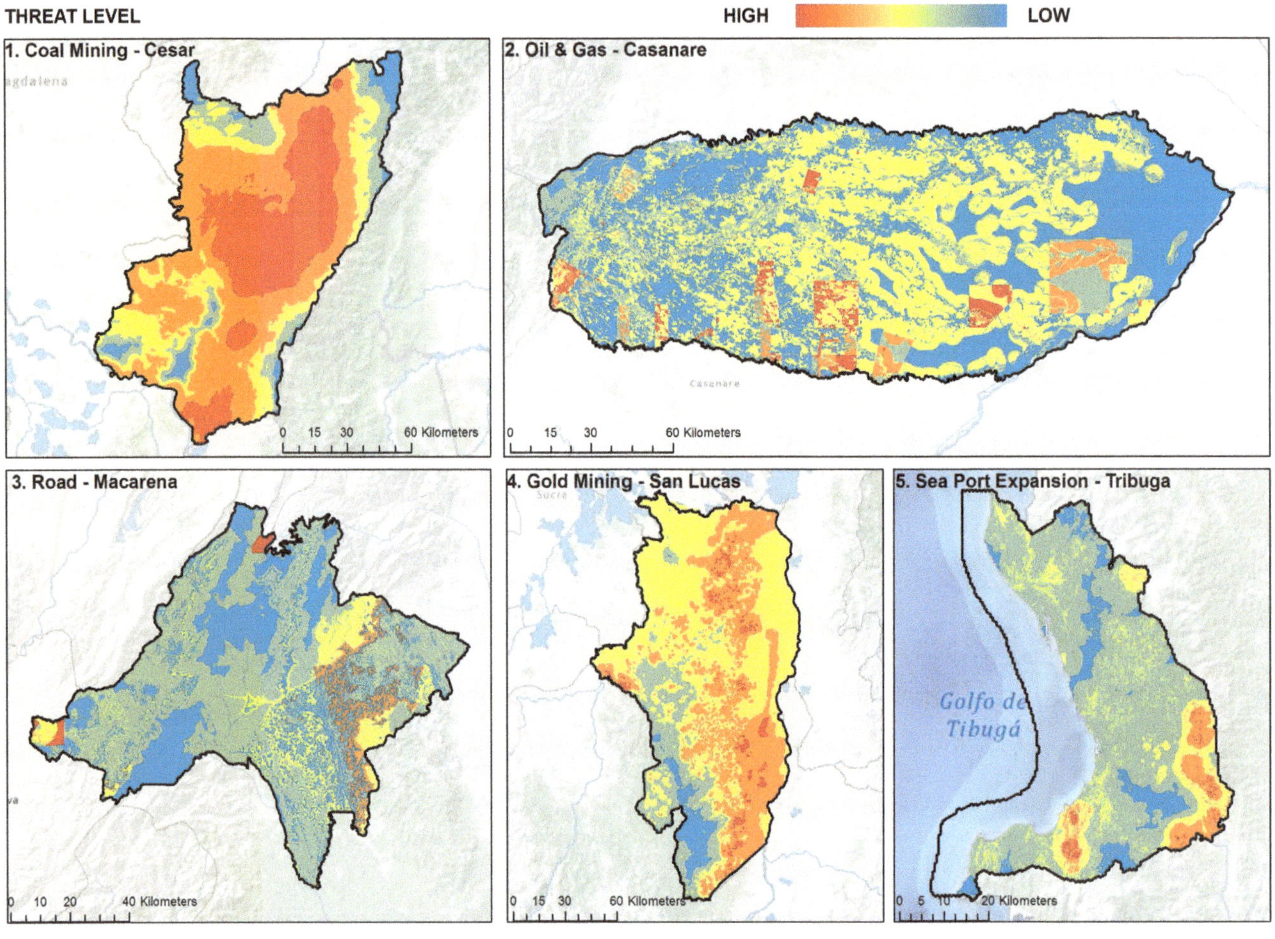

Figure 5. Threat Level or cost (= current condition) within each pilot project area.

Using selected conservation areas to guide mitigation decisions

We assessed the cumulative impacts of future development projects within a regional context to better guide which step of the mitigation hierarchy should be applied (i.e. avoidance versus offsets). We examined areas selected as part of our conservation portfolios in the context of expected future development for each of the five pilot landscapes (Figure 6). We took a fairly simple and conservative approach when making decisions about which proposed developments should be avoided. Based on the amount of habitat already converted to other uses and the desire to emphasize the need for avoidance, the working group decided that any overlap between the conservation portfolio and proposed development should be avoided and permits should not be granted for these leases. By assessing how future development impacts may affect biodiversity goals in a regional context, there is an opportunity to be proactive in avoiding conflict between development and conservation goals. Decision-makers can seek solutions that support development and maintenance of natural systems.

Results

The results of our analysis show the potential of conservation planning to proactively identify potential conflicts between conservation goals and development objectives and guide application of the mitigation hierarchy. The areas selected as part of the conservation portfolio represented from 38 percent to 75 percent of the total area of the pilot landscapes, providing some opportunity to resolve conflicts by simply re-designing development objectives to be met in areas that are not part of the conservation portfolio (Table 1, Figure 6). In general the overlap between the conservation portfolio and areas of future development represents only a small conflict for meeting conservation goals. With the exception of the gold mining activities in the Sur de Bolivar region (~7 percent%) and the oil and gas development in the Casanare region (~30 percent), the intersection between future development and conservation portfolio represents less than 3 percent of the conservation portfolio. However, the conflict appears more significant if we consider the proportion of the total area proposed for development that has been selected as part of the conservation portfolio. Approximately ~17 to 94 percent of the total area proposed for development (depending on the region) overlaps with the conservation portfolio, mainly because of the smaller size of proposed development areas compared to the area needed to meet conservation goals (Figure 6, Table 1). This overlap will represent a significant challenge for development in some of the landscapes.

Discussion

Our results illustrate how a portfolio of sites selected as part of an ecoregional assessment can be used to guide application the mitigation hierarchy identifying areas where development may

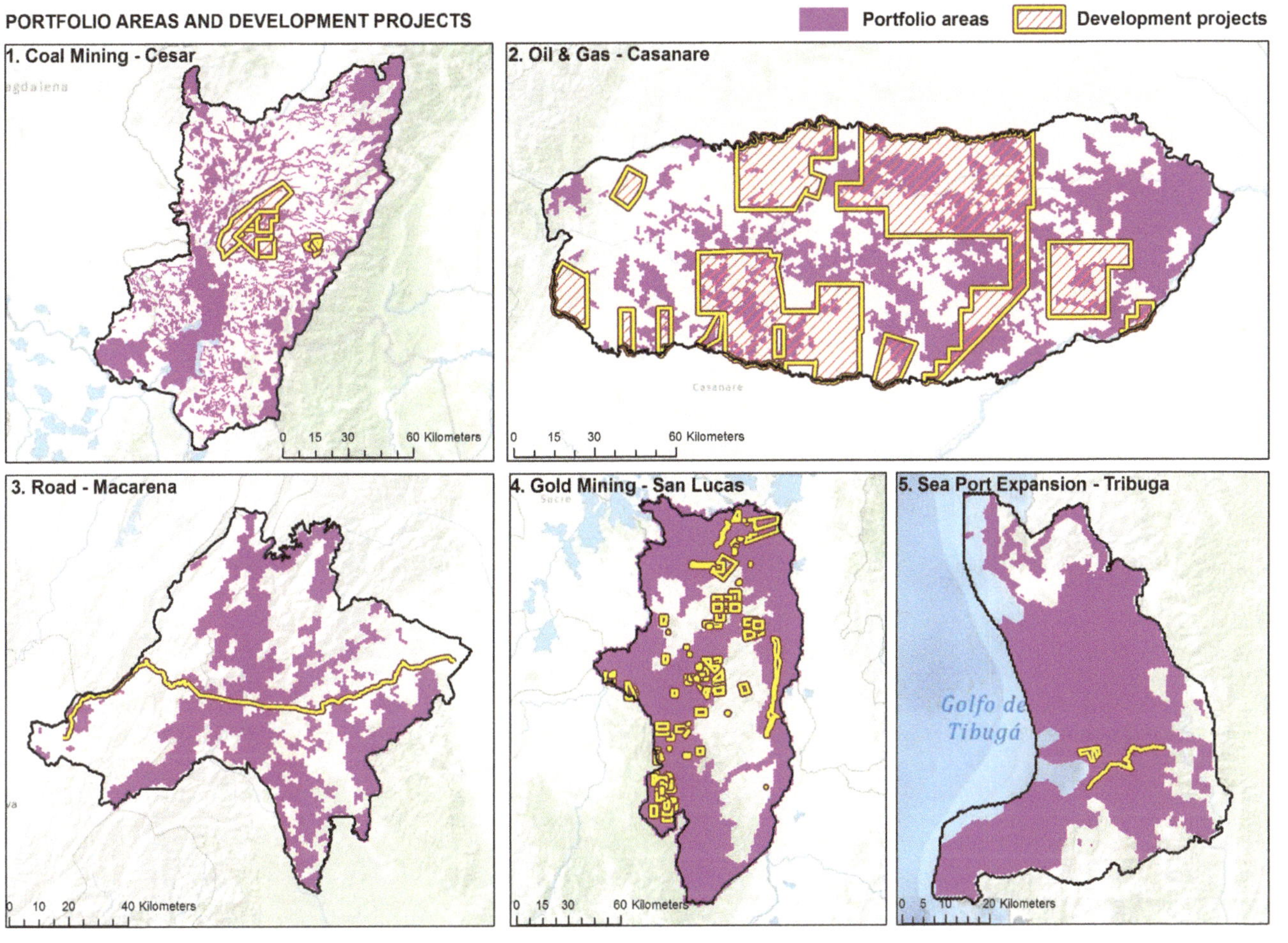

Figure 6. Landscape-level recommendations for the application of the mitigation hierarchy for each pilot project area. Portfolio of conservation sites selected by the ecoregional assessment in purple. Development potential outlined in yellow hash marks showing overlap between potential development and conservation priorities.

and may not be compatible with landscape level conservation goals. Our results also demonstrate the flexibility that exists in applying the mitigation hierarchy when tradeoffs are examined at an appropriate landscape scale. Using a landscape scale assessment allows the flexibility to design a conservation portfolio that reduces potential conflicts between current and future potential development. By considering current impacts in the portfolio design, and incurring a higher cost for selection of sites with development potential, we steer conservation priorities away from potential conflicts where possible. Our conservation portfolio reflects these principles. Although the conservation portfolio represents 38 to 75 percent of the landscape, less than three percent of the conservation portfolio would be affected by future development activities, with two exceptions – the Casanare region (~30 percent) and Sur de Bolivar region (~7 percent) (Figure 6, Table 1). Moreover, the conflicts in the Casanare and Sur de Bolivar regions could be reduced through local siting and/or development approaches that minimize impacts to the degree compatible with biodiversity goals. For example, in the case of oil and gas development in the Casanare, there is significant opportunity for reducing impacts in the process of siting and designing well pads, roads and pipelines [47]. In addition, directional drilling may allow access to oil and gas resources in a manner the avoids or minimizes impacts [48].

As a proportion of the total area proposed for development the conflict with the conservation portfolio is considerable, ranging from about 17 to 94 percent depending on the region (Figure 6, Table 1). Prohibiting this proposed development could potentially result in significant economic losses. But it is important to recognize that our analysis represented development potential rather coarsely using concession boundaries, which likely overestimate the potential conflict between development and conservation. As more specific development planning information for the concessions in question becomes available, this information can be taken into account with the geospatial data available from our analysis to better inform and guide decisions. Moreover our representation of biodiversity features is relatively coarse in scale (1:100,000 scale) and as development plans are refined it likely that biodiversity assessments will be refined as well. These refinements may make it possible to design development activities that minimizes impacts or avoids impacts to biodiversity features completely.

Given both the considerable conflict the conservation portfolio represents for development goals and the high economic value the leases likely represent, conservation planning could be adapted to further reduce conflicts with development objectives by designing a portfolio to avoid conflict with potential development [10,19,20]. Proposed developments can be mapped and assessed relative to the conservation portfolio and the minimum viability needs of the targets. Overlap between the portfolio and proposed development may result in a "re-drawing" of the portfolio to re-capture habitat needed to meet biodiversity goals impacted by development in areas not slated for development [10,19,20]. Alternatively proposed development activities can be incorporated into the cost surface incurring a higher cost for selection of sites where proposed development activities may occur, ultimately steering the selection conservation of the portfolio away from potential conflicts. Given the small overall loss the majority of these leases represent to the conservation portfolio it seems unlikely that the Colombian government will forego the revenue in favor of conservation goals. In our pilot studies we could have examined more complicated scenarios to avoid conflicts between conservation and development objectives. However, it was the decision of the working group to use the portfolio design process in a simplistic manner, using it to highlight leases that should be considered for avoidance. Future assessments seeking to either update the analysis within our focal landscapes or expand the application of this concept to other landscapes should seek to incorporate more complicated scenarios to avoid conflicts between development and conservation objectives [10,19,20].

Our framework attempts to address some of the key deficiencies with current mitigation practices. It provides decision makers and stakeholders the opportunity to proactively understand potential losses to biodiversity resources associated with proposed development plans, supporting informed decisions. In too many places mitigation is still carried out on a project-by-project basis, with piecemeal actions taken on-site or nearby. There is little or no consideration about how these actions contribute to wider goals for the landscape. Traditional mitigation also ignores the future. Too often mitigation is implemented without considering the projected cumulative impacts for the region from planned energy, mining, and infrastructure. More broadly, by integrating conservation and development planning at a landscape scale, it supports application of mitigation at a more appropriate ecological scale. Projecting cumulative impacts at this scale moves mitigation beyond a project-by-project approach to one that can support a dynamic vision consistent with systematic conservation planning. While we choose not to consider future potential cumulative impacts of development in our portfolio design our analysis can still help decision makers incorporate cumulative impacts into their decision making. For example a regulator considering an impact to a conservation target at one site can examine how that decision will contribute to overall potential impacts. Moreover these regulators can consider how that impact may or may not restrict the ability to meet conservation goals. If impacts represent a significant portion of the conservation goal and there is limited flexibility to meet goals elsewhere they can recommend more avoidance and minimization be utilized. The framework we present here may also be a way to improve the application of conservation planning exercises. Over the last three decades a great deal of research, funding, and effort have been put into the development of systematic conservation planning, yet these methods are used infrequently by those charged with managing landscapes [49]. By blending landscape level conservation planning with mitigation we have the potential to move these analyses out of the academic world and into the hands of the regulatory agencies responsible for the decisions that drive the majority of land-use change [10,20].

Landscape-level plans can also improve the conservation benefits of applying the mitigation hierarchy, in particular how offsets are designed and sited [9,10]. Most biodiversity offset legislation and policies presume "like-for-like" or "in-kind" offsets (i.e. offsets that conserve biodiversity of a similar kind to that affected by the development) [4]. At times, however, better conservation results may be obtained by placing the offset in an ecosystem of higher conservation priority [10,50,51]. A regional landscape perspective can provide opportunities for identifying situations where "trading up" or "out-of-kind" offsets may offer valuable alternatives. Consider, for example, development that results in impacts to a widely distributed or highly conserved target. Requiring in-kind offsets could limit the potential benefit that an offset might provide. For example, losses of a particular common habitat type could be offset in a habitat of higher priority in the region, because it is under great threat (i.e. vulnerable) or because it is the last remaining example of its kind, and is therefore irreplaceable [10,20]. Out-of-kind offsets may also be preferable where there is an opportunity to take advantage of existing conservation management to locate the offset, or consolidating several offsets in one location.

The goal of our analyses was to illustrate a way in which gaps in the existing siting and mitigation regulatory framework for Colombia could be improved using available data and tools. The pilot sites selected to illustrate these concepts were chosen jointly by The Nature Conservancy and the Colombian Ministry of Environment and Sustainable Development (MADS) because they are expected to experience significant increases in development pressure [25]. As a result of this work MADS adopted a resolution (MADS 2010 –Resolución 1503 de 2010) and a methodology to incorporate the principles of biodiversity offsets outlined in our analyses into its licensing process for terrestrial projects [52]. For the first time, companies will be required to compensate for impacts to biodiversity in accordance with an explicit science-based framework. It will also encourage MADS to place impacts of development into a landscape perspective: highlighting the cumulative impacts of development, revealing the potential losses and making clear the need for mitigation, including both avoidance and the compensation of impacts. Prior to these changes Colombia's approach to impact mitigation focused primarily on impacts to forested systems. In some cases forest clearing resulted in the planting of fruit trees as a way to compensate for impacts (Saenz & Walschburger unpublished data). There was also a need for more structured decision-making framework to determine when projects could proceed or should be avoided. Now the MADS has guidelines for proactively evaluating the compatibility of proposed development with conservation goals and determining when impacts should be avoided [50,52]. In addition to decisions about avoidance and minimization, the framework will support MADS in determining ecologically equivalent offset opportunities, locations where these offsets can best contribute to landscape conservation goals, and the amount of compensatory mitigation needed to address impacts [50,51]. While only recently signed into law on August 31, 2012 and becoming effective on January1, 2013, this change in the licensing process should drive both a significant increase in, and more effective use of, funding for biodiversity conservation across Colombia [50,51].

Despite the progress MADS's regulatory change represents, there are a number of issues that will need to be addressed to effectively implement the guidance especially as it relates to the avoidance of impacts. Since MADS's engagement in the siting and mitigation of development occurs as part of the licensing process, it may be too late in the process to require a project's impacts be avoided. In situations where direct avoidance of impacts is not possible, MADS may encourage companies to avoid impacts by calling for high rates of compensation [50,51]. In addition, other agencies and sectors of government may have authority to make decisions regarding avoidance of impacts. For example, local municipalities may use their territorial land use planning process to guide and shape which resources are made available for development. The Colombian government has started to define a process to identify areas necessary to maintain ecological structure for the entire country, asking what is the minimum area needed to maintain biodiversity and ecosystem services [25]. The framework we have presented here could help guide that process.

Conclusion

By avoiding or minimizing impacts to irreplaceable biodiversity features, then ensuring that damaged ecosystems are restored on site, using the best available technology, and finally offsetting any remaining residual impacts, we can provide a framework that is consistent with sustainable development. A landscape vision is essential, because it helps us to move beyond a project by project, business-as-usual approach to addressing conflicts between development and conservation goals. Given the extensive amount of development projected for Colombia, a requirement that development projects achieve no-net-loss outcomes for biodiversity could be the impetus needed to conserve biodiversity across the country.

Acknowledgments

We thank, Vice Minister of Environment Adriana Soto, Xiomara Sanclemente, Aurelio Ramos, Jose Yunis, Julio Carcamo, Diana Zapata, Claudia Mora, Juan Sebastian Lozano, Economic Analysis Group of the Ministry of Environment, Sofia Montenegro, Joseph Tribbiani, Michael Heiner and Collin O'Neil, for providing technical assistance, and the people from Café Havana of Cartagena and Choice Butchers of Fort Collins, Colorado for helpful discussions.

Author Contributions

Conceived and designed the experiments: SS TW JL JCG BM JK. Performed the experiments: SS TW JL. Analyzed the data: SS TW JL. Wrote the paper: SS TW JK.

References

1. World Bank (2007) Global Economic Prospects 2007: Managing the Next Wave of Globalization. Washington DC.
2. International Energy Agency (2007) World Energy Outlook 2007. Paris, France.
3. Environmental Law Institute (2007) Mitigation of Impacts to Fish and Wildlife Habitat: Estimating Costs and Identifying Opportunities. Washington, DC: Environmental Law Institute.
4. McKenney BA, Kiesecker JM (2010) Policy Development for Biodiversity Offsets: A Review of Offset Frameworks. Environ Manag 45:165–176.
5. Clare S, Krogman N, Foote L, Lemphers N (2011) Where is the avoidance in the implementation of wetland law and policy? Wetlands Ecol Manage 19:165–182. DOI 10.1007/s11273-011-9209-3
6. Maron M, Dunn PK, McAlpine CA, Apan A (2010) Can offsets really compensate for habitat removal? The case of the endangered red-tailed black-cockatoo. J Appl Ecol 47, 348–355.
7. Council on Environmental Quality (2000) Protection of the Environment (under the National Environment Policy Act)(40 CFR 1500–1517).
8. MAVDT (Ministerio de Ambiente, Vivienda y Desarrollo Territorial) (2008) Convenio de Asociación No. 09 de 2008., The Nature Conservancy, World Wildlife Fund y Conservación Internacional. Colombia
9. Kiesecker JM, Copeland H, Pocewicz A, Nibbelink N, McKenney et al. (2009) A Framework for Implementing Biodiversity Offsets: Selecting Sites and Determining Scale. Bioscience 59:77–84.
10. Kiesecker JM, Copeland H, Pocewicz A, McKenney B (2010) Development by design: blending landscape-level planning with the mitigation hierarchy. Frontiers in Eco Environ 8:261–266.
11. Margules CR, Pressey RL (2000) Systematic conservation planning. Nature 405, 243–253.
12. Pressey RL, Bottrill MC (2008) Opportunism, threats, and the evolution of systematic conservation planning. Conser Biol 22:1340–1345.
13. Noss RF, Carroll C, Vance-Borland K, Wuerthner G (2002) A multicriteria assessment of the irreplaceability and vulnerability of sites in the greater Yellowstone ecosystem. Conser Biol 16: 895–908.
14. Ball IR, Possingham HP (2000). MARXAN (V1.8.2): Marine Reserve Design Using Spatially Explicit Annealing, a Manual
15. Pressey RL, Possingham HP, Day JR (1997) Effectiveness of alternative heuristic algorithms for identifying indicative minimum requirements for conservation reserves. Biol Conser 80:207–19.
16. Lovejoy TE (1980) Discontinuous wilderness: minimum areas for conservation. Parks 5:13–15.
17. Armbruster P, Lande R (1993) A population viability analysis for African Elephant (*Loxodonta Africana*)—how big should reserves be? Conserv Biol 7:602–10.
18. Doncaster CP, Micol T, Jensen SP (1996) Determining minimum habitat requirements in theory and practice. Oikos 75:335–39.
19. Kiesecker JM, Copeland HC, McKenny B, Pocewicz A, Doherty K (2011a) Energy by design: Making mitigation work for conservation and development. Pages 159–182 *in* D. E. Naugle, editor. Energy Development and Wildlife Conservation in Western North America. Washington DC, Island Press.
20. Heiner M, Davaa G, Kiesecker J, McKenney B, Evans J, et al. (2011) Identifying Conservation Priorities in the Face of Future Development: Applying Development by Design in the Grasslands of Mongolia. Arlington, Virginia, The Nature Conservancy.
21. Stotz DF, Fitzpatrick JW, Parker TE III, Moskowitz DK (1996) Neotropical Birds, Ecology and Conservation. Chicago: University of Chicago Press. 502 p.
22. Erwin TL, Pimienta MC, Murillo OE, Aschero V (2004) Mapping patterns of beta diversity for beetles across the western Amazon Basin: A preliminary case for improving conservation strategies. Proc Calif Acad Sci 56: 72–85.
23. Ceballos G, Ehrlich PR (2006) Global mammal distributions, biodiversity hotspots, and conservation. Proc Natl Acad Sci U S A 103: 19374–19379.
24. Finer M, Jenkins CN, Pimm SL, Keane B, Ross C (2008) Oil and Gas Projects in the Western Amazon: Threats to Wilderness, Biodiversity, and Indigenous Peoples. PLoS ONE 3(8): e2932. doi:10.1371/journal.pone.0002932
25. Departamento nacional de Planeación (DNP). (2011) Plan Nacional de Desarrollo 2010–2014. Colombia.
26. Kiesecker JM, Evans JS, Fargione J, Doherty K, Foresman KR et al. (2011) Win-Win for Wind and Wildlife: A Vision to Facilitate Sustainable Development. PLoS ONE 6: e17566. doi:10.1371/journal.pone.0017566
27. Groves CR (2003). Drafting a Conservation Blueprint: A Practioner's Guide to Planning for Biodiversity. WashingtonDC: Island Press, 457 pp.
28. Dobson A (1996) Conservation and Biodiversity. Scientific American Library, New York. page 66.
29. León J, Saenz S, Walschburger T, Porras M (2010) Portafolio de Áreas Importantes para la Conservación de la Biodiversidad del Sur de Bolívar. Producto V. Convenio de Asociación No. 123 de 2008. Ministerio de Ambiente, Vivienda y Desarrollo Territorial, The Nature Conservancy y World Wildlife Fund. Colombia. 33p
30. León J, Cardenas JJ, Walschburger T, Porras M, Saenz S (2010) Portafolio de Áreas Importantes para la Conservación de la Biodiversidad de Bahía Tribugá. Producto V. Convenio de Asociación No. 123 de 2008. Ministerio de Ambiente, Vivienda y Desarrollo Territorial, The Nature Conservancy y World Wildlife Fund. Colombia. 32p
31. Romero M, Rodríguez N, Cárdenas K (2008) Portafolio de Áreas Importantes para la Conservación de la Biodiversidad del Norte del Casanare. Producto V. Convenio de Asociación No. 123 de 2008. Ministerio de Ambiente, Vivienda y Desarrollo Territorial, The Nature Conservancy y World Wildlife Fund. Colombia. 59p.
32. Rodríguez N, Romero M, Cárdenas K, Castellanos L (2008) Portafolio de Áreas Importantes para la Conservación de la Biodiversidad del Meta - Huila. Producto V. Convenio de Asociación No. 123 de 2008. Ministerio de Ambiente, Vivienda y Desarrollo Territorial, The Nature Conservancy y World Wildlife Fund. Colombia. 59p
33. Cabrera E, Rodriguez J, Otero J, Pedraza C, Ruiz F (2009) Portafolio de áreas prioritarias de conservación en la zona central del Cesar. Producto 16. Convenio de Asociación No. 09 de 2008. Ministerio de Ambiente, Vivienda y Desarrollo Territorial, The Nature Conservancy, World Wildlife Fund y Conservación internacional. Colombia. 55p
34. IDEAM IGAC, IAvH Invemar, Sinchi IIAP, (2007) Mapa de Ecosistemas Continentales, Costeros y Marinos de Colombia.
35. Galindo G, Pedraza C, Betancourt F, Moreno R, Cabrera E (2007) Planeación ambiental del sector hidrocarburos para la conservación de la biodiversidad en los llanos de Colombia. Convenio de cooperación 05–050. Instituto de

Investigación de Recursos Biológicos Alexander von Humboldt. Bogotá. Colombia.
36. Galindo G, Cabrera E, Otero J, Bernal NR, Palacios S (2009) Planificación ecoregional para la conservación de la biodiversidad en los Andes y en el Piedemonte amazónicos colombianos. Instituto de investigación de recursos biológicos Alexander von Humboldt, Agencia Nacional de Hidrocarburos, The Nature Conservancy e Instituto de Meteorología y Estudios Ambientales. Bogotá D.C. Colombia. 24p
37. NatureServe (2007) InfoNatura: Birds, mammals, and amphibians of Latin America. NatureServe'. Arlington, Virginia (USA). Base de datos en línea. Version 4.1> URL: http://www.natureserve.org/infonatura(2007/05/20).
38. Corzo S (2008) Áreas prioritarias para la conservación de la biodiversidad continental en Colombia. Sistema Nacional de áreas protegidas.
39. Pressey RL, Taffs KH (2001) Scheduling conservation action in production landscapes: Priority areas in western New South Wales defined by irreplaceability and vulnerability to vegetation loss. Biol Conser 100: 355–376.
40. Alonso D, Ramírez L, Segura-Quintero C, Castillo-Torres P, Diaz JM, et al. (2008) Prioridades de Conservación in situ para la biodiversidad marina y costera de la plataforma continental del Caribe y Pacifico colombiano. Instituto de Investigaciones Marinas y Costeras INVEMAR, The Nature Conservancy y Unidad Administrativa Especial del Sistema de parques Nacionales Naturales-UAESPNN. Santa Marta, Colombia, 20 p.
41. Osaragi T (2002) Classification methods for spatial data representation. (CASA Working Papers 40). Centre for Advanced Spatial Analysis (UCL): London, UK.
42. Possingham HP, Ball IR, Andelman S (2000) Mathematical methods for identifying representative reserve networks. In: S. Ferson and M. Burgman (Eds.), Quantitative methods for conservation biology (pp. 291–305). New York: Springer-Verlag
43. Ardron JA, Possingham HP, Klein CJ (eds). (2010) Marxan Good Practices Handbook, Version 2. Pacific Marine Analysis and Research Association, Victoria, BC, Canada. 165 pages. www.pacmara.org.
44. White D, Kimerling AJ, Overton WS. (1992) Cartographic and geometric components of a global sampling design for environmental monitoring. Carto and Geog Info Syst 19: 5–22.
45. Copeland H, Ward J, Kiesecker JM (2007) Threat, cost, and biological value: Prioritizing conservation within Wyoming ecoregions. J Conser Planning 2007: 1–16
46. Salafsky N. Margoluis R (1999) Greater than the Sum of Their Parts: Designing Conservation and Development Programs to Maximize Results and Learning. Washington (DC): Biodiversity Support Program.
47. Rosenfeld AB, Gordon D, Guerin-McManus M (2001) Reinventing the well: approaches to minimizing the environmental and social impact of oil development in the tropics. In: Bowles IA, Prickett GT, editors. Footprints in the jungle. New York: Oxford University Press.
48. The Energy & Biodiversity Initiative (2003) Integrating biodiversity conservation into oil and gas development. http://www.theebi.org/pdfs/ebi_report.pdf
49. Prendergast JR, Quinn RM, Lawton JH (1999) The gaps between theory and practice in selecting nature reserves. Conser Biol 13:484–492.
50. Saenz S, Walschburger T, León J, Gonzalez J (2010) Manual para asignación de compensaciones por pérdida de biodiversidad. Convenio de Asociación No. 09 de 2008. Ministerio de Ambiente, Vivienda y Desarrollo Territorial, The Nature Conservancy, World Wildlife Fund, Conservación Internacional. Colombia. 45p Documento sin publicar
51. Saenz S, Walschburger T, León J, Gonzalez J, Kiesecker JM et al. (In Review) A Framework for Implementing and Valuing Offsets in Colombia: Can Development Deliver Net Gains for Nature?
52. MAVDT (Ministerio de Ambiente, Vivienda y Desarrollo Territorial) (2010) Metodología general para la presentación de estudios ambientales / Zapata P., Diana M, Zamira Lozano; Carlos A. Londoño, B; es ál (eds) Claudia V. Gonzalez H; Jorge Idarraga; Amanda Poveda G; et ál (textos). Bogota D.C. 2010. 97 p

5

Towards Monitoring Biodiversity in Amazonian Forests: How Regular Samples Capture Meso-Scale Altitudinal Variation in 25 km² Plots

Darren Norris[1,3]*, Marie-Josée Fortin[2], William E. Magnusson[3]

1 Programa de Pós-Graduação em Biodiversidade Tropical, Universidade Federal do Amapá, Macapá, Amapá, Brazil, **2** Department of Ecology & Evolutionary Biology, University of Toronto, Ontario, Canada, **3** Coordenação de Biodiversidade, Instituto Nacional de Pesquisas da Amazônia, Manaus, Amazonas, Brazil

Abstract

Background: Ecological monitoring and sampling optima are context and location specific. Novel applications (e.g. biodiversity monitoring for environmental service payments) call for renewed efforts to establish reliable and robust monitoring in biodiversity rich areas. As there is little information on the distribution of biodiversity across the Amazon basin, we used altitude as a proxy for biological variables to test whether meso-scale variation can be adequately represented by different sample sizes in a standardized, regular-coverage sampling arrangement.

Methodology/Principal Findings: We used Shuttle-Radar-Topography-Mission digital elevation values to evaluate if the regular sampling arrangement in standard RAPELD (rapid assessments ("RAP") over the long-term (LTER ["PELD" in Portuguese])) grids captured patters in meso-scale spatial variation. The adequacy of different sample sizes ($n = 4$ to 120) were examined within 32,325 km²/3,232,500 ha (1293×25 km² sample areas) distributed across the legal Brazilian Amazon. Kolmogorov-Smirnov-tests, correlation and root-mean-square-error were used to measure sample representativeness, similarity and accuracy respectively. Trends and thresholds of these responses in relation to sample size and standard-deviation were modeled using Generalized-Additive-Models and conditional-inference-trees respectively. We found that a regular arrangement of 30 samples captured the distribution of altitude values within these areas. Sample size was more important than sample standard deviation for representativeness and similarity. In contrast, accuracy was more strongly influenced by sample standard deviation. Additionally, analysis of spatially interpolated data showed that spatial patterns in altitude were also recovered within areas using a regular arrangement of 30 samples.

Conclusions/Significance: Our findings show that the logistically feasible sample used in the RAPELD system successfully recovers meso-scale altitudinal patterns. This suggests that the sample size and regular arrangement may also be generally appropriate for quantifying spatial patterns in biodiversity at similar scales across at least 90% (≈5 million km²) of the Brazilian Amazon.

Editor: Bruno Hérault, Cirad, France

Funding: The authors thank the Brazilian National Counsel of Technological and Scientific Development ("Conselho Nacional de Desenvolvimento Científico e Tecnológico" -CNPq) for the financial support provided through the Science Without Borders Program ("Ciência sem Fronteiras" - process 401370/2012-7). WM held a Bolsa de Produtividade scholarship from CNPq. DN thanks CNPq (process 159806/2012-7 & 164999/2013-2) and CAPES (PNPD20130074) for the scholarships granted. The funders had no role in study design, data collection and analysis, decision to publish, or preparation of the manuscript.

Competing Interests: The authors have declared that no competing interests exist.

* Email: dnorris75@gmail.com

Introduction

Gathering sample data that reliably reflects the underlying spatial and temporal variability in target measurements is a fundamental requisite for ecological studies. Bias (sampling bias, estimation bias, etc.) occurs when the sampling design used induces errors and artificial differences in the values among samples [1,2]. These first principles form the basis of many an undergraduate statistics course but beyond lecture halls obtaining reliable and representative sample data is one of the primary challenges for any biodiversity monitoring program [3,4], where the necessary requirement is that it is sensible and meaningful to compare and contrast sample values [1,3,4].

There is no panacea in ecological sampling because sampling optima are context and location specific. Novel applications (e.g., monitoring of biodiversity for payments of environmental services) present further challenges and call for renewed efforts to establish general monitoring guidelines and frameworks, especially in biodiversity rich areas [5,6,7]. A particular challenge to establishing payments for biodiversity services is that reliable and robust long-term monitoring of biodiversity indicators is required to ensure that services are and will continue to be provided. Such long-term monitoring of biodiversity requires decisions today about the sampling distribution to recover information about changes in response to future, and often unknown, threats.

While the adequacy of sampling for a specific threat might not be known, we do know that most analyses require knowledge of the spatial structure, and hence the distribution patterns of response variables. If the distribution of sampling is inadequate to capture the spatial patterns in biodiversity variables, any subsequent analyses are likely to be inefficient and/or biased. Design of an optimal spatial sampling scheme requires a careful balance between sampling locations that are too close to one another, thus not providing enough new information (highly autocorrelated data), and sampling locations that are too sparse, so that processes at other spatial scales introduce too much variability [8]. Within biodiversity monitoring programs, sample quality (representativeness) and sample detail (resolution) often require separate investment in time, money and expertise (p251 of [4]). Finite resources mean that any biodiversity monitoring program faces a tension between securing samples that are of both sufficient quantity (to ensure that results can be extrapolated across scales that are relevant to management) and quality (to ensure they reliably capture the underlying variability in the measurement of interest) for addressing the research objective at hand [3,4].

Field (on-the-ground) data collection provides the empirical basis upon which our understanding of biodiversity conservation is built. When a non-contiguous "sparse" [9] sampling strategy is used (be it regular, stratified, random or clustered), the extent is not completely surveyed and information is missing about the spatial pattern [9]. Sample pattern (geometrical configuration of the sample elements in space) and sample density (number of samples per unit area) are recognized as two important determinants for optimizing sparse sampling designs [10]. However, there is no information on the distribution of most biodiversity variables across the Amazon basin with the precision necessary to test the adequacy of sampling density and configuration. Faced with such uncertainty, ecologists often turn to rules of thumb, such as a minimum of 30 sample locations are required to detect significant spatial autocorrelation [9], and the reliable estimation of spatial structure and spatial model parameters may require 100 or more sampling locations [9]. If other factors remain unchanged, sparse sample patterns ranked by decreasing level of logistical efficiency are regular, stratified, random, and clustered [10].

Advances in statistical methodologies and computational power mean that there are a wide variety of approaches that can be applied to deal with sampling designs that were traditionally considered as suboptimal/inappropriate due to lack of spatial independence between samples [11]. However, for the statistics to be meaningful and useful, it is vital to understand how well the sampling captures the variation in the target variable. Although the substantial body of geostatistical literature enables us to generate well informed expectations of how well a sample captures patterns in known abiotic and biotic attributes, it is hard to establish informed expectations/guidelines when there is little or no knowledge regarding the response of interest [12].

Although there is little information on the distribution of most biodiversity variables across the Amazon basin, altitude is a driver and modulator of species distribution patterns from microhabitat to biogeographic scales [13,14,15,16]. Altitude not only affects soil, water availability, climate and a myriad of other abiotic and biotic variables [16], it is also a key determinant of Amazonian biodiversity [17,18,19,20,21]. As such, spatial variation in altitude and biodiversity are expected to be strongly correlated, especially at local (10 m–1 km [19]) to meso-scales (1 km–100 km [13]).

We can identify at least three reasons why altitude is also perhaps the only variable available that can be used to provide a baseline evaluation of sample arrangements for biodiversity monitoring. Firstly, freely available digital elevation models (such as those obtained from ASTER/SRTM satellite images) provide a continuous global coverage of known altitude values at resolutions <100 m (resolutions vary slightly with distance from the equator). Secondly, not only is the variation in many response variables associated with the variation in altitude, but in the majority of situations, it will also be necessary to model and/or control for the effects of altitude in both within and between site comparisons. For example, altitude is used to predict soil characteristics and both can then be modeled as covariates for numerous response variables [22,23,24,25,26]. Thirdly, on the scales relevant to biodiversity monitoring (i.e. excluding changes that occur on geological timescales) meso-scale altitude values are dependent simply on the geographical location. Therefore it should be possible to accurately predict altitude using only spatial predictors, i.e. by modeling geographic coordinates. These three reasons make altitude a useful metric for evaluating the adequacy of sampling arrangements to represent meso-scale spatial variation.

Altitude can be particularly informative for evaluating sample arrangements for biodiversity monitoring in cases (i) when strata are not identifiable; (ii) there is uncertainty in the expected changes in the spatial arrangement of responses [12] (i.e. biodiversity indicator variables) to future, and often unknown, threats and/or (iii) where questions involve investigating responses of multiple indicators [27,28]. Here, we use data from the most widespread system of standardized biodiversity monitoring in the Amazon (RAPELD [3,22,29]) to investigate whether regular sampling at 1-km intervals is sufficient to accurately recover spatial patterns in altitude within 25 km^2 sampling grids across the Brazilian Amazon. To examine the accuracy of both samples and associated predictions within the survey area we focus on both sample and interpolated values within grids. Specifically we address the following questions:

1) How do sample size and sample standard deviation influence the representativeness of values from regularly arranged samples and the associated spatially interpolated altitude values?
2) How does sample size and sample standard deviation influence the similarity and accuracy of spatially interpolated altitude values?
3) What are the lower thresholds in sample size and sample standard deviation needed to obtain representative, similar and accurate values?
4) What are the practical implications of these findings for the use of regular meso-scale sample arrangements in Amazon forest biodiversity monitoring programs?

Materials and Methods

Study areas and sample arrangements

To investigate how well a regular sample design captures spatial variation in altitude, we used a series of 5×5 km study areas overlaid on SRTM altitude values (version 2 "void filled", downloaded from http://earthexplorer.usgs.gov/) distributed across the legal Brazilian Amazon (Figure S1). The legal Brazilian Amazon is a 5.06 million km^2 area defined by the boundaries of nine Brazilian States: Acre, Amapá, Amazonas, Pará, Rondônia, Roraima, Tocantins, Mato Grosso, and Maranhão. The 5×5 km study areas included seven areas (hereafter "active research areas", Figure 1, Table S1, Figure S3) that are part of the Brazilian Program for Biodiversity Research ("Programa de Pesquisa em Biodiversidade" – hereafter PPBio) [3,22] and 1286 randomly selected areas (Figure S1, Figure S2). The randomly selected areas provided a representative sample of altitude across the entire legal

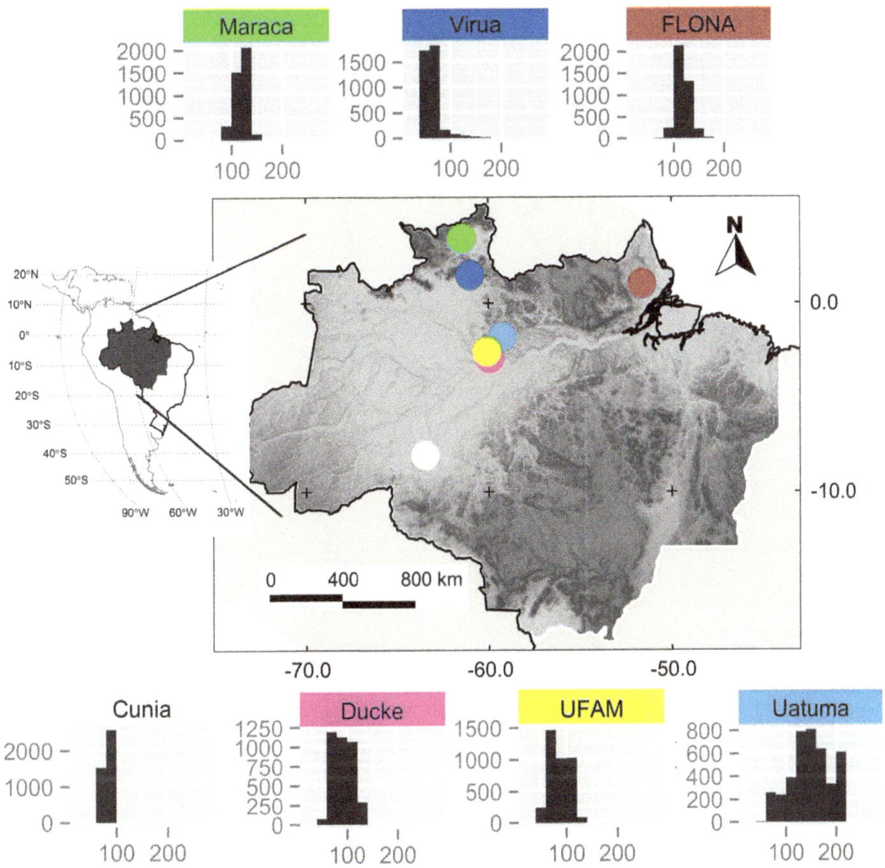

Figure 1. Distribution of seven active research sites within the Brazilian Amazon. Site locations (circles) are overlaid on a 30 arc-second shaded relief map derived from mean altitude values [45] distributed by the U.S. Geological Survey (USGS) Earth Resources Observation & Science (EROS) Center (http://topotools.cr.usgs.gov/gmted_viewer/). Histograms show the frequency distribution of SRTM altitude values (90 m resolution) within the study sites, with counts (y-axis) grouped in 20 meter altitude bins (x-axis).

Brazilian Amazon (Figure S1) and enabled us to establish that conclusions from the seven active research areas were more generally applicable.

The active research areas were included to provide a real-world sample of sites established to address research and biodiversity monitoring questions. PPBio has adopted a standardized survey system ("RAPELD" [3,22,27]) that uses regularly spaced plots along 5 km trails as its basic sample unit. One of the sample arrangements used within PPBio [3] is a 25 km^2 area delimited by a series of 5 km trails that form a survey "grid" with sample plots regularly spaced at 1 km intervals (Figure S3). At the time of writing PPBio and partners has established ten 25 km^2 research areas across Brazilian Amazonia. We included 7 of the 10 active research areas in our analysis based on two criteria: (i) with an established regular arrangement of sample plots ($n = 30–31$) distributed along a trail system, 1 km distant from each other (Figure S3), and (ii) with active research and data freely available (http://ppbio.inpa.gov.br/sitios).

In addition to the regular sample plots, the RAPELD grids also have "riparian plots" [3]. These riparian plots are installed wherever a trail crosses a waterbody and are used to survey the distinctive species found in riparian zones [30,31,32]. The distance between riparian plots within grids is not standardized, and only one grid ("Ducke") has established riparian plots. Therefore, to enable comparison between grids we derived the probable locations of riparian plots within the other grids using standardized GIS processes (Figure S4).

Data analysis

All statistical analyses were undertaken with the R language and environment for statistical computing [33], using base functions and functions available in the following packages: "ggplot2" [34], "gstat" [35], "mgcv" [36], "raster" [37], and "sp" [38].

We evaluated the adequacy of samples and their associated spatially interpolated values across the random and active research areas. Active research areas included three sample sizes (Table S1): (i) the existing arrangement of regular 1 km interval plots ($n = 30–31$), (ii) the same regular 1 km interval plots plus riparian plots ($n = 45–52$), and (iii) to examine the expected improvements generated by increasing sample size we also included locations if plots were to be established at regularly spaced 500 m intervals along the existing trails ($n = 96$),which also included the locations of the existing 1 km interval regular plots. Within the randomly selected areas, we simulated regularly arranged samples across a wider range of sample sizes ($n = 4$, 8, 16, 30, 60, 120). As the locations of the regular samples generated within each random area can differ (depending on the starting point), metrics for the samples and associated interpolations were generated for 10 iterations of each sample size within each random area. The mean value of the metrics for samples and spatial interpolations from the 10 iterations was used in all subsequent analysis.

Inverse Distance Weighted (IDW) spatial interpolations were calculated from the SRTM altitude values at the different sample sizes within each of the random and active research areas. Interpolations were generated for each area using a 90 m grid size that corresponded approximately to that of the original SRTM altitude values. The IDW approach was adopted as this spatial interpolation technique can be used with small sample sizes [9,22,26] and in our case provided similar results to Ordinary Kriging and Generalized Additive Model (GAM) interpolations (Figure S5, Figure S6).

To answer questions 1 and 2, the adequacy of samples and associated spatial interpolations was evaluated using the following methods. For question 1, we used the Kolmogorov-Smirnov (KS-test) as a measure of sample representativeness [26]. This was done by comparing the distribution of both sample and spatially interpolated altitude values with the original SRTM altitude values in both randomly selected and active research areas using the KS-test. For question 2 we examined the interpolated values from samples in random areas, using correlation as a measure of similarity between sample and population (i.e. original SRTM values) values, and the root mean square error (RMSE) to represent the accuracy of parameter estimation.

For question 3, the thresholds in all metrics were evaluated based on whether the parameter of interest reached stability and/or the chosen level of precision [39]. This was achieved using complementary approaches. Firstly, Generalized Additive Models (GAMs) [36] were used to model trends in the response of the different metrics to sample size and standard deviation (SD). Secondly, to identify thresholds (i.e. break-points) where sample size and SD influenced the responses we used the conditional inference tree method [40,41], implemented in the "partykit" package [42]. Although various regression-tree methods are available (e.g. CART, random forest, etc.) we assume that results from different methods will be similar for simple cases such as ours.

Results

Sample and Interpolation Representativeness

Mean SRTM altitude within 5×5 km areas of the 5.06 million km^2 legal Brazilian Amazon was 159.5 m (range 0–1216 m). Mean altitude in the randomly selected areas (159.3) was similar to this overall mean, but the greatest mean value in the randomly selected areas was lower at 683 m. Altitude heterogeneity (SD) ranged from 0 to 466 across 5×5 km areas of the 5.06 million km^2 legal Brazilian Amazon (Figure S1). However, 90% of areas had SD values $<=40$, which corresponds to a range in altitude of ≈210 meters [mean of range values from the randomly selected areas with SD of 40 ($n = 7$)]. Our sample of seven active research areas (SD ranging from 4.8 to 40.2, Table S1) therefore represents the meso-scale altitudinal heterogeneity found across an area of approximately 5 million km^2.

The Kolmogorov-Smirnov (KS) tests showed that a regular arrangement of samples could represent the distribution of altitude values within our study areas (Figure 2, Figure S7, Figure S8). For the active research areas, the KS-test showed that the distribution of altitude values in the samples from two areas [FLONA do Amapá ($n = 30$), and Virua ($n = 30$ and $n = 49$)] were (marginally) significantly different from the distribution of altitude values within grids (Figure 2a, Figure S8). Although statistically different, visual inspection of back-to-back histograms showed that even these smaller samples ($n = 30$, $n = 49$), appeared to capture the distribution of altitude values in these two areas with long-tailed distributions i.e. differences appear to be caused by variation in the distribution of the more extreme values (Figure S8).

Spatial interpolation: similarity and accuracy

Both sample size and standard deviation (SD) strongly influenced the similarity and accuracy of spatial interpolations (Figure 3). The similarity between interpolations and the original altitude values increased rapidly with sample size (Pearson correlation mean values increased from 0.35 to 0.57, at $n = 4$ to $n = 16$ samples respectively) after which correlations continued to increase but at a lesser rate, reaching a mean value of 0.78 at a sample size of 120.

Thresholds

Conditional inference tree analysis revealed differences in the influence of sample size and sample standard deviation on thresholds (break-points) in sample accuracy (representativeness, similarity and accuracy - Figures S9–S13). This analysis showed that sample size was the most important determinant of sample representativeness (KS-test p-values, Figure S9), with the first, second and third splits all due to sample size. The first split of this tree was at $n = 16$, grouping the smaller sample sizes ($n = 4$, 8, 16), which, independent of SD, were less representative (lower p-values) compared with the larger sample sizes ($n = 30$, 60, 120). The GAM model (Figure 2a, $p<0.00001$, deviance explained $= 65.2\%$) showed that the sample KS-test p-values did not stabilize within the range of sample sizes examined, but did stabilize at SD values ≥9.

Generally, the representativeness of interpolated values showed a similar pattern to sample values (Figure 2). This similarity was also reflected in the first two splits of the interpolated KS-test p-value classification tree (Figure S10), which were also based on sample size, with the first split again at $n = 16$. Overall, 5 of the 9 sample KS-test tree inner nodes were based on sample size, and 5 of the 11 interpolation KS-test tree inner nodes were based on sample size ($p<0.001$ for all sample and interpolation inner nodes). However, GAM models reinforced differences between sample and interpolated values, with interpolated values statistically representative (on average) even at the smallest sample size ($n = 4$, Figure 2b, $p<0.00001$, deviance explained $= 49.3\%$).

With respect to the similarity and accuracy of spatially interpolated values, the conditional inference trees (Figures S11, S12) agreed with GAM results (Figure 3), showing that sample size was a more important determinant of correlations than for RMSE values. The GAM model (Figure 3, $p<0.00001$, deviance explained $= 58.5\%$), showed that correlations started to stabilize after $n = 16$ (increasing little after $n = 60$) and stabilized at SD values >15. As was found for the KS-test p-values (both sample and interpolation), the first split of the correlation classification tree was at $n = 16$ (Figure S11), whereas the first split of the RMSE tree was at SD = 19.77. The GAM model (Figure 3, $p<0.00001$, deviance explained $= 86.1\%$) showed that RMSE values stabilized after $n = 16$, but continued to increase across the range of SD values examined. Although sample size was not as important for RMSE values, a sample size of 16 was also identified at the 3rd and 4th splits of the RMSE tree (Figure S12). Overall, 5 of the 10 correlation tree inner nodes were based on sample size, whereas only 3 of the 11 RMSE tree inner nodes were based on sample size ($p<0.001$ for all correlation and RMSE inner nodes).

The difference in the importance of sample sizes and SD for correlations and RMSE was also confirmed by visual inspection (Figure 3), which showed that RMSE values tended to increase with SD at any given sample size, whereas this relationship was not as clear for correlations. Although there was a significant association between SD and interpolation similarity and accuracy, visual inspection of the IDW estimate maps showed that the meso-scale trends in altitude were represented well across the research

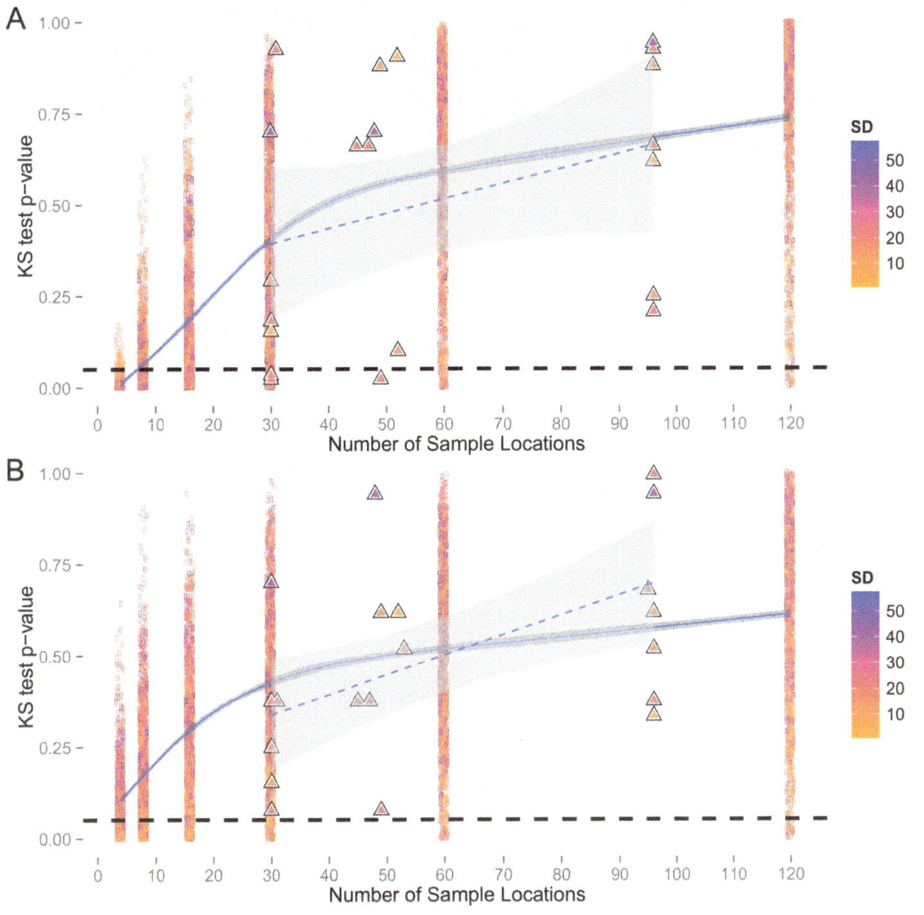

Figure 2. Comparison of sample representation across 25 km² of Amazonian forest. The sample representativeness of altitude values was determined in two sets of 5×5 km sample areas [1286 randomly selected (rectangles) and seven active research areas (triangles)] across the Brazilian Amazon. Within each area, the representativeness of (a) samples and (b) their corresponding IDW interpolations were evaluated. The distribution of values was compared with that of the area altitude (SRTM DEM) values using the Kolmogorov-Smirnov (KS) test. Increasing *p*-values denote more similar distributions (i.e. samples/interpolations are more representative). The blue lines (solid = randomly selected and dashed = active research areas) and shaded areas are the mean value and 95% confidence intervals from a GAM model illustrating the trend in representativeness with increasing sample size. The filled colors show the standard deviation (SD) of altitude values within each of the different sample areas. The black dashed horizontal line shows a *p*-value of 0.05.

grids (with SD values between 4 and 40) using a regularly arranged sample size of 30 (Figure S13).

Discussion

Our findings indicate that a regular arrangement of 30 sparse samples represent meso-scale altitude in 25 km² Amazonian areas. We also found that the associated interpolations are statistically similar to values at a much more detailed resolution (≈90 m). We discuss these findings and their potentially wide reaching applications in relation to obtaining samples necessary for monitoring biodiversity changes.

Sample and Spatial Interpolation Representativeness

Accurate and unbiased samples are needed to enable derivation of any of the myriad metrics used by researchers to examine ecological patterns and processes and to monitor and predict changes in biodiversity [1]. Our understanding of biological diversity within tropical biomes usually comes from sparse samples [1,3,4]. The continued debate regarding species richness [43], perhaps the most commonly used indicator in studies investigating biodiversity and conservation (not to mention ecological and macroecological research) highlights how sample adequacy affects conclusions even for the most simple biodiversity metrics.

Finding that regularly arranged samples are representative of meso-scale altitude values was expected as this arrangement is known to be efficient [10]. What was surprising was that even relatively small sample sizes ($n = 8$) were statistically representative (i.e. mean KS-test p value >0.05). However the considerable variation around the mean values suggests that such low sample sizes may not be generally reliable. This variation in sample representativeness was only partly explained by differences in sample size and standard deviation (65.2% of the GAM model deviance explained). As shown by numerous previous studies [9,10], it therefore seems likely that additional factors such as pre-existing gradients and directionality (i.e. anisotropy) will also influence sample representativeness [9].

A more detailed examination of the research grids showed that statistical representativeness appeared to be strongly influenced by the frequency distribution of the sample values. Samples from research grids with longer tailed distributions (i.e. with "extreme"/ "unusual" values) were less representative. This finding was

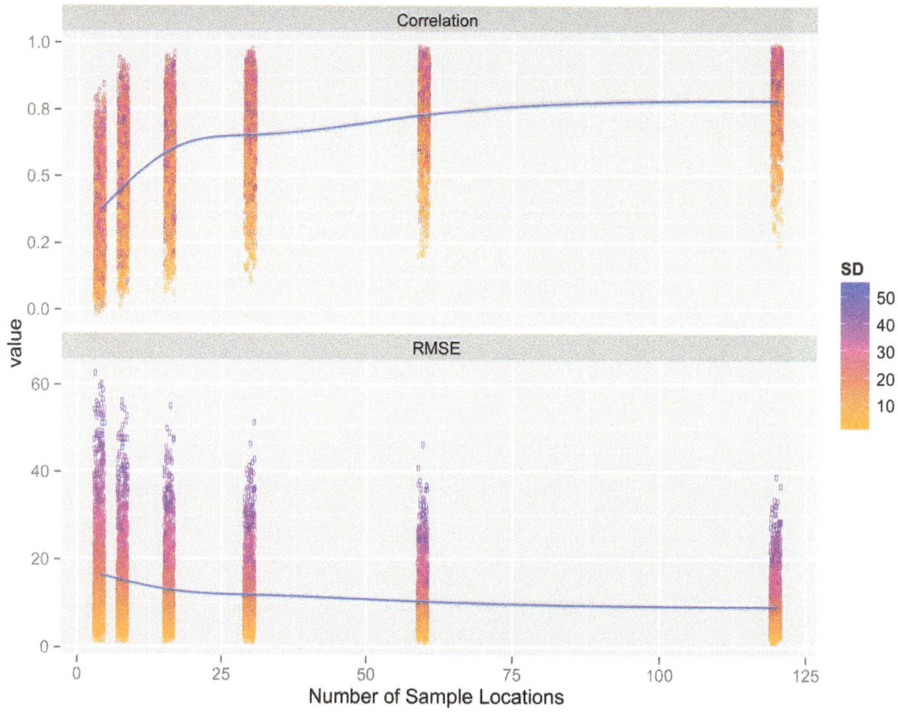

Figure 3. Trends in interpolation adequacy with increasing sample size. IDW (Inverse Distance Weighted) models were used to interpolate altitude (SRTM DEM) based on values from regularly distributed sample locations (*n*: 4, 8, 16, 30, 60 and 120) within 1286 areas (5×5 km) across the Brazilian Amazon. Interpolation adequacy was evaluated using root mean square error (RMSE) and correlations calculated from the interpolated values in relation to the original SRTM altitude values. Lines and shaded areas are mean values and 95% confidence intervals from GAM models illustrating trends with increasing sample size. The filled colors show the standard deviation (SD) of altitude values within each of the different sample areas.

expected as it is widely recognized that studies focusing on rare/extreme events/values require the development of specific sample arrangements and/or analytic techniques [44,45]. Nevertheless, visual examination of mapped values and the histograms comparing sample and original distributions suggested that the regular samples captured meso-scale patterns in at least some of these statistically "unrepresentative" cases.

The mapped research grid interpolations also illustrated how the statistical representativeness of the samples translated into predicting the real-world local-scale (90 m) spatial patterns in altitude. Indeed our findings suggest that a regular arrangement of ≥4 samples can (on average) be considered as statistically representative within the randomly selected areas. However, there was considerable variation around the mean. This variation shows that such small sample sizes do not necessarily translate into samples that are appropriate for the detection of either local or meso-scale patterns.

Spatial interpolation: similarity and accuracy

We show that regular meso-scale samples generate interpolations representative of a response measured at the local scale (90 m). Even at small sample sizes (<30) the interpolations were accurate (mean RMSE <20 m) and representative (mean KS-test p value >0.05). However, correlations with the original SRTM values were weak (r<0.5) at these smaller sample sizes. Additionally the accuracy of interpolated samples was strongly dependent on the sample standard deviation at all sample sizes. Such strong dependence on sample standard deviation means that even with the most intensive sampling ($n = 120$) there was a degree of uncertainty in the accuracy of such local scale predictions in the more heterogeneous areas (SD values greater than 10). While it is clearly important to acknowledge such uncertainty, visual examination of the maps generated showed that interpolations from sample sizes of 30 captured the meso-scale distribution in altitude even in the more heterogeneous areas (SD>10).

The representativeness, similarity and accuracy of interpolated values means that it is theoretically possible to compare metrics derived from these interpolations with those obtained from any other sample size/arrangement at similar scales that adequately represent the variable of interest. The ability to generate accurate interpolations also means that we are able to adopt the (typically more powerful) statistics used for continuously sampled areas [9]. The use of interpolated values can therefore provide an option to integrate/compare/contrast results from studies that adopt different sample designs and sizes. As demonstrated by recent reviews [46,47] and case studies [48], we would expect the integration of results from different studies to generate substantial improvements in relation to the insights and potential applications of the data generated. Such improvements are also likely to generate financial economies [46,48]. The economies arising from the ability to integrate data from different sources are particularly relevant in the case of payments for ecosystem services where indicators and monitoring protocols are likely to vary spatially with context and location and temporally with the emergence of future threats [46,49,50,51].

Thresholds

Classification-tree analysis showed that sample sizes equal to or greater than 30 were an important threshold for representativeness, accuracy and similarity. In combination with the GAM results, these findings show that a regularly arranged sample size of 30 presents values that are representative of, accurate, and similar to the original "population" (i.e. meso-scale altitude). The thresholds identified broadly agree with established rules of thumb for a sample size of 30 in spatial ecology [9] and a sample size of 50 as the computational minimum for geostatistical analysis [26].

But how representative are these sample size thresholds for biodiversity monitoring? Our findings are based on the spatial variation in altitude and the underlying assumption is that there will be similar variation in meso-scale "biodiversity" patterns. Defining biodiversity was one of the questions addressed in a classic issue of the Philosophical Transactions of the Royal Society B | Biological Sciences (July 29, 1994, vol. 345 (1311): http://rstb.royalsocietypublishing.org/content/345/1311.toc). After 20 years, the increasing number and complexity of biodiversity metrics (e.g. [7,48]) suggests that we are not much closer to a unifying definition for quantifying biodiversity. Lacking a clear understanding of how to quantify biodiversity means that extrapolating conclusions from meso-scale altitude to meso-scale tropical forest "biodiversity" is challenging. However, to examine whether the thresholds identified are likely to be generally applicable for biodiversity monitoring we can compare the meso-scale variation in altitude with variation in some of the species and community responses documented from tropical forests.

Studies within one of the PPBIO research grids ("Ducke") provide a detailed multi-taxa assessment of biodiversity [52]. Although there are exceptions, previous studies (e.g. [53,54,55,56]), and the more general overview provided in [3]), show that meso-scale patterns in the vast majority of species and communities vary at similar scales to altitude within this area (altitude mean ± SD (range): 91±18.9 (53–127)). In other words, 2 to 3 fold variations in meso-scale richness and diversity are relatively common within Ducke but 100 fold variations are rare. The fact that variation in the community diversity of several groups is not explained by topographic variation i.e. slope (see Table 1 of [52]), is not necessarily critical in this case. The more important point is that at meso-scales the spatial variation of species and communities is often similar to or less than the variation in altitude [53,54,55,56]. Therefore, we can expect that the same sample size thresholds we identified for representing altitude should also be representative for quantifying numerous meso-scale biodiversity responses.

There are obvious limitations to the thresholds identified. We only considered the sample representativeness in areas with a standard deviation ≤56, corresponding to 95% of the legal Brazilian Amazon. The thresholds identified are unlikely to be applicable in areas where altitude varies more, such as the Andean highlands. Although it is theoretically possible to establish regular 25 km^2 sample grids in highly heterogeneous areas, this is unlikely to be a cost effective approach in locations with strong preexisting gradients/strata (e.g. highly heterogeneous topographic variation/land cover types). In such situations we would expect that some sort of stratified sample scheme should be the most cost effective [4,57].

Practical implications

But what is the optimum sample design for monitoring Amazon biodiversity? The answer to this question is location and context specific and therefore lies with individual researchers/project managers. Within the context of biodiversity monitoring, study designs must provide the most information for the lowest price [1,4] i.e. meet statistical requirements and be practically and logistically possible [1,4]. Some consider a regular arrangement to be unnecessarily "demanding" in terms of sampling effort. For example, Gardner [4] highlights that "the major disadvantage of a uniform-grid approach is that it demands a high level of sampling effort", but the same author describes a stratified-random approach using considerably more effort than 30 sample points [57]. It is clear that a regular design would probably cost more to implement than the stratified approach to answer the questions in [57]. But the efforts described in [57] suggest that a sample size of 30 across a 25 km^2 area is well within the logistic and financial "norm" of Amazon research projects. An often overlooked benefit of standardized samples that adequately represent the target response is that it is possible to directly compare and contrast values of interest. For example, it is possible to directly compare intuitive and easily communicable metrics, such as number of individuals (N) or number of species (S) to represent biodiversity. The ability to communicate and inform results is one of the advantages of standardized samples that is, however, rarely included in comparisons of sampling designs [1].

In cases where strata can be identified (the logical pre-requisite for stratification), there will probably be sufficient information (e.g. describing magnitude of expected response, magnitude of possible confounding effects) to develop robust sampling designs to meet research objectives. Land use change coupled with the presence of multi-jurisdictional landscapes, (where it is often impossible to assume a regular design as some land-owners will not permit access (T. Gardner pers. com.)) means that there is a large area within the legal Brazilian Amazon where a 25 km^2 regular sampling grid is unlikely to be practical. Yet at the same time, Amazon deforestation is decreasing and there remain vast tracts of forest cover across the Brazilian Amazon [58]. For example the Amazon Region Protected Areas (ARPA) program is a network that includes over 90 protected areas, and covers 51 million hectares [59]. Therefore, there remains a substantial area where knowledge to inform sample stratification for biodiversity monitoring is unavailable (e.g. unknown future threats, poorly described response/predictor variables, multiple response variables) and stratification is therefore unlikely to be a viable option [3,27,28].

We are not advocating one arrangement over another, and are merely demonstrating the adequacy of regularly arranged sample sizes. Indeed, a regular arrangement is unlikely to be suitable for a variety of relevant ecological questions, such as studies examining rare species or the density of large canopy trees [60]. However, it is a relatively simple task to develop more detailed studies using a subsample of the logistics provided by a broader scale sample arrangement. For example, many researchers use only a fraction of a 25 km^2 PPBio grid to conduct a wide variety of relevant studies [3]. It is also worth noting that a recent study showed that common indicators are more likely to be appropriate compared with rare species indicators within the context of monitoring biodiversity for environmental service payments [5].

Sommerville et al. (2011) [5] present a framework for effectively monitoring environmental services where it is necessary to make three major decisions: (i) which indicators to monitor (e.g. a species of concern/threats/positive actions), (ii) how to monitor the chosen indicators (e.g. ground based, remote sensing or a combination) and (iii) how to use the monitoring results to determine payments (e.g. presence-absence of an indicator/the trend of the indicator over time/detecting differences between monitored sites/achievement of targets). These choices will dictate the most adequate sampling design for monitoring biodiversity within any given context/locality. Decades of research present detailed evaluations of how to derive spatially optimal sampling strategies [10,61,62,63] and rather than seek to optimize for the general "biodiversity monitoring" case (which never exists) we show that metrics obtained from a regular sample are likely to be both widely applicable and comparable across ≈5 million km^2 of the legal Brazilian Amazon.

Conclusions

Although not all biodiversity indicators are directly related to altitude in the Amazon, in the majority of situations, it will at least be necessary to model and/or control for the effects of altitude in both within and between site comparisons. The ability to represent meso-scale altitudinal variation with a regular arrangement of 30 samples provides strong support for the idea that such an arrangement and sample size will be adequate for a wide range of present and future biodiversity monitoring challenges. Knowing that samples/interpolations provide a reasonable representation of meso-scale altitude values means that we are able to reliably use such samples to examine whether within and between-site differences in patterns of biodiversity indicators are or are not driven by altitudinal/topographical variation. The capability to compare interpolated values with those derived from other representative sampling designs is also likely to facilitate both comparison across studies and communication between researchers and managers.

Supporting Information

Figure S1 Random sample areas.

Figure S2 Correlation and RMSE of interpolated altitude values.

Figure S3 Distribution of sample plots within seven active research areas.

Figure S4 Obtaining locations of river-trail intersections.

Figure S5 Mapped comparison of IDW, Kriging and GAM interpolations.

Figure S6 Comparison of IDW, Kriging and GAM interpolations.

Figure S7 Sample "representativeness."

Figure S8 Sample "unrepresentativeness."

Figure S9 Sample KS-test conditional inference tree.

Figure S10 IDW KS-test conditional inference tree.

Figure S11 IDW correlation conditional inference tree.

Figure S12 IDW error conditional inference tree.

Figure S13 Mapped IDW estimates.

Table S1 Active research areas.

Acknowledgments

Data were provided by the Programa de Pesquisa em Biodiversidade (PPBio) and the Instituto Nacional de Ciência, Tecnologia e Inovação para Biodiversidade Amazônica (INCT-CENBAM). We thank Toby Gardner, Bruno Hérault and two anonymous reviewers for comments that improved previous versions of the text.

Author Contributions

Conceived and designed the experiments: DN MF WEM. Analyzed the data: DN. Wrote the paper: DN MF WEM. Designed and produced the figures: DN.

References

1. Bonar SA, Fehmi JS, Mercado-Silva N (2011) An overview of sampling issues in species diversity and abundance surveys. In: Magurran AE, McGill BJ, editors. Biological diversity: frontiers in measurement and assessment. Oxford: Oxford University Press. pp. 11–24.
2. Groves RM (2004) Survey errors and survey costs. New York: Wiley-Blackwell. 616 p.
3. Magnusson W, Braga-Neto R, Pezzini F, Baccaro F, Bergallo H, et al. (2013) Biodiversity and integrated environmental monitoring (Biodiversidade e monitoramento ambiental integrado). Manaus: Áttema Editorial :: Assessoria e Design. 352 p. Downloadable from: http://ppbio.inpa.gov.br/livros.
4. Gardner TA (2010) Monitoring forest biodiversity: Improving conservation through ecologically responsible management. London: Earthscan. 360 p.
5. Sommerville MM, Milner-Gulland E, Jones JP (2011) The challenge of monitoring biodiversity in payment for environmental service interventions. Biological Conservation 144: 2832–2841.
6. Dickson B, Kapos V (2012) Biodiversity monitoring for REDD+. Current Opinion in Environmental Sustainability 6: 717–725.
7. Jones JP, Asner GP, Butchart SH, Karanth KU (2013) The 'why','what'and 'how'of monitoring for conservation. Key Topics in Conservation Biology 2: 327–343.
8. Haining RP (1990) Spatial data analysis in the social and environmental sciences. Cambridge: Cambridge University Press. 432 p.
9. Fortin M-J, Dale MRT (2005) Spatial analysis: a guide for ecologists. New York: Cambridge University Press. 365 p.
10. Olea RA (1984) Sampling design optimization for spatial functions. Journal of the International Association for Mathematical Geology 16: 369–392.
11. Skidmore AK, Franklin J, Dawson TP, Pilesjo P (2011) Geospatial tools address emerging issues in spatial ecology: a review and commentary on the Special Issue. International Journal of Geographical Information Science 25: 337–365.
12. Magnusson W (2014) Uncertainty and the design of in-situ biodiversity-monitoring programs. Nature Conservation In press.
13. Hughes C, Eastwood R (2006) Island radiation on a continental scale: Exceptional rates of plant diversification after uplift of the Andes. Proceedings of the National Academy of Sciences of the United States of America 103: 10334–10339.
14. Svenning JC (1999) Microhabitat specialization in a species-rich palm community in Amazonian Ecuador. Journal of Ecology 87: 55–65.
15. Hothorn T, Mueller J, Schroeder B, Kneib T, Brandl R (2011) Decomposing environmental, spatial, and spatiotemporal components of species distributions. Ecological Monographs 81: 329–347.
16. Körner C (2007) The use of 'altitude'in ecological research. Trends in Ecology & Evolution 22: 569–574.
17. Laurance SG, Laurance WF, Andrade A, Fearnside PM, Harms KE, et al. (2010) Influence of soils and topography on Amazonian tree diversity: a landscape-scale study. Journal of Vegetation Science 21: 96–106.
18. Tuomisto H, Ruokolainen K, Kalliola R, Linna A, Danjoy W, et al. (1995) Dissecting amazonian biodiversity. Science 269: 63–66.
19. Malhi Y, Silman M, Salinas N, Bush M, Meir P, et al. (2010) Introduction: elevation gradients in the tropics: laboratories for ecosystem ecology and global change research. Global Change Biology 16: 3171–3175.
20. Hoorn C, Wesselingh F, Ter Steege H, Bermudez M, Mora A, et al. (2010) Amazonia through time: Andean uplift, climate change, landscape evolution, and biodiversity. Science 330: 927–931.
21. Costa FRC, Magnusson WE (2010) The need for large-scale, integrated studies of biodiversity– the experience of the Program for Biodiversity Research in Brazilian Amazonia. Natureza & Conservacao 8: 3–12.
22. Yao XL, Fu BJ, Lu YH, Sun FX, Wang S, et al. (2013) Comparison of Four Spatial Interpolation Methods for Estimating Soil Moisture in a Complex Terrain Catchment. Plos One 8: e54660.
23. Greve MH, Kheir RB, Greve MB, Bocher PK (2012) Quantifying the ability of environmental parameters to predict soil texture fractions using regression-tree model with GIS and LIDAR data: The case study of Denmark. Ecological Indicators 18: 1–10.
24. Vanwalleghem T, Meentemeyer RK (2009) Predicting Forest Microclimate in Heterogeneous Landscapes. Ecosystems 12: 1158–1172.

25. Renschler CS, Lee T (2005) Spatially distributed assessment of short- and long-term impacts of multiple best management practices in agricultural watersheds. Journal of Soil and Water Conservation 60: 446–456.
26. Hengl T (2011) A practical guide to geostatistical mapping of environmental variables. Amsterdam: University of Amsterdam. 291 p.
27. Johnson DH (2012) Monitoring that matters. In: Gitzen RA, Millspaugh JJ, Cooper AB, Licht DS, editors. Design and Analysis of Long-term Ecological Monitoring Studies. Cambridge: Cambridge University Press. pp. 54–73.
28. McDonald T (2012) Spatial sampling designs for long-term ecological monitoring. In: Gitzen RA, Millspaugh JJ, Cooper AB, Licht DS, editors. Design and Analysis of Long-term Ecological Monitoring Studies. Cambridge: Cambridge University Press. pp. 101–125.
29. Magnusson WE, Lima AP, Luizão R, Luizão F, Costa FR, et al. (2005) RAPELD: a modification of the Gentry method for biodiversity surveys in long-term ecological research sites. Biota Neotropica 5: 19–24.
30. Naiman RJ, Décamps H (1997) The ecology of interfaces: riparian zones. Annual review of Ecology and Systematics 28: 621–658.
31. Sabo JL, Sponseller R, Dixon M, Gade K, Harms T, et al. (2005) Riparian zones increase regional species richness by harboring different, not more, species. Ecology 86: 56–62.
32. Drucker DP, Costa FRC, Magnusson WE (2008) How wide is the riparian zone of small streams in tropical forests? A test with terrestrial herbs. Journal of Tropical Ecology 24: 65–74.
33. R Core Team (2013) R: A language and environment for statistical computing. 2.15.3. Vienna, Austria: R Foundation for Statistical Computing. Available: http://www.R-project.org.
34. Wickham H (2009) ggplot2: elegant graphics for data analysis. New York: Springer.
35. Pebesma EJ (2004) Multivariable geostatistics in S: the gstat package. Computers & Geosciences 30: 683–691.
36. Wood SN (2011) Fast stable restricted maximum likelihood and marginal likelihood estimation of semiparametric generalized linear models. Journal of the Royal Statistical Society (B) 73: 3–36.
37. Hijmans RJ, van Etten J (2013) raster: Geographic data analysis and modeling. R package version 2.1-0/r2632. Available: http://CRAN.R-project.org/package=raster.
38. Pebesma EJ, Bivand R (2005) Classes and methods for spatial data in R. R News 2: 9–13.
39. Pillar V (1998) Sampling sufficiency in ecological surveys. Abstracta Botanica 22: 37–48.
40. Hothorn T, Hornik K, Zeileis A (2006) Unbiased recursive partitioning: A conditional inference framework. Journal of Computational and Graphical Statistics 15: 651–674.
41. De'ath G, Fabricius KE (2000) Classification and regression trees: a powerful yet simple technique for ecological data analysis. Ecology 81: 3178–3192.
42. Hothorn T, Zeileis A (2013) partykit: A toolkit for recursive partitioning. R package version 0.1–6. Available: http://CRAN.R-project.org/package=partykit.
43. Hortal J, Borges PA, Gaspar C (2006) Evaluating the performance of species richness estimators: sensitivity to sample grain size. Journal of Animal Ecology 75: 274–287.
44. Gutschick VP, BassiriRad H (2003) Extreme events as shaping physiology, ecology, and evolution of plants: toward a unified definition and evaluation of their consequences. New Phytologist 160: 21–42.
45. Katz RW, Brush GS, Parlange MB (2005) Statistics of extremes: Modeling ecological disturbances. Ecology 86: 1124–1134.
46. Lindenmayer DB, Likens GE (2010) The science and application of ecological monitoring. Biological Conservation 143: 1317–1328.
47. Gitzen RA, Millspaugh JJ, Cooper AB, Licht DS, editors (2012) Design and Analysis of Long-term Ecological Monitoring Studies. Cambridge: Cambridge University Press. 590 p.
48. Certain G, Skarpaas O, Bjerke J-W, Framstad E, Lindholm M, et al. (2011) The Nature Index: A general framework for synthesizing knowledge on the state of biodiversity. Plos One 6: e18930.
49. Engel S, Pagiola S, Wunder S (2008) Designing payments for environmental services in theory and practice: An overview of the issues. Ecological Economics 65: 663–674.
50. Nelson E, Mendoza G, Regetz J, Polasky S, Tallis H, et al. (2009) Modeling multiple ecosystem services, biodiversity conservation, commodity production, and tradeoffs at landscape scales. Frontiers in Ecology and the Environment 7: 4–11.
51. Rands MRW, Adams WM, Bennun L, Butchart SHM, Clements A, et al. (2010) Biodiversity conservation: Challenges beyond 2010. Science 329: 1298–1303.
52. Landeiro VL, Bini LM, Costa FR, Franklin E, Nogueira A, et al. (2012) How far can we go in simplifying biomonitoring assessments? An integrated analysis of taxonomic surrogacy, taxonomic sufficiency and numerical resolution in a megadiverse region. Ecological Indicators 23: 366–373.
53. Kinupp V, Magnusson WE (2005) Spatial patterns in the understorey shrub genus Psychotria in central Amazonia: effects of distance and topography. Journal of Tropical Ecology 21: 363–374.
54. Costa FRC, Guillaumet J-L, Lima AP, Pereira OS (2009) Gradients within gradients: The mesoscale distribution patterns of palms in a central Amazonian forest. Journal of Vegetation Science 20: 69–78.
55. Cintra R, Naka LN (2011) Spatial variation in bird community composition in relation to topographic gradient and forest heterogeneity in a Central Amazonian rainforest. International Journal of Ecology 2012.
56. de Oliveira Freitas R, Buscardo E, Nagy L, dos Santos Maciel AB, Carrenho R, et al. (2014) Arbuscular mycorrhizal fungal communities along a pedo-hydrological gradient in a Central Amazonian terra firme forest. Mycorrhiza 24: 21–32.
57. Gardner TA, Ferreira J, Barlow J, Lees AC, Parry L, et al. (2013) A social and ecological assessment of tropical land uses at multiple scales: the Sustainable Amazon Network. Philosophical Transactions of the Royal Society B: Biological Sciences 368: 20120166.
58. Davidson EA, de Araújo AC, Artaxo P, Balch JK, Brown IF, et al. (2012) The Amazon basin in transition. Nature 481: 321–328.
59. Hance J (2014) Happy Amazon: $215 million raised for world's largest protected area network. Available http://news.mongabay.com/2014/0521-hance-arpa-funding.html#S6PweWmXmzpJRt7J.99. Accessed 03 June 2014.
60. Druce DJ, Shannon G, Page BR, Grant R, Slotow R (2008) Ecological thresholds in the savanna landscape: Developing a protocol for monitoring the change in composition and utilisation of large trees. PLoS ONE 3: e3979.
61. Legendre P, Dale MR, Fortin MJ, Gurevitch J, Hohn M, et al. (2002) The consequences of spatial structure for the design and analysis of ecological field surveys. Ecography 25: 601–615.
62. Van Groenigen J, Stein A (1998) Constrained optimization of spatial sampling using continuous simulated annealing. Journal of Environmental Quality 27: 1078–1086.
63. Stevens DL Jr, Olsen AR (2004) Spatially balanced sampling of natural resources. Journal of the American Statistical Association 99: 262–278.

6

Spatial Heterogeneity in Ecologically Important Climate Variables at Coarse and Fine Scales in a High-Snow Mountain Landscape

Kevin R. Ford[1]*, Ailene K. Ettinger[1], Jessica D. Lundquist[2], Mark S. Raleigh[2], Janneke Hille Ris Lambers[1]

1 Department of Biology, University of Washington, Seattle, Washington, United States of America, **2** Department of Civil and Environmental Engineering, University of Washington, Seattle, Washington, United States of America

Abstract

Climate plays an important role in determining the geographic ranges of species. With rapid climate change expected in the coming decades, ecologists have predicted that species ranges will shift large distances in elevation and latitude. However, most range shift assessments are based on coarse-scale climate models that ignore fine-scale heterogeneity and could fail to capture important range shift dynamics. Moreover, if climate varies dramatically over short distances, some populations of certain species may only need to migrate tens of meters between microhabitats to track their climate as opposed to hundreds of meters upward or hundreds of kilometers poleward. To address these issues, we measured climate variables that are likely important determinants of plant species distributions and abundances (snow disappearance date and soil temperature) at coarse and fine scales at Mount Rainier National Park in Washington State, USA. Coarse-scale differences across the landscape such as large changes in elevation had expected effects on climatic variables, with later snow disappearance dates and lower temperatures at higher elevations. However, locations separated by small distances (~20 m), but differing by vegetation structure or topographic position, often experienced differences in snow disappearance date and soil temperature as great as locations separated by large distances (>1 km). Tree canopy gaps and topographic depressions experienced later snow disappearance dates than corresponding locations under intact canopy and on ridges. Additionally, locations under vegetation and on topographic ridges experienced lower maximum and higher minimum soil temperatures. The large differences in climate we observed over small distances will likely lead to complex range shift dynamics and could buffer species from the negative effects of climate change.

Editor: Francesco de Bello, Institute of Botany, Czech Academy of Sciences, Czech Republic

Funding: Research was supported by the University of Washington Department of Biology (KF, AE), the American Alpine Club (KF), the Seattle chapter of the ARCS Foundation (KF), an NSF Doctoral Dissertation Improvement Grant (DEB-1010787) (AE), the University of Washington Royalty Research Foundation (JH) and the US Department of Energy (DOE#DE-FC02 06ER64159) (JH). This material is based on work supported by the NSF Graduate Research Fellowship Program (DGE-0718124- KF, AE), NSF grant no. CBET-0931780 (JL) and by NASA Headquarters under the NASA Earth and Space Science Fellowship Program (NNX09AO22H) (MR). The funders had no role in study design, data collection and analysis, decision to publish, or preparation of the manuscript.

Competing Interests: The authors have declared that no competing interests exist.

* E-mail: kford10@gmail.com

Introduction

Biologists have long recognized the fundamental role climate plays in determining the geographic distributions of species and biomes [1–3]. As a result, climate change is expected to induce shifts in the geographic ranges of species. This prediction is supported by the many observations of upward or poleward range shifts over the last 100 years consistent with observed warming [4] as well as range shifts inferred from the fossil record [5,6]. Alarmingly, models of the impacts of future anthropogenic climate change on species ranges have forecasted widespread extinction risks as the climatic niche of many species disappears or shifts faster than species can likely migrate [7,8].

However, these projections of climate change-induced range shifts (and subsequent extinction risks) are sensitive to the spatial scale at which the analyses are conducted [9]. Most range shift assessments rely on gridded maps of climate variables with grid cell sizes ranging from 800×800 m (e.g. PRISM [10] and WorldClim [11]) up to 50×50 km (e.g. [12]). The finer scale maps (800×800 m grid cells) capture a wide variety of climatic patterns, but the scales of these maps are still far coarser than the scales at which organisms experience their environment. Thus, these climate maps may hide fine-scale differences in climate that are important for organism distributions [13]. For example, north and south facing slopes separated by tens of meters may receive different amounts of solar radiation and experience very different temperature regimes [14,15], which could lead to differences in species composition within these microhabitats.

The implication of significant fine-scale climatic heterogeneity that is not captured by coarse-scale climate maps is that projections based on these maps could fail to capture important range shift dynamics. For example, cool microhabitats (such as north-facing slopes in the Northern Hemisphere) near the contracting edge or core of a species' distribution may allow populations of that species to persist if individuals can disperse to them from warmer microhabitats (such as south-facing slopes in the Northern Hemisphere), even if most of the surrounding landscape becomes unsuitably warm (as long as these microhabitat types comprise a total area large enough to support a population [16]). At the same time, warm microhabitats beyond the advancing edge of a species' range may provide the first sites of colonization that allow that species to migrate to new locations. Thus, instead of needing to move hundreds of meters upward or hundreds of kilometers poleward to track suitable climate, many species may only need to move tens of meters from one microhabitat to another and could be buffered from the negative effects of climate change [17].

For such fine-scale climatic heterogeneity to strongly influence range dynamics, however, fine-scale differences in climate must be large relative to coarse-scale differences. We addressed this issue by examining the magnitude of fine-scale heterogeneity relative to coarse-scale heterogeneity in snow disappearance date and growing season soil temperature. Specifically, we deployed 284 microclimate sensors across a ~1500 m elevation gradient spanning forest, subalpine and alpine biomes at Mount Rainier National Park. Our objectives were to 1) quantify snow disappearance date and soil temperature as a function of coarse-scale differences in elevation and exposure to storm tracks (i.e. being on the windward or leeward side of the mountain) and fine-scale differences in vegetation structure or topography, 2) compare fine-scale differences in climatic variables (that would be missed by climate models) to coarse-scale differences (that would be captured by climate models), and 3) determine whether fine-scale patterns in climatic variables related to topography (but not vegetation structure) are correlated with fine-scale patterns in vegetation characteristics. We focus on snow disappearance date and growing season soil temperature because snow disappearance date influences the length of the growing season (especially important in this region where the growing season can be very short due to the persistence of large winter snowpacks) while soil temperature strongly influences plant growth rates and other physiological processes [18]. These variables have also been shown to be strongly associated with patterns of distribution, abundance, productivity and diversity of plant species, at our sites and others [19–21]. Additionally, both variables are likely to change in the coming decades as a result of anthropogenic climate change, with rising temperatures and declining snowpacks leading to warmer and longer growing seasons [22,23].

It has long been known that climate can vary dramatically at fine spatial scales (reviewed in [14] and [24]), but these patterns have only recently begun to be studied explicitly and systematically with respect to the impacts of climate change on species distributions. Studies have found that locations separated by only tens of meters experienced mean seasonal soil temperatures that differed by 3–7°C, equivalent to the average temperature difference experienced in locations separated by hundreds of meters in elevation or hundreds of kilometers in latitude [15,21,25–28]. Moreover, such large differences in temperature are known to strongly influence organismal performance [18,29] and are greater than the expected increase in temperature due to climate change in many parts of the globe [30]. Similarly large differences were also observed in air temperature, snow cover duration or snow disappearance date over fine spatial scales in these studies. Our paper builds on these case studies and is notable for its large sample size of 284 sensors (important for assessing microclimate patterns in a statistically rigorous way), its explicit comparison of coarse- and fine-scale climatic heterogeneity (important for assessing the biases of coarse-scale models) and the broad environmental gradients covered (important for assessing how widespread these biases may be).

Methods

Ethics Statement

We obtained the necessary scientific research permits from Mount Rainier National Park, where all data collection occurred, before conducting this study. We did not sample any protected species.

Study Area

Mount Rainier National Park encompasses 95,354 ha of land in the western Cascade Mountains in Washington State, USA (Figure 1). The region experiences a temperate, maritime climate with mild, dry summers and cool, wet winters that produce large snowpacks. Elevation ranges from 518 m in the deep valley floors to 4392 m at the peak of Mount Rainier, the volcano located in the middle of the Park. The mountainous terrain produces steep climatic gradients: temperature decreases and precipitation increases with elevation, while the rainshadow effect produced by the volcano leads to lower precipitation on the east side of the Park. There are two primary climate stations in the Park. At the station located at 842 m elevation, mean annual temperature is 6.6°C and mean annual precipitation is 2030 mm; at the 1654 m station, mean annual temperature is 3.7°C and mean annual precipitation is 3005 mm (1981–2010 normals, NOAA National Climate Data Center – www.wrcc.dri.edu/Climsum.html).

The large climatic gradients create three major biomes in the Park. The forest biome extends from the lowest elevations of the Park up to about 1600–2000 m and is dominated by coniferous trees. The subalpine biome typically extends about 300 m above the upper limit of the forest and is a mosaic of conifer tree patches and subalpine meadows. The alpine biome occupies the highest elevations, stretching from 1900–2300 m to the summit of Mount Rainier, and consists of large patches of mostly continuous alpine meadows (dominated by forbs, grasses and dwarf shrubs) near the lower limit of the biome, with exposed rock, glaciers, bare soil, and cryptogams (mostly mosses, lichens, algae and cryptobiotic soil crusts) predominating above.

Study Design

From September 2009 through October 2010, we deployed 284 soil temperature sensors (HOBO Pendants made by the Onset Computer Corporation and iButtons made by Maxim Integrated Products) across Mount Rainier at elevations ranging from 638 m to 2164 m as part of two different plant ecology studies where microclimate was measured as an explanatory variable. The first study took place in the forest biome and spanned the elevational range of forests on the south side of Mount Rainier. The second study took place in the subalpine and alpine biomes, with study sites set up at the lower limit of the subalpine biome and the upper limit of alpine meadows on three sides of the mountain (Figure 1). Microclimate data from these studies were ideally suited for our questions as they covered large elevational gradients with sensors at each location stratified by vegetation or topographic features expected to influence microclimate. Due to differences in study design (described below), we analyzed the microclimate data from the two studies separately. The sensors remained in place and logged data until we collected them in September/October 2010.

For each sensor, we calculated the values of four climatic variables: snow disappearance date, and average daily mean, maximum and minimum soil temperature. We could assess snow cover from soil temperature measurements because snow insulates soil from fluctuations in air temperature so that temperatures beneath the snowpack in this region remain constant around 0°C. Thus, the soil temperature data allowed us to determine whether

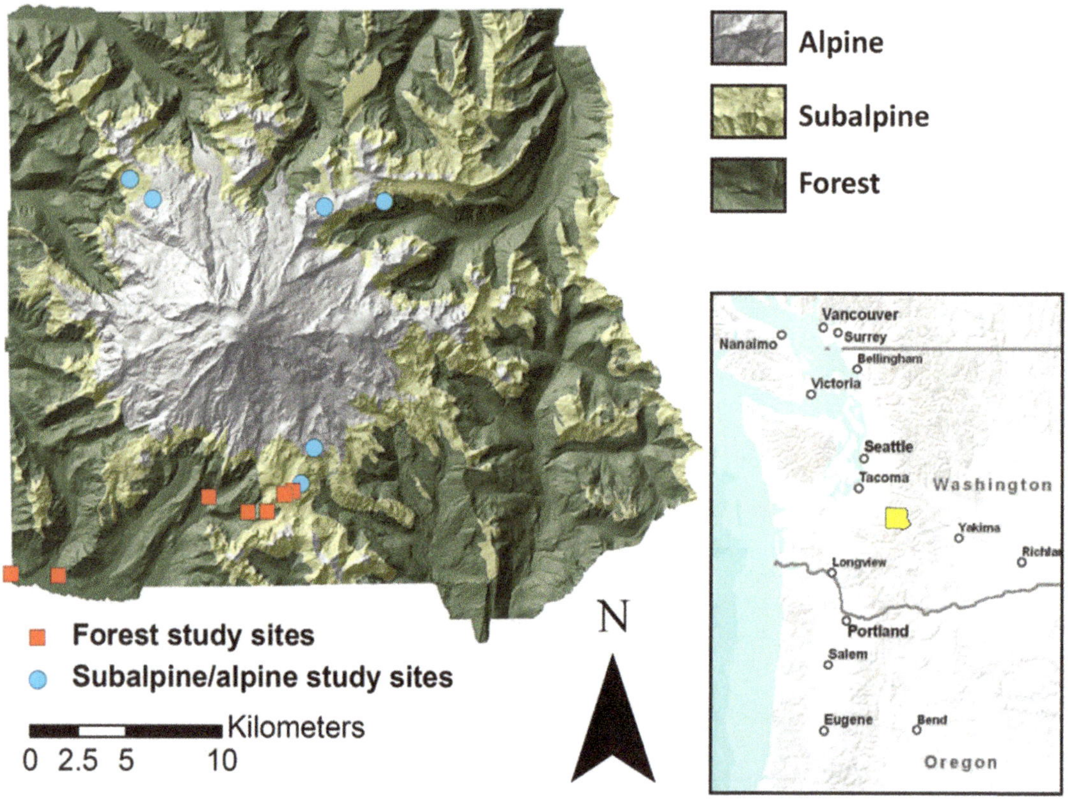

Figure 1. Study area. Mount Rainier National Park and its three major biomes, along with study site locations. Shading depicts topographic relief.

snow was covering the sensor for each day the sensor was deployed using an algorithm that considers daily temperature ranges and maxima [31,32]. We calculated average daily mean, maximum and minimum soil temperature for periods in Summer/Fall 2010 when all sensors in a study experienced snow-free conditions and reported data. This period was August 14 through October 3, 2010 for the sensors in the forest study and August 11 through August 18, 2010 for the sensors in the subalpine/alpine study. This meant we only used a portion of the snow-free temperature data for some sensors, even though temperatures outside this period are likely to also be ecologically relevant. However, it was necessary to use the same time period for all sensors in a study so that temporal differences in the snow-free period between locations did not confound our analysis of spatial differences in temperature.

Arrays of sensors were deployed at 13 sites throughout the Park. For the forest study, we established three study sites along an elevation gradient in Summer 2009, allowing us to calculate snow disappearance date in 2010. We quantified growing season soil temperature in Summer/Fall 2010 at these three sites plus an additional four sites along the same elevation gradient. Within each site, we deployed sensors under gaps in the forest canopy caused by tree falls ("gaps") and in locations under intact canopy within 20 m of one of the gaps ("non-gaps"). Each study site contained five of these gap/non-gap pairs. Gaps were ~130 m^2 on average. Within each gap or non-gap location, we placed one sensor in a 5.5×1.5 m area where all understory vegetation up to 2 m tall had been experimentally removed since early Summer 2009 (the "removed" plot) and one sensor at an adjacent location 2 m away where the vegetation had been left undisturbed (the "control" plot) (Figure 2A, Table 1).

For the subalpine/alpine study, we quantified microclimate at study sites on three sides of the mountain (south, northwest and northeast) which have different exposures to storm tracks and experience different precipitation regimes. On each side, we established two study sites, one close to the lower limit of the subalpine biome and one close to the upper limit of continuous alpine meadows, (below this limit, the ground is mostly vegetated while above it is almost entirely rock, glaciers and bare soil). These sites were about 200–300 m apart in elevation on each side of the mountain. Within each site, we established six linear transects that ran from a depression in the landscape up to a ridge (transects parallel to the slope) and were about 20 m in length. Within each transect, two sensors were located in the depression and two sensors were located on the ridge (Figure 2B, Table 1).

At each of the sensors in the subalpine/alpine biomes (where fine-scale sensor placement was stratified by topographic position and not vegetation structure), we measured vegetation characteristics in order to compare patterns in microclimate to ecological patterns. At the study sites near the lower limit of the subalpine biome, where closed-canopy forests transition to open meadows with increasing elevation, we measured percent cover by tree canopy above each sensor using a spherical densiometer (a gridded, hemisphere-shaped mirror used to estimate percent cover by foliage above a point on the ground), allowing us to assess the density of trees (higher values of tree canopy cover implies more or bigger trees). At the study sites near the upper limit of alpine meadows, where meadows transition to bare ground with increasing elevation, we estimated the percent of ground covered by vegetation at each sensor using a square-shaped PVC frame (1×1 m) placed on the ground adjacent to the sensor. String tied to the PVC frame created 100 evenly spaced grid points, allowing us

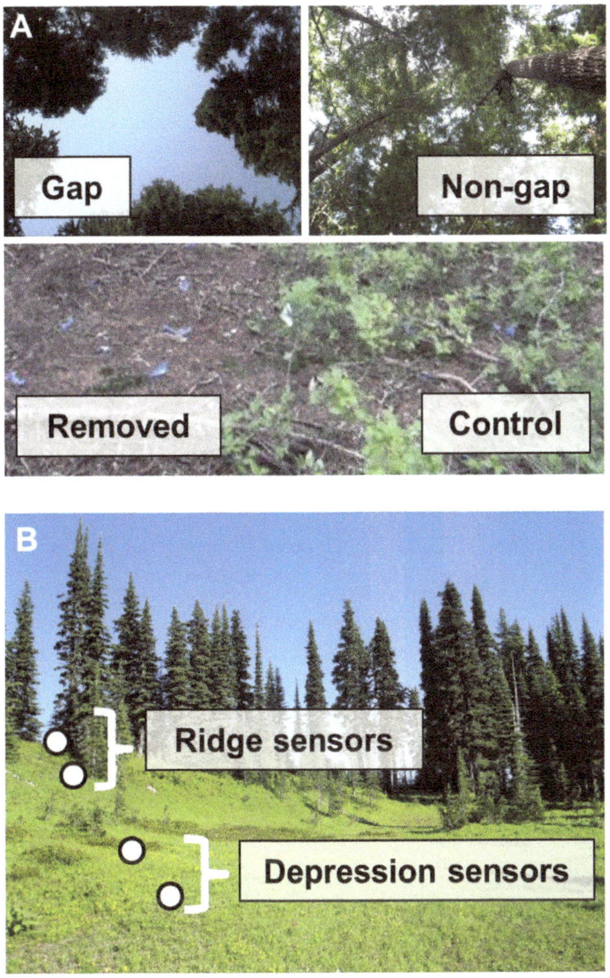

Figure 2. Temperature sensor deployment. Sensor deployment in (A) forest and (B) subalpine/alpine biomes. At each elevation in the forest biome (A), sensors were placed in gaps in the forest canopy (top left) and non-gaps with intact forest canopy (top right). Within each of these canopy types, sensors were located in plots where understory vegetation was removed (bottom left) and control plots where it was left undisturbed (bottom right). In the subalpine/alpine biomes (B), temperature sensors were located along transects running from depressions in the landscape to ridges.

to count the number of grid points overlaying vegetation in the area within the frame.

Data Analysis

We used linear mixed effects models (LMMs) to characterize the relationships between the potential drivers of climate and each of the four climatic response variables [33]. The LMMs allowed us to estimate the effects of explanatory variables and their two-way interactions on the response variable ("fixed effects") while statistically controlling for the effects of randomly selected experimental units on the response variable ("random effects"). At the forest sites, the drivers of climate were elevation, canopy structure (gap vs. non-gap) and understory structure (removed vs. control treatments), with gap/non-gap pair designated as a random effect. At the subalpine/alpine sites, the drivers of climate were side of the mountain (south vs. northwest vs. northeast), elevation (upper limit of alpine meadows vs. lower limit of the

Table 1. Details of temperature sensor deployment.

Biome(s)	Time span of deployment	# sites	# sensors per site	Total # sensors	Type of sensor	Sensor accuracy	Data logging interval	Sensor location
Forest	Summer 2009– Fall 2010	3	8	24	HOBO Pendants made by the Onset Computer Corporation	±0.53°C from 0° to 50°C	2 hours	Soil surface
Forest	Summer–Fall 2010	7*	20	140	HOBO Pendants made by the Onset Computer Corporation	±0.53°C from 0° to 50°C	1 hour	Soil surface
Subalpine/alpine	Summer 2009– Summer 2010	6	24	144	iButtons made by Maxim Integrated Products	±0.5°C from −10° to 65°C, or ±1°C from −30° to 70°C†	1, 2 or 4 hours†	3 cm below soil surface

*These sites include the three sites with sensors deployed in Summer 2009.
†Differences in accuracy and logging intervals were due to differences in the specific model of iButton sensor used.

subalpine biome, treated as a categorical variable since there were only two values of elevation on each side of the mountain) and topographic position (depression vs. ridge), with sensor transect designated as a random effect. We verified that the residuals of these models were normally distributed, to validate our use of linear mixed effects models (rather than generalized linear mixed effects models).

For each model, we used Akaike's information criterion (AIC) to select the most parsimonious combination of fixed and random effects to derive the "best-fit" model. Specifically, we used a three-step process following [33] where we (1) used AIC to determine the optimal random effects structure, selecting amongst several LMMs (fit with restricted maximum likelihood) that had different random effect terms (no random effects, random intercepts, random slopes or both), but the same fixed effect terms (which included all main effect and two-way interaction terms for each explanatory variable); (2) determined the optimal combination of fixed effect terms by using AIC to select amongst models (fit with maximum likelihood) with all possible combinations of fixed effect terms (but sharing the optimal random effects structure selected in the first step); and (3) fit a model with the random effects structure selected in the first step (which could be no random effects) and the fixed effects structure selected in the second step and considered this model to be our final "best-fit" model. This final model was fit with restricted maximum likelihood if it included random effects or maximum likelihood if it did not. All models were fit in R version 2.12.0 using the lme4 package for the LMMs [34,35]. See Appendix S1 for more details of the model fitting and selection procedure.

We assessed the significance of the model coefficients using Markov chain Monte Carlo sampling implemented with the MCMCglmm package in R [36] or *t*-tests (when no random effects were included in the best-fit model). We then used the explanatory variable coefficients of the best-fit models to calculate the magnitude of differences in microclimate response variables relative to differences in the explanatory variables. For example, if the coefficients related to topographic position in the model of snow disappearance date at the subalpine/alpine sites indicated that the difference between ridges and depressions was 20 days, controlling for differences in other variables, then the effect of topographic position on snow disappearance date would be equal to 20 days. In order to compare the effects of elevation (which we consider a coarse-scale driver of climate) to the effects of other explanatory variables, we calculated the difference in snow disappearance date or temperature between two points 100 m apart in elevation for each model (controlling for differences in other variables). Like differences in climate amongst different sides of the mountain, differences in climate caused by a 100 m difference in elevation can typically be captured by coarse-scale climate models (e.g. PRISM), while differences caused by vegetation structure and fine-scale topography cannot. If one of the explanatory variables was not included in the best-fit model, we included the main effect of that variable in the final model for comparative purposes. This happened for one explanatory variable in one model (the understory structure variable in the snow disappearance date model in the forest study).

For sites in the subalpine/alpine biomes, we also fit linear models (LMs) to characterize the relationships between each of the four microclimate variables and percent cover by tree canopy at the lower elevation sites, and the relationships between each of the four microclimate variables and percent cover by ground vegetation at the higher elevation sites, for a total of eight LMs. In these models, the response variable was the vegetation characteristic while the explanatory variables were the microclimate variable, side of the mountain (included as a covariate) and their interaction. Using LMMs with sensor transect designated as a random effect did not improve model fit for any of the relationships, so we used the simpler LMs for all of these analyses. Next, we used AIC to select the best-fit LM. With one exception, this best-fit model included both the microclimate variable and side of the mountain, but not their interaction, as explanatory variables. In these cases, we used *t*-tests to assess the significance of the microclimate variable coefficient in the best-fit model in order to assess the significance of that microclimate variable. When modeling percent tree canopy cover at the lower elevation sites as a function of average daily minimum temperature and side of the mountain, neither explanatory variable nor their interaction was included in the best-fit model (i.e. the best-fit model was the null model with only an intercept). For this situation, we performed a *t*-test on the minimum temperature coefficient in a model that included both minimum temperature and side of the mountain, but not their interaction, as explanatory variables (i.e. a model with the same structure as the best-fit model for all the other vegetation-climate analyses) in order to assess the significance of minimum temperature.

We calculated the proportion of variance in the response variable explained by variance in the fixed effect explanatory variables (r^2) for all models, following [37].

Results

Variations in climate were explained by both coarse- and fine-scale drivers, with best-fit models having r^2 values ranging between 0.20 and 0.94 (Table 2). As expected, higher elevations experienced later snow disappearance dates and lower average daily mean and minimum temperatures (Figure 3A, B, D). However, the relationship between elevation and average daily maximum temperature was weak, and variability in this parameter was dominated by variability amongst locations at similar elevations (Figure 3C). At the subalpine/alpine sites, snow disappearance date and temperature varied depending on what side of the mountain sensors were on – e.g. the south side experienced later snow disappearance dates on average, probably because this side of the mountain is the most exposed to winter storms and receives the largest amount of winter precipitation. However, there were also substantial differences amongst locations at similar elevations for each of these variables that could be attributed to vegetation structure or topographic position (Figures 3, 4). We also assessed heterogeneity in growing degree days (GDD – calculated as the sum of daily mean soil temperatures for all days where the daily mean soil temperature was over 5°C), which showed patterns very similar to those of snow disappearance date (results not shown due to limitations of the data – sensors were not deployed long enough to estimate GDD for the full year or growing season, which could bias comparisons of GDD amongst locations).

Forest Biome: Stratification by Vegetation Structure at Fine Scales

Snow disappearance date was later at higher elevations and in canopy gaps, while understory vegetation structure had little effect (Figure 3A). We also found that the effect of fine-scale differences in canopy structure (gaps vs. non-gaps) on snow disappearance date was similar to the effects of coarse-scale differences in elevation (100 m elevation differences) (Figure 4A). Thus, snow disappearance date differed as much at fine scales (where locations differed in forest canopy structure) as it did over coarse spatial scales.

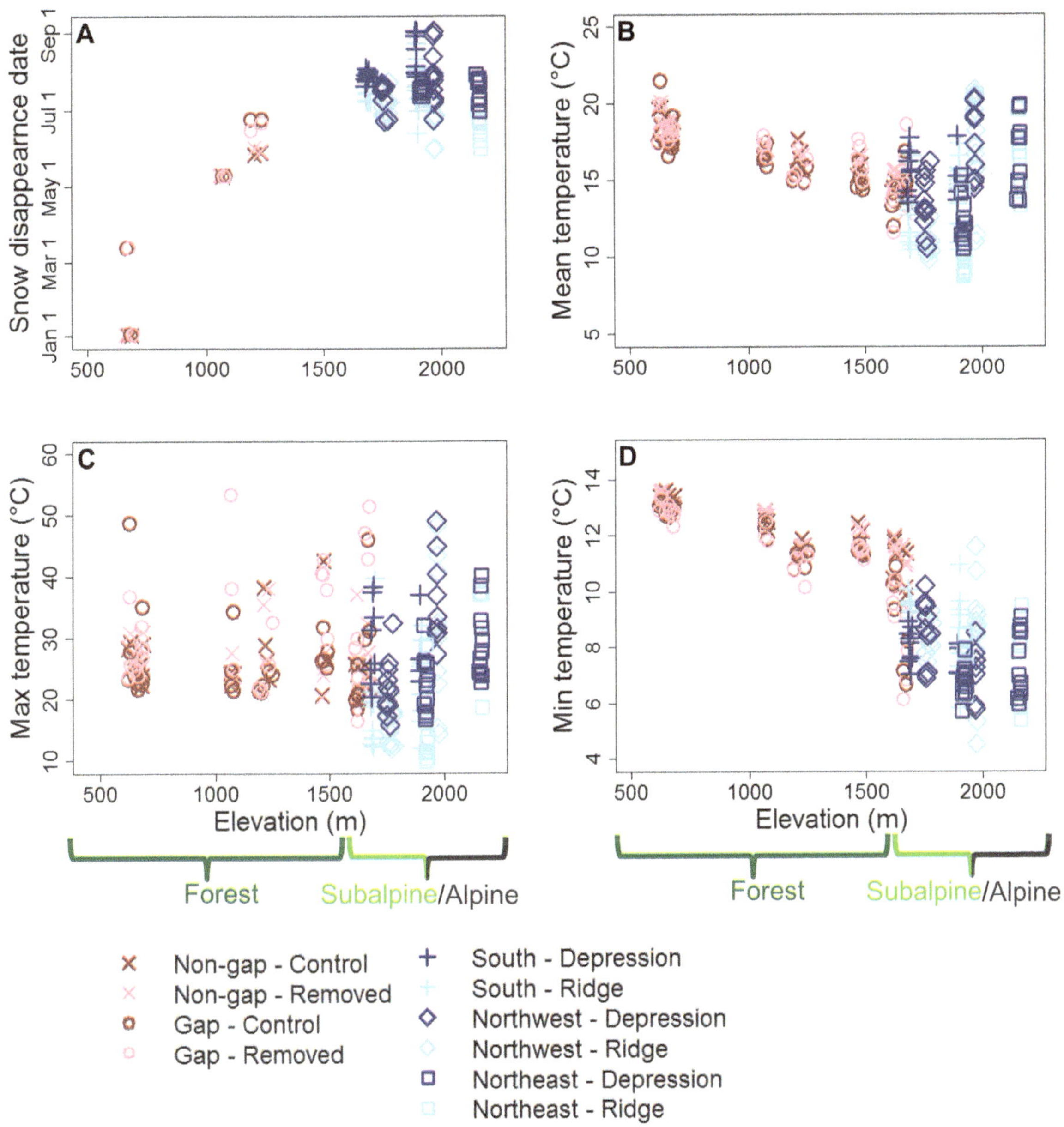

Figure 3. Patterns in climate. (A) Snow disappearance date in 2010 and average daily (B) mean, (C) maximum and (D) minimum soil temperature for a representative week during the growing season (August 11–18, 2010) plotted against elevation. Note the differences in scale on the axes showing temperature values. Points represent individual sensors with symbol type and color designating sampling stratification for forest (dark and light red) and subalpine/alpine sites (dark and light blue). "Non-gap"/"gap" refer to canopy structure categories while "control"/"removed" refer to understory structure categories (forest sites). "South"/"northwest"/"northeast" refer to sides of the mountain while "ridge"/"depression" refer to topographic positions (subalpine/alpine sites). Approximate biome ranges are shown below the elevation axes.

As expected, growing season soil temperature generally declined with increasing elevation. Canopy gaps had higher maximum temperatures, but lower minimum and mean temperatures relative to non-gaps (Figure 3B–D). Canopy structure had a similar or greater effect on temperature than a 100 m change in elevation for average daily maximum and minimum temperature (Figure 4C, D). Locations where understory vegetation was removed experienced higher maximum and mean temperatures, but lower minimum temperatures, relative to control plots where vegetation was undisturbed (Figure 3B–D). Understory structure had a greater effect on average daily maximum temperature than a 100 m change in elevation, but had weaker effects on average daily mean and minimum temperature (Figure 4B–D). Overall, there was about as much heterogeneity in temperature at fine scales (differing vegetation structure) as there was at coarse scales (100 m differences in elevation).

Subalpine/alpine Biomes: Stratification by Topographic Position at Fine Scales

Snow disappearance date was later on the south side of the mountain, at higher elevations and in topographic depressions (Figure 3A). Furthermore, the effect of fine-scale topographic differences (depressions vs. ridges) on snow disappearance date was

Table 2. Best-fit models for the climatic response variables.

<table>
<tr><th>Study</th><th>Climatic response variable</th><th>Model formula*</th><th>r^2</th></tr>
<tr><td>Forest: stratification by vegetation structure</td><td>Snow disappearance date</td><td>SDD = f { **elev**+**canopy**+understory[†]+ (1|pair) }</td><td>0.94</td></tr>
<tr><td>Forest: stratification by vegetation structure</td><td>Average daily mean temperature</td><td>Tmean = f { **elev**+canopy+understory+elev:canopy+**elev:understory**+**canopy:understory**+(0+canopy|pair) }</td><td>0.35</td></tr>
<tr><td>Forest: stratification by vegetation structure</td><td>Average daily maximum temperature</td><td>Tmax = f { elev+**canopy**+understory+**elev:canopy**+**elev:understory**+**canopy:understory**+(0+canopy|pair) }</td><td>0.40</td></tr>
<tr><td>Forest: stratification by vegetation structure</td><td>Average daily minimum temperature</td><td>Tmin = f { **elev**+canopy+**understory**+**elev:canopy**+**canopy:understory**+(0+canopy|pair) }</td><td>0.20</td></tr>
<tr><td>Subalpine/alpine: stratification by topographic position</td><td>Snow disappearance date</td><td>SDD = f { **side**+**elev**+**topo**+**elev:topo**+**side:elev**+(1|tran) }</td><td>0.60</td></tr>
<tr><td>Subalpine/alpine: stratification by topographic position</td><td>Average daily mean temperature</td><td>Tmean = f { **side**+elev+topo+**elev:topo**+**side:elev** }</td><td>0.60</td></tr>
<tr><td>Subalpine/alpine: stratification by topographic position</td><td>Average daily maximum temperature</td><td>Tmax = f { **side**+elev+topo+**elev:topo**+**side:elev** }</td><td>0.52</td></tr>
<tr><td>Subalpine/alpine: stratification by topographic position</td><td>Average daily minimum temperature</td><td>Tmin = f { side+elev+**topo**+side:elev }</td><td>0.30</td></tr>
</table>

*Parameters in bold have significant coefficients ($p<0.05$). For the forest biome study, elev = elevation; canopy = forest canopy. structure, gap or non-gap; understory = understory vegetation structure, removed or control; pair = gap/non-gap pairings. For the subalpine/alpine biomes study, side = side of the mountain, south or northwest or northeast; elev = elevation; topo = topographic position, ridge or depression; tran = sensor deployment transect. The colon indicates an interaction effect between two explanatory variables. The parentheses indicate the term is a random effect – all other terms are fixed effects. If *fe* is a particular fixed effect and *re* is a particular random effect, then (1|*re*) indicates the intercept was allowed to vary randomly with respect to *re* while (0+*fe*|*re*) indicates the interaction of *fe* and *re* was allowed to vary randomly. The case of both the intercept and interaction being allowed to vary randomly was not included in any of the best-fit models.

[†]Understory was not included in the best-fit model, based on AIC, but was retained for comparative purposes.

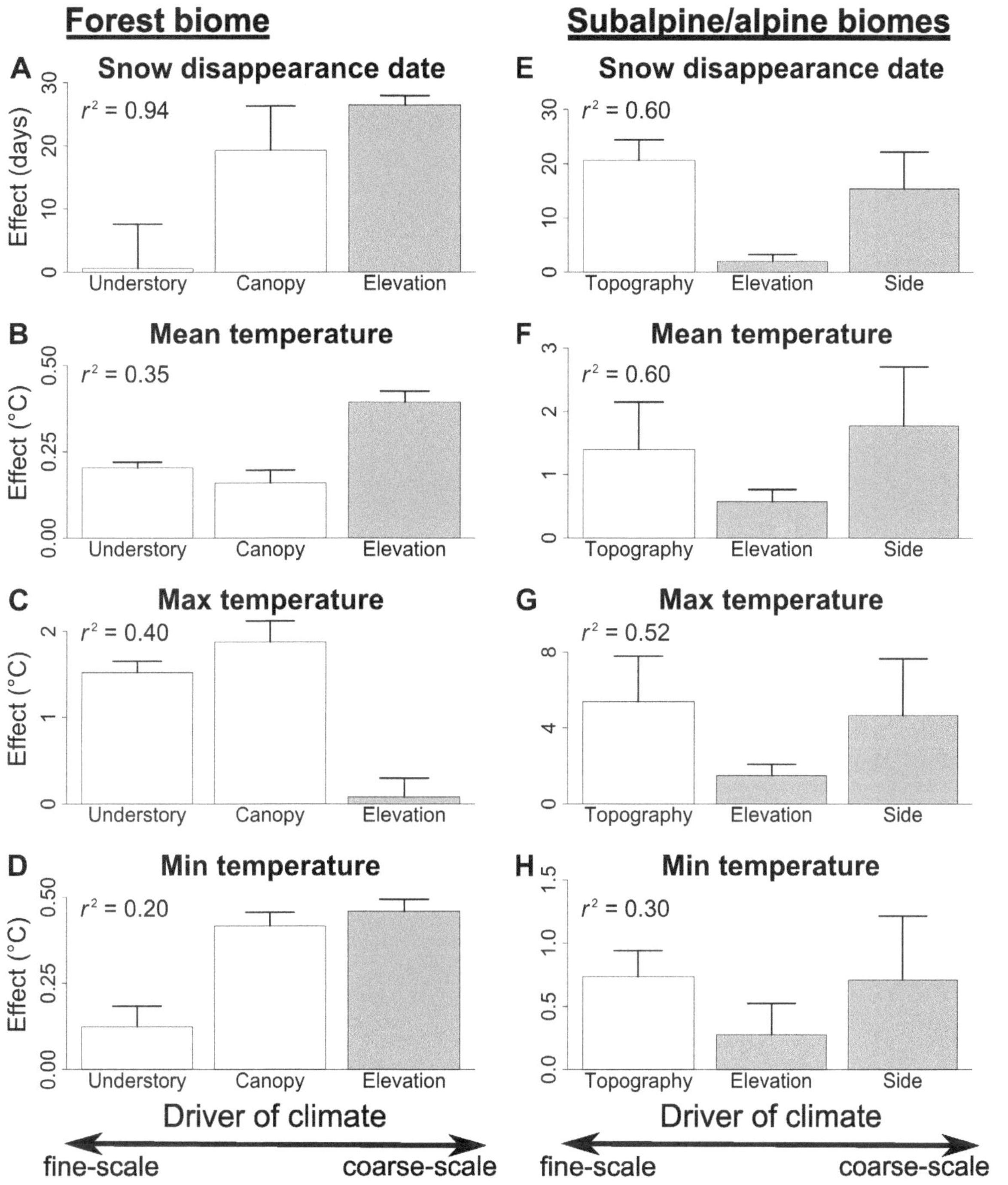

Figure 4. Effects of fine- and coarse-scale drivers of climate. The effects of fine- and coarse-scale drivers of climate on snow disappearance date and the average daily values of mean, maximum and minimum growing season soil temperature. Bars show differences in snow disappearance date or temperature attributed to the effect of different drivers of climate by the best-fit model, with standard error. The effect of elevation was standardized to the effect of a 100 m difference in elevation. Bars filled with gray represent drivers that are coarse enough in scale to be captured by typical climate models (>1 km) while unfilled bars represent drivers too fine in scale to be captured by these models (≤20 m). Fine-scale drivers of climate often had a greater effect on snow or soil temperature than coarse-scale drivers.

similar to the effects of coarse-scale differences in elevation (100 m difference in elevation) and side of the mountain (Figure 4E). In other words, snow disappearance date differed as much over fine spatial scales as it did over coarse spatial scales.

Growing season soil temperatures during our sampling period were lower on the northeast side of the mountain than on the northwest and south sides, potentially because the meadows are at higher elevations on this side of the mountain. On a given side of the mountain, higher elevations (the upper limit of alpine meadows) had higher mean and maximum temperatures, but lower minimum temperatures, than lower elevations (the lower limit of the subalpine biome). Compared to ridges, depressions had

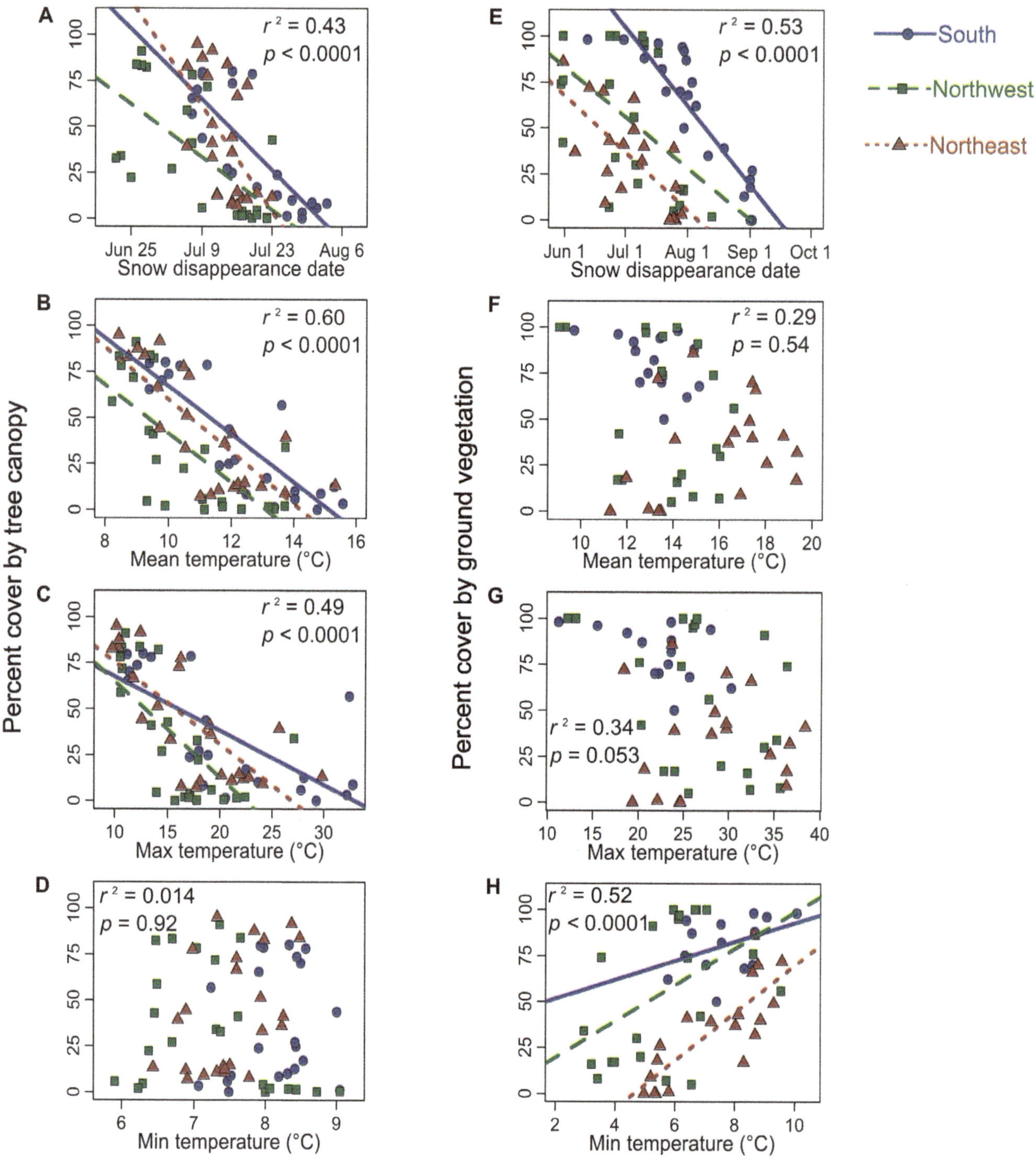

Figure 5. Relationships between vegetation characteristics and microclimate. (A–D) Percent cover by tree canopy at sites near the lower limit of the subalpine biome and (E–H) percent cover by ground vegetation at sites near the upper limit of alpine meadows plotted against the four microclimate variables (snow disappearance date and average daily mean, maximum and minimum soil temperature) on each of the three sides of the mountain. The r^2 values are for models that included the microclimate variable and side of the mountain as explanatory variables, while the p values indicate the significance of the microclimate variable in these models. Regression lines are shown for significant p values (<0.05).

higher mean and maximum temperatures but lower minimum temperatures (Figure 3B–D). We found that the effects of fine-scale topographic differences (depressions vs. ridges) were similar to the effects of coarse-scale differences in elevation and side of the mountain for average daily mean, maximum and minimum temperature (Figure 4F–H). Overall, there was as much heterogeneity in temperature at fine scales as there was at coarse scales.

Several microclimate variables were significantly correlated with vegetation characteristics (Figure 5). At study sites near the lower limit of the subalpine biome, percent cover by tree canopy was lower where snow disappearance date was later and average daily mean and maximum soil temperatures were higher ($p<0.0001$). At study sites near the upper limit of alpine meadows, percent cover by ground vegetation was lower where snow disappearance date was later and average daily minimum soil temperature was lower ($p<0.0001$).

Discussion

Our study suggests that climatic heterogeneity at the fine spatial scales most organisms experience their environment is substantial, implying that projections based on coarse-scale climate models will not capture the full complexity of range shifts in response to climate change. Specifically, we found large differences in snow disappearance date and growing season soil temperatures over small distances (Figures 3, 4), differences that were sometimes as large as those experienced when travelling hundreds of meters upward in elevation or several kilometers to a different side of the mountain. These microclimate variables have been shown to be strongly associated with plant species distributions and abundances [20,21], suggesting that the microclimate heterogeneity we observed is important for plant communities. We also found that vegetation characteristics (canopy and ground vegetation cover) can be strongly correlated with the microclimate variables influenced by fine-scale topographic features (Figure 5), further suggesting that the fine-scale climatic heterogeneity we observed is ecologically important. However, because we did not measure species distributions or abundances in this study, we cannot conclusively state that the microclimate heterogeneity we observed is linked to species distributions or abundances at Mount Rainier. Nonetheless, understanding fine-scale climatic heterogeneity will likely be critical for management, as cool or snowy microhabitats could provide an important buffer against the negative effects of climate change on biodiversity. Thus, when assessing potential species range shifts in response to climate change, it is critical for ecologists to consider fine-scale patterns in climate in addition to other important factors such as broad-scale climate patterns, dispersal constraints, biotic interactions and evolutionary dynamics.

Explanations of Fine-scale Climatic Heterogeneity

In the forest biome, a complex interplay between elevation and vegetation structure is likely responsible for the heterogeneous patterns in snow disappearance date and soil temperature we observed. For example, locations under tree canopy gaps likely experienced later snow disappearance dates than locations under an intact canopy (Figure 3A) because tree canopies intercept snowfall where it can rapidly sublimate or melt instead of being incorporated into the snowpack on the ground [38]. Tree canopies also increase incoming longwave radiation (which increases ablation rates) and this effect can sometimes be greater than the effect of canopies decreasing incoming shortwave radiation by shading the snowpack (which reduces ablation rates), leading to a net effect of canopies increasing ablation rates [39]. Although the presence of trees has also been shown to lead to longer snow persistence by shading the snowpack and decreasing wind speeds (reducing incoming sensible and latent heat fluxes) [38], these effects appear to be relatively weak at our study sites. Increased shading from tree canopies and understory vegetation in forest locations probably led to substantially lower maximum soil temperatures (Figure 3C). But these low sky exposure locations also experienced higher minimum soil temperatures (Figure 3D), probably due to vegetation emitting more longwave radiation (which warms the surface) than the night sky [14]. Differences in mean soil temperatures appeared to be the net effect of these two counteracting influences of sky exposure, with mean soil temperatures being higher in the shadier non-gap locations, but lower in the shadier undisturbed understory vegetation locations (Figure 3B).

Similarly, in the subalpine/alpine biomes we found that both coarse- and fine-scale features had large effects on climate. For example, snow disappeared substantially later from depressions in the landscape than from ridges only ~20 m away, likely because snow typically collects in these depressions while it is blown off of ridges and because shading from surrounding slopes can reduce ablation rates [24]. Feedbacks between vegetation and climate are also likely to influence fine-scale climatic variability. At the lower elevation sites, for example, patches of trees with trunks sticking out above the snowpack emit substantial amounts of longwave radiation which quickens the ablation of snow next to the tree patch and can lead to earlier snow disappearance dates. Trees can also intercept snowfall, reducing snowpack accumulation under canopy and resulting in earlier snow disappearance [38]. These effects can lead to a positive feedback, where trees establish in microsites with earlier snow disappearance dates (e.g. ridges), and the established trees lead to even earlier snow disappearance dates and more tree establishment. This result is consistent with previous studies from subalpine meadows in the region that have documented increased tree establishment on ridges that tend to have earlier snow disappearance dates [40,41].

A striking pattern to emerge from our data was that mean and maximum soil temperatures were greater at higher elevations within the subalpine/alpine biomes (though minimum soil temperatures were lower). Feedbacks between climate and vegetation likely play important roles in producing this temperature pattern. First, tree cover declines with increasing elevation, leading to less shading and potentially higher daytime soil temperatures, especially during the sunny growing season when our data were collected. This explanation is supported by the negative correlation we observed between percent canopy cover (a measure of tree density) and mean/maximum soil temperature in the subalpine/alpine biomes (Figure 5B, C). Second, ground vegetation density declines with increasing elevation, which can lead to lower organic matter content in the soil and lower soil moisture levels. The lower moisture levels probably cause the soil to have a lower heat capacity, leading to greater temperature change per unit of energy input and hence higher maximum temperatures and lower minimum temperatures. This second explanation is supported by the pattern of low soil organic matter content and soil water holding capacity at high elevations in Mount Rainier's subalpine/alpine biomes (Appendix S2). Soil characteristics also have important effects on vegetation in subalpine/alpine environments [42], creating the possibility for complex feedbacks amongst soil, vegetation and climate. These two hypotheses are not mutually exclusive, and further study is needed to assess the importance of each. Regardless, our results suggest that even if patterns in climate are ultimately responsible for patterns in vegetation, the feedback effect of vegetation on soil temperature (either directly, or indirectly through the effects of vegetation on soil characteristics which then affect temperature) appears to at times be stronger than the original forcing of physiographic effects on soil temperature.

Implications of Fine-scale Climatic Heterogeneity for Species Distributions in a Warming World

Since snow disappearance date and growing season temperature vary dramatically over short distances, species whose distributions are primarily constrained by these climate variables may not need to migrate long distances to remain in suitable habitat even when there are large changes in climate. For example, in the subalpine/alpine biomes we found that the average difference in snow disappearance date between depressions and ridges separated by only ~20 m was often one month or more. This is an especially large difference considering the ground is typically only free of snow for 3–5 months out of the year at these elevations. Snow manipulation experiments have shown that differences of this magnitude have large impacts on the phenology, species composition, diversity and productivity of plant communities [20]. Thus, these snowy microhabitats have the potential to serve as refugia for species in a warmer world and provide linkages to new areas of suitable climate, implying that fine-scale climatic heterogeneity could buffer species from climate change [27,43,44], as it may have done during past periods of rapid climate change [45]. Given that we did not stratify our sensors along all gradients likely to produce fine-scale differences in climate (e.g. wind direction, aspect – [14,24]), our results may even be an underestimate of the magnitude of fine-scale heterogeneity.

The importance of topographic heterogeneity for creating climatic heterogeneity shown in this study and others [15,21,25–28] also suggests that mountainous regions will be important for providing climatic refugia in a warming world. However, mountain biotas will still likely face unique challenges. For example, organisms currently living on or near summits will not be able to shift upwards to track suitable climate, and deep valleys between mountains will likely pose serious obstacles to poleward shifts [7]. Broad-scale modeling will continue to be important for addressing these problems. Furthermore, fine-scale environmental heterogeneity does not guarantee that biodiversity will be buffered from climate change. It is possible for heterogeneity to produce small, isolated patches of habitat that cannot support many species, producing a negative effect on diversity [16]. Thus, whether the net effect of heterogeneity on maintaining diversity will be positive during a period of rapid climate change remains an open question.

Different kinds of cool or snowy microhabitats will likely differ in their effectiveness as refugia in a warming world. First, the abundance of microhabitat types will strongly influence how effective they can be as refugia. For example, depressions in the landscape in the subalpine/alpine biomes may have a high likelihood of serving as refugia because they are a common topographic feature. Second, the longevity of microhabitat types will affect their ability to act as refugia. For example, canopy gaps may disappear relatively quickly as trees establish in them, forcing species that might use gaps as refugia to migrate amongst gaps, which may not be possible for some species (though others may be adapted to this migration). In contrast, depressions in the landscape could provide more long-term refugia. Third, the temporal climatic heterogeneity experienced in microhabitat types may affect how well they can serve as refugia. For example, gaps had lower daily minimum and higher daily maximum temperatures compared to non-gaps, showing that these microhabitats experience a wide variety of temperatures. This heterogeneity may favor some species but not others. A final complicating factor influencing how and whether microhabitat types can serve as climatic refugia are the non-climatic conditions associated with them. For example, depressions may differ from other topographic positions in soil characteristics, which could prevent some species from using them as snowy microrefugia.

An important caveat to these findings is that they are based on one year of data. Spatial patterns in climate can change from year to year due to differences in prevailing synoptic weather patterns [46–48], so the patterns we observed in the year we conducted this study may not represent typical spatial patterns. However, the year of our study was a fairly typical year in terms of snow disappearance date and in terms of growing season air temperature for the past few decades (Appendix S3). And although spatial patterns in climate can vary year to year, the patterns are generally constant from one year to the next, especially in terms of snow [49–52]. For example, locations with later snow disappearance dates in one year tend to have later snow disappearance dates in other years, even though the spatially averaged snow disappearance date varies from year to year.

Challenges and Opportunities for Management

To best protect biodiversity in a period of rapid climate change, conservation biologists and resource managers will require realistic assessments of future species distributions [53]. Thus, incorporating fine-scale climatic heterogeneity is essential for improving projections of species range shifts and extinction risks. Current coarse-scale models of the relationships between climate and species distributions ignore fine-scale heterogeneity and may therefore overestimate the distance species must migrate to track suitable climate (because forecasted range shifts are necessarily at the resolution of the model), and overpredict habitat loss and extinction risks ([54–57], but see [58]). Ecologists have previously criticized these bioclimate envelope models for only predicting where the climate that is currently associated with a species distribution will shift to and failing to account for biotic factors that could affect a species' ability to track these climate shifts (dispersal limitations, biotic interactions and evolutionary changes) [59]. However, the limitation of model spatial resolution could undermine predictions not only of species' abilities to track shifts in climate but also of the climate shifts themselves.

In addition to more realistically forecasting range shifts, knowledge of fine-scale climatic heterogeneity may also allow managers to increase species and ecosystem resilience to climate change. For example, protecting microhabitats with cooler temperatures or later snow disappearance dates could become increasingly important as climate change occurs because these microhabitats may provide critical refugia for species. Additionally, our results suggest that planting seedlings and sowing seeds at a variety of microhabitats when restoring degraded sites is an important bet-hedging strategy because it could increase the probability that species establish in microsites that remain suitable as climate change progresses, even if those microsites are only marginally suitable now. Thus, information on fine-scale climate heterogeneity has the potential to be useful for natural area protection and restoration when taken together with other important factors such as edaphic constraints, biotic interactions, genetic diversity and financial costs [60]. More detailed and longer term studies are needed to assess whether microclimate heterogeneity can contribute substantially to plant establishment and restoration efforts in a warming world.

Conclusions

We have shown that snow disappearance date and growing season soil temperature vary dramatically over small distances due to differences in vegetation structure and topography. In fact, fine-scale features such as gaps in the forest canopy or small depressions in the landscape can produce differences in snow disappearance date and soil temperature as large as those produced by shifting hundreds of meters up a mountain slope. This large degree of fine-scale spatial heterogeneity may provide an important buffer against the negative effects of rapid climate change, as many species may only need to migrate tens of meters from one microhabitat to another in order to track suitable climate, as opposed to shifting hundreds of meters upward in elevation or hundreds of kilometers poleward. Climate change will undoubtedly pose serious threats to biodiversity, but knowledge of fine-scale climatic heterogeneity may allow managers to better assess and potentially alleviate some of these threats.

Supporting Information

Appendix S1 Model fitting and selection procedure.

Appendix S2 Soil characteristics in the subalpine and alpine biomes.

Appendix S3 Climatological context of the study.

Acknowledgments

Thanks to Erin Curtis, Rachel Konrady, Tony Krueger, Jack Lee, Gerald Lisi, Anna O'Brien, Mitch Piper, Elli Jenkins Theobald and Courtney Wenneborg for field assistance. We also want to thank Lou Whiteaker and Bret Christoe at Mount Rainier National Park for advice and help with permitting as well as Josh Lawler, Jeremy Littell, Regina Rochefort, Catharine Copass Thompson, Pete Wyckoff and two anonymous reviewers for providing helpful feedback on the manuscript.

Author Contributions

Conceived and designed the experiments: KF AE JHRL. Performed the experiments: KF AE. Analyzed the data: KF. Contributed reagents/materials/analysis tools: KF AE MR JHRL. Wrote the paper: KF AE JL MR JHRL.

References

1. Humboldt Av, Bonpland A (1807) Essay on the geography of plants. Paris: Levrault Schoell. 155 p.
2. Merriam CH, Steineger L (1890) Results of a biological survey of the San Francisco Mountain region and the desert of the Little Colorado, Arizona. North American Fauna Report 3. Washington, DC: US Department of Agriculture, Division of Ornithology and Mammalogy. 136 p.
3. Whittaker RH (1975) Communities and ecosystems. New York: Macmillan. 385 p.
4. Chen I, Hill JK, Ohlemueller R, Roy DB, Thomas CD (2011) Rapid range shifts of species associated with high levels of climate warming. Science 333: 1024–1026. doi: 10.1126/science.1206432.
5. Davis M, Shaw R (2001) Range shifts and adaptive responses to Quaternary climate change. Science 292: 673–679. doi: 10.1126/science.292.5517.673.
6. Huntley B (2005) North temperate responses. In: Lovejoy TE, Hannah L, editors. Climate change and biodiversity. New Haven, Connecticut: Yale University Press. pp. 109–124.
7. Fischlin A, Midgley GF, Price JT, Leemans R, Gopal B, et al. (2007) Ecosystems, their properties, goods, and services. In: Parry ML, Canziani OF, Palutikof JP, Linden PJvd, Hanson CE, editors. Climate change 2007: Impacts, adaptation and vulnerability. Contribution of Working Group II to the Fourth Assessment Report of the Intergovernmental Panel on Climate Change. Cambridge, UK and New York, NY, USA: Cambridge University Press. pp. 211–272.
8. Loarie SR, Duffy PB, Hamilton H, Asner GP, Field CB, et al. (2009) The velocity of climate change. Nature 462: 1052–1057. doi: 10.1038/nature08649.
9. Levin SA (1992) The problem of pattern and scale in ecology. Ecology 73: 1943–1967.
10. Daly C, Halbleib M, Smith JI, Gibson WP, Doggett MK, et al. (2008) Physiographically sensitive mapping of climatological temperature and precipitation across the conterminous United States. Int J Climatol 28: 2031–2064. doi: 10.1002/joc.1688.
11. Hijmans RJ, Cameron SE, Parra JL, Jones PG, Jarvis A (2005) Very high resolution interpolated climate surfaces for global land areas. Int J Climatol 25: 1965–1978. doi: 10.1002/joc.1276.
12. Thuiller W, Lavorel S, Araujo MB, Sykes MT, Prentice IC (2005) Climate change threats to plant diversity in Europe. Proc Natl Acad Sci U S A 102: 8245–8250. doi: 10.1073/pnas.0409902102.
13. Guisan A, Zimmermann NE (2000) Predictive habitat distribution models in ecology. Ecol Model 135: 147–186.
14. Geiger R, Aron RH, Todhunter P (2009) The climate near the ground. Lanham, MD: Rowman & Littlefield Publishers, Inc. 623 p.
15. Ackerly DD, Loarie SR, Cornwell WK, Weiss SB, Hamilton H, et al. (2010) The geography of climate change: Implications for conservation biogeography. Divers Distrib 16: 476–487. doi: 10.1111/j.1472-4642.2010.00654.x.
16. Tamme R, Hiiesalu I, Laanisto L, Szava-Kovats R, Partel M (2010) Environmental heterogeneity, species diversity and co-existence at different spatial scales. Journal of Vegetation Science 21: 796–801. doi: 10.1111/j.1654-1103.2010.01185.x.
17. Willis KJ, Bhagwat SA (2009) Biodiversity and climate change. Science 326: 806–807. doi: 10.1126/science.1178838.
18. Larcher W (2003) Physiological plant ecology: Ecophysiology and stress physiology of functional groups. Berlin: Springer-Verlag. 513 p.
19. Franklin JF, Moir WH, Hemstrom MA, Greene SE, Smith BG (1988) The forest communities of Mount Rainier National Park. Scientific Monograph Series No. 19. Washington, DC: US Department of the Interior, National Park Service. 194 p.
20. Wipf S, Rixen C (2010) A review of snow manipulation experiments in Arctic and alpine tundra ecosystems. Polar Res 29: 95–109. doi: 10.1111/j.1751-8369.2010.00153.x.
21. Scherrer D, Körner C (2011) Topographically controlled thermal-habitat differentiation buffers alpine plant diversity against climate warming. J Biogeogr 38: 406–416. doi: 10.1111/j.1365-2699.2010.02407.x.
22. Mote P, Salathé EP (2009) Future climate in the Pacific Northwest. In: Elsner MM, Littell J, Binder LW, editors. The Washington climate change impacts assessment. Seattle, WA: Climate Impacts Group, Center for Science in the Earth System, Joint Institute for the Study of the Atmosphere and Oceans, University of Washington. pp. 21–43.
23. Elsner MM, Cuo L, Voisin N, Deems JS, Hamlet AF, et al. (2009) Implications of 21st century climate change for the hydrology of Washington State. In: Elsner MM, Littell J, Binder LW, editors. The Washington climate change impacts assessment. Seattle, WA: Climate Impacts Group, Center for Science in the Earth System, Joint Institute for the Study of the Atmosphere and Oceans, University of Washington. pp. 69–106.
24. Clark MP, Hendrikx J, Slater AG, Kavetski D, Anderson B, et al. (2011) Representing spatial variability of snow water equivalent in hydrologic and land-surface models: A review. Water Resources Research 47: doi:10.1029/2011WR010745.
25. Fridley JD (2009) Downscaling climate over complex terrain: High finescale (<1000 m) spatial variation of near-ground temperatures in a montane forested landscape (Great Smoky Mountains). Journal of Applied Meteorology and Climatology 48: 1033–1049. doi: 10.1175/2008JAMC2084.1.
26. Millar CI, Westfall RD (2010) Distribution and climatic relationships of the American pika (*Ochotona princeps*) in the Sierra Nevada and western Great Basin, USA; periglacial landforms as refugia in warming climates. Arctic Antarctic and Alpine Research 42: 76–88. doi: 10.1657/1938-4246-42.1.76.
27. Scherrer D, Körner C (2010) Infra-red thermometry of alpine landscapes challenges climatic warming projections. Global Change Biol 16: 2602–2613. doi: 10.1111/j.1365-2486.2009.02122.x.
28. Wundram D, Pape R, Loeffler J (2010) Alpine soil temperature variability at multiple scales. Arctic Antarctic and Alpine Research 42: 117–128. doi: 10.1657/1938-4246-42.1.117.
29. Deutsch CA, Tewksbury JJ, Huey RB, Sheldon KS, Ghalambor CK, et al. (2008) Impacts of climate warming on terrestrial ectotherms across latitude. Proc Natl Acad Sci U S A 105: 6668–6672. doi: 10.1073/pnas.0709472105.
30. Meehl GA, Stocker TF, Collins WD, Friedlingstein P, Gaye AT, et al. (2007) Global climate projections. In: Solomon S, Qin D, Manning M, Chen Z, Marquis M, et al, editors. Climate change 2007: The physical science basis. Contribution of Working Group I to the Fourth Assessment Report of the Intergovernmental Panel on Climate Change. Cambridge, UK and New York, NY, USA: Cambridge University Press. pp. 747–845.
31. Lundquist JD, Lott F (2008) Using inexpensive temperature sensors to monitor the duration and heterogeneity of snow-covered areas. Water Resour Res 44: W00D16. doi: 10.1029/2008WR007035.
32. Raleigh MS, Rittger K, Moore C, E., Henn B, Lutz J, A., etal. (2013) Ground-based testing of MODIS fractional snow cover in subalpine meadows and forests of the Sierra Nevada. Remote Sensing of Environment 128: 44–57.
33. Zuur AF, Ieno EN, Walker NJ, Saveliev AA, Smith GM (2009) Mixed effects models and extensions in ecology with R. New York: Springer. 574 p.
34. R Development Core Team (2010) R: A language and environment for statistical computing.
35. Bates D, Maechler M, Bolker B (2011) lme4: Linear mixed-effects models using S4 classes. R package version 0.999375-39.
36. Hadfield JD (2010) MCMC methods for multi-response generalized linear mixed models: The MCMCglmm R package. Journal of Statistical Software 33: 1–22.
37. Xu RH (2003) Measuring explained variation in linear mixed effects models. Stat Med 22: 3527–3541. doi: 10.1002/sim.1572.
38. Varhola A, Coops NC, Weiler M, Moore RD (2010) Forest canopy effects on snow accumulation and ablation: An integrative review of empirical results. Journal of Hydrology 392: 219–233. doi: 10.1016/j.jhydrol.2010.08.009.
39. Sicart JE, Pomeroy JW, Essery RLH, Hardy J, Link T, et al. (2004) A sensitivity study of daytime net radiation during snowmelt to forest canopy and

atmospheric conditions. J Hydrometeorol 5: 774–784. doi: 10.1175/1525-7541(2004)005<0774:ASSODN>2.0.CO;2.
40. Rochefort RM, Peterson DL (1996) Temporal and spatial distribution of trees in subalpine meadows of Mount Rainier National Park, Washington, USA. Arct Alp Res 28: 52–59.
41. Zald HSJ, Spies TA, Huso M, Gatziolis D (2012) Climatic, landform, microtopographic, and overstory canopy controls of tree invasion in a subalpine meadow landscape, Oregon Cascades, USA. Landscape Ecol 27: 1197–1212. doi: 10.1007/s10980-012-9774-8.
42. Körner C (2003) Alpine plant life: Functional plant ecology of high mountain ecosystems, 2nd edition. Berlin: Springer. 344 p.
43. Dobrowski SZ (2011) A climatic basis for microrefugia: The influence of terrain on climate. Global Change Biol 17: 1022–1035. doi: 10.1111/j.1365-2486.2010.02263.x.
44. Keppel G, Van Niel KP, Wardell-Johnson GW, Yates CJ, Byrne M, et al. (2012) Refugia: Identifying and understanding safe havens for biodiversity under climate change. Global Ecol Biogeogr 21: 393–404. doi: 10.1111/j.1466-8238.2011.00686.x.
45. Hof C, Levinsky I, Araujo MB, Rahbek C (2011) Rethinking species' ability to cope with rapid climate change. Global Change Biol 17: 2987–2990. doi: 10.1111/j.1365-2486.2011.02418.x.
46. Lundquist JD, Cayan DR (2007) Surface temperature patterns in complex terrain: Daily variations and long-term change in the central Sierra Nevada, California. Journal of Geophysical Research-Atmospheres 112: D11124. doi: 10.1029/2006JD007561.
47. Bennie JJ, Wiltshire AJ, Joyce AN, Clark D, Lloyd AR, et al. (2010) Characterising inter-annual variation in the spatial pattern of thermal microclimate in a UK upland using a combined empirical-physical model. Agric For Meteorol 150: 12–19. doi: 10.1016/j.agrformet.2009.07.014.
48. Lundquist JD, Minder JR, Neiman PJ, Sukovich E (2010) Relationships between barrier jet heights, orographic precipitation gradients, and streamflow in the northern Sierra Nevada. J Hydrometeorol 11: 1141–1156. doi: 10.1175/2010JHM1264.1.
49. Erickson T, Williams M, Winstral A (2005) Persistence of topographic controls on the spatial distribution of snow in rugged mountain terrain, Colorado, United States. Water Resour Res 41: W04014. doi: 10.1029/2003WR002973.
50. Deems JS, Fassnacht SR, Elder KJ (2008) Interannual consistency in fractal snow depth patterns at two Colorado mountain sites. J Hydrometeorol 9: 977–988. doi: 10.1175/2008JHM901.1.
51. Sturm M, Wagner AM (2010) Using repeated patterns in snow distribution modeling: An Arctic example. Water Resour Res 46: W12549. doi: 10.1029/2010WR009434.
52. Egli L, Jonas T, Gruenewald T, Schirmer M, Burlando P (2012) Dynamics of snow ablation in a small alpine catchment observed by repeated terrestrial laser scans. Hydrol Process 26: 1574–1585. doi: 10.1002/hyp.8244.
53. Hannah L, Midgley G, Andelman S, Araujo M, Hughes G, et al. (2007) Protected area needs in a changing climate. Frontiers in Ecology and the Environment 5: 131–138.
54. Austin MP, Van Niel KP (2011) Impact of landscape predictors on climate change modelling of species distributions: A case study with *Eucalyptus fastigata* in southern New South Wales, Australia. J Biogeogr 38: 9–19. doi: 10.1111/j.1365-2699.2010.02415.x.
55. Barrows CW, Murphy-Mariscal ML (2012) Modeling impacts of climate change on Joshua trees at their southern boundary: How scale impacts predictions. Biol Conserv 152: 29–36. doi: 10.1016/j.biocom.2012.03.028.
56. Luoto M, Heikkinen RK (2008) Disregarding topographical heterogeneity biases species turnover assessments based on bioclimatic models. Global Change Biol 14: 483–494. doi: 10.1111/j.1365-2486.2007.01527.x.
57. Randin CF, Engler R, Normand S, Zappa M, Zimmermann NE, et al. (2009) Climate change and plant distribution: Local models predict high-elevation persistence. Global Change Biol 15: 1557–1569. doi: 10.1111/j.1365-2486.2008.01766.x.
58. Trivedi MR, Berry PM, Morecroft MD, Dawson TP (2008) Spatial scale affects bioclimate model projections of climate change impacts on mountain plants. Global Change Biol 14: 1089–1103. doi: 10.1111/j.1365-2486.2008.01553.x.
59. Pearson RG, Dawson TP (2003) Predicting the impacts of climate change on the distribution of species: Are bioclimate envelope models useful? Global Ecol Biogeogr 12: 361–371.
60. Rochefort RM, Kurth LL, Carolin TW, Riedel JL, Mierendorf RR, et al. (2006) Mountains. In: Apostol D, Sinclair M, editors. Restoring the Pacific Northwest: The art and science of ecological restoration in Cascadia. Washington, DC: Island Press. pp. 241–275.

7

Local and Regional Effects on Community Structure of Dung Beetles in a Mainland-Island Scenario

Pedro Giovâni da Silva*, Malva Isabel Medina Hernández

Programa de Pós-Graduação em Ecologia, Departamento de Ecologia e Zoologia, Universidade Federal de Santa Catarina, Florianópolis, Santa Catarina, Brazil

Abstract

Understanding the ecological mechanisms driving beta diversity is a major goal of community ecology. Metacommunity theory brings new ways of thinking about the structure of local communities, including processes occurring at different spatial scales. In addition to new theories, new methods have been developed which allow the partitioning of individual and shared contributions of environmental and spatial effects, as well as identification of species and sites that have importance in the generation of beta diversity along ecological gradients. We analyzed the spatial distribution of dung beetle communities in areas of Atlantic Forest in a mainland-island scenario in southern Brazil, with the objective of identifying the mechanisms driving composition, abundance and biomass at three spatial scales (mainland-island, areas and sites). We sampled 20 sites across four large areas, two on the mainland and two on the island. The distribution of our sampling sites was hierarchical and areas are isolated. We used standardized protocols to assess environmental heterogeneity and sample dung beetles. We used spatial eigenfunctions analysis to generate the spatial patterns of sampling points. Environmental heterogeneity showed strong variation among sites and a mild increase with increasing spatial scale. The analysis of diversity partitioning showed an increase in beta diversity with increasing spatial scale. Variation partitioning based on environmental and spatial variables suggests that environmental heterogeneity is the most important driver of beta diversity at the local scale. The spatial effects were significant only at larger spatial scales. Our study presents a case where environmental heterogeneity seems to be the main factor structuring communities at smaller scales, while spatial effects are more important at larger scales. The increase in beta diversity that occurs at larger scales seems to be the result of limitation in species dispersal ability due to habitat fragmentation and the presence of geographical barriers.

Editor: Judi Hewitt, University of Waikato (National Institute of Water and Atmospheric Research), New Zealand

Funding: This study is part of the Project "Biodiversidade de artrópodes terrestres e aquáticos em diferentes gradientes ambientais da Mata Atlântica do estado de Santa Catarina," and was supported by Coordenação de Pessoal de Nível Superior (Proc. 001/2010 MEC/CAPES/PNPD). PGS received a post-graduate fellowship from Coordenação de Pessoal de Nível Superior. MIMH received a research grant from Conselho Nacional de Desenvolvimento Científico e Tecnológico (Research Productivity/PQ/2010, Proc. 303800/2010-0, Project "Comportamento de besouros Scarabaeinae e sua função no ciclo de decomposição da matéria orgânica"). The funders had no role in study design, data collection and analysis, decision to publish, or preparation of the manuscript.

Competing Interests: The authors have declared that no competing interests exist.

* Email: pedrogiovanidasilva@yahoo.com.br

Introduction

Community ecology aims to understand and explain the processes that influence the patterns of distribution, abundance and composition of species [1,2] over space and time [3], both locally and regionally [4,5]. Community structure may be influenced by several ecological processes that involve biotic and abiotic factors operating at different spatial and temporal scales [6–8]. When we consider large scales, historical, evolutionary and stochastic processes become critical to the understanding of these patterns [5,9–11]. Further, studying only local processes may not be sufficient for understanding how communities are structured locally and regionally [5], because local and regional processes may act in different ways in relation to the increase or decrease in species diversity [11].

Despite the large number of mechanisms (theories and models) proposed as drivers of patterns of species distribution, only four processes are fundamentally involved: selection, drift, speciation and dispersal [2]. Three main hypotheses are proposed to explain the origin of beta diversity (i.e., variation in the identities and relative abundance of species among sites) with respect to these processes [12]. The first hypothesis suggests that the species composition may be stable over large areas, and that biological interactions (e.g., competition inter- and/or intraspecific) play an important role in maintaining beta diversity [12]. The second hypothesis states that species composition varies in a random and autocorrelated way, emphasizing spatially limited dispersal [12]. The last hypothesis suggests that species distribution is driven by environmental conditions, and that landscapes are mosaics in which local environmental drivers control species composition [12]. These hypotheses seem to be somewhat related regardless of the organismal group or ecosystem, and testing them is crucial for elucidating issues on ecosystem functioning and biodiversity conservation initiatives [12].

In community ecology there exists a variety of concepts and methodologies commonly employed by ecologists to measure beta diversity and to identify the processes related to its generation [12–17]. Recent approaches have been based on the dependence of environmental, spatial and random processes, with the goal of explaining which processes have more influence on beta diversity, e.g., by using variation partitioning methods [18] on composition or abundance community data among groups of explanatory variables (e.g., environmental and spatial) [12]. These methods are used to attempt to explain how beta diversity is influenced by environmental and/or spatial factors, or by random factors [19]. Despite being criticized [20,21], variation partitioning has long been used in the context of metacommunity theory [1] and it highlights the importance of increasing the spatial scale in understanding the ecological processes structuring biological communities locally and regionally [22].

A metacommunity is defined as a set of communities connected by the dispersal of multiple interacting species [1,22]. There are four theoretical paradigms (models) to explain metacommunity dynamics (species sorting, mass effects, patch dynamics and neutral) and they take into account three (drift, selection and dispersal) of the four basic processes aforementioned [2]; differences in species dispersal ability and environmental characteristics are important factors for determining which model best describes the metacommunity [1,22]. Mass effects (high dispersal) and patch dynamics (low dispersal) would be variations of species sorting (efficient dispersal), as there are different levels of dispersal ability of species in each metacommunity model [23]. The adoption of metacommunity theory has led to substantial changes in the way that ecologists interpret ecological phenomena at both local and metacommunity (regional) scales [1].

A key point in assessing the relative importance of proposed metacommunity processes is the identification and use of environmental and/or spatial gradients as study scenarios [12]. Direct gradient ordination techniques (e.g., redundancy analysis) followed by variation partitioning [18] allows determination of the fraction of beta diversity explained solely by environmental or spatial predictors, and by shared effects of both sets of predictors [12]. The prevalence of environmental effects indicates species sorting, the predominance of spatial effects indicates neutral processes, historical events and/or dispersal limitation, and shared effects of both environmental and spatial predictors indicate species sorting, dispersal limitation or a combination of both (mass effects and patch dynamics sensu [1]) [23,24]. The relative importance of metacommunity paradigms is still dependent on spatial scale, spatial extent or spatial distances between sites [25,26], and varies between environments and groups of species due to inherent differences of ecosystem type and species dispersal ability [27]. Recent techniques have also allowed the identification of species and sites that may contribute to beta diversity along an ecological gradient by using community dataset total variance as an estimate of beta diversity [15].

The Brazilian Atlantic Forest is one the most threatened biomes in terms of biodiversity conservation [28]. About 12% of its original size, it is highly fragmented with a high degree of isolation, existing primarily in intermediate successional state [29]. Less than 2% of Atlantic Forest areas are located in protected zones [29], despite being considered global biodiversity hotspots [30]. Historically, the coast of Brazil has always showed the highest population and industrial concentration, and thus, the Atlantic Forest has been affected by the growth and development of the country [29]. An understanding of how species respond to anthropogenic modifications to the structure or complexity of habitats is fundamental for the development of future conservation initiatives, especially for organisms that play key roles in the maintenance and/or restoration of ecosystems, such as dung beetles (Coleoptera: Scarabaeidae: Scarabaeinae).

Dung beetles feed on decaying organic matter (e.g., mammalian feces, animal carcasses, rotting vegetation) [31] and they play several ecosystem services [32]. In tropical ecosystems they are used as indicators of diversity, as well as for monitoring environmental changes, because they respond quickly in terms of species composition, richness, abundance and biomass to the effects caused by habitat destruction, fragmentation and/or isolation [33–37]. These beetles are easily sampled using standardized, efficient and low-cost sampling methods [34]. They have wide distribution and are correlated with other taxa (e.g., mammals) [33,34,38]. Therefore, community changes have potential to affect ecological functions performed by dung beetles and hence, ecosystem function [35,39–42]. As such, dung beetles are an excellent model system [31] with which to investigate the main processes that influence community structure in Atlantic Forest regions.

Therefore, the aim of our study was to investigate the effect of spatial scales on the patterns of species diversity of dung beetles in Brazilian Atlantic Forest and to identify the mechanisms that drive these patterns applying aspects of metacommunity theory. We tested the hypothesis that the distribution of dung beetles in the Atlantic Forest is associated with differences in forest structure and that high levels of beta diversity will be found with increasing spatial scale due to dispersal limitation. Our predictions are as follows: (i) because dung beetles are sensitive to environmental changes, environmental gradients should result in high beta diversity among sites via species sorting, (ii) due to differences in habitat structure of each site, environmental characteristics and dung beetle species distribution are spatially structured, (iii) because of dispersal limitation among areas (mainland-island and fragmented landscape), the spatial effect has high importance in structuring communities at increased spatial scales.

Materials and Methods

Study area

The study was conducted at four large Atlantic Forest areas in the state of Santa Catarina, Brazil, two on the island of Santa Catarina (Florianópolis city) and two on the mainland, both on the east coast (Figure 1). The island of Santa Catarina is approximately 54 km north-south and maximally 18 km wide, with a total land area of 424.4 km^2. The distance between the mainland and the island varies greatly, with minimum of 500 m and maximum around 10 km.

On the island, the study areas were within the Lagoa do Peri Municipal Park (PER, 27°43′30″S, 48°32′18″W) and the Permanent Protection Area of Ratones (RAT, 27°31′52″S, 48°30′45″W). On the mainland, the areas sampled were within the Environmental Protection Area of Anhatomirim in Governador Celso Ramos city (ANH, 27°25′1″S, 48°34′25″W), and in Permanent Protection Area in Itapema city (ITA, 27°05′13″S, 48°35′54″W). According to the Brazilian Forest Code (Law n°. 12.651/2012), permanent protection areas are sites with characteristics that have the environmental function of preserving water, biodiversity resources, and landscape and geological stability, and for facilitation floral and faunal gene flow. The distance among areas is approximately 21 km between PER and RAT, 34 km between PER and ANH, 71 km between PER and ITA, 13.5 km between ANH and RAT, 50 km between ITA and RAT, and 37 km between ANH and ITA. The altitude of the sampling sites ranged between 28 and 265 m. All sites sampled are near the Brazilian Atlantic coastline and have dense rain forest vegetation within the Atlantic Forest biome, with various levels of vegetation succession [43]. According Köeppen classification, the climate in the eastern region of Santa Catarina is Cfa, humid subtropical (mesothermal) without dry season, with hot summers (average of 25°C) and well

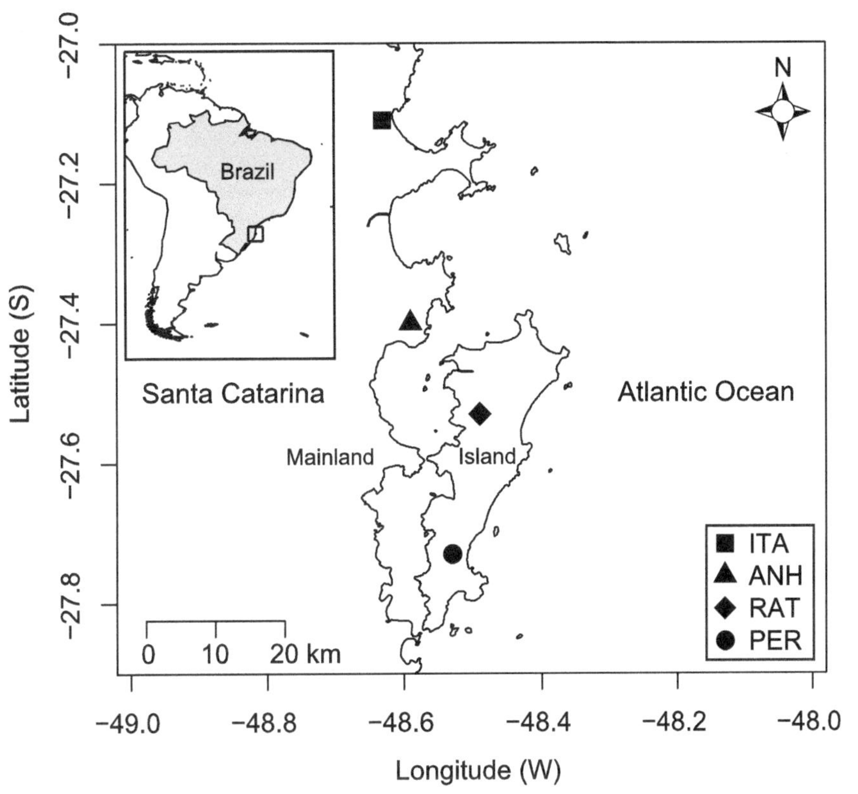

Figure 1. Map of the study region. Location of the four areas sampled in eastern Santa Catarina state, Brazil. ANH: Environmental Protection Area of Anhatomirim; ITA: Permanent Protection Area of Itapema; PER: Lagoa do Peri Municipal Park; RAT: Permanent Protection Area of Ratones.

distributed rainfall throughout the year, with annual average of 1500 mm approximately [43]. Santa Catarina shows four seasons of the year well defined.

Scarabaeinae sampling

We sampled dung beetles using baited pitfall traps, as they are a highly efficient method to capture this group [44]. The traps were made with plastic containers (15 cm diameter×20 cm depth) buried with the top edge at ground level, allowing insects to fall in. To prevent overflow, the traps were protected against rain using a small sheet supported by wooden sticks, placed approximately 10 cm above the trap. A mixture (300 ml) of water and detergent (neutral) was added to each container to catch and kill insects. We used human feces and rotting flesh (aged in plastic container at room temperature three days prior to sampling) as bait to attract dung beetles, as both satisfy the two main eating habits of dung beetles – coprophagy and necrophagy, respectively [31]. Approximately 30 g of both baits were wrapped in thin cloth and tied in the central part of the rain protection above the traps.

The insects collected were sorted, mounted on entomological pins and dried in an oven (60°C for 72 h), then weighed on a precision balance (0.0001 g). Species identities were confirmed by experts. The beetles were deposited in the Entomological Collection of the Centro de Ciências Biológicas at the Universidade Federal de Santa Catarina, Brazil. We thank the Instituto Chico Mendes de Conservação da Biodiversidade (ICMBio/MMA) and Fundação do Meio Ambiente (FATMA-SC) for permission to collect (permit #32333-3 to MIMH). The field study did not involve endangered or protected species. Dataset S1 provides the database of values for abundance and biomass of dung beetle species across the study sites.

Sampling design

Samples were taken at five different forested (hillside) sites within each sampling area. Each site contained five pairs of traps spaced 5–10 m apart, each pair containing the two kinds of bait. The pairs were spaced 50 m apart, as a minimum distance of 50 m decreases the influence between sets of traps in sampling Scarabaeinae [45]. Each pair of traps was regarded as a sampling point, and remained in the field for 48 h prior to collection.

The samplings were carried out during the summer of 2012 (January and February), because of high temperatures, and it being the period of greatest dung beetle abundance in subtropical regions in Brazil [46,47]. Due to the spatial configuration of our sampling design, the large distance among the four areas, and the effect of spatial discontinuity between mainland-island, the sampling sites showed a hierarchical distribution. Thus, it was possible to investigate the variation in dung beetle communities at three spatial scales, i.e. mainland-island, areas, and sites. Sites represent the local spatial scale, i.e., the smallest spatial extent in our study that encompasses five sampling points. Areas represent the intermediate spatial scale with five sites per area. Mainland-island represents the regional spatial scale, i.e., the largest spatial extent in our study that encompasses two areas each one. Variation in dung beetle species composition, number of individuals, and dry biomass was used to assess the influence of environmental and spatial factors at each spatial scale.

Environmental variables

We measured 20 environmental variables related to habitat structure to test their influence on dung beetle distribution. Differences in environmental conditions (environmental variables measured) among sampling sites is defined as environmental heterogeneity. Measurement was performed using the adapted point-centered quarter method [48,49]. This method was chosen for its simplicity and common use in phytosociological surveys [50]. Briefly, a plastic pipes crossing in an x-shape were placed in the center of each pair of traps (i.e., at each sampling point), dividing the sampling point into four quadrants (northwest, southwest, southeast and northeast). Tree, shrub and soil environmental variables were measured in each quadrant as follows: (1) circumference at breast height when diameter at breast height >5 cm), (2) height, (3) top diameter and (4) distance away from the nearest tree to the center of cross, (5–8) same measures for the greater tree distant up to 10 m, (9–12) similar measures for shrubs (circumference at ankle height when <5 cm and with a minimum height of 1 m), (13) land slope, (14) canopy cover, (15) percentage of leaf litter cover, (16) green cover and (17) exposed soil, (18) height and (19) dry biomass of leaf litter, and (20) altitude.

The height of trees and shrubs was visually estimated with a ruler of 4 m length. Circumference and distance were measured with a millimeter tape measure. The percentage of litter, green cover, and exposed soil coverage in each quadrant was estimated in different classes (0–5, 6–25, 26–50, 51–75, 76–95, 96–100%) using a square of 1 m plastic pipes, placed about 20 cm away from the cross. Land slope was obtained at the center of the square using an inclinometer. Litter height was measured using a mm ruler at five points inside the square (near each corner and in the center). A five-inch square was constructed in the center of the 1 m square, and a portion of litter was removed. Litter was later dried in an oven (60°C for 72 hours) and weighed to obtain dry biomass. Using the same classes described above, the percentage of canopy cover was visually estimated using a hollow square of 10 cm side length, placed at a distance of 60 cm from the eye of the observer at a 20° angle in relation to the zenith [50]. Altitude was obtained using a hand-held GPS at ground level. The basal area of trees and shrubs was calculated from the trunk circumference (based on the area of the circle). For each variable, a measure of central tendency was calculated based on the Shapiro-Wilk normality test. Thus, each environmental variable represented a central value (mean or median, as appropriate) of the four measures of each point; this was done to minimize the effects of visual estimation. A subset of the variables analyzed (three basal area, three heights, DBH) is used by the Conselho Nacional do Meio Ambiente, the Brazilian Council of Environmental issues, to characterize successional stages of Atlantic Forest in the state of Santa Catarina [51]. Dataset S1 provides the database of values for each environmental variable across the study sites.

Spatial variables

Spatial predictors were created using a method called Principal Coordinates of Neighbour Matrices [19], which is part of a set of spatial eigenfunction analyses called Moran's Eigenvector Maps [52]. The creation of spatial predictors was performed using *create.MEM.model* function [25] for the program R [53], because the sampling sites in our study showed a spatial hierarchical structure [54] with large distances between sites in different areas. This function produces a staggered matrix arranged in blocks from the geographical coordinates, generating information on the number of blocks (or groups) and sampling sites in each block [54]. Each block represents the hierarchical spatial distribution of sampling points, and in the staggered matrix the blocks from another hierarchy receive value of zero (0) for each spatial variable created. These variables represent the spatial variation at different spatial scales and may be used as predictors in gradient analysis to model the spatial relationship of the community data [25]. To create the spatial variables, we used data from geographic coordinates (Universal Transverse Mercator) obtained at each sampling point using a hand-held GPS. Dataset S1 provides the database of geographic coordinates for each study site.

Data analysis

Beta diversity across spatial scales. A recent approach called "true diversity" [55] has been used to partition diversity into its different components in an additive or multiplicative way [14]. The additive partitioning approach ($\gamma = \alpha+\beta1+\beta2+\beta3$) was used to estimate the beta diversity at the different spatial scales. Alpha (α) is the average species richness in local communities, while gamma (γ) refers to the total richness observed in the entire set of samples. Each component of beta diversity refers to different spatial scales studied: $\beta1$ = between sampling sites, $\beta2$ = between areas, $\beta3$ = between mainland-island. We used data on species richness and individual abundance (i.e., true Shannon diversity) for the hierarchical analysis of diversity partitioning. Partitioned components of diversity based on abundance were natural log-transformed to make them additive (i.e., Shannon entropy [55]). These analyses were performed in Partition 3.0 program [56].

Species and local contributions to beta diversity. The total beta diversity (BD_{Total}) was analyzed by calculating the total variance of the species matrix using ***beta.div*** function [15] for R program [53]. This method calculates the total sum of squares of the species matrix and from it, one may obtain an index of the total data variance that represents the total beta diversity, and it may be compared among sampling units. The BD_{Total} may then be partitioned in Species Contribution to Beta Diversity (SCBD, or degree of variation of the species along the studied area) and Local Contribution to Beta Diversity (LCBD, or comparative indicators of ecological uniqueness of the sampling sites) [15]. The values of LCBD were tested using random and independent permutations (in columns) of the species matrix, testing whether species are randomly and independently distributed between sampling sites [15]. This approach was used to identify the species and sites that contributed most to the beta diversity index throughout the spatial gradient. Before running the analysis, species data (composition, abundance and biomass) were Hellinger-transformed, after which Euclidean distance was used in the execution of the analysis. We used Spearman correlation to assess the association between the values of LCBD and species richness, abundance and biomass, in order to determine whether sampling site contribution is related to the number of species, number of individuals, and total biomass. Analyses were performed in R 3.0.1 program [53].

Variation partitioning explained by explanatory variables. Double stopping criterion [57] was used as forward selection procedure of explanatory variables in order to avoid type I error, and to avoid overestimating the amount of explained variance in the species matrix before variation partitioning [54,57]. Variation partitioning allows partitioning the variation in the species data explained by pure environmental effects [a], spatially structured environmental variables [b], pure spatial effects [c], and unexplained variation (i.e., residuals or fraction [d]) [18,58]. This method estimates and tests the percentage of variation (R^2_{adj}) attributed to each unique set of explanatory variables [18]. Three steps were necessary to perform the variation partitioning: (1) implementation of a redundancy analysis (RDA) with sets of environmental and spatial variables, (2) a second RDA with environmental data, controlling for spatial effects (E | S), (3) a third RDA with spatial data, controlling for environmental effects (S | E) [18]. Variation partitioning was performed for the composition, abundance and biomass of dung beetles at each spatial scale studied. Before running RDAs, species datasets were

Hellinger-transformed in order to eliminate the disparity between values [59]. The proportion of variance explained by each set of explanatory variables is described by R^2_{adj} according to the Ezekiel correction [18], and significance levels are calculated by permutation tests (N = 999) [54]. We tested for a linear spatial trend and found a significant longitudinal trend for dung beetle composition data ($F = 3.34$; $df = 2$; $p<0.01$), abundance ($F = 6.77$; $df = 2$; $p<0.01$), and biomass ($F = 7.35$; $df = 2$; $p<0.01$). Thus, all datasets were detrended prior to the analyses [54]. R^2_{adj} values were indicated in percentage format in the text and tables. The analyses were conducted in R 3.0.1 program [53] using Packfor [60] and Vegan [61] packages.

Results

Species richness, abundance and biomass across spatial scales

Regionally, we collected a total of 3,004 individuals of Scarabaeinae, belonging to 21 species (Table S1). The mainland and the island had the same total number of species (16), sharing 11. On the island, the number of individuals was 2.5 higher, and total biomass was 2 times higher compared to the mainland. Among areas, RAT had the greatest number of species and individuals, and greatest biomass, followed by PER (both island areas). Only eight species (38.1%) were shared by all four areas.

The number of species per site ranged between five and 14 (Table S1). Only one species occurred in all sampling sites (*Canthon rutilans cyanescens*). Three species were sampled at least 19 sites (*Deltochilum morbillosum*, *Deltochilum multicolor*, and *Dichotomius sericeus*). Five species were responsible for 92.8% of the total dung beetle biomass (*D. sericeus*, *Coprophanaeus saphirinus*, *C. rutilans cyanescens*, *D. multicolor* and *D. morbillosum*) (see Figure S1 for a spatial comparison of species richness, abundance and biomass).

Beta diversity across spatial scales

The hierarchical partitioning analysis of diversity based on species richness data showed a large contribution of regional ($\beta3 = 5$ species) and local ($\beta1 = 4.4$ species) spatial scales to gamma diversity (Figure 2). Beta diversity among areas ($\beta2 = 2.2$ species) was relatively small when compared to other spatial scales. A similar pattern was observed for Shannon entropy based on species abundance. The hierarchical partitioning of diversity analysis also indicated a small contribution of $\beta2$, and a large relative contribution of $\beta3$ and $\beta1$, respectively.

Our results show that there were five species found only on the mainland (*Bdelyrus braziliensis*, *Coprophanaeus dardanus*, *Deltochilum furcatum*, *Dichotomius quadrinodosus*, and *Eurysternus cyanescens*), and five only found on the island (*Dichotomius* sp., *Eurysternus parallelus*, *Paracanthon* aff. *rosinae*, *Uroxys* sp. 1, and *Uroxys* sp. 2). *Bdelyrus braziliensis* and *Eurysternus cyanescens* were found only in ANH, on the mainland. *Coprophanaeus dardanus*, *Deltochilum furcatum* and *Dichotomius quadrinodosus* occurred only in ITA, on the mainland. *Dichotomius* sp. and *Uroxys* sp. 2 occurred only in RAT, on the island. *Paracanthon* aff. *rosinae* and *Uroxys* sp. 1 were only shared between RAT and PER.

Species and local contributions to beta diversity index

The partitioning of the total variance in components of the contribution of species and sites to beta diversity showed different results when data on composition, abundance and biomass of dung beetles were analyzed. For composition, the total sum of squares (SS_{Total}) was 38.183 and the index of variance of beta diversity (BD_{Total}) was 0.395 for dung beetle data across all sampling sites. SCBD values ranged between 0.002 and 0.145, and 10 species contributed above the mean (0.047) to beta diversity (Table 1 left). The values of LCBD ranged between 0.003 and 0.032, indicating the uniqueness of the dung beetle community at each sampling site. Six sampling points were statistically significant to beta diversity (Figure 3A), all within two ITA sites and one PER site. LCBD values were negatively correlated with species richness ($r = -0.45$, $p<0.001$) indicating that, in general, sites with unique species composition have a low number of species.

For abundance data, the SS_{Total} was 35.691 and the BD_{Total} was 0.360. SCBD values ranged between 0.0003 and 0.179, and five species contributed above the mean (0.047) to beta diversity (Table 1 center). LCBD values ranged between 0.002 and 0.040, and seven sampling points were statistically significant (Figure 3B), all occurring in the same two ITA sites sampled for composition data. LCBD values were negatively correlated with abundance at each sampling point ($r = -0.32$, $p<0.001$), demonstrating that sites with unique species composition, in general, have low abundance.

For biomass data, the SS_{Total} was 35.275 and the BD_{Total} was 0.356. SCBD ranged between $6.915e^{-06}$ and 0.235, and six species contributed above the mean (0.047) to beta diversity (Table 1 right). LCBD values ranged between 0.002 and 0.042, and eight sampling points were statistically significant (Figure 3C). LCBD values were negatively correlated with biomass of each sampling point ($r = -0.49$, $p<0.001$), and as well as to species richness and abundance, suggesting that sites with unique species composition, in general, have low dung beetle biomass.

Only four species (*C. rutilans cyanescens*, *C. saphirinus*, *D. multicolor* and *D. morbillosum*) contributed to the beta diversity index taking into account the composition, abundance and biomass of dung beetles. Although these species are very common among the sampled sites, this result suggests that they had strong local spatial variation in terms of occurrence, number of individuals and total biomass between sites.

Environmental and spatial effects on community variation at different spatial scales

The variation partitioning based on community composition, abundance and biomass showed different responses at each spatial scale when we analyzed each species dataset. The variation in species composition at mainland-island scale showed a greater and significant environmental effect (Table 2A left). Altitude was the only environmental variable selected to compose the environmental model and it explained 4.5% of variation at this scale. When the spatial configuration was removed from the environmental model the explanation decreased to 4.4%. The spatial effect was not significant for variation in species composition, and after the environmental effect was removed the spatial model was still not significant. The variation explained by spatially structured environmental variables [b] explained only 0.02%.

At the scale of areas, spatial effects were stronger than environmental effects, and it explained 13.2% of the variation in species composition (Table 2B left). After environmental effects were removed, the spatial model explained 9.9% of the variation in the data. The environmental model, which was composed of altitude only, explained only 1.2% after spatial effects were removed. Spatially structured environmental variables [b] explained 3.3% of the variation in the data. At the smallest scale, i.e. sites, only the environmental model was significant and explained 9.0% of the data variation after spatial effects were removed (Table 2C left). At this scale, the spatial model showed no

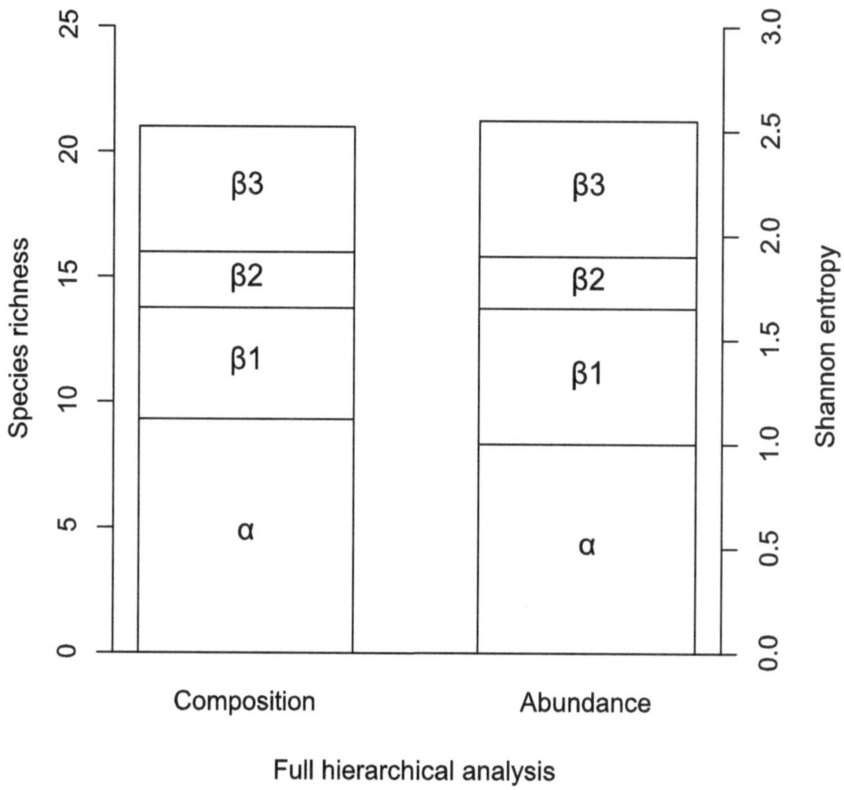

Figure 2. Full hierarchical analysis of diversity partitioning. The partitioning was performed for species richness and Shannon entropy of dung beetles. α = local diversity, β1 = diversity among sites, β2 = diversity among areas, β3 = diversity among mainland-island.

significant patterns. The variation explained by spatially structured environmental variables [b] showed negative values.

Almost 10% of the variation in composition (using species abundance) at the scale of mainland-island was attributed to the environmental model, which included altitude, green cover and land slope (Table 2A center). After spatial effects were removed, the environmental model explained 10.0% of the variation in the data. The spatial model was also significant, but explained only 1.4%. Both models were significant when only the pure effects were analyzed. Spatially structured environmental variables [b] showed negative values.

At the area scale, the spatial effect was significant (explaining 17.4% of the variation) and greater than the environmental effect. Both models were significant when only the pure effects were analyzed, in which the spatial model explained 14.8% and the environmental model explained 7.3% of the variation (Table 2B

Table 1. Partitioning of the total variance in species contribution to beta diversity (SCBD) based on the beta diversity index (BD_{Total}) and the total sum of squares (SS_{Total}).

Species	Composition	Abundance	Biomass
	$SS_{Total} = 38.183$	$SS_{Total} = 35.691$	$SS_{Total} = 35.275$
	$BD_{Total} = 0.395$	$BD_{Total} = 0.360$	$BD_{Total} = 0.356$
Canthidium aff. *trinodosum*	0.121	0.141	
Canthon luctuosus	0.061		
Canthon rutilans cyanescens	0.055	0.153	0.113
Coprophanaeus saphirinus	0.100	0.123	0.235
Deltochilum multicolor	0.145	0.116	0.155
Deltochilum morbillosum	0.109	0.101	0.067
Deltochilum rubripenne	0.068		
Dichotomius sericeus	0.059		0.230
Phanaeus splendidulus	0.063		0.49
Uroxys sp. 1	0.053		

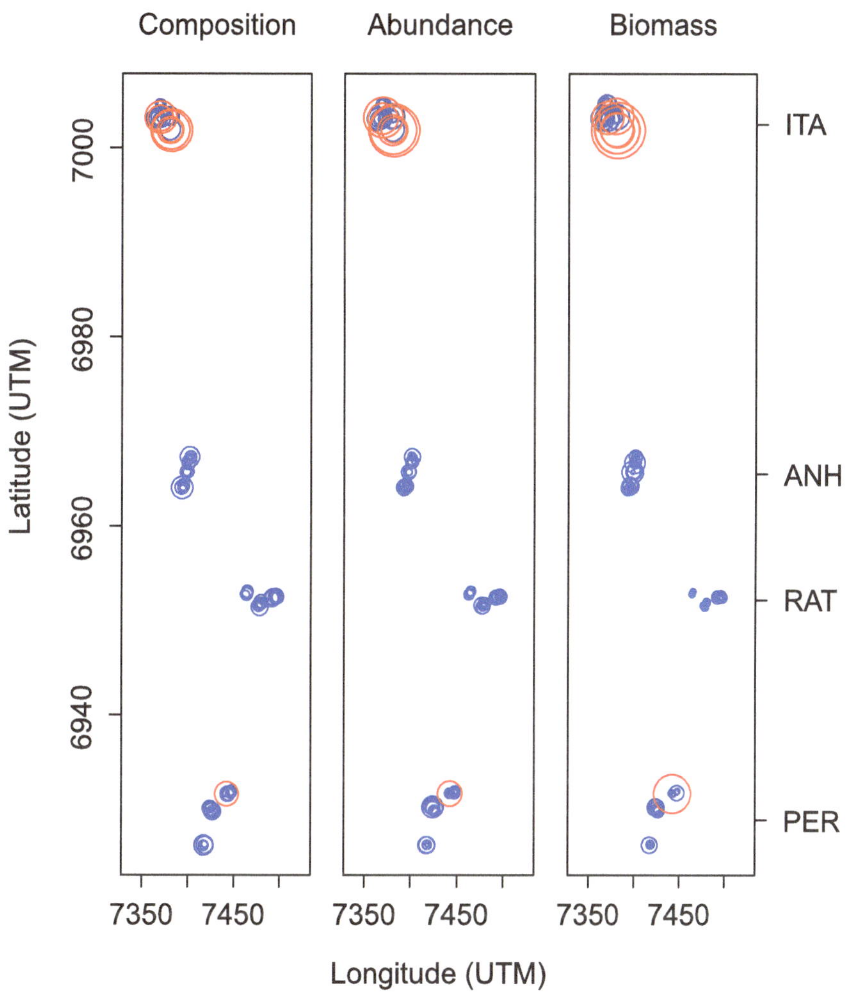

Figure 3. Map of the sampling points showing significant values (red) of the local contribution to beta diversity (LCBD). LCBD analysis used composition, abundance and dry biomass data. ANH: Environmental Protection Area of Anhatomirim; ITA: Permanent Protection Area of Itapema; PER: Lagoa do Peri Municipal Park; RAT: Permanent Protection Area of Ratones. The circles are proportional to the total value of LCBD for each analysis.

center). The variation explained by spatially structured environmental variables [b] explained 2.5% of the variation in the data. At the site scale, the environmental model had greatest relative importance for dung beetle abundance (Table 2C center). After spatial effects were removed, the variables that composed the environmental model explained 15.8% of the variation in abundance data. The spatial model showed no significant spatial patterns at this scale. Spatially structured environmental variables [b] showed negative values.

At the mainland-island scale, analysis of the variation in species composition based on biomass showed that both environmental and spatial effects were significant (Table 2A right). The environmental model composed of altitude, green cover and land slope explained 9.9% of the variation, and spatial effects explained 2.8% of the variation in the data (after corrections). The variation explained by spatially structured environmental variables [b] explained 0.002% of the variation in the data.

At the area scale, the spatial model was significant and explained the greatest amount of the variation in the biomass data (16.6%) followed by the environmental model (9.9%) (Table 2B right). After correction, the spatial and environmental models explained 15.1% and 8.4%, respectively, of the variation in the biomass data. Spatially structured environmental variables [b] explained 1.5%. At the local scale, the environmental model explained 9.9% of variation, and when spatial effects were removed the proportion increased to 17.2%. The spatial model showed no significance at this scale. Values for the variation explained by spatially structured environmental variables [b] were negative.

Discussion

In recent decades, there has been increased interest in understanding scale-dependence of the structuring processes of biological communities, including studies of protozoa [62], zooplankton [25,63], ichthyoplankton [64], dragonflies [65], coral reefs [66], reef-fishes [67], freshwater fishes [65], plants [68,69], frogs [65], birds [70], and mammals [71], covering a wide variety of ecosystems. The unique biology of dung beetles makes them excellent models with which to explore general concepts in ecology [31], including new approaches suggested by metacommunity theory. Our results represent a first step towards a better understanding of the relative importance of ecological processes

Table 2. Partitioning of variation in dung beetle communities at three spatial scales using redundancy analysis on composition, abundance and biomass.

		Composition				Abundance				Biomass			
		R^2_{adj}	DF	F	P	R^2_{adj}	DF	F	P	R^2_{adj}	DF	F	P
A) Mainland-island													
E	[a + b]	**4.5**[a]	1	5.62	0.001	**9.9**[e]	3	4.62	0.001	**9.9**[e]	3	4.61	0.001
S	[b + c]	0.4[b]	2	1.19	0.268	**1.4**[f]	1	2.40	0.016	**2.8**[b]	1	3.88	0.003
E | S	[a]	**4.4**	1	5.53	0.001	**10.0**	3	4.70	0.001	**9.9**	3	4.69	0.001
S | E	[c]	0.4	2	1.19	0.242	**1.6**	1	2.68	0.006	**2.8**	1	4.11	0.001
B) Areas													
E	[a + b]	**4.5**[a]	1	5.62	0.001	**9.9**[e]	3	4.62	0.001	**9.9**[e]	3	4.61	0.001
S	[b + c]	**13.2**[c]	5	4.02	0.001	**17.4**[g]	7	3.98	0.001	**16.6**[h]	6	4.28	0.001
E | S	[a]	**1.2**	1	2.27	0.012	**7.3**	3	3.97	0.001	**8.4**	3	4.45	0.001
S | E	[c]	**9.9**	5	3.27	0.001	14.8	7	3.69	0.001	**15.1**	6	4.21	0.001
C) Sites													
E	[a + b]	**4.5**[a]	1	5.62	0.001	**9.9**[e]	3	4.62	0.001	**9.9**[e]	3	4.61	0.001
S	[b + c]	−11.2[d]	40	0.75	0.999	−20.2[d]	40	0.58	1.000	−22.9[d]	40	0.53	1.000
E | S	[a]	**9.0**	1	6.19	0.001	**15.8**	3	3.98	0.001	**17.2**	3	4.20	0.001
S | E	[c]	−6.7	40	0.83	0.976	−14.2	40	0.67	0.999	−15.5	40	0.64	0.999

E: environmental model, S: spatial model, constructed from MEM variables, E | S: environmental model without spatial patterns within each spatial scale, S | E: spatial model without environmental patterns within each spatial scale, R^2_{adj}: data variation explained by the model (values are in percentage), DF: degrees of freedom of model. Significant models are in bold.

[a]Environmental model constructed from the altitude variable; [b]Spatial model constructed from the MEM1 and MEM2 variables; [c]Spatial model constructed from the MEM4, MEM9, MEM5, MEM3, and MEM1 variables; [d]Spatial model constructed from all MEM variables; [e]Environmental model constructed from the altitude, green cover and land slope variables; [f]Spatial model constructed from the MEM1 variable; [g]Spatial model constructed from the MEM4, MEM9, MEM3, MEM7, MEM1, MEM2, and MEM5 variables; [h]constructed from the MEM4, MEM3, MEM2, MEM5, MEM7, and MEM1 variables.

on dung beetle community structure in a coastal mainland-island landscape across three different spatial scales.

In this study, the environmental heterogeneity had greater importance at smaller scales, and may be the cause of high beta diversity in terms of species richness and abundance (i.e., Shannon entropy) found among sampling sites. Local environmental factors seem to be crucial in the structuring of local communities; such factors may be responsible for high beta diversity at the local scale, as has been demonstrated for several groups of organisms in a variety of ecosystems [25,72–77]. Thus, the ecological gradient evaluated here appears to have a distribution defined by spatially structured environmental heterogeneity, which may have strong effects on dung beetle community structure locally.

Beta diversity at the area scale was lower than at the site scale, despite the increase in geographic distance among the sampling points. At area scale, we found a significantly greater importance of spatial effects compared to environmental effects, even after the analysis of individual effects of the models. Beta diversity among areas appears to be mainly related to the spatial patterns of the sampling sites. The occurrence of shared environmental and spatial effects as drivers of beta diversity are very common with increasing spatial scale [24,75], and these shared effects may suggest significant limitations in species dispersal ability between site and area scales. Besides environmental effects, spatial limitation may be related to geographic distance, lack of connectivity caused by fragmentation, or the landscape structure between the mainland and the island.

Between the mainland and the island, beta diversity showed the highest values and at this scale only the environmental model was significant for species composition, while for abundance the environmental and spatial models were significant. The high beta diversity found between the mainland and the island has its origin at site and area scales, where environmental and spatial patterns have high relative importance. Thus, we observed that the distribution of dung beetles along an ecological gradient occurs in a spatially structured environment, where such patterns may be generated due to dispersal limitation at intermediate scales, and due to environmental heterogeneity at local scales.

The distribution pattern of dung beetle species composition was associated with the altitude gradient. This variable was significant at all scales studied after spatial effects were removed, demonstrating its strong influence on the species composition of dung beetle communities. Altitude ranged between 28 and 265 m among sampling sites. A study performed in the Colombian Andes demonstrated that dung beetle composition varied along an altitude gradient between 1,000 and 2,250 m at intervals of roughly 250 m [78]; the differences found in this study were associated with different environmental adaptations of the species. Environmental and climatic differences are also important for dung beetle distribution at low altitudes. The proximity to the sea and the effect of wind on humidity [46], and soils with higher salt concentration, although not measured in our study, could also affect the relative success of some species. Thus, the environmental and spatial configuration of sampling sites evaluated in the mainland-island landscape may influence the distribution of dung beetle species.

Except for at the mainland-island scale, in general species composition and abundance showed similar responses to the ecological gradient studied. However, the relative importance of the models was greater for abundance data. Although abundance may not sufficiently explain patterns of species distribution (i.e., due to confounding effects caused by highly abundant species), it may help to explain the responses of species across the environmental gradients, because it reflects changes in the relative success of each species against these gradients [25]. In our study, abundance and biomass data were explained by the same set of environmental variables, and showed very similar responses to the ecological gradients. In general, dung beetle biomass was more influenced by individual spatial effects than abundance data. Thus, biomass may be an important descriptor of changes in the relative success of dung beetles along ecological gradients, because it is mainly derived from nutrients obtained from mammal feces [79], and availability of this resource may also be affected by environmental heterogeneity.

In addition to altitude, the percentage of green cover and land slope were part of the environmental model describing the distribution of dung beetle abundance and biomass. Green cover has also been found to explain the distribution of dung beetles species in different-sized Atlantic forest fragments [80]. Variation in the percentage of green cover illustrates the differences among sites with greater or fewer small plants and shrubs covering the soil. Sites with greater spacing between trees and less tree cover allow more sunlight, which may influence the microclimate and soil moisture, as shown to occur in forest edges [81]. Land slope ranged between five and 36° degrees, and having some degree of slope is a common characteristic among our sampling sites, due to the fact that Atlantic Forest is typically located on hillsides with a large altitudinal range [29]. In another study of Atlantic Forest in Serra do Japi, located in the western region of São Paulo state's Atlantic plateau, Brazil, dung beetle composition was shown to vary between the tops, hillsides and valleys, which are associated with differences in environmental structure [46]. Sites with high degrees of land slope may be most affected by rainfall, and may present unfavorable soil features for some dung beetle species. These environmental characteristics may influence the distribution of dung beetles, and may have greater power to affect relative species success.

Changes in the structural complexity of forested areas may modify the entire community associated with these habitats, diminishing the species richness of some taxonomic groups and increasing the others [33]. For example, the structure of the environment was more important in determining dung beetle community composition than resource availability in areas occupied by cattle in Mexico [82]. The distribution of dung beetles along different environmental characteristics may show discrete associations typical to particular biotypes within the landscape [83]. Species richness, abundance and biomass of dung beetles were negatively affected in disturbed habitats (e.g., secondary forests and *Eucalyptus* plantations in the Brazilian Amazon) when compared to primary forest habitats [35]. Microclimatic variations in tropical forests related to canopy height and opening affected dung beetle communities in French Guiana [84]. Thus, many species of dung beetles have relationships with certain habitat characteristics, likely to facilitate finding mates and/or food, or could be directly related to the presence of organisms that produce their food resource.

High inter- and intraspecific competition, random distribution, and ephemeral nature of food resources together suggest, a priori, that dung beetles are probably good dispersers [85]. However, relatively few quantitative descriptions of dispersal in these beetles exist [85]. The dispersal ability of *Canthon acutus* was investigated in Venezuela using capture-mark-recapture technique [45]. The authors installed pitfall traps baited with feces at different distances in a semi-deciduous tropical forest, and they found that 95% of individuals were collected up to 25 m. In contrast, using similar techniques, other authors [86] evaluated the dispersal ability of *Canthon cyanellus cyanellus* across a Mexican landscape that contained different components such as forest fragments, hedgerows and pastures. They found a maximum movement distance among the different landscape components of 1,560 m for males (average 390 m) and 860 m for females (average 290 m), suggesting that landscape type change is not a barrier to dispersal for some species. In fact, some species from continuous Amazonian forest fragments do not extend their activities to adjacent open areas, and this effect is reduced when there is presence of secondary forest in these areas [40]. We may expect a similar pattern in the Brazilian Atlantic Forest.

Besides differences in dispersal ability, several species of dung beetles are associated with certain habitat types [35,46,47,87–91] due to microclimatic factors [84] or resource availability [31]. Data on differences in dispersal ability in the species sampled in this study are still lacking. Based on our results, it is possible that the high beta diversity found among sites reflects low dispersal ability due to environmental and spatial effects. Many species of dung beetles that inhabit forests tend to not extend their range to open areas [40,92], which influences their ability to disperse and colonize new habitats when the matrix is not favorable. However, species that live in forest edges or in the matrix [83,86] may have a stronger ability to disperse and colonize new habitats compared to those living inside the forests.

The high beta diversity of dung beetle communities found among our sampling sites throughout the ecological gradient could still be related to historical events [5,11] or neutral processes [10]. In a biogeographical context, the island of Santa Catarina shows similar physiographic and structural features to those of the mainland, since the island and the mainland were connected during past periods when the Atlantic Ocean level was low [93]. During that time, the small minimum distance between mainland and island (minimum of 500 m) and low maximum depth between them (about 30 m) may have allowed a favorable environment for dispersal of the species from the mainland to the island. Site "C" of ITA is unique in that is separated from the continuous forest that occurs in each area due to fragmentation caused by a highway; it had the lowest values for species richness (five) and for number of individuals (49), and was the exclusive site of occurrence of *Coprophanaeus dardanus*. This site also showed the lowest average altitude, and like others, this site has significant contribution to the negative relationship between LCBD values and community descriptors (species richness, abundance and biomass).

Due to the current fragmented structure of the landscape and the negative impacts on many coastal environments that urbanization has caused [93,94], the Atlantic Forest landscape is highly fragmented and in different stages of succession, with each functioning as "islands". The communities are isolated and dispersal and colonization rates are low [95] due to insertion in a matrix of inhospitable environments [96] for forest-inhabiting dung beetle species. Our results show that there were five species found only on the mainland, and five only found on the island. On the mainland, three species occurred only in ITA, and two only in ANH. On the island, two species occurred only in RAT, and two others were shared between RAT and PER. These results, as well as the analysis of the local contribution to beta diversity (significant sampling points occurred only near the ends of the spatial gradient, Figure 3) reflect the high importance of large-scale spatial effects in structuring dung beetle communities. The distinct occurrence of species between mainland-island may be result of isolation processes [95,97], or local extinction due to lack of certain food sources (e.g., presence of certain mammals [38]) or simple inability to colonize [98]. Alternatively, species may persist at a given location due to biotic and/or abiotic conditions in spatially structured environmental conditions [99]. We propose that the processes listed above act as joint drivers of the current distribution of dung beetle species in the landscape studied, and our result suggest that the relative importance of each process depends on the spatial scale.

Environmental control (i.e., the species sorting paradigm) seems to be the dominant structuring process in the metacommunity at the local scale. However, environment was also important at larger scales, and environmental factors were spatially structured along the ecological gradient studied. Spatial effects were more important at larger scales, where there was an increase in beta diversity that appears to be due to limitation in dispersal ability of the species due to geographic barriers and fragmented landscape. Contrary to what was found in other studies [25], our results suggest that the increase in the spatial scale was related to increased environmental heterogeneity, although only mildly, agreeing with the general findings for stream insect communities [26]. We believe that our results, extrapolated with caution, represent general patterns that serve as the basis for other organisms with similar characteristics and requirements.

Appropriate management of spatially heterogeneous ecosystems requires an understanding of both local and regional processes by which beta diversity is created and maintained, in order to preserve the spatial organization or species-environment relationships on which beta diversity is dependent [12,100]. Due to current scenario of fragmentation and isolation of remaining fragments of the Brazilian Atlantic Forest [29], knowing these answers is of great interest to managers and decision makers to plan appropriate conservation strategies in an increasingly human-modified world.

Supporting Information

Figure S1 Map of the sampling sites showing the distribution of species richness, abundance and total biomass of dung beetles. ANH: Environmental Protection Area of Anhatomirim; ITA: Permanent Protection Area of Itapema; PER: Lagoa do Peri Municipal Park; RAT: Permanent Protection Area of Ratones. The circles represent the proportion to the total for each analysis.

Table S1 Dung beetle species sampled in Atlantic Forest in eastern Santa Catarina, Brazil. ANH: Environmental Protection Area of Anhatomirim in Governador Celso Ramos; ITA: Permanent Protection Area of Itapema; PER: Lagoa do Peri Municipal Park, Florianópolis; RAT: Permanent Protection Area of Ratones, Florianópolis. Letters A to E depict the sampled sites in each area. T: total.

Dataset S1 Dataset of abundance and dry biomass of dung beetle species, environmental variables, and geographical coordinates. Samplings were performed in Brazilian Atlantic Forest, Santa Catarina, Brazil using baited pitfall traps from January to February 2012.

Acknowledgments

We are grateful to Ana Siegloch, Franciéle Garcês, Javier Toso, Jéssica Pires, Juliano Bogoni, Rafael Schmitt, Renata Calixto and Tamyris Gregolini for help during the fieldwork. We thank Fernando Vaz de Mello and David Edmonds for dung beetle identification. We thank the Instituto Chico Mendes de Conservação da Biodiversidade (ICMBio/MMA) and Fundação do Meio Ambiente (FATMA-SC) for permission to collect.

Author Contributions

Conceived and designed the experiments: PGS MIMH. Performed the experiments: PGS MIMH. Analyzed the data: PGS MIMH. Contributed reagents/materials/analysis tools: PGS MIMH. Contributed to the writing of the manuscript: PGS MIMH.

References

1. Leibold MA, Holyoak M, Mouquet N, Amarasekare P, Chase JM, et al. (2004) The metacommunity concept: a framework for multi-scale community ecology. Ecol Lett 7: 601–613.
2. Vellend M (2010) Conceptual synthesis in community ecology. Q Rev Biol 85: 183–206.
3. Legendre P, De Cáceres M, Borcard D (2010) Community surveys through space and time: testing the space-time interaction in the absence of replication. Ecology 91: 262–272.
4. Simberloff D (2004) Community ecology: is it time to move on? Am Nat 163: 787–799.
5. Ricklefs RE (2008) Disintegration of the ecological community. Am Nat 172: 741–750.
6. Howeth JG, Leibold MA (2010) Species dispersal rates alter diversity and ecosystem stability in pond metacommunities. Ecology 91: 2727–2741.
7. Tack A, Ovaskainen O, Harrison P, Roslin T (2009) Competition as a structuring force in leaf miner communities. Oikos 118: 809–818.
8. Wilson DS (1992) Complex interactions in metacommunities, with implications for biodiversity and higher levels of selection. Ecology 73: 1984–2000.
9. Davis ALV, Scholtz CH (2001) Historical vs. ecological factors influencing global patterns of Scarabaeine dung beetle diversity. Divers Distrib 7: 161–174.
10. Hubbell SP (2001) The unified neutral theory of biodiversity and biogeography. Princeton: Princeton University Press. 448 p.
11. Ricklefs RE, Schluter D (1993) Species diversity: regional and historical influences. In: Ricklefs RE, Schluter D, editors. Species diversity in ecological communities. Chicago: The University of Chicago Press. 350–363.
12. Legendre P, Borcard D, Peres-Neto PR (2005) Analyzing beta diversity: partitioning the spatial variation of community composition data. Ecol Monogr 75: 435–450.
13. Anderson MJ, Crist TO, Chase JM, Vellend M, Inouye BD, et al. (2011) Navigating the multiple meanings of beta diversity: a roadmap for the practicing ecologist. Ecol Lett 14: 19–28.
14. Jost L, DeVries P, Walla T, Greeney H, Chao A, et al. (2010) Partitioning diversity for conservation analyses. Divers Distrib 16: 65–76.
15. Legendre P, De Cáceres M (2013) Beta diversity as the variance of community data: dissimilarity coefficients and partitioning. Ecol Lett 16: 951–963.
16. Tuomisto H (2010) A diversity of beta diversities: straightening up a concept gone awry. Part 1. Defining beta diversity as a function of alpha and gamma diversity. Ecography 33: 2–22.
17. Tuomisto H (2010) A diversity of beta diversities: straightening up a concept gone awry. Part 2. Quantifying beta diversity and related phenomena. Ecography 33: 23–45.
18. Peres-Neto PR, Legendre P, Dray S, Borcard D (2006) Variation partitioning of species data matrices: estimation and comparison of fractions. Ecology 87: 2614–2625.
19. Borcard D, Legendre P, Avois-Jacquet C, Tuomisto H (2004) Dissecting the spatial structure of ecological data at multiple scales. Ecology 85: 1826–1832.
20. Gilbert B, Bennett JR (2010) Partitioning variation in ecological communities: do the numbers add up? J Appl Ecol 47: 1071–1082.
21. Smith TW, Lundholm JT (2010) Variation partitioning as a tool to distinguish between niche and neutral processes. Ecography 33: 648–655.
22. Holyoak M, Leibold MA, Holt RD (2005) Metacommunities: Spatial dynamics and ecological communities. Chicago: The University of Chicago Press. 520 p.
23. Winegardner AK, Jones BK, Ng IS, Siqueira T, Cottenie K (2012) The terminology of metacommunity ecology. Trends Ecol Evol 27: 253–254.
24. Cottenie K (2005) Integrating environmental and spatial processes in ecological community dynamics. Ecol Lett 8: 1175–1182.
25. Declerck SAJ, Coronel JS, Legendre P, Brendonck L (2011) Scale dependency of processes structuring metacommunities of cladocerans in temporary pools of High-Andes wetlands. Ecography 34: 296–305.
26. Heino J, Peckarsky BL (2014) Integrating behavioral, population and large-scale approaches for understanding stream insect communities. Curr Opin Insect Sci 2: 7–13.
27. Logue JB, Mouquet N, Peter H, Hillebrand H, The Metacommunity Working Group (2011) Empirical approaches to metacommunities: a review and comparison with theory. Trends Ecol Evol 26: 482–491.
28. Myers N, Mittermeier RA, Mittermeier CG, Fonseca GAB, Kent J (2000) Biodiversity hotspots for conservation priorities. Nature 403: 853–845.
29. Ribeiro MC, Metzger JP, Martensen AC, Ponzoni FJ, Hirota MM (2009) The Brazilian Atlantic Forest: How much is left, and how is the remaining forest distributed? Implications for conservation. Biol Conserv 142: 1141–1153.
30. Tabarelli M, Pinto LP, Silva JMC, Hirota MM, Bedê LC (2005) Desafios e oportunidades para a conservação da biodiversidade na Mata Atlântica brasileira. Megadiversidade 1: 132–138.
31. Simmons LW, Ridsdill-Smith TJ (2011) Reproductive competition and its impact on the evolution and ecology of dung beetles. In: Simmons LW, Ridsdill-Smith TJ, editors. Ecology and evolution of dung beetles. Oxford: Blackwell Publishing. 1–20.
32. Nichols E, Spector S, Louzada J, Larsen T, Amezquita S, et al. (2008) Ecological functions and ecosystem services provided by Scarabaeinae dung beetles. Biol Conserv 141: 1461–1474.
33. Barlow J, Gardner TA, Araujo IS, Avila-Pires TC, Bonaldo AB, et al. (2007) Quantifying the biodiversity value of tropical primary, secondary, and plantation forests. Proc Natl Acad Sci USA 104: 18555–18560.
34. Gardner TA, Barlow J, Araujo IS, Avila-Pires TC, Bonaldo AB, et al. (2008) The cost-effectiveness of biodiversity surveys in tropical forests. Ecol Lett 11: 139–150.
35. Gardner TA, Hernández MIM, Barlow J, Peres CA (2008) Understanding the biodiversity consequences of habitat change: the value of secondary and plantation forests for neotropical dung beetles. J Appl Ecol 45: 883–893.
36. Halffter G, Favila ME (1993) The Scarabaeinae (Insecta: Coleoptera) an animal group for analysing, inventorying and monitoring biodiversity in Tropical Rainforest and modified landscapes. Biol Int 27: 15–21.
37. Nichols E, Larsen T, Spector S, Davis AL, Escobar F, et al. (2007) Global dung beetle response to tropical forest modification and fragmentation: A quantitative literature review and meta-analysis. Biol Conserv 137: 1–19.
38. Culot L, Bovy E, Zagury Vaz-de-Mello F, Guevara R, Galetti M (2013) Selective defaunation affects dung beetle communities in continuous Atlantic rainforest. Biol Conserv 163: 79–89.
39. Andresen E (2003) Effect of forest fragmentation on dung beetle communities and functional consequences for plant regeneration. Ecography 26: 87–97.
40. Klein BC (1989) Effects of forest fragmentation on dung and carrion beetle communities in Central Amazonia. Ecology 70: 1715–1725.
41. Slade EM, Mann DJ, Villanueva JF, Lewis OT (2007) Experimental evidence for the effects of dung beetle functional group richness and composition on ecosystem function in a tropical forest. J Anim Ecol 76: 1094–1104.
42. Vulinec K (2002) Dung beetle communities and seed dispersal in primary forest and disturbed land in Amazonia. Biotropica 34: 297–309.
43. Veloso HP, Rangel-Filho ALR, Lima JCA (1991) Classificação da vegetação brasileira, adaptada a um sistema universal. Rio de Janeiro: Instituto Brasileiro de Geografia e Estatística. 124 p.
44. Lobo JM, Martín-Piera F, Veiga CM (1988) Las trampas pitfall con sebo, sus posibilidades en el estudio de las comunidades coprófagas de Scarabaeoidea (Col.). I. Características determinantes de su capacidad de captura. Rev Ecol Biol Sol 25: 77–100.
45. Larsen T, Forsyth A (2005) Trap spacing and transect design for dung beetle biodiversity studies. Biotropica 37: 322–325.
46. Hernández MIM, Vaz-de-Mello FZ (2009) Seasonal and spatial species richness variation of dung beetle (Coleoptera, Scarabaeidae *s. str.*) in the Atlantic Forest of southeastern Brazil. Rev Bras Entomol 53: 607–613.
47. Silva PG, Vaz-de-Mello FZ, Di Mare RA (2013) Diversity and seasonality of Scarabaeinae (Coleoptera: Scarabaeidae) in forest fragments in Santa Maria, Rio Grande do Sul, Brazil. An Acad Bras Cienc 85: 679–697.
48. Brower JE, Zar JH (1984) Field and laboratory methods for general ecology. Dubuque: Wm. C. Brown Company. 226 p.
49. Cottam G, Curtis JT (1956) The use of distance measures in phytosociological sampling. Ecology 37: 451–460.
50. Ramos FA (2000) Nymphalid butterfly communities in an amazonian forest fragment. J Res Lepid 35: 29–41.
51. Conselho Nacional do Meio Ambiente (1994) Resolução CONAMA n° 4, de 4 de maio de 1994. Publicada no DOU n° 114, de 17 de junho de 1994, Seção 1, páginas 8877–8878. Brasília: Ministério do Meio Ambiente. 2 p.
52. Dray S, Pélissier R, Couteron P, Fortin MJ, Legendre P, et al. (2012) Community ecology in the age of multivariate multiscale spatial analysis. Ecol Monogr 82: 257–275.
53. R Core Team (2013) R: A language and environment for statistical computing, Vienna, Austria. Available: http://www.R-project.org. Accessed 2013 Apr 22.
54. Borcard D, Gillet F, Legendre P (2011) Numerical ecology with R. New York: Springer. 306 p.
55. Jost L (2007) Partitioning diversity into independent alpha and beta components. Ecology 88: 2427–2439.
56. Veech JA, Crist TO (2009) Partition: software for hierarchical partitioning of species diversity. Program version 3.0, <http://www.users.muohio.edu/cristto/partition.htm>.
57. Blanchet FG, Legendre P, Borcard D (2008) Forward selection of explanatory variables. Ecology 89: 2623–2632.
58. Borcard D, Legendre P, Drapeau P (1992) Partialling out the spatial component of ecological variation. Ecology 73: 1045–1055.
59. Legendre P, Gallagher E (2001) Ecologically meaningful transformations for ordination of species data. Oecologia 129: 271–280.

60. R-Forge Administration and Development Team (2013) Spatial ecological data analysis with R. R Development Page, <http://r-forge.r-project.org/R/?group_id=195>.
61. Oksanen J, Blanchet FG, Kindt R, Legendre P, O'Hara RG, et al. (2013) Vegan: community ecology package. R package version 2.0–7. http://cran.r-project.org/web/packages/vegan/index.html.
62. Cadotte MW (2006) Metacommunity influences on community richness at multiple spatial scales: a microcosm experiment. Ecology 87: 1008–1016.
63. Forbes AE, Chase JM (2002) The role of habitat connectivity and landscape geometry in experimental zooplankton metacommunities. Oikos 96: 433–440.
64. Macedo-Soares LC, Garcia CA, Freire AS, Muelbert JH (2014) Large-scale ichthyoplankton and water mass distribution along the South Brazil Shelf. PLoS ONE 9: e91241.
65. Heino J (2011) A macroecological perspective of diversity patterns in the freshwater realm. Freshwater Biol 56: 1703–1722.
66. Cornell HV, Karlson RH, Hughes TP (2007) Scale-dependent variation in coral community similarity across sites, islands, and island groups. Ecology 88: 1707–1715.
67. MacNeil MA, Graham NAJ, Polunin NVC, Kulbicki M, Galzin R, et al. (2009) Hierarchical drivers of reef-fish metacommunity structure. Ecology 90: 252–264.
68. Giladi I, Ziv Y, May F, Jeltsch F (2011) Scale-dependent determinants of plant species richness in a semi-arid fragmented agro-ecosystem. J Veg Sci 22: 983–996.
69. Meynard CN, Lavergne S, Boulangeat I, Garraud L, Van Es J, et al. (2013) Disentangling the drivers of metacommunity structure across spatial scales. J Biogeogr 40: 1560–1571.
70. Heikkinen RK, Luoto M, Virkkala R, Rainio K (2004) Effects of habitat cover, landscape structure and spatial variables on the abundance of birds in an agricultural-forest mosaic. J Appl Ecol 41: 824–835.
71. Kelt DA, Meserve PL, Patterson BD, Lang BK (1999) Scale dependence and scale independence in habitat associations of small mammals in southern temperate rainforest. Oikos 85: 320–334.
72. Cottenie K, De Meester L (2004) Metacommunity structure: synergy of biotic interactions as selective agents and dispersal as fuel. Ecology 85: 114–119.
73. Davies KF, Melbourne BA, Margules CR (2001) Effects of within- and between-patch processes on community dynamics in a fragmentation experiment. Ecology 82: 1830–1846.
74. Davies KF, Melbourne BA, Margules CR, Lawrence JF (2005) Metacommunity structure influences the stability of local beetle communities. In: Holyoak M, Leibold MA, Holt RD, editors. Metacommunities: Spatial dynamics and ecological communities. Chicago: The University of Chicago Press. 170–188.
75. Gianuca AT, Bastazini VAG, Dias RA, Hernández MIM (2013) Independent and shared effects of environmental features and space driving avian community beta diversity across a coastal gradient in southern Brazil. Austral Ecol 38: 864–873.
76. Siqueira T, Bini LM, Roque FO, Marques Couceiro SR, Trivinho-Strixino S, et al. (2012) Common and rare species respond to similar niche processes in macroinvertebrate metacommunities. Ecography 35: 183–192.
77. Tews J, Brose U, Grimm V, Tielbörger K, Wichmann MC, et al. (2004) Animal species diversity driven by habitat heterogeneity/diversity: the importance of keystone structures. J Biogeogr 31: 79–92.
78. Escobar F, Lobo JM, Halffter G (2005) Altitudinal variation of dung beetle (Scarabaeidae: Scarabaeinae) assemblages in the Colombian Andes. Glob Ecol Biogeogr 14: 327–337.
79. Spector S, Forsyth AB (1998) Indicator taxa for biodiversity assessment in the vanishing tropics. In: Mace GM, Balmford A, Ginsberg JR, editors. Conservation in a changing world. Cambridge: Cambridge University Press. 181–209.
80. Campos RC, Hernández MIM (2013) Dung beetle assemblages (Coleoptera, Scarabaeinae) in Atlantic forest fragments in southern Brazil. Rev Bras Entomol 57: 47–54.
81. Didham RR, Lawton JH (1999) Edge structure determines the magnitude of changes in microclimate and vegetation structure in tropical fragments. Biotropica 31: 17–30.
82. Halffter G, Arellano L (2002) Response of dung beetle diversity to human-induced changes in a tropical landscape. Biotropica 34: 144–154.
83. Davis AJ, Holloway JD, Huijbregts H, Krikken J, Kirk-Spriggs AH, et al. (2001) Dung beetles as indicators of change in the forests of northern Borneo. J Appl Ecol 38: 593–616.
84. Feer F (2013) Variations in dung beetles assemblages (Coleoptera: Scarabaeidae) within two rain forest habitats in French Guiana. Rev Biol Trop 62: 753–768.
85. Roslin T, Viljanen H (2011) Dung beetle populations: structure and consequences. In: Simmons LW, Ridsdill-Smith TJ, editors. Ecology and evolution of dung beetles. Oxford: Blackwell Publishing. 220–244.
86. Arellano L, León-Cortés JL, Ovaskainen O (2008) Patterns of abundance and movement in relation to landscape structure: a study of a common scarab (*Canthon cyanellus cyanellus*) in Southern Mexico. Landsc Ecol 23: 69–78.
87. Almeida S, Louzada J, Sperber C, Barlow J (2011) Subtle land-use change and tropical biodiversity: dung beetle communities in Cerrado grasslands and exotic pastures. Biotropica 43: 704–710.
88. Davis AJ, Sutton SL (1998) The effects of rainforest canopy loss on arboreal dung beetles in Borneo: implicantions for the measurement of biodiversity in derived tropical ecosystems. Divers Distrib 4: 167–173.
89. Davis ALV (1996) Habitat associations in a South African, summer rainfall, dung beetle community (Coleoptera: Scarabaeidae, Aphodiidae, Staphylinidae, Histeridae, Hydrophilidae). Pedobiologia 40: 260–280.
90. Filgueiras BKC, Iannuzzi L, Leal IR (2011) Habitat fragmentation alters the structure of dung beetle communities in the Atlantic Forest. Biol Conserv 144: 362–369.
91. Larsen T, Lopera A, Forsyth A (2006) Extreme trophic and habitat specialization by Peruvian dung beetles (Coleoptera: Scarabaeidae: Scarabaeinae). Coleopts Bull 60: 315–324.
92. Spector S, Ayzama S (2003) Rapid turnover and edge effects in dung beetle assemblages (Scarabaeidae) at a Bolivian Neotropical forest-savanna ecotone. Biotropica 35: 394–404.
93. Klein AHF, Menezes JT, Diehl FL, Abreu JGN, Polette M, et al. (2006) Erosão e progradação do litoral brasileiro - Santa Catarina. In: Muehe D, editor editors. Erosão e progradação do litoral brasileiro. Brasília: Ministério do Meio Ambiente. 401–436.
94. Caruso MML (1983) O desmatamento da Ilha de Santa Catarina de 1500 aos dias atuais. Florianópolis: Editora da Universidade Federal de Santa Catarina. 158 p.
95. May F, Giladi I, Ristow M, Ziv Y, Jeltsch F (2013) Metacommunity, mainland-island system or island communities? Assessing the regional dynamics of plant communities in a fragmented landscape. Ecography 36: 842–853.
96. Nouhuys S, Hanski I (2005) Metacommunities of butterflies, their hosts and their parasitoids. In: Holyoak M, Leibold MA, Holt RD, editors. Metacommunities: Spatial dynamics and ecological communities. Chicago: The University of Chicago Press. 99–121.
97. MacArthur RH, Wilson EO (1967) The Theory of Island Biogeography. Princeton: Princeton University Press. 224 p.
98. Andrewartha HG, Birch LC (1954) The distribution and abundance of animals. Chicago: The University of Chicago Press. 793 p.
99. Legendre P (2008) Studying beta diversity: ecological variation partitioning by multiple regression and canonical analysis. J Plant Ecol 1: 3–8.
100. Mouquet N, Carr MC, Cottenie K (2005) The world is patchy and heterogeneous! Trade-off and source-sink dynamics in competitive metacommunities. In: Holyoak M, Leibold MA, Holt RD, editors. Metacommunities: Spatial dynamics and ecological communities. Chicago: The University of Chicago Press. 237–262.

8

Deforestation and Benthic Indicators: How Much Vegetation Cover Is Needed to Sustain Healthy Andean Streams?

Carlos Iñiguez-Armijos[1,2]*, Adrián Leiva[1], Hans-Georg Frede[2], Henrietta Hampel[3], Lutz Breuer[2]

1 Departamento de Ciencias Naturales, Universidad Técnica Particular de Loja (UTPL), Sección de Ecología, Loja, Ecuador, **2** Institute for Landscape Ecology and Resources Management (ILR), Research Centre for Bio Systems, Land Use and Nutrition (IFZ), Justus-Liebig-University Giessen, Giessen, Germany, **3** Universidad de Cuenca, Quinta Balzaín, Cuenca, Ecuador

Abstract

Deforestation in the tropical Andes is affecting ecological conditions of streams, and determination of how much forest should be retained is a pressing task for conservation, restoration and management strategies. We calculated and analyzed eight benthic metrics (structural, compositional and water quality indices) and a physical-chemical composite index with gradients of vegetation cover to assess the effects of deforestation on macroinvertebrate communities and water quality of 23 streams in southern Ecuadorian Andes. Using a geographical information system (GIS), we quantified vegetation cover at three spatial scales: the entire catchment, the riparian buffer of 30 m width extending the entire stream length, and the local scale defined for a stream reach of 100 m in length and similar buffer width. Macroinvertebrate and water quality metrics had the strongest relationships with vegetation cover at catchment and riparian scales, while vegetation cover did not show any association with the macroinvertebrate metrics at local scale. At catchment scale, the water quality metrics indicate that ecological condition of Andean streams is good when vegetation cover is over 70%. Further, macroinvertebrate community assemblages were more diverse and related in catchments largely covered by native vegetation (>70%). Our results suggest that retaining an important quantity of native vegetation cover within the catchments and a linkage between headwater and riparian forests help to maintain and improve stream biodiversity and water quality in Andean streams affected by deforestation. This research proposes that a strong regulation focused to the management of riparian buffers can be successful when decision making is addressed to conservation/restoration of Andean catchments.

Editor: Tomoya Iwata, University of Yamanashi, Japan

Funding: This research was funded through a scholarship of the Secretaria Nacional de Educación Superior, Ciencia, Tecnología e Innovación of Ecuador (SENACYT), a DFG (Deutsche Forschungsgemeinschaft) grant (FR615/28-1/-2) and by the Universidad Técnica Partucular de Loja (UTPL). The support from Henrietta Hampel was thanks to the PROMETEO Program of SENESCYT. The funders had no role in study design, data collection and analysis, decision to publish, or preparation of the manuscript.

Competing Interests: The authors have declared that no competing interests exist.

* Email: cainiguez@utpl.edu.ec

Introduction

Tropical deforestation is one of the most important topics worldwide associated with biodiversity loss and ecosystem degradation [1–3]. However, the effects of deforestation on ecological condition of tropical streams have been rarely investigated despite evidence that forest loss negatively affects these streams [4–7]. In Ecuador, deforestation for agriculture and/or urbanization is a critical threat for Andean ecosystems [8], reaching annual deforestation rates of 2.7% between 1989 and 2008 in the montane forests of southern Andes (Tapia Armijos, pers. communication). However many of these ecosystems provide essential environmental services including biodiversity conservation, carbon storage and water supply [9]. Several studies focused along the Ecuadorian Andean region have demonstrated that deforestation affects habitat structure and macroinvertebrate community assemblages in streams [10], litter breakdown rates [11] and hydrological processes [12]. Clearly the Ecuadorian montane streams are influenced by the surrounding landscape and several research groups have recently deployed considerable efforts to assess these relationships. Nonetheless, information gaps concerning the influence of land cover/use and spatial scale on stream condition remains, despite they affect water quality and aquatic diversity [13,14]. These data are crucial for management and conservation of developed catchments to insure long term sustainability [15].

In other regions, organizations such as the European Environment Agency (EEA), the United States Environmental Protection Agency (EPA) and national agencies in New Zealand have identified, and are currently considering within their monitoring programs, the impacts of land cover/use changes on stream ecological conditions based on responses of macroinvertebrate communities [16–19]. In addition, they have developed catchment and/or riparian rehabilitation projects in order to enhance stream condition [20]. In South and Central America such extensive monitoring or restoration programs by governmental initiatives are rare, despite their potential to reduce the impact of deforestation on stream condition [13]. In this regard, a Costa

Rican regulation has showed the effectiveness of protecting riparian buffers to reduce the impacts of deforestation on macroinvertebrate communities and sensitive taxa in tropical streams [4].

Several authors [10,21–23] have applied a variety of macroinvertebrate metrics (e.g. Fisher's index; EPT taxa) to assess the impact of land cover/use on the ecological condition of streams in South America focusing on the structure and composition of benthic communities. These studies have confirmed changes to aquatic biodiversity and water quality because of deforestation around the streams. Nevertheless, how does vegetation cover influences stream ecological condition? In streams running through forested catchments the macroinvertebrate communities are more heterogeneous, diversity of intolerant taxa to environmental stress is higher, and abundance and variety of habitats is greater [10,14,21]. Likewise, streams with forested riparian areas have low variation in water temperature and less nutrients and sediment loads along the entire length [4,13]. At the catchment scale, there is also an influence of vegetation cover on controlling hydrology and the entire diversity of the streams [12]. In this regard, estimating the extent of native vegetation cover in a catchment and the importance of riparian management and headwater forest protection to maintain and improve biodiversity and water quality of streams affected by deforestation is a pressing task for management and conservation [13,14,24].

Geographic Information Systems (GIS) allow researchers to explore correlations in spatial patterns with ecological features in the landscape. Allan [13] describes the ability to apply GIS tools with remote sensing data to assess the influence of land cover/use on stream ecological condition. Studies in temperate areas have described the influence of land cover and spatial scale on stream communities [25,26] and on stream water quality [14] by applying GIS based evaluation techniques to derive land cover data and to relate them with physical, chemical and biological data.

In this study, our objective was to investigate the influence of vegetation cover at three spatial scales (local, riparian buffer and catchment) on structure and composition of macroinvertebrate communities and on water quality of headwater streams in the southern Ecuadorian Andes. These streams are located in two contrasting regions of varying intensities of anthropogenic pressure (i.e. deforestation, population growth, and farming). The questions we addressed were: what spatial scale and how much vegetation cover are associated with healthy streams based on macroinvertebrate metrics? In order to address these questions, we calculated several diversity and water quality metrics from macroinvertebrate communities, derived spatial information on land cover from GIS, and related these metrics statistically to three spatial scales and stream parameters. Finally, we addressed management and conservation problems in order to help mitigating impacts of deforestation on stream ecosystems.

Methodology

For this study, the Ministry of Environment of Ecuador issued us a general research permit for every sampling site and to collect water samples and macroinvertebrate specimens. In the case of the Zamora upstream catchment (Zu), the Municipality of Loja granted a specific permission to work at sites Zu1–2, 4–5, 8 because these streams are used for the water supply of the city. To work at the San Francisco catchment (SF), the administration of the San Francisco Research Station and Nature and Culture International allowed us to access sites SF1–7. For all other sites we used the general research permit.

Most of macroinvertebrates collected in this study were identified at genus level and we do not know if any of these genera are endangered or protected, because of the lack of information for Ecuador.

Study area

We studied 23 streams of two headwater catchments of the Zamora River basin, located in the mountain forests in Ecuadorian southern Andes (Figure 1). The Zamora upstream catchment covers an area of 276.2 km^2 and ranges between 2,020 to 3,250 m a.s.l. It is the most populated catchment within the Zamora River basin with approximately 120, 000 inhabitants in the city of Loja. The San Francisco catchment is located in a sparsely populated area of the Zamora River basin with an area of 75.3 km^2 and an elevation range from 1,800 to 3,250 m a.s.l. The natural vegetation in both study catchments is evergreen mountain rain forests, humid shrub, cloud forest and páramo [27,28]. Anthropic activities in both catchments have substituted the original vegetation mainly to pasture, and to a lesser extent to cropland, pine (*Pinus*) and eucalyptus (*Eucalyptus*) plantations [29]. Urban development has only taken place in the Zamora upstream catchment and has spread along its main river. Both catchments have similar climatic conditions with a wet season from May to September and a dry season from October to April. The average annual precipitation and temperature in the Zamora upstream and San Francisco catchments are about 2,200 mm and 15.2°C respectively.

Stream measurements and macroinvertebrates sampling

We sampled 16 streams in the Zamora upstream catchment and 7 streams in the San Francisco catchment. Sampling focused on riffles with an altitudinal variation of 350 m among sites. Every site consisted of a 30–60 m reach placed at the outlet of each stream in order to capture the influences of land use upstream on macroinvertebrate communities. According to the sampling frequency applied in other studies with benthic macroinvertebrates both in Ecuador and in other Neotropical regions [4,10,11,30–34], we collected physical–chemical and macroinvertebrate data during both dry (November 2010; December 2012) and wet seasons (May 2011; April 2012) at the same stream reach. Physical–chemical parameters such as pH, temperature (*T*), specific conductance (SC), dissolved oxygen (DO), were assessed in–stream in each of the sampling periods (Table S1) with a portable YSI 556 Multi Probe System and were averaged from several measurements during the sampling campaigns. Grab samples were taken for measuring further water chemistry at the laboratory of the Universidad Técnica Particular de Loja (UTPL) using standard methods [35]: nitrate (NO_3^-), phosphate (PO_4^{3-}), biochemical oxygen demand during decomposition in a period of 5 days (BOD_5), total coliforms (TC) and turbidity (Table S1).

For sampling of macroinvertebrate communities several methods exist and each one provides different results, nevertheless a combination of sampling methods is preferable [30]. In this study, qualitative macroinvertebrates samples were taken along the stream reach applying two sampling methods: 1) collecting macroinvertebrates directly in their stream habitats from stones and leaves, and 2) using a D–frame net (500 μm mesh) for kick sampling at 3–6 points selected randomly. Collection method 1 was applied during 10 min along the reach to remove all visible macroinvertebrates from stones and collect leaves respectively. Selected stones collected were totally submerged and had an average diameter of 20 cm approximately. Leaves were stored in a plastic bag and rinsed with tap water to remove macroinvertebrates in the laboratory. For collection method 2, macroinverte-

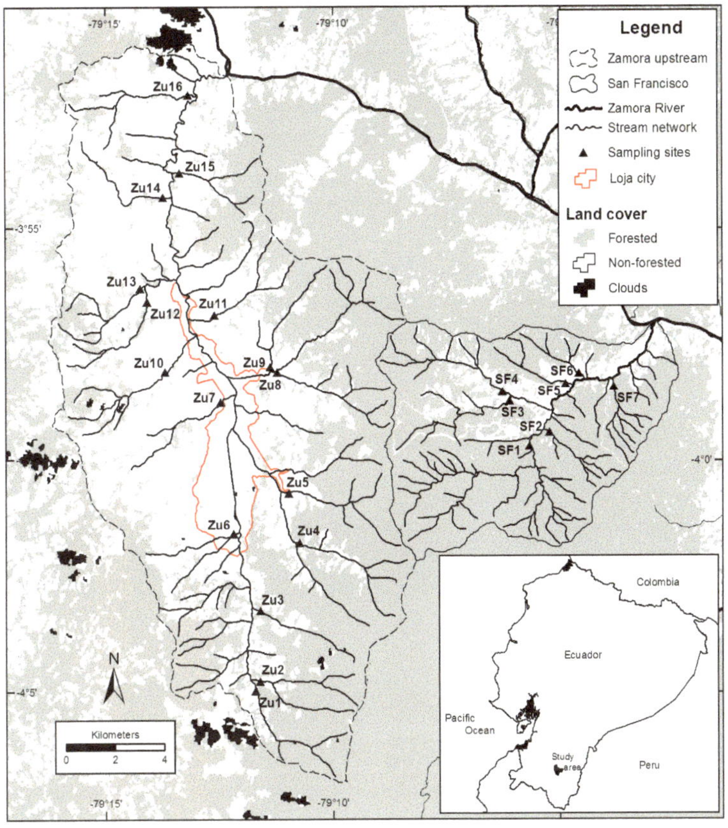

Figure 1. Study area and location of the sampling sites. Codes represent the sampling sites in the Zamora upstream (Zu1 to Zu16) and San Francisco (SF1 to SF7) catchments. Native vegetation cover was classified into forested and non-forested. Only the complete stream network for the study area is shown. Catchment characteristics are further shown in Table S1.

brates were sampled by kicking the substrate while holding the D–frame net downstream at each point. The application of this technique was based on its suitability for sampling in a variety of stream substrates and depths [36,37]. Both collections were combined into a single collection per site in order to avoid differences on macroinvertebrate communities due to seasonality of the Ecuadorian Andes [38,39]. All macroinvertebrates were preserved in 95% ethanol and brought to the laboratory to be sorted and identified to the lowest taxonomical level possible using identification keys for South American macroinvertebrates [40,41].

GIS data and vegetation cover

The Geographic Information System software ArcGIS 9.3 was used to evaluate spatial digital information. Catchments were delineated from a Digital Elevation Model (DEM) with a10 m resolution, available at the Research Unit 816 (RU 816) database (www.tropicalmountainforest.org/) of the Deutsche Forschungsgemeinschaft (DFG). Stream networks were obtained from the national cartography created by the National Mapping Agency (IGM; Instituto Geográfico Militar).

Based on Allan [13], the influence of vegetation cover (VC) on stream condition was evaluated at three spatial scales: 1) catchment scale representing the entire area upstream of the sampling site; 2) riparian scale, using a buffer of 30 m on each bank along the entire length of the stream; and 3) local scale, comprised by the same buffer width along a 100 m reach upstream of the sampling site. Buffer width chosen was based on regulations by local governments (called Municipal Ordinances) in Ecuador. These ordinances reflect several law articles of the Ecuadorian Government in terms of water resources protection, but it does not specify or recommend an amount of riparian vegetation.

A 2010 land use map provided by M.F. Tapia–Armijos (pers. communication) was used to quantify VC within each catchment. This map was generated for southern Ecuador from ASTER images (http://asterweb.jpl.nasa.gov/) with a minimum mapping unit of 1 ha and classified into eight main land use types: páramo, forest, shrub, pasture, crop, plantation, urban and un–vegetated. In Table 1 we show the land use and percentage area covered by each land use type within the 23 studied catchments, with forest and pasture classes as the most dominating ones. Urban land use was absent or very poorly represented. For this study we aggregated this classification to two land cover categories, i.e. forested and non–forested. Forest class included all types of native vegetation and non–forest class all types of disturbed areas (pasture, cropland, plantation, urban, un–vegetated). The forest class was used to determine the VC only at catchment scale, because resolution and accuracy may limit the interpretation of riparian land cover in mountainous areas [13,25,42]. In order to address the riparian and local scales, the VC within the 30 m buffer was digitized and interpreted from 2011 rectified colored orthophotographs (0.30 m resolution). These orthophotographs were obtained and provided by the Ecuadorian project SIG-TIERRAS (Ministerio de Agricultura, Ganadería, Acuacultura y

Table 1. Land use characteristics and area (%) of each land use type of the 23 studied catchments.

Catchment	Páramo	Forest	Shrub	Pasture	Plantation	Crop	Urban	Un-vegetated
Zu1	0.8	69.4	13.4	13.6	0.3	2.3	0.0	0.2
Zu2	9.3	61.1	14.4	12.5	0.5	2.1	0.0	0.0
Zu3	1.4	59.4	10.7	25.0	0.7	2.6	0.0	0.1
Zu4	2.8	46.9	19.3	25.2	0.5	5.3	0.0	0.2
Zu5	4.4	54.9	27.9	10.4	0.9	1.4	0.1	0.1
Zu6	0.3	8.8	17.6	66.0	2.2	2.9	1.5	0.7
Zu7	0.0	4.6	13.2	71.1	0.8	9.0	0.0	1.2
Zu8	2.7	39.0	25.6	30.9	0.7	1.0	0.0	0.2
Zu9	0.5	41.0	12.4	35.9	1.5	8.3	0.2	0.1
Zu10	0.4	8.0	28.0	47.1	2.5	13.4	0.0	0.6
Zu11	0.4	33.7	9.1	42.0	2.7	11.9	0.0	0.2
Zu12	0.2	17.6	4.7	57.3	2.9	16.3	0.0	0.9
Zu13	0.4	16.0	19.2	48.9	1.2	13.5	0.0	0.7
Zu14	0.0	6.5	0.1	73.2	1.9	15.0	0.0	3.3
Zu15	0.0	51.9	9.9	32.9	1.1	4.0	0.0	0.2
Zu16	0.0	15.6	0.2	72.4	2.5	7.0	0.0	2.3
SF1	0.0	82.3	5.3	5.6	0.0	0.0	0.0	6.8
SF2	2.0	93.5	4.0	0.0	0.0	0.0	0.0	0.5
SF3	0.0	78.0	5.0	15.8	0.3	0.0	0.0	0.9
SF4	1.9	87.6	3.7	6.3	0.1	0.0	0.0	0.3
SF5	0.4	81.8	3.0	14.1	0.4	0.0	0.0	0.3
SF6	0.4	76.3	2.8	18.9	1.2	0.0	0.0	0.4
SF7	2.0	92.0	3.9	2.0	0.0	0.0	0.0	0.1

Pesca; Proyecto Sistema Nacional de Información y Gestión de Tierras Rurales e Infraestructura Tecnológica).

Data analysis

We calculated nine indices (also referred as metrics) to describe ecological conditions of the streams. We calculated three structural, two compositional and four water quality indices (Table 2) that allowed us assessing stream conditions in response to increasing VC. The structural indices were richness, Fisher's index, and evenness (Pielou's index). For compositional indices we classified macroinvertebrates by functional feeding groups (FFG) according to [43,44], but we only used percentage of scrappers and shredders because they have an expected decreasing response to perturbation [36], in our case deforestation.

Water quality was assessed with the Water Quality Index (WQI), the EPT index, the Biological Monitoring Working Party index adapted to Colombia (BMWP/Col), and the %5 Dominant taxa. WQI is a single score composed of several measurements of various physical, chemical and bacteriological parameters, where low WQI values show high influence of environmental stress [45]. EPT index shows the taxa richness within the macroinvertebrate groups which are considered sensitive to stressors, and therefore should increase with increasing water quality [46]. BMWP/Col index provides a single value at the family level, representative of the organism's tolerance to stressors, where high tolerance to pollution corresponds to low BMWP scores [40]. %5 Dominant taxa is a measure of the abundance of the five most dominant taxa to the total number of organisms in the sample at each site [36]. When a community is dominated by few taxa it will have high dominance values, thus indicating the community is under the influence of environmental stress.

Two statistical approaches have been used to evaluate relationships between vegetation cover/land use and ecological indices; these are regression [14,42,47] and ordination analyses [4,10,48]. Prior to analyses, macroinvertebrate abundance data (x) were transformed by $\log_{10}(x)$ to reduce the effect of large numbers in single taxa (e.g. *Baetodes*). Rare taxa, i.e. taxa represented by three or less individuals in a sampling site, were excluded in order to avoid the influence of these taxa in the ordination analysis [49]. Because explanatory variables did not have a uniform scale, stream parameters were $\log_{10}(x+1)$ transformed in order to avoid the problem with 0 values and reduce the range of data. All of these variables were analyzed by pairs in scatterplot matrices, and through an exploratory Redundancy Analysis (RDA) to identify the least correlated and most informative variables (PO_4^{3-}, DO, TC and turbidity).

An important issue was to identify whether ecological condition of the streams was attributable to a land use type and/or a stream parameter. We used a stepwise multiple linear regression (SMLR) with calculated macroinvertebrate metrics as dependent variables against land use and PO_4^{3-}, DO, TC and turbidity to determine which of those independent variables affect our response variables. SMLR was applied with the 'stepAIC' function of 'MASS' package in R [50] using backward elimination of variables and Akaike Information Criterion for variables or predictors selection. In addition, we estimated the relative importance of predictors based on their contribution to R^2 using boostrap ('relaimpo' package). All p-values were adjusted with a Bonferroni correction for multiple comparisons during regression analyses. These analyses indicated that forest class (i.e. VC) was consistently more informative to explain macroinvertebrate metrics than other land use types or stream parameter (see Table S2 for SMLR results). Therefore, to examine the linear dependency of each of our response variables with increasing VC, macroinvertebrate metrics (dependent) were regressed against percentages of VC at each spatial scale (independent) using the method of least mean squares ('lm' function of 'stats' package).

We used a RDA ('rda' function of 'vegan' package) to identify macroinvertebrate community assemblages associated with percentages of VC, PO_4^{3-}, DO, TC and turbidity. RDA evaluates multivariate response data and is a powerful tool of constrained ordination in community ecology [51]. RDA works on a matrix Y of response data (e.g. macroinvertebrate data) and a matrix X of explanatory variables (e.g. water chemistry, VC). It identifies groups and shows the vector fitting of variables used in the analysis. While the vector fitting is commonly used for environmental interpretation in ordination diagrams, this implies a linear relationship between ordination scores and the environment

Table 2. Results of regression analysis between calculated macroinvertebrate metrics and percentage of vegetation cover at multiple spatial scales.

Spatial scale	Catchment		Buffer		Local	
Metric	R^2	p	R^2	p	R^2	p
Structural						
Richness	0.30	<0.01	0.34	<0.001	0.10	0.13
Fisher's index	0.50	<0.001	0.38	<0.001	0.01	0.30
Evenness	0.24	<0.01	0.11	0.11	0.00	0.93
Compositional						
% Scrappers	0.02	0.57	0.01	0.72	0.02	0.55
% Shredders	0.42	<0.001	0.46	<0.001	0.12	0.09
Water Quality						
WQI†	0.56	<0.001	0.50	<0.001	0.12	0.11
EPT index	0.53	<0.001	0.59	<0.001	0.03	0.39
BMWP/Col	0.44	<0.001	0.46	<0.001	0.17	0.05
%5 Dominant taxa	0.85	<0.001	0.60	<0.001	0.02	0.50

†WQI is the unique metric derived from water chemistry data showed in Table S1.

variables and may not always be appropriate if it is desired to understand the ordination of studied sites or species in relation to one or more variables. Thus, we fitted the surface of the best predictor variable, identified with an ANOVA of the RDA results, to detect how macroinvertebrate communities are grouped and spread in relation to this variable. We examined whether macroinvertebrate communities were different between those groups using an Analysis of Similarities (ANOSIM; 'anosim' function of 'vegan' package). ANOSIM is a nonparametric procedure that provides a test of whether there is a significant difference between two or more groups of sampling units.

Results

Stream condition and vegetation cover

Relationships between VC and all metrics are generally stronger at catchment and buffer scale than at local scale where associations are not apparent (Table 2). Significant correlations are found on catchment and buffer scale in 90% and 80% of all cases. Within the structural metrics, only evenness does not show significant relationship with the percentage of VC at buffer scale ($R^2 = 0.11$; $p = 0.11$). In general, correlations decrease with smaller spatial scales, disappearing completely at local scale.

As for compositional metrics, only one FFG shows a strong relationship with VC at the two larger scales ($p<0.001$). The percentage of shredders is more related to VC at catchment and buffer scales (Table 2). Scrappers show weaker regressions at all spatial scales ($R^2<0.02$). At local scale no significant correlation existed at all.

All water quality metrics have stronger relationships at catchment and buffer scale than at local scale ($R^2>0.44$ and $p<0.001$, Table 2). Within water quality metrics, the %5 Dominant taxa depicts the highest correlation across all scales with 0.85 on catchment scale. Regression slopes indicate significant differences in water quality when VC increases at both spatial scales. However, there are some streams with a large area covered by forest at catchment and buffer scale which present low values in water quality indices (Figure 2B–C; F–G). For example, in both spatial scales the EPT index increase with VC ($R^2 = 0.53; p<0.001$ and $R^2 = 0.59; p<0.001$), but there is one site with approximately 70% of forest cover whose index value is less than 50, indicating fairly poor water quality. In addition, the strongest relationship is inversely correlated between the %5 Dominant taxa and the percentage of VC in the catchment ($R^2 = 0.85$; $p = 0.001$). As a general pattern most of the macroinvertebrate metrics indicate good water quality when the percentage of native VC is higher than 70% within the catchment and the riparian buffer (see Figure 2).

Macroinvertebrate communities and vegetation cover

All structural indices are related with VC at catchment scale rather than at other spatial scales (Table 2). Thus, we decided to further analyze the community assemblage patterns related to catchment forest cover. A correlation analysis run for those variables showed scatterplot matrices and correlation values from 0 to 1.We defined a decision threshold of 0.50 to remove those variables which are correlated with more than one variable. Remaining variables include VC, DO, PO_4^{3-}, TC and turbidity. RDA reveal that 51% of the variation in the data can be explained by these five selected variables and the ANOVA identified that catchment VC is the best predictor for macroinvertebrate composition in the ordination ($p<0.001$). In Figure 3A, most variables are spread along negative axis 1 and negative axis 2, with VC, PO_4^{3-} and turbidity being the most influential factors for axis 1, while DO and TC are more associated with axis 2. Figure 3B depicts the fitted surface of the best predictor variable (i.e. VC) to detect how macroinvertebrate communities are grouped and spread in relation to VC. Thus the ordination plot illustrates two visibly distinguishable groups; one group associated to catchments with a forest cover above 70% which is even more aggregated than the other group related to lower VC. The ANOSIM indicates that both groups of macroinvertebrate assemblages were strongly dissimilar between both ranges of VC ($R = 0.55$; $p = 0.001$). Both RDA plot and Bray–Curtis distance of the ANOSIM suggest that macroinvertebrate assemblages differ in taxonomical composition when there is a decline in the percentage of VC within the catchments.

Overall, macroinvertebrate communities in the 23 study streams present specific patterns. They are composed of a total of 13,512 individuals belonging to 11 orders, 33 families and 53 genera (data available at PANGAEA – Data Publisher for Earth & Environmental Science; doi:10.1594/PANGAEA.831687). Only two taxa (taxa–1 = Empididae; taxa–2 = Pyralidae) were not identified to genera level.

Comparing occurrence and abundance patterns between both groups of catchments (i.e. less than 70% and greater than 70% VC) we detected several singularities. In catchments with a VC between 70 to 100% we collected 4,061 individuals of 11 orders. In contrast 9,451 individuals of nine orders were found in catchments with less VC (Basommatophora and Lepidoptera were absent). We found 33 families, all of them present in catchments highly forested; whilst 10 families were absent in catchments with less than 70% forest cover (Calamoceratidae, Ceratopogonidae, Coenagrionidae, Dolichopodidae, Gyrinidae, Odontoceridae, Physidae, Psephenidae, Ptilodactylidae, and Pyralidae). With regard to the 53 genera found, only *Scirtes* (Coleptera) and *Molophilus* (Diptera) were absent in catchments above 70% forest cover, and 25 genera were absent in catchments with lower forest cover showing the greatest difference between both groups of catchments at this taxonomic level. Here genera absent at <70% catchment VC were: *Physa* (Basommatophora); *Noelmis*, *Phanacerus*, *Dineutus*, *Anchytarsus*, *Psephenops*, *Elodes* (Coleoptera); *Probezzia*, *Aphrosytus*, *Hexatoma* (Diptera); *Andesiops*, *Moribaetis*, *Haplohyphes*, *Tricorythodes*, *Farrodes* (Ephemeroptera); taxa–2 (Lepidoptera); *Argia* (Odonata); *Phylloicus*, *Leptonema*, *Macrostemum*, *Grumichella*, *Nectopsyche*, *Triplectes*, *Marilia*, *Cernotina* (Tricoptera).

A trend can be observed here as well, confirmed by the ANOSIM, as there are differences in macroinvertebrate compositions between both groups of catchments (Table 3). We detect an increase for almost all metrics of the macroinvertebrate communities with an increase of the percentage of VC in the catchment, except for evenness where no changes are observed.

Discussion

Stream condition and vegetation cover

Our results show a stronger relationship between VC and the macroinvertebrate metrics at catchment and buffer scales. This suggests an important linkage between headwater and riparian forests to conserve diversity and water quality of the streams in Ecuadorian Andean catchments. We observe that stream condition begins to improve when native VC within the catchment is over 70% (see Figure 2). Therefore our suggestion is to retain at least 70% native VC within the catchment in order to keep sufficient ecological condition of the streams. Such a percentage of VC can contribute significantly to riparian management for protection or restoration of watersheds where forest has been

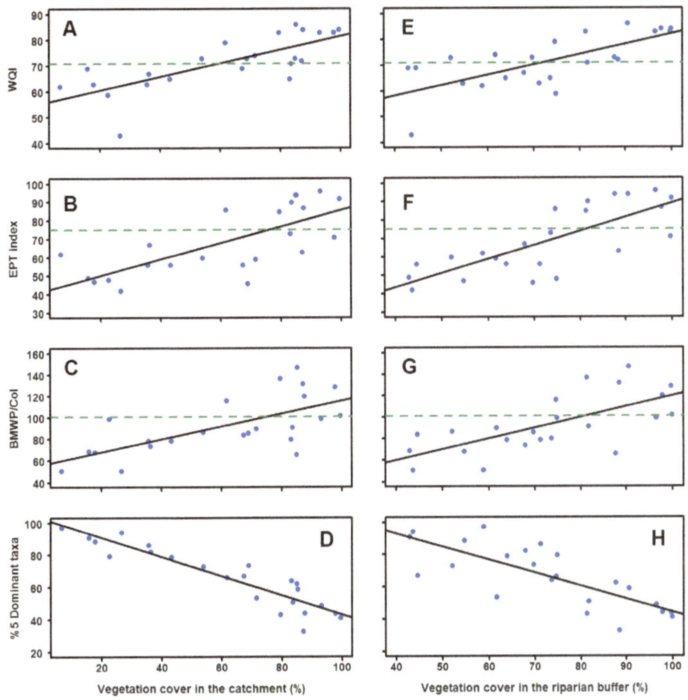

Figure 2. Water quality indices plotted against vegetation cover (%) at catchment and buffer scales. Horizontal green dashed lines represent thresholds of good water quality defined for each metric in the literature. Only the %5 Dominant index does not have ranges of tolerance, but high percentages indicate stress in the communities. For both spatial scales R^2 values are over 0.44 with a p–value <0.001, see Table 2 for specific values of each regression.

replaced by grasslands. Further, our findings show that local scale strategies (e.g. conserving small forest patches in riparian areas) do not improve water quality and are unlikely to improve the overall ecological condition of streams. According to [13,47], the maintenance of ecological conditions may require conservation of more than only a few native forest patches, and efforts should

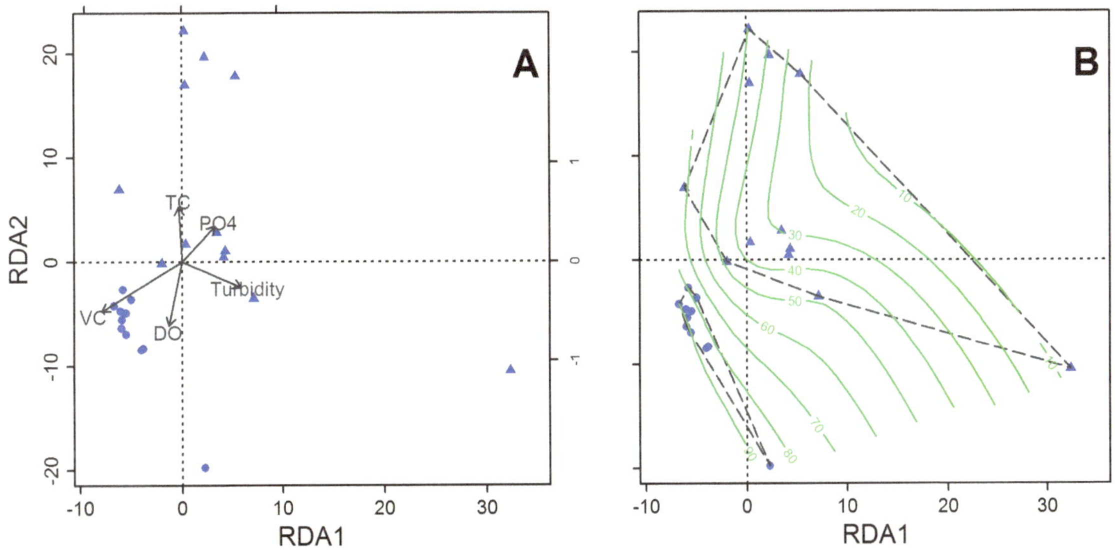

Figure 3. Redundancy Analysis (RDA) of macroinvertebrate community assemblages from catchments with different vegetation covers (VC). This ordination analysis detected two groups of macroinvertebrate assemblages: solid circles represent macroinvertebrate communities in catchments with 70–95% in VC and triangles indicate macroinvertebrate communities in catchments with less than 70% in VC. Each symbol represents the community of a single catchment, i.e. macroinvertebrate community at catchment SF3. A Shows the RDA ordination with vector fitting of five variables selected during analysis. DO = dissolved oxygen, TC = total coliforms, PO_4 = phosphate. B Displays in green curves the fitted surface only for the percentage of VC, which is the most explicative variable of the ordination (65%). Dashed lines enclose and separate the detected groups.

Table 3. Diversity patterns of macroinvertebrate communities in both groups of catchments identified in RDA ordination plot (Figure 3B). VC = catchment vegetation cover.

Vegetation cover	<70% VC		>70% VC		Response to increasing VC
Metric	**Mean**	**SD**	**Mean**	**SD**	
Structural					
Richness	14.3	2.1	20.1	6.2	Increase
Fisher's index	2.5	0.4	4.9	1.3	Increase
Evenness	0.7	0.1	0.7	0.1	Variable
Compositional					
% Scrappers	26.0	16.3	30.6	17.3	Increase
% Shredders	0.3	0.8	3.1	2.6	Increase

focus on considering larger spatial patterns that have more important consequences for stream conservation.

Macroinvertebrates diversity is more related to native VC at catchment and buffer scale than at a local scale. Many studies have documented the influence of upstream vegetation on stream ecological condition downstream, especially on water chemistry and habitat conditions [13,14,52]. For example, DO increases as the stream enters the forest [53]. Likewise, the similar relationships shown between water quality metrics and VC at catchment and buffer scale is not surprising given the strong longitudinal connectivity in a riverine landscape [13,54]. Riparian forests can influence water chemistry and habitat quality, which are directly linked to the macroinvertebrate metrics and help to minimize the effects of catchment deforestation on streams [4,42,55].

In this study, there were two particular sampling points in streams of catchments Zu4 and Zu8 (see Figure 1 and Table S1). Both had low values in water quality indices but a relatively high VC (60–68%). We assume that livestock and minor roads which are located parallel to the streams influence these metrics. For example, in [4] the authors found a stream affected by inputs of sediments and nutrients derived from roads and livestock, despite having a wide riparian buffer. We believe those elements are affecting the macroinvertebrate composition of very well forested catchments as discussed below.

Macroinvertebrate communities, indicators and vegetation cover

Analyzing the composition of macroinvertebrate communities of 23 Andean streams, we are able to identify direct relationships with the coverage of the catchment's native forest. The difference between macroinvertebrate assemblages increases when decreasing the percentage of native VC. We find a group with a more aggregated distribution when VC is large in the RDA plot. This pattern was also observed by [14] in a larger sample of streams in New Zealand. It supports the premise that deforestation is affecting the macroinvertebrate community assemblages in Ecuadorian Andean streams as in other Neotropical streams [4,21,56].

The difference between both groups of macroinvertebrate communities detected in the RDA is also revealed in the ANOSIM. Bücker et al. [10] found similar patterns in the San Francisco catchment having less variation at structural and compositional level and higher water quality metric values in forested than in deforested streams. Comparable results were also found in other Andean streams [11] and Amazon streams [21] of northern Ecuador. Nonetheless, among all metrics used here and mostly applied in similar studies, %5 Dominant taxa reveals a superior fitting to the regression slope despite it has not been used to reflect deforestation effects on streams before. In [36], authors denoted that high values of this metric indicate a community under environmental stress.

In terms of the FFG composition, we observe a direct response to increasing VC in both groups of macroinvertebrates (Table 3). In average, the abundance of shredders present in streams with > 70% VC in catchment is always higher than in less forested catchments. Shredders abundance is typically linked to leaf litter from riparian vegetation and it is sensitive to riparian zone impacts [43,57,58]. Shedders could be considered as an indicator of deforestation in Andean streams, despite proportion of shredders in macroinvertebrate communities in our catchments are always small (see Table 3). According to [4,5] this seems to be a general pattern in tropical streams. On average, scrapers abundance is slightly higher in more forested catchments in contrast to other Neotropical streams where their abundance is higher in less forested streams [4]. Additionally no association between scrappers and VC was detected.

Management and conservation

Allan et al. [13,47] have recommended that it is preferable to retain as much forest cover as possible in a catchment to preserve natural stream condition. However, in tropical regions like Ecuadorian Andes, deforestation has been increasing throughout the years affecting a large part of catchments along the Andes [29,59,60]. Deforestation has stopped in the upper Andean catchments because of topography and accessibility, preserving headwater forests and/or páramo [29]. However, this native vegetation in the headwaters is not enough to sustain healthy streams. In Andean countries, human activities are especially concentrated on midstream sections and valleys of catchments affecting water/habitat quality downstream and other rivers after the confluence. For example, within Guayas river basin in Ecuador there are several cities and in combination with intensive agricultural land use they are affecting the entire river network and its water quality [61]. Our results suggest that retaining over 70% native forest cover in Andean catchments now dominated by grasslands and keeping the riparian buffers, the water quality and stream biodiversity could be enhanced or at least preserved. However, this task is not so easy to achieve if there is not a strong regulation at national level.

The Ecuadorian constitution defines the right to water and promotes the protection, restoration and management of basins and water resources; but it does not mention how much forest needs to be retained in a catchment or a recommended width for riparian buffers. Best conservation actions that intend maintaining as much forest as possible have been focused on catchments for drinking water supply from initiatives of local governments (Municipalities) supported by non-governmental organizations, through the establishment of trust funds for catchment conservation. Also, local governments through Municipal Ordinances have decided to protect riparian strips with different buffer widths (15–30 m), depending if the stream is running in the city or outside the urban boundary. Riparian buffers could considerably reduce the effects of deforestation on benthic communities and water quality, but there is need for a proper management of riparian areas [4]. Unfortunately, Municipal Ordinances in Ecuador do not include the management of riparian buffers. Here, we assessed the effect of a 30 m buffer width on stream condition showing strong associations with the water quality metrics. Therefore, we also recommend protection and an effective management of native riparian forest buffers in Andean streams to assist headwater forests to sustain ecological condition of the streams.

We did not evaluate any management approaches of riparian areas, but we would like to recommend some strategies that governments can include in national regulations. We believe that the most important problems that managers will face in Andean streams are livestock density, residential areas and parallel roads along riparian zones which affect water quality and macroinvertebrate communities in forested catchments. Therefore, these problems must be addressed; for example, livestock exclusion jointly with a long-term research program to assess its effectiveness [62]; or an in-depth monitoring to further study road effects on stream and riparian vegetation [63]. Another important issue to be considered is the restoration of riparian vegetation, mainly because it enhances water quality, as our study indicates. Restoration positively affects habitat and macroinvertebrate community heterogeneity [48] and can be ecologically successful by using native plant species and avoiding the use of invasive plant species, which can affect structure and functionally of streams [64]. Finally, as mentioned by [4] we would like to emphasize that "riparian forest buffers are not a replacement for continuous forest, but can promote biodiversity and stream ecosystem integrity in tropical catchments affected by deforestation".

Supporting Information

Table S1 List of environmental variables considered in the 23 studied Andean streams. Data of water parameters are means obtained from several measurements in two sampling campaigns. Altitude was determined once at the sampling sites. Catchment area and percentage of vegetation cover (VC) at three spatial scales were calculated in ArcGIS 9.3 Geographic Information System software. T = Temperature, SC = Specific Conductance, DO = Dissolved Oxygen, NO_3^- = Nitrate, PO_4^{3-} = Phosphate, BOD_5 = Biochemical Oxygen Demand in 5 days. Turbidity in Nephelometric Turbidity Units or NTU and Total Coliforms (TC) as Colony Forming Units or CFU.

Table S2 Stepwise multiple linear regression analysis between macroinvertebrate metrics, land use and stream parameters for 23 Andean streams. Degrees of freedom for all regressions were 1, 20. AIC = Akaike Information Criterion used for selection of predictor variables. DO = Dissolved Oxygen, PO_4^{3-} = Phosphate, TC = Total Coliforms.

Acknowledgments

Special thanks are due to the personnel of the San Francisco Research Station, Nature and Culture International, Municipality of Loja and the Ministry of Environment for the provision of research facilities. Many thanks are due to C. Prieto to provide us additional information on stream parameters and macroinvertebrates of the Zamora catchment; and to L.M. Cun for his support with field sampling and stream measurements. The help of C. Romero, S. Aguilar and M. Villa with water analyses is appreciated. We would like to thank S. Bourret for proof reading, C. Espinosa for his assistance with R functions and routines, and P.F. Quintana-Ascencio and two anonymous reviewers for constructive comments on previous drafts of the manuscript.

Author Contributions

Conceived and designed the experiments: CIA HGF LB. Performed the experiments: CIA AL. Analyzed the data: CIA LB HH. Contributed reagents/materials/analysis tools: CIA HGF LB. Wrote the paper: CIA HH LB.

References

1. Laurance WF (1999) Reflections on the tropical deforestation crisis. Biol Conserv 91: 109–117.
2. Achard F, Eva HD, Stibig H-J, Mayaux P, Gallego J, et al. (2002) Determination of deforestation rates of the world's humid tropical forests. Science (80-) 297: 999–1002.
3. Lambin EF, Geist HJ, Lepers E (2003) Dynamics of l and-use and land -cover change in tropical regions. Annu Rev Environ Resour 28: 205–241.
4. Lorion CM, Kennedy BP (2009) Relationships between deforestation, riparian forest buffers and benthic macroinvertebrates in neotropical headwater streams. Freshw Biol 54: 165–180.
5. Wantzen KM, Wagner R (2006) Detritus processing by invertebrate shredders: a neotropical – temperate comparison. J North Am Benthol Soc 25: 216–232.
6. Iwata T, Nakano S, Inoue M (2003) Impacts of past riparian deforestation on stream communities in a tropical rain forest in Borneo. Ecol Appl 13: 461–473.
7. Ordóñez MV (2011) Influencia del uso del suelo y la cobertura vegetal natural en la integridad ecológica de los ríos altoandinos al noreste del Ecuador Universidad San Francisco de Quito.
8. Sierra R, Campos F, Chamberlin J (2002) Assessing biodiversity conservation priorities: ecosystem risk and representativeness in continental Ecuador. Landsc Urban Plan 59: 95–110.
9. Buytaert W, Cuesta-Camacho F, Tobón C (2011) Potential impacts of climate change on the environmental services of humid tropical alpine regions. Glob Ecol Biogeogr 20: 19–33.
10. Bücker A, Sondermann M, Frede H-G, Breuer L (2010) The influence of land-use on macroinvertebrate communities in montane tropical streams – a case study from Ecuador. Fundam Appl Limnol 177: 267–282.
11. Encalada AC, Calles J, Ferreira V, Canhoto CM, Graça MAS (2010) Riparian land use and the relationship between the benthos and litter decomposition in tropical montane streams. Freshw Biol 55: 1719–1733.
12. Crespo PJ, Feyen J, Buytaert W, Bücker A, Breuer L, et al. (2011) Identifying controls of the rainfall–runoff response of small catchments in the tropical Andes (Ecuador). J Hydrol 407: 164–174.
13. Allan JD (2004) Landscapes and riverscapes: the influence of land use on stream ecosystems. Annu Rev Ecol Evol Syst 35: 257–284.
14. Death RG, Collier KJ (2009) Measuring stream macroinvertebrate responses to gradients of vegetation cover: when is enough enough? Freshw Biol 55: 1447–1464.
15. Van Sickle J, Baker J, Herlihy A, Bayley P, Gregory S, et al. (2004) Projecting the biological condition of streams under alternative scenarios of human land use. Ecol Appl 14: 368–380.
16. Collier KJ, Hamer M (2012) The ecological condition of Waikato wadeable streams based on the Regional Ecological Monitoring of Streams (REMS) Programme. Waikato Regional Council, Technical Report, Waikato.
17. European Environment Agency (2012) European waters - assessment of status and pressures. EEA Report No 8. Copenhagen.
18. Environmental Protection Agency (2013) National Rivers and Streams Assessment 2008–2009: A Collaborative Survey. Draft February 28. Washington.
19. Perrie A, Morar S, Milne JR, Greenfield S (2012) River and stream water quality and ecology in the Wellington region: State and trends. Greater Wellington Regional Council, Publication No. GW/EMI-T-12/143, Wellington.
20. Parkyn SM, Davies-Colley RJ, Halliday NJ, Costley KJ, Croker GF (2003) Planted Riparian Buffer Zones in New Zealand: Do They Live Up to Expectations? Restor Ecol 11: 436–447.
21. Bojsen BH, Jacobsen D (2003) Effects of deforestation on macroinvertebrate diversity and assemblage structure in Ecuadorian Amazon streams. Arch für Hydrobiol 158: 317–342.

22. Miserendino ML, Casaux R, Archangelsky M, Di Prinzio CY, Brand C, et al. (2011) Assessing land-use effects on water quality, in-stream habitat, riparian ecosystems and biodiversity in Patagonian northwest streams. Sci Total Environ 409: 612–624.
23. Villamarín C, Rieradevall M, Paul MJ, Barbour MT, Prat N (2013) A tool to assess the ecological condition of tropical high Andean streams in Ecuador and Peru: The IMEERA index. Ecol Indic 29: 79–92.
24. Gergel SE, Turner MG, Miller JR, Melack JM, Stanley EH (2002) Landscape indicators of human impacts to riverine systems. Aquat Sci 64: 118–128.
25. Stewart JS, Wang L, Lyons J, Horwatich JA, Bannerman R (2001) Influences of watershed, riparian-cprridor and reach-scale characteristics on aquatic biota in agricultural watersheds. J Am Water Resour Assoc 37: 1475–1487.
26. Sutherland AB, Meyer JL, Gardiner EP (2002) Effects of land cover on sediment regime and fish assemblage structure in four southern Appalachian streams. Freshw Biol 47: 1791–1805.
27. Göttlicher D, Obregón A, Homeier J, Rollenbeck R, Nauss T, et al. (2009) Land-cover classification in the Andes of southern Ecuador using Landsat ETM+ data as a basis for SVAT modelling. Int J Remote Sens 30: 1867–1886.
28. Thies B, Meyer H, Nauss T, Bendix J (2014) Projecting land-use and land-cover changes in a tropical mountain forest of Southern Ecuador. J Land Use Sci 9: 1–33.
29. Torrachi S, Tapia MF, Escudero A, de la Cruz M (2013) Deforestación en una región montañosa megadiversa en los Andes: Dinámica del paisaje en el sur de Ecuador. In: de la Cruz M, Maestre FT, editors. Avances en el Análisis Espacial de Datos Ecológicos: Aspectos Metodológicos y Aplicados. Móstoles: ECESPA–Asociación Española de Ecología Terrestre. pp. 275–289.
30. Stein H, Springer M, Kohlmann B (2008) Comparison of two sampling methods for biomonitoring using aquatic macroinvertebrates in the Dos Novillos River, Costa Rica. Ecol Eng 34: 267–275.
31. De Cáceres M, Legendre P, Moretti M (2010) Improving indicator species analysis by combining groups of sites. Oikos 119: 1674–1684.
32. Sites RW, Willig MR, Linit MJ (2003) Macroecology of Aquatic Insects: A Quantitative Analysis of Taxonomic Richness and Composition in the Andes Mountains of Northern Ecuador. Biotropica 35: 226–239.
33. Jacobsen D (2004) Contrasting patterns in local and zonal family richness of stream invertebrates along an Andean altitudinal gradient. Freshw Biol 49: 1293–1305.
34. Tarras-Wahlberg N., Flachier A, Lane S., Sangfors O (2001) Environmental impacts and metal exposure of aquatic ecosystems in rivers contaminated by small scale gold mining: the Puyango River basin, southern Ecuador. Sci Total Environ 278: 239–261.
35. American Public Health Association (1995) Standard Methods for the Examination of Water and Wastewater. 19th ed. American Public Health Association, editor Washington DC.
36. Barbour MT, Gerritsen J, Snyder BD, Stribling JB (1999) Rapid Bioassessment Protocols for Use in Streams and Wadeable Rivers: Periphyton, Benthic Macroinvertebrates, and Fish. 2nd ed. Environmental Protection Agency.
37. Hauer FR, Resh VH (2007) Macroinvertebrates. In: Hauer FR, Lamberti GA, editors. Methods in Stream Ecology. San Diego CA: Academic Press. pp. 435–454.
38. Ríos-Touma B, Encalada AC, Prat Fornells N (2011) Macroinvertebrate Assemblages of an Andean High-Altitude Tropical Stream: The Importance of Season and Flow. Int Rev Hydrobiol 96: 667–685.
39. Jacobsen D, Encalada AC (1998) The Macroinvertebrate fauna of Ecuadorian high–land streams in the wet and dry season. Arch für Hydrobiol 142: 53–70.
40. Roldán Pérez G (2003) Bioindicación de la calidad del agua en Colombia. Uso del método BMWP/Col. 1st ed. Medellín: Editorial Universidad de Antioquia.
41. Fernádez HR, Domínguez E (2001) Guía para la determinación de los artrópodos bentónicos sudamericanos. 1st ed. Tucumán: Universidad Nacional de Tucumán.
42. Sponseller R, Benfield E, Valett H (2001) Relationships between land use, spatial scale and stream macroinvertebrate communities. Freshw Biol.
43. Cummins KW, Merritt RW, Andrade PCN (2005) The use of invertebrate functional groups to characterize ecosystem attributes in selected streams and rivers in south Brazil. Stud Neotrop Fauna Environ 40: 69–89.
44. Tomanova S, Goitia E, Helešic J (2006) Trophic levels and functional feeding groups of macroinvertebrates in neotropical streams. Hydrobiologia 556: 251–264.
45. Brown R (1970) A Water Quality Index – Do We Dare? Water Sew Work 11: 339–343.
46. Lenat DR (1988) Water quality assessment of streams using a qualitative collection method for benthic macroinvertebrates. J North Am Benthol Soc 7: 222–233.
47. Allan JD, Erickson DL, Fay J (1997) The influence of catchment land use on stream integrity across multiple spatial scales. Freshw Biol 37: 149–161.
48. Harding JS, Benfield EF, Bolstad P V, Helfman GS, Jones III EBD (1998) Stream biodiversity: The ghost of land use past. Proc Natl Acad Sci U S A 95: 14843–14847.
49. Legendre P, Gallagher E (2001) Ecologically meaningful transformations for ordination of species data. Oecologia 129: 271–280.
50. R Development Core Team (2011) R: A language and environment for statistical computing. R Foundation for Statistical Computing, Vienna. http://www.r-project.org/.
51. Bocard D, Gillet F, Legendre P (2011) Numerical Ecology with R. 1st ed. New York: Springer.
52. Harding JS, Claassen K, Evers N (2006) Can forest fragments reset physical and water quality conditions in agricultural catchments and act as refugia for forest stream invertebrates? Hydrobiologia 568: 391–402.
53. Suga CM, Tanaka MO (2012) Influence of a forest remnant on macroinvertebrate communities in a degraded tropical stream. Hydrobiologia 703: 203–213.
54. Shandas V, Alberti M (2009) Exploring the role of vegetation fragmentation on aquatic conditions: Linking upland with riparian areas in Puget Sound lowland streams. Landsc Urban Plan 90: 66–75.
55. Rios SL, Bailey RC (2006) Relationship between Riparian Vegetation and Stream Benthic Communities at Three Spatial Scales. Hydrobiologia 553: 153–160.
56. Couceiro SRM, Hamada N, Luz SLB, Forsberg BR, Pimentel TP (2006) Deforestation and sewage effects on aquatic macroinvertebrates in urban streams in Manaus, Amazonas, Brazil. Hydrobiologia 575: 271–284.
57. Lecerf A, Dobson M, Dang CK, Chauvet E (2005) Riparian plant species loss alters trophic dynamics in detritus-based stream ecosystems. Oecologia 146: 432–442.
58. Lecerf A, Richardson JS (2010) Litter decomposition can detect effects of high and moderate levels of forest disturbance on stream condition. For Ecol Manage 259: 2433–2443.
59. James LA, Marcus WA, Harden CP (2006) Human impacts on headwater fluvial systems in the northern and central Andes. Geomorphology 79: 249–263.
60. Molina A, Vanacker V, Balthazar V, Mora D, Govers G (2012) Complex land cover change, water and sediment yield in a degraded Andean environment. J Hydrol 472: 25–35.
61. Borbor-Córdova MJ, Boyer EW, McDowell WH, Hall CA (2006) Nitrogen and phosphorus budgets for a tropical watershed impacted by agricultural land use: Guayas, Ecuador. Biogeochemistry 79: 135–161.
62. Sarr D (2002) Riparian livestock exclosure research in the Western United States: A critique and some recommendations. Environ Manage 30: 516–526.
63. Jones JA, Swanson FJ, Wemple BC, Snyder KU (2000) Effects of roads on hydrology, geomorphology, and disturbance patches in stream networks. Conserv Biol 14: 76–85.
64. Hladyz S, Kajsa A, Giller PS, Woodward G (2011) Impacts of an aggressive riparian invader on community structure and ecosystem functioning in stream food webs. J Appl Ecol 48: 443–452.

9

Species Distribution Models and Impact Factor Growth in Environmental Journals: Methodological Fashion or the Attraction of Global Change Science

Lluís Brotons[1,2]*

1 CEMFOR-CTFC, Forest Sciences Center of Catalonia, Solsona, Catalonia, Spain, 2 CREAF, Centre for Ecological Research and Forestry Applications, Cerdanyola del Vallès, Catalonia, Spain

Abstract

In this work, I evaluate the impact of species distribution models (SDMs) on the current status of environmental and ecological journals by asking the question to which degree development of SDMs in the literature is related to recent changes in the impact factors of ecological journals. The hypothesis evaluated states that research fronts are likely to attract research attention and potentially drive citation patterns, with journals concentrating papers related to the research front receiving more attention and benefiting from faster increases in their impact on the ecological literature. My results indicate a positive relationship between the number of SDM related articles published in a journal and its impact factor (IF) growth during the period 2000–09. However, the percentage of SDM related papers in a journal was strongly and positively associated with the percentage of papers on climate change and statistical issues. The results support the hypothesis that global change science has been critical in the development of SDMs and that interest in climate change research in particular, rather than the usage of SDM per se, appears as an important factor behind journal IF increases in ecology and environmental sciences. Finally, our results on SDM application in global change science support the view that scientific interest rather than methodological fashion appears to be the major driver of research attraction in the scientific literature.

Editor: Diego Fontaneto, Consiglio Nazionale delle Ricerche (CNR), Italy

Funding: This study has received financial support from the projects CGL2011-29539/BOS and Consolider Montes CSD2008-00040 granted by the Spanish Ministry of Education and Science (MEC) and it is also a contribution to the FP7-226852 project SCALES. Additional funding to LB was received from the Catalan government through project 2010 BE 272. The funders had no role in study design, data collection and analysis, decision to publish, or preparation of the manuscript.

Competing Interests: The author has declared that no competing interests exist.

* Email: lluis.brotons@ctfc.cat

Introduction

Science is under continuous change and the appearance and development of new methodologies and approaches often has profound impact on the research panorama [1]. Species distribution models (SDM) exploring the association of environmental and species location data have rapidly developed over the last 15 years and appear to have had a great influence on environmental sciences and ecology in particular. SDM applications to climate change have been identified as the broadest research front in ecology and environment from Thomson ISI according to the clustering of the co-citing highly cited papers on this topic [2].

The popularity of SDM may be rooted in a range of different factors. Since understanding species distributions is a fundamental goal of ecology, the appearance of SDMs may have provided an efficient methodological approach to estimate species distributions and allowed the use of model outputs in a wide range of ecological applications (from species-energy relationships to niche conservationism [3]). The great availability of location data sets and environmental information in digital format (GIS) and the rapid development of statistical methods allowing efficient use of available information may have influenced the successful adoption of these techniques and their rapid spread in the ecological literature [3]. Being easy to implement using widely available GIS and distributional data coming from existing databases, SDMs may have benefited from a combination of fashion and ease of implementation. Alternatively, the popularity of SDMs may be related to the application of these techniques to expanding new ecological disciplines derived from an increasing interest in the effects of global change on biodiversity. SDMs allow a rapid estimation of the spatially explicit effects of drivers such as climate change on biodiversity at large spatial scales. Some seminal applications using SDMs have been instrumental in setting baselines of potential future impacts of climate change on a range of species [4].

In this work, I want to evaluate the impact of SDMs on the current status of ecological journals by asking the question to which degree of the use of SDMs in the literature is related to recent changes in the impact factors of ecological journals. The hypothesis evaluated derives from the idea that research fronts are likely to attract attention and drive research developments in a given discipline. Therefore, journals with a stronger focus on the research front should concentrate higher attention and receive more citations, thus benefiting from faster increases in their impact on the ecological literature. If this holds true, we predict that journals publishing more SDM-related articles should have benefited from the interest of this prolific field and show stronger increases in their citation rates and impact factors. However, if

SDM usage is related to increases in citations rates, two main mechanisms may be identified as potential explanations to the observed patterns. First, SDM-related articles may be associated with studies on global change impacts on biodiversity (climate changes, land use changes and the impact of invasive species), and therefore, one should expect that the number of papers on these topics and not on SDM per se should better explain journal citation patterns. Alternatively, SDM may have influenced journal citations rates through of their intrinsic attraction as methodological novelty allowing the easy estimation of species distributions. In this case, I expect the number of papers on SDM to be associated to changes in the journal impact factor independently of the range of global change topics included in environmental and ecology journals.

Methods

I used data from ISI web of science and test the prediction that the proportion of articles in a journal containing a larger number of SDMs related articles is related to the journal changes in impact factor during the period 2000–09. I used an objective method to select journals publishing a minimum number of articles related to SDM. This method included a general search for SDM related articles and the selection of a subset of articles included in non-multidisciplinary journals with more than 5 SDM articles published in the 2000–2009 period. Multidisciplinary journals were discarded because they included a much broader number of topics than thematic journals thus leading to potential biases in our bliblimetric estimators. First, I searched the ISI web of science for articles containing the words "predictive species distribution model" "niche model" or "habitat suitability model" or a combination of these [3]. I identified a total of 2.118 articles leading to a total of 37.854 citations. Second, I selected a subset of articles published in currently active, non-multidisciplinary journals (according to the ISI categories, Thompson Scientific) with more than 5 SDM articles published during the period 2000–09 from ISI categories accounting for at least 2% of the total SDM references. These articles accounted for 1305 of the articles above. Although this subset, which accounts for over 60% of the SDM related articles included in our search, may not represent a comprehensive compilation of articles in the literature dealing with SDMs, I believe that due to the wide range of journals included, is representative of their distribution in the ecology, environmental sciences and biodiversity journals panorama. For this subset of journals mostly within the three subject areas mentioned above, I estimated impact factor trajectories and compiled the number of articles published per year during the period 2000–09. With the information derived from the databases, I was able to derive for, each of the 56 journals selected (Table S1), a measure of SDM relevance, *SDMr*, as the proportion of SDM related articles from the total number of articles published by the journal during the study period (range 0.01 to 12%). For each of the journals in this subset, I also obtained the number of articles published on different topics related to global change by searching for different combination of key words ("climate change*" (1505 articles), "land use change* or fragmentation" (603 articles) and "invasive species*" (1128 articles) in biology and ecology ("biolog* and ecology*")) and calculated the proportion of articles for each topic in each journal (Table S1). Finally, I also used two additional different controls searches to account for general patterns in general ecological studies searching for the words "population and species" in biology and ecology ("biolog* and ecology*") (2511 articles) and methodological biases searching for the word "statistics" in biology and ecology ("biolog* and ecology*") (373 articles).

Impact factor (IF) is generally recognised as the primary measure of journal "quality" [6], but see [7]. Changes in the impact of the articles published in each journal (absolute increase) were quantified by calculating the slope of the regression of the journal's impact factor [5] and the respective year with positive slopes for journals with increasing impact factors and negative slopes for journals with decreasing impact. Finally, I also included, for each journal, the number of published articles during the study period and the year of the journal first issue (journal age) to account for general differences in article production and antiquity between journals [8]. I tested the role of journal descriptors on IF change and SDMr by means of linear models and forward variable selection using information theory based criteria (Bayesian Information Criteria, BIC) in R (package "MASS", [9]). Both variables, IF change and SDMr, were log transformed to ensure normality. To deal with collinearity problems, I also used an analytical method named hierarchical partitioning (HP hereafter). HP reduces collinearity problems by determining the independent contribution of each explanatory variable to the response variable (I) and separates it from the joint contribution (J), resulting from correlation with other variables (for a detailed explanation of how HP works, see [10]. This allows ranking the importance of the covariates in explaining the response variable independently of the others covariates. Given its usefulness for complementing multiple regression analysis, I applied HP using the "hier.part package" in R [11].

Results

The number of articles on SDM in the literature has rapidly increased during the period 2000–09 (Fig. 1) with the number of citations these papers are receiving also increasing rapidly with one third of the total amount of citations received in 2010. Ecology journals with a higher percentage of SDM related articles showed higher increases in their IFs between 2000 and 2009 (Fig. 2, Table S2). The total number of papers published by a journal, its IF at the beginning of the study period or the journal age were not significant factors behind changes in IF for the set of journals analysed and were thus discarded from the final model. The three ecology journals with values of SDMr larger than 5% (Diversity & Distributions, Global Ecology & Biogeography and Ecography) showed increases during the study period larger than 200% in their IF (Table S1). The relationship between changes in IF and SDMr was stronger and accounted for up to 16% of the variability when the journal Ecology Letters (experiencing a spectacular increase in IF during this period) was excluded from the analyses ($\beta = 11.05$, $t = 3.13$ $d.f. = 51$, $p<0.005$).

SDMr was highly predictable from the combination of topics analysed and contained in a given journal. SDMr was in particular strongly and positively related to the number of articles on global change topics (climate change, $\beta = 5.53$, $t = 4.74$, $d.f. = 51$, $p< 0.0001$, Table S2) and to the percentage of articles on statistics published by a journal ($\beta = 29.69$, $t = 4.62$, $d.f. = 51$, $p<0.0001$, Table S2). I also found a minor tendency for journals with lower IF in the year 2000 ($\beta = -0.00044$, $t = 2.44$, $d.f. = 51$, $p<0.05$) and a lower number of total articles published ($\beta = -0.0042$, $t = 2.13$, $d.f. = 51$, $p<0.0001$) to include a largest percentage of SDM related articles. The final model predicted 60% of variability in journal SDMr.

The effect of SDMr on IF change disappeared after the inclusion of variables accounting for the thematic scope of the journals. IF change was strongly and positively related to the

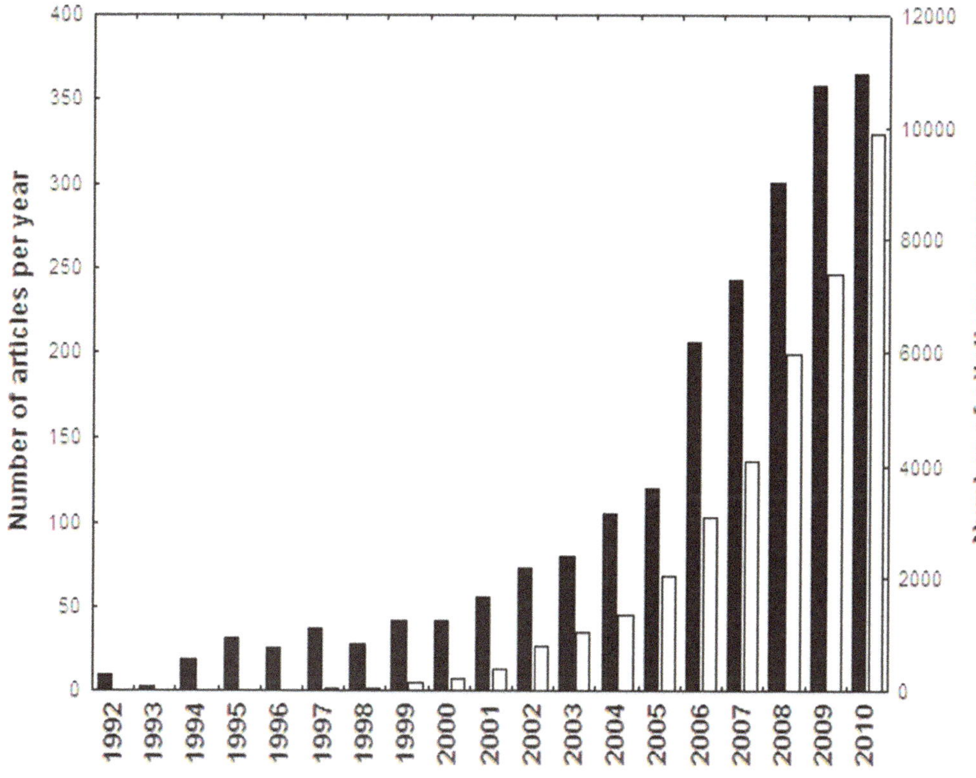

Figure 1. Number of published SDM related articles in the ecological literature (black bars) and number of citations received by these articles per year (white bars) during the period 1992–2010.

number of climate change papers published (CLIr) in a journal during the study period (Fig. 2, Table S2). In fact, CLIr showed a much higher both, independent and joint explanatory power than any other variable included in the assessments, suggesting that it is a much stronger candidate to drive IF changes than SDMr or the others bibliometric descriptors used (Table S2).

Discussion

My results indicate a positive relationship between the numbers of SDM related articles published in a journal and its IF increases during the period 2000–09. However, given the strong association between SDMr and the number of global change articles published in a journal, the role of SDM on IFs is likely to be an indirect effect of the increases in the journal IF being associated with a larger number of climate change articles published. The results support the hypothesis that global change science has been critical in the development of SDM and that climate change research in particular appears as an important factor behind increases of IF in ecological and environmental journals devoting larger attention to this topic.

Recent studies have found a positive trend in the number of articles cited by ecological journals in recent years leading to a potential for general increase in citation rates and thus impact factors [12,13]. However, it seems that increases in IF are not evenly distributed with some journals getting a disproportionally larger share of the IF growth leading to changes in the potential impact of these journals [13,14]. Our results indicate that journals with an overall higher percentage of SDMr, but specially those with more articles on climate change topics have grown at relatively faster rates than others [15]. However, SDM use in the journals included in the analyses did not directly drive IF increases in the set of journals analysed. The application of SDMs has been described as a one of the biggest emerging fronts in ecology in the last years [2], with a large number of highly impacting articles, and appears to be rapidly growing. SDM articles often use already existing and readily available environmental data and therefore are less constrained than more traditional ecology works based on field data thus opening the way to faster and more widespread publication on different issues of general interest to ecologists. Furthermore, available software makes SDM applications very easy and potentially articles using these methodologies may be easier to write than in other ecology areas. Other studies have indeed described increases in citations rates of ecological journals associated to the number of articles published on SDMs (i.e. invasion biology, [16]). However, our results do not support the view that journals publishing more SDM related articles receive more citations per se.

Rather, the role of SDM on IF trends disappeared when climate change was included in the analyses. This indicates that journals with a larger numbers of climate change related papers have indeed grown larger IFs in the 10-year period of our study. Journals with a higher proportion of SDMs also published more articles on hotter topics such as climate change or invasion biology [17] suggesting that the increase in impact factor is not a direct consequence of the number of SDM articles published. Overall, these results show that specific topics disproportionally drive changes in research attention and appeared to influence journal citation patterns [17]. The finding that climate change research contributes to the variability in recent IF increases of environ-

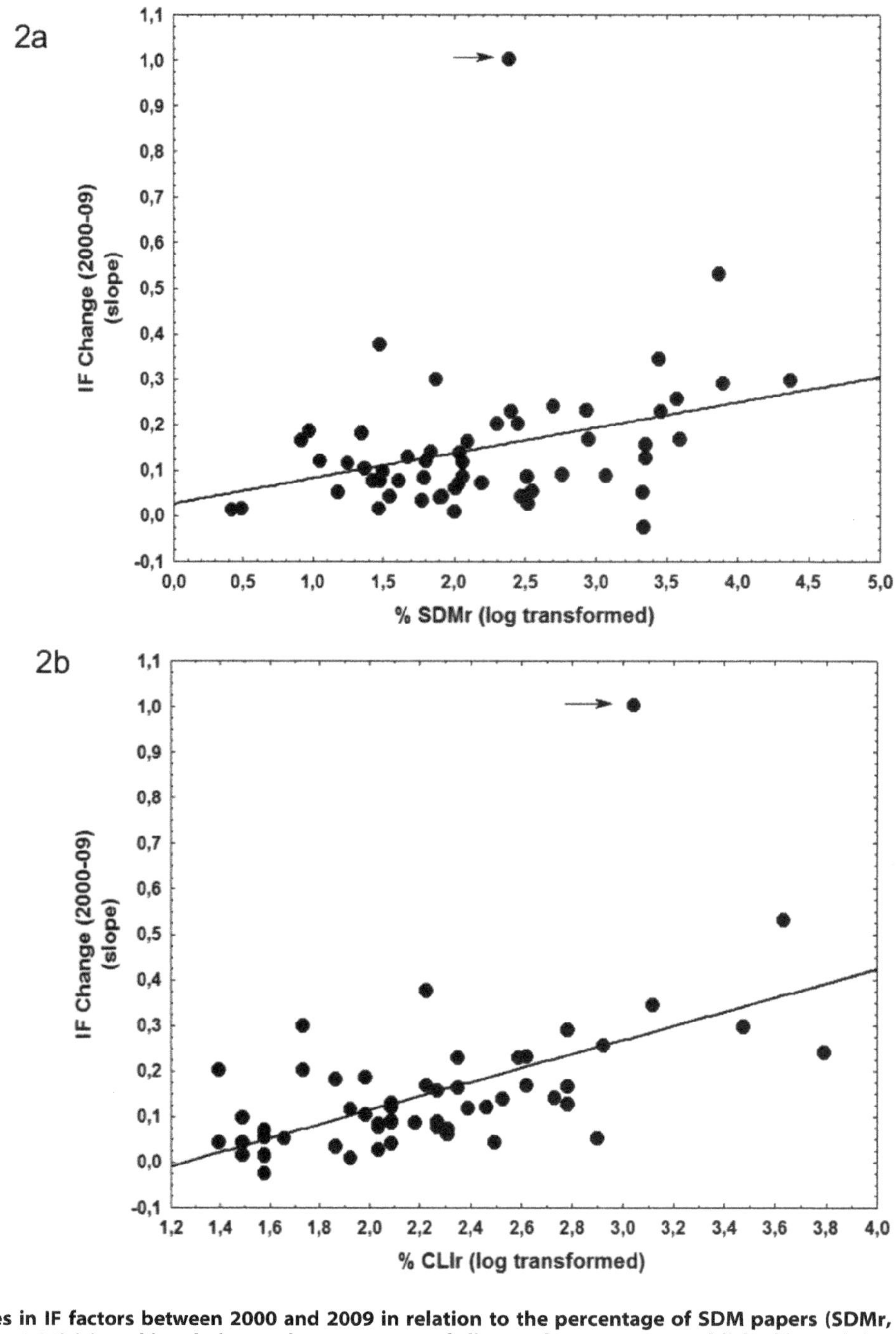

Figure 2. Increases in IF factors between 2000 and 2009 in relation to the percentage of SDM papers (SDMr, $R^2 = 0.12$, $\beta = 10.68$, $t = 2.74$, $d.f. = 54$, $p < 0.01$) (a), and in relation to the percentage of climate change papers published in each journal (CLIr, $R^2 = 0.31$, $\beta = 17.82$, $t = 4.98$, $d.f. = 54$, $p < 0.001$) (b). The arrow identifies the journal Ecology Letters.

mental and ecology journals supports the view that scientific interest and not methodological fashion appears to be a major driver of research attraction [18].

SDM usage appeared therefore related to changes in journal IF most likely because their development has been largely driven by applications in climate change science. SDM development has been instrumental in moving global change science forward due to the capability of the models to be used for a large number of species over large spatial scales [4,16]. The real impact of SDM in ecology may be therefore be better interpreted as one of the foundation stones of global change science applications in ecology. SDMs have allowed the environmental research community to efficiently integrate the extensive availability of large-scale biological data, appropriate tools and environmental data sets into the growing needs of spatially explicit biodiversity assessments. SDMs may continue to play a significant role in the future panorama of ecology and environmental sciences as long as they remain as key methodological approaches in global change science. Spatial models allowing the projection of species distributions to future environmental conditions such as climate change are still required and tend to progressively become more complex to overcome the limitations of the correlative nature of

SDMs [19]. However, I think that the challenges faced by model building in global change science will require flexible, integrative approaches allowing the use of extensively available data, and SDMs are likely to continue playing a significant role in this context.

Acknowledgments

I thank Jane Elith and Miguel Clavero for fruitful discussions on impact factors in ecology.

Author Contributions

Conceived and designed the experiments: LB. Performed the experiments: LB. Analyzed the data: LB. Contributed reagents/materials/analysis tools: LB. Wrote the paper: LB.

References

1. Neff MW, Corley E (2009) 35 years and 160,000 articles: A bibliometric exploration of the evolution of ecology. Scientometrics 80: 657–682.
2. Institute for Scientific Information® (2005) Thomson ISI Essential Science IndicatorsSM Special Topics: Research Front Methodology [online]. Philadelphia: ISI. (http://www.esi-topics.com/RFmethodology.html).
3. Elith J, Leathwick J (2009) Species distribution models: ecological explanation and prediction across space and time. Ann Rev Ecol Sys 40: 677–697.
4. Thomas C, Cameron A, Green RE, Bakkenes M, Beaumont LJ, et al. (2004) Extinction risk from climate change. Nature 427: 145–148.
5. Garfield E (2006) The history and meaning of the journal impact factor. J Am Med Asso 295: 90–93.
6. Olden JD (2007) How do ecological journals stack-up? Ranking of scientific quality according to the h index. Ecoscience 14: 370–376.
7. Leimu R, Koricheva J (2005) What determines the citation frequency of ecological papers. TREE 20:28:32.
8. Wilson AE (2007) Journal Impact Factors are inflated. Bioscience 57: 550–551.
9. Venables WN, Ripley BD (2002) Modern Applied Statistics with S. Fourth edition. Springer.
10. Mac Nally R (2002) Multiple regression and inference in ecology and conservation biology: further comments on identifying important predictor variables. Biodivers Conserv 11: 1397–1401.
11. Walsh C, Mac Nally R (2005) "The hier.part Package" version 1.0-1. Hierarchical Partitioning. Documentation for R: A language and environment for statistical computing. R Foundation for statistical Computing, Vienna, Austria
12. Neff BD, Olden JD (2010) Not so fast: Inflation in Impact factors contributes to apparent improvents in journal quality. Bioscience 60: 455–459.
13. Althouse BM, West JD, Bergstrom CR, Bergstrom T (2009) Differences in impact factor across fields and over time. J Am Soc Inf Sci Tech 60: 27–34 http://dx.doi.org/10.1002/asi.20936.
14. Groesser SN (2012) Dynamics of journal impact factors. Sys Res Behav Sci 29: 624–644
15. Ioannidis JPA (2006) Concentration of the most-cited papers in the scientific literature: analyses of journal ecosystems. PLOS ONE 1: art. no.-e5.
16. Barbosa FG, Schneck F, Melo AS (2012) Use of ecological niche models to predict the distribution of invasive species: a scientometric analysis. Braz J Biol 72: 821–829.
17. Pysek P, Richardson DM, Vojtech J (2006) Who cites who in the invasion zoo: insights from an analysis of the most highly cited papers in invasion ecology. Preslia 78: 437–468
18. Lortie CJ, Aarssen LW, Budden AE, Leimu R (2013) Do citations and impact factors relate to the real numbers in publications? A case study of citation rates, impact, and effect sizes in ecology and evolutionary biology. Scientometrics 94: 675–682
19. Dormann CF, Schymanski SJ, Cabral J, Chuine I, Graham C, et al. (2012). Correlation and process in species distribution models: bridging a dichotomy. J. Biogeogr, 39: 2119–2131.

10

Landscape Genetics for the Empirical Assessment of Resistance Surfaces: The European Pine Marten (*Martes martes*) as a Target-Species of a Regional Ecological Network

Aritz Ruiz-González[1,2,3]*, Mikel Gurrutxaga[2,4], Samuel A. Cushman[5], María José Madeira[1,2], Ettore Randi[3,6], Benjamin J. Gómez-Moliner[1,2]

1 Department of Zoology and Animal Cell Biology, University of the Basque Country, UPV/EHU, Vitoria-Gasteiz, Spain, **2** Systematics, Biogeography and Population Dynamics Research Group, Lascaray Research Center, University of the Basque Country, UPV/EHU, Vitoria-Gasteiz, Spain, **3** Conservation Genetics Laboratory, National Institute for Environmental Protection and Research, ISPRA, Ozzano dell'Emilia, Bologna, Italy, **4** Department of Geography, University of the Basque Country, UPV/EHU, Vitoria-Gasteiz, Spain, **5** U.S. Forest Service, Rocky Mountain Research Station, Flagstaff, AZ, United States of America, **6** Department 18/Section of Environmental Engineering, Aalborg University, Aalborg, Denmark

Abstract

Coherent ecological networks (EN) composed of core areas linked by ecological corridors are being developed worldwide with the goal of promoting landscape connectivity and biodiversity conservation. However, empirical assessment of the performance of EN designs is critical to evaluate the utility of these networks to mitigate effects of habitat loss and fragmentation. Landscape genetics provides a particularly valuable framework to address the question of functional connectivity by providing a direct means to investigate the effects of landscape structure on gene flow. The goals of this study are (1) to evaluate the landscape features that drive gene flow of an EN target species (European pine marten), and (2) evaluate the optimality of a regional EN design in providing connectivity for this species within the Basque Country (North Spain). Using partial Mantel tests in a reciprocal causal modeling framework we competed 59 alternative models, including isolation by distance and the regional EN. Our analysis indicated that the regional EN was among the most supported resistance models for the pine marten, but was not the best supported model. Gene flow of pine marten in northern Spain is facilitated by natural vegetation, and is resisted by anthropogenic landcover types and roads. Our results suggest that the regional EN design being implemented in the Basque Country will effectively facilitate gene flow of forest dwelling species at regional scale.

Editor: Jesus E. Maldonado, Smithsonian Conservation Biology Institute, United States of America

Funding: This study has been funded by the Basque Government through the Research group on "Systematics, Biogeography and Population Dynamics" (Ref. IT317-10; GIC10/76; IT575/13) and by the University of the Basque Country (UPV-EHU) and the Department of Environment, Territorial Planning, Agriculture and Fisheries (Basque Government) through IKT S.A under the University-Enterprise research program (Ref. UE07/02). Ruiz-González holds a Post doc fellowship awarded by the Dept. of Education Universities and Research of the Basque Government (Ref. DKR-2012-64). Several samples analysed in this study have been obtained in the framework of different carnivore surveys funded by regional or national administrations (Spanish Ministry of Environment, Regional Governments of Navarre and Aragon, Alava Provincial Council). The funders had no role in study design, data collection and analysis, decision to publish, or preparation of the manuscript.

Competing Interests: The authors have declared that no competing interests exist.

* Email: aritz.ruiz@ehu.es

Introduction

Long-term biodiversity conservation requires the preservation of ecological and evolutionary processes, such as gene flow, dispersal movements and population range shifts [1]. The ability of individuals to move across changing landscapes is crucial for maintaining regional populations [2,3]. The preservation of these processes requires, in turn, that landscape connectivity be preserved, especially when we take into account the synergetic effects of habitat fragmentation and climate change [1]. Landscape connectivity is defined as the degree to which landscape facilitates or impedes movement of organisms among resource patches [4]. Connectivity is species-specific and reflects the response of individuals to landscape features and the patterns of dispersal and gene flow that result from these individual responses [5]. Thus, landscape connectivity depends to a large extent on how the spatial configuration of habitat and land use interact with the movement ecology of particular species [6].

Ecological networks have been promoted as coherent systems composed of core areas linked by ecological corridors capable of facilitating the dispersal, migration and gene flow of wild species in landscapes and regions [7–9]. They are configured and managed with the objective of maintaining ecological functions and conserving biodiversity [7]. Although the development of ecological networks is based on the precautionary principle and on ecological theory [8], the absence of empirical evidence regarding their effectiveness and the difficulty in obtaining this evidence has been a focus of criticism about the extent to which they have in fact ensured landscape connectivity and increased biodiversity conservation [10,11].

In the design of ecological networks there is a need to predict regional ecological corridors and to quantify the degree of expected landscape connectivity between specific areas [3,9,11–13]. 'Least-cost modeling' is one commonly employed approach for designing ecological corridors [9,14], in which resistance values are assigned to distinct habitat or land use types and the least-cost paths (LCP) between specific locations are calculated using a geographical information system (GIS). How landscape influences effective distances between locations is calculated as the accumulated cost through the least cost paths [14,15]. However, for most organisms, setting the resistance values is a difficult process in which expert judgment and data available in the literature play an important role [16–19].

Accurate identification of the potential factors that drive gene flow in heterogenous landscapes and the scales at which they are acting is a foundation of reliable mapping of corridors [9,18]. Thus, reliable development of corridors must be based on a correct representation of the local resistance relative to the movement ecology of the organism of focus [9,18]. Landscape genetics, a research area that integrates landscape ecology, population genetics and spatial statistics, provides a valuable framework for testing the influence of landscape structure and composition on dispersal and gene flow [20,21]. It facilitates quantification of the resistance to gene flow a given landscape element poses [12,22]. Thus, one of the principal applications of landscape genetics in landscape planning and conservation biology is to empirically test and optimize resistance maps [23–26]. This facilitates the optimal design of ecological corridors [3,16,23], the detection of barriers to gene flow [27–29] and the identification of the landscape features which favour or impede dispersal [30–35].

Landscape genetics has shifted towards individual-based sampling and analysis, especially when organisms are continuously distributed [12,22]. However, sufficient sample collection for this purpose is a difficult task, especially in rare and elusive species in which sampling is a limiting factor [36]. In this context, non-invasive genetic sampling allows us to address studies of wildlife species without the need to capture or even observe them [37–40].

In 2005 a regional ecological network was established in the Basque Country (North Spain) by delimiting the ecological corridors linking forest protected areas [41]. A functional group of forest mammal species was selected to guide the development of a generic resistance map, which would, in turn, serve as a basis least-cost modeling of the network of ecological corridors linking these core areas. These mammals were considered suitable target species due to their sensitivity to recent fragmentation and homogenization dynamics in the regional landscape, such as road construction, urbanization and agrarian intensification [41,42]. The resistance map was parameterized through bibliographical review and expert opinion and was based on the assignment of different resistance levels to each land use [41]. The regional government of the Basque country incorporated that coherent ecological network as a reference for the environmental assessment of plans, programs and projects in 2005 [41]. In addition to its intrinsic internal relevance, the Basque country has been chosen for its crucial role in the regulation of biotic flows in south-western Europe [43]. This is because of its strategic location between two important biodiversity reservoirs in south-western Europe, the mountain chains of the Pyrenees and the Cantabrian Range [43–45]. Consequently, the preservation and restoration of connectivity in this transitional area between mountain ranges requires reliable knowledge about ecological responses of organisms to landscape composition and structure [45].

Among the set of functional forest mammals used in the design of the coherent regional ecological network, the European pine marten (*Martes martes*) is the most forest dependent species [46]. The pine marten is generally associated with forest habitats, mainly mature forests [46–48]. Deforestation and forest fragmentation limit the distribution and density of pine martens [48–50], which are believed to need a minimum woodland area to survive (ca. $2km^2$) [47] and tend to avoid treeless areas [48,51,52]. Their occurrence patterns are affected by forest patch size, percentage of woodland cover, food abundance, sex, age class and habitat fragmentation levels [47,51]. Given their strong associations with high forest structural complexity, the species is particularly sensitive to human influences on their habitats, including habitat loss and landscape-scale effects of habitat fragmentation [48,51,53]. Nonetheless, they have also been recently reported in fragmented landscapes characterized by isolated, small forest fragments within an agricultural landscape matrix [48,51,54], suggesting they are not as obligately interior-forest dependent as previously described [51]. However, in such landscapes, linear features, such as hedgerows and small woods, play a key role to connect adjacent forest patches [48,54,55].

Consequently, the pine marten is a species which is well suited to studies focused on the effects of forest fragmentation on genetic structure and gene flow [56]. However, whether habitat characteristics that predict marten occupancy act as barriers to dispersal, influencing gene flow and population genetic structure across the landscape, is largely unknown [56].

The main objective of this research is to evaluate a large suite of alternative resistance hypotheses for the pine marten and compare the most supported empirical model with the expert-derived landscape resistance model used to parameterize the corridor network for the Basque Country. Specifically, we aim to evaluate (1) different binary landscape resistance maps which cover a gradient from greater to lesser preference of the pine marten for forest environments in order to identify which land uses favour or impede genetic interchange in the study area; and secondly (2) whether or not the resistance map with which the regional ecological network was originally designed in the Basque Country was correctly parametrized to reflect European pine marten gene flow.

If there is no effect of landscape structure on dispersal and gene flow in martens, then we expected either: (a) panmixia, where there is no genetic pattern, or (b) isolation-by-distance, where genetic differences increase with geographic distance [57]. If landscape structure influences marten dispersal, then we expected (c) isolation by landscape resistance [30]. Given that the most consistent marten-habitat relation appears to be a general association with forest habitats, and avoidance of open, non-forested habitats [47,48,51,52], we expected that open and human altered landscapes would act as a barrier for martens, and hence that landscape structure would have an effect on gene flow. In addition, we hypothesize that the intervening landscape features between forest patches (i.e., matrix) could also play a key role to substantially affect pine marten dispersal, and consequently the connectivity between forest environments [48,55,58].

Methods

Study area and spatial data

The region of the Basque Country is located in the northern Iberian Peninsula (Fig. 1) within the Atlantic and Mediterranean biogeographical regions. It comprises an area of 7,235 km^2 and has an average human population density of 298 inhabitants per square kilometer. Forests cover 28%, forestry plantations 29%,

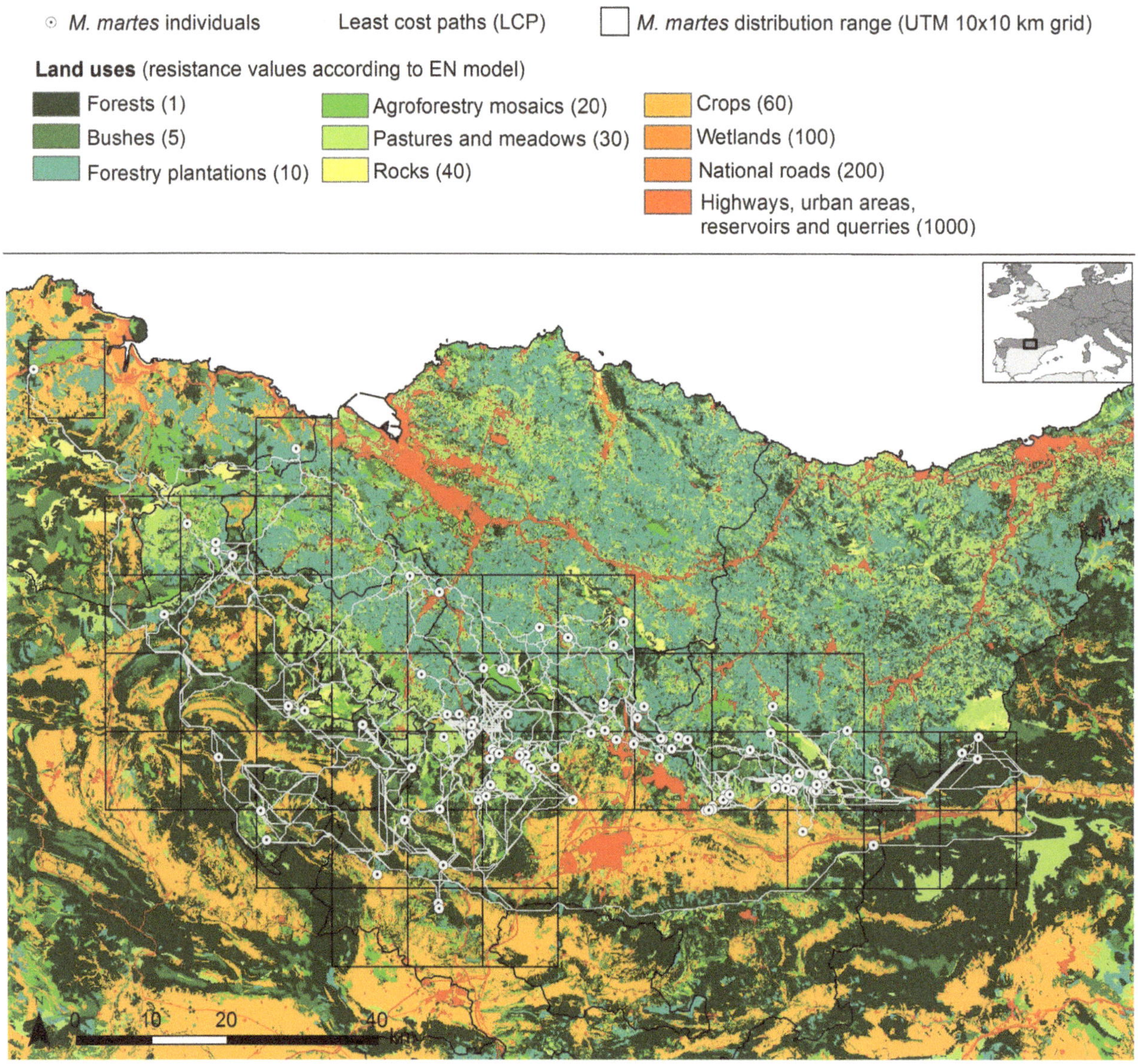

Figure 1. Ecological network resistance map (EN) and LCP analysis between European pine marten individuals in the study area. Least cost paths (LCP) obtained between the 101 pine marten individuals in accordance with the EN resistance map, analogous to that used in the design of the corridors in the ecological network of the Basque Country (North Spain) [41]. Resistance values for each land use are indicated in brackets.

non-wooded mountains 24%, cultivated land 14%, and urban land and infrastructures 5.7% of the land area, respectively.

Land use information was obtained in vector format from the most recent forest map of Spain [59] and from national road network maps [60].

Non-invasive genetic sampling and species identification

We used non-invasive scat sampling to collect genetic samples from the *Martes* sp. (*Martes martes* and *Martes foina*) in the study area between 2004 and 2010. Thus, no specific permissions were required for faecal sampling purposes, as the sampling was carried out without needing to intervene directly in the species in focus. We conducted a multi-stage sampling scheme, in which samples from a pilot study were used to assess the appropriateness of the sampling with respect to the research questions. Thus, two scat-based surveys were conducted between 2004 and 2010 across the sympatric range of both species in the study area. The first survey, conducted in 2004–2005, was used to initially estimate the distribution range of the two sympatric species of the genus *Martes* in the study area and isolate genetic samples of the focal species (*M. martes*). The second, conducted in 2006–2010, was used to refine species distribution information and to obtain a higher number of *M. martes* samples for microsatellite genotyping after a genetic species identification process [52] Aiming to homogenously cover the wide study area and obtain the highest number of different individuals we prioritized our sampling to faecal samples that were separated a minimum of 1km apart (i.e. potentially avoiding re-samplings of the same individual). We also prioritized fresh scat samples to increase genotyping success [61]. Additionally, fresh tissue specimens from road-killed pine martens were included in the data base, when possible. Tissue specimens were collected by authorized veterinarian personnel of the Wildlife

Rehabilitation Centre of Martioda (Alava Regional Council. Department of Environment. Biodiversity section), in line with the laws and ethical protocols governing wildlife management (Law 42/2007) and were submitted to Department of Zoology and Animal Cell Biology (UPV/EHU) for further DNA analyses. No animals were sacrificed for the only purposes of this study. Therefore, a formal approval by an Institutional Animal Care and Use Committee was not necessary. Universal Transversal Mercator (UTM) coordinates were recorded for all the samples collected using a global positioning system (Garmin eTtrex) [61]. The faecal samples were stored in autoclaved tubes containing ethanol 96% and frozen at $-20°C$ until processed [52]. DNA was isolated from tissues and scat using the Qiagen DNeasy Tissue DNA (Qiagen, Hombrechtikon, Switzerland) and DNA Stool MiniKit (Qiagen, Hombrechtikon, Switzerland) according to the manufacturer's instructions, respectively. As pine marten faeces cannot be distinguished from those of the sympatric stone marten (*M. foina*), which is widespread in the study area, and can also be easily confused with those of other carnivores [62], a molecular technique was applied for the identification of faecal samples. Species identification was accomplished by a polymerase chain reaction – restriction fragment length polymorphism (PCR-RFLP) method, providing for an effective genetic identification of sympatric marten species following the method described in Ruiz-González et al. [52].

Microsatellite analyses and individual identification

Identification of individual pine martens used nuclear DNA following methods in Ruiz-González et al. [61]. All the faecal samples identified by the PCR-RFLP method [52] as pine marten were genotyped at 15 variable microsatellite loci (Table S1) using a multiplex protocol specifically designed for degraded faecal DNA analysis [61] and following a modified multitube-approach [63]. The multitube-approach of 4 independent replicates followed by a stringent criteria to construct consensus genotype (i.e. accepting heterozygotes if the two alleles were seen at least in two replicates and homozygotes if a single allele was seen at least in three replicates) is a commonly used approach in non invasive genetic studies leading to a low probability of retaining a false homozygote or false allele error (e.g. [64–66]). Briefly, DNA quality was initially screened by PCR-amplifying each DNA sample four times at four loci (Multiplex 1: MP0188; MP0059; Gg-7; Ma-1), since the results obtained for this four loci are indicative of the genotyping success for the full panel of 15 microsatellites [61].

Only samples showing> 50% positive PCRs were further amplified four times at the remaining 11 loci. Samples with ambiguous results after four amplifications per locus or with < 50% successful amplifications across loci were removed from further analysis as they were not considered reliable genotypes. Multiplex PCR products were run on an ABI (Foster City, CA) 3130XL automated sequencer (Applied Biosystems), with the internal size standard GS500 LIZ (Applied Biosystems). Fragment analyses were conducted using the ABI software Genemapper 4.0.

RELIOTYPE software [67] was used to assess genotype reliability obtained by 4 independent replicates. Samples that were not reliably typed at all loci after 4 replicates (at score threshold $R = 0.95$) were discarded from the analysis. GIMLET software v 1.3.4 [68] was used to calculate the probabilities of identity (PID and PID-sibs) so as to quantify the efficacy in discriminating the fifteen loci in combination. Consensus genotypes from four replicates were reconstructed using GIMLET, accepting heterozygotes if the two alleles were seen at least in two replicates and homozygotes if a single allele was seen at least in three replicates (e.g. [64–66]). GIMLET was also used to estimate genotyping errors: allelic dropout (ADO) and false alleles (FA) [63,69].

The raw microsatellite data and geographic coordinates of the 101 pine marten individuals are included in Table S2.

Genetic diversity and pairwise individual genetic distances

We summarized genetic variation through the number of alleles per locus (A), expected (HE) and observed (HO) heterozygosities using GENETIX v 4.05.2 [70]. Estimates of pairwise linkage disequilibria for each pair of loci and deviation from Hardy Weinberg equilibrium (HWE) genotypic proportions at each locus were tested using the exact test implemented in GenePop version 4.0 [71]. Statistical significance was evaluated by running a Markov Chain Monte Carlo (MCMC) consisting of 10,000 batches of 10,000 iterations each, with the first 10,000 iterations discarded before sampling [72]. Significance levels were adjusted with sequential Bonferroni correction in order to correct for the effect of multiple tests [73], (i.e. $\alpha = 0.05$/number markers). MICRO-CHECKER software [74] was used to check for potential scoring errors and the presence of null alleles. The Rousset's a_r inter-individual genetic distance [75] was computed using the program SPAGeDI [76] since this parameter of relatedness does not rely on a reference population [77] and has been successfully applied to infer the effect of landscape on genetic structure of continuously distributed vertebrates [32,78–80].

Construction of landscape resistance models

We produced different resistance maps representing 59 different hypotheses about the resistance of different land use types using ARCGIS version 9.3 [81], with a raster cell size set to 50 m (Table 1; File S1). As suggested by Anderson et al. [82] the sampling grain selected (i.e. 50×50 m) is adequate to infer landscape effects on gene flow as is smaller than the average home-range size of the study species (i.e.> 0.5Km^2, [47]). In addition, this resolution allows representation of small landscape patches, but also those smaller elements in the landscape that will be crucial for the resulting effective distances, including linear elements such roads and highways [14].

1) Isolation by distance: Our first hypothesis and null model was a test of isolation by distance across a uniform resistance landscape [57,83]. In this model we assumed movement could occur with equal facility in any direction, with all raster cell values equal in resistance (i.e. resistance value 1).

2) Binary Landscape resistance maps: Our second set of hypotheses propose that some land uses promote genetic connectivity for forest dependant species, such as the pine marten, that specialize in such habitats [47,48], while others resist gene flow. Thus, different binary resistance maps were developed to evaluate the specific land uses which were favourable and unfavourable to the dispersal movements of the martens (i.e. habitat *vs* non-habitat-model). As pine martens are believed to need a minimum woodland area to survive (ca. 2km^2) [47] and tend to avoid treeless areas [48,49,51], we expected to find a positive effect of closed-canopy forest habitats and negative effect of open areas and human transformed landscapes on gene flow. Thus, the different binary resistance maps created (Land_A, Land_B, Land_C, Land_D, Land_E, Land_F and Land_G) covered a gradient from greater to lesser preference of the focal species for forest environments, ranging from strictly forest land (Land_A) up to and including open spaces (Land_G) (Table 1, Fig. 2). Therefore, we classified land use data as habitat *vs.* non-habitat and parameterized the models according to a range of plausible resistance values. As there is not a general rule for the

Table 1. Resistance values corresponding to the resistance maps taken evaluated.

		Binary landscape resistance maps														**Ecological Network**	
		Land_A$_x$ to Land_G$_x$							**Land_A$_x$ to Land_G$_x$**								
Land uses	IBD	A_x	B_x	C_x	D_x	E_x	F_x	G_x	Ab_x	Bb_x	Cb_x	Db_x	Eb_x	Fb_x	Gb_x	EN	ENnb
Forests	1	1	1	1	1	1	1	1	1	1	1	1	1	1	1	1	1
Forestry plantations	1	X	1	1	1	1	1	1	X	1	1	1	1	1	1	10	10
Scrubland	1	X	X	1	1	1	1	1	X	X	1	1	1	1	1	5	5
Agroforestry mosaics	1	X	X	X	1	1	1	1	X	X	X	1	1	1	1	20	20
Pastures and meadows	1	X	X	X	X	1	1	1	X	X	X	X	1	1	1	30	30
Rocks	1	X	X	X	X	X	1	1	X	X	X	X	X	X	1	40	40
Crops	1	X	X	X	X	X	X	1	X	X	X	X	X	X	1	60	60
Wetlands	1	X	X	X	X	X	X	X	100	100	100	100	100	100	100	100	100
National roads	1	X	X	X	X	X	X	X	200	200	200	200	200	200	200	200	50
Highways, urban areas, reservoirs and quarries	1	X	X	X	X	X	X	X	1000	1000	1000	1000	1000	1000	1000	1000	50

Binary landscape resitance maps: 1) *Land_A-Land_G*: Binary resistance maps, on a gradient from greater to lesser preference of the focal species in relation to forest environment; 2) *Land_Ab-Land_Gb*: Maps with letter "b" include the barrier effect of national roads, highways, urban areas, reservoirs and quarries. All the models were evaluated for 4 different resistance values (X = 5, 25, 50, 100) (e.g. Land_A_5 correspond to Land_A model with resistance value of 5). *Ecological Network resistance map*: 1) *EN*: a resistance map analogous to that used in the design of the ecological network of the Basque country; *ENnb*: a variant of the latter which diminishes the barrier effect of national roads, highways, urban areas, water reservoirs and quarries.

Table 2. Results of causal modeling of landscape resistance on genetic distance in European pine marten according to Mantel and partial mantel tests.

	Model/ Resistance Values	1) Simple mantel		Rank	2) Partial mantel test		Rank	3) Partial mantel		CMS?		1) Simple mantel test		Rank	2) Partial mantel test		Rank	3) Partial mantel		CMS?
		G*L	G*L		G*L/Dis	G*L/Dis		G*Dis/L	G*Dis|L			G*L	G*L		G*L/Dis	G*L/Dis		G*Dis/L	G*Dis/L	
		R	p		R	p		R	p			R	p		R	p		R	p	
Binary Landscape Resistance Models (Land_A to Land G)	**Land_A** *Forest*										**Binary Landscape Resistance Models (Land_Ab to Land_Gb)**	**Land_Ab**								
	5	0.221	0.0001	49	0.057	0.202	54	−0.007	0.877	**N**		0.243	0.0001	28	0.126	0.0005	26	−0.045	0.256	**Y**
	25	0.219	0.0001	53	0.066	0.215	51	0.047	0.317	**N**		0.243	0.0001	29	0.117	0.0138	32	0.008	0.861	**Y**
	50	0.217	0.0001	56	0.074	0.170	49	0.066	0.156	**N**		0.238	0.0001	34	0.112	0.0279	36	0.038	0.421	**Y**
	100	0.211	0.0001	58	0.080	0.138	46	0.089	0.053	**N**		0.229	0.0001	42	0.107	0.0445	38	0.067	0.159	**Y**
	Mean (±SE)	**0.2169 (±0.0044)**			**0.0692 (±0.0099)**							**0.2381 (±0.0063)**			**0.1152 (±0.0078)**					
	Land_B *Forest + Forestry plantations)*											**Land_Bb**								
	5	0.216	0.0001	57	0.027	0.463	58	0.011	0.764	**N**		0.242	0.0001	30	0.125	0.0001	27	−0.050	0.161	**Y**
	25	0.224	0.0001	47	0.068	0.134	50	0.009	0.838	**N**		0.245	0.0001	23	0.124	0.0022	28	−0.028	0.511	**Y**
	50	0.227	0.0001	46	0.082	0.085	45	0.024	0.612	**N**		0.245	0.0001	25	0.121	0.0083	29	−0.004	0.938	**Y**
	100	0.232	0.0001	40	0.099	0.051	41	0.040	0.380	**N**		0.241	0.0001	32	0.117	0.0195	31	0.028	0.539	**Y**
	Mean (±SE)	**0.2247 (±0.0069)**			**0.0689 (±0.0309)**							**0.2432 (±0.0018)**			**0.1220 (±0.0037)**					
	Land_C *Forest + forestry plantations + Scrublands*											**Land_Cb**								
	5	0.218	0.0001	54	0.041	0.128	56	−0.006	0.863	**N**		0.245	0.0001	24	0.135	0.0002	21	−0.059	0.105	**Y**
	25	0.228	0.0001	45	0.080	0.043	47	−0.009	0.840	**Y**		0.250	0.0001	14	0.140	0.0005	18	−0.045	0.305	**Y**
	50	0.232	0.0001	41	0.091	0.033	43	0.008	0.868	**Y**		0.251	0.0001	11	0.137	0.0022	20	−0.024	0.587	**Y**
	100	0.239	0.0001	33	0.111	0.018	37	0.021	0.648	**Y**		0.248	0.0001	17	0.129	0.0085	25	0.009	0.859	**Y**
	Mean (±SE)	**0.2291 (±0.0088)**			**0.0806 (±0.0292)**							**0.2487 (±0.0028)**			**0.1349 (±0.0045)**					
	Land_D *Forest + forestry plantations + scrublands + agroforestry mosaics*											**Land_Db**								
	5	0.220	0.0001	52	0.052	0.161	55	−0.019	0.608	**N**		0.246	0.0001	22	0.137	0.0001	19	−0.062	0.090	**Y**
	25	0.229	0.0001	44	0.083	0.041	44	−0.012	0.801	**Y**		0.251	0.0001	12	0.142	0.0006	15	−0.047	0.280	**Y**
	50	0.235	0.0001	37	0.100	0.038	40	−0.001	0.981	**Y**		0.253	0.0001	8	0.140	0.0024	17	−0.027	0.564	**Y**
	100	0.242	0.0001	31	0.116	0.026	33	0.017	0.721	**Y**		0.251	0.0001	13	0.134	0.0064	22	0.005	0.915	**Y**
	Mean (±SE)	**0.2313 (±0.0095)**			**0.0877 (±0.0272)**							***0.2501 (±0.0031)***			***0.1382 (±0.0036)***					
	Land_E *Forest + forestry plantations + scrublands + agroforestry mosaics + pastures*											**Land_Eb**								
	5	0.220	0.0001	51	0.062	0.049	53	−0.035	0.265	**Y**		0.248	0.0001	19	0.149	0.0001	13	−0.078	0.016	**Y**

Table 2. Cont.

Model/ Resistance Values	1) Simple mantel		Rank	2) Partial mantel test		Rank	3) Partial mantel		CMS?	1) Simple mantel test		Rank	2) Partial mantel test		Rank	3) Partial mantel		CMS?
	G*L	G*L		G*L/Dis	G*L/Dis		G*Dis/L	G*Dis|L		G*L	G*L		G*L/Dis	G*L/Dis		G*Dis/L	G*Dis/L	
	R	p		R	p		R	p		R	p		R	p		R	p	
25	0.234	0.0001	39	0.113	0.009	35	−0.061	0.132	Y	0.255	0.0001	7	0.167	0.0001	6	−0.089	0.015	Y
50	0.243	0.0001	27	0.133	0.004	24	−0.064	0.142	Y	0.259	0.0001	**4**	0.171	0.0001	**2**	−0.086	0.032	Y
100	0.253	0.0001	10	0.150	0.002	11	−0.061	0.176	Y	0.261	0.0001	**2**	0.170	0.0002	**4**	−0.074	0.097	Y
Mean (±SE)	**0.2373 (±0.0138)**			**0.1146 (±0.0380)**						**0.2556 (±0.0060)**			**0.1638 (±0.0103)**					
Land_F *Forest + forestry plantations + scrublands + agroforestry mosaics + pastures + rocky areas*										**Land_Fb**								
5	0.220	0.0001	50	0.062	0.044	52	−0.035	0.266	Y	0.248	0.0001	20	0.149	0.0001	12	−0.078	0.016	Y
25	0.234	0.0001	38	0.114	0.008	34	−0.062	0.134	Y	0.255	0.0001	6	0.167	0.0001	**5**	−0.089	0.014	Y
50	0.243	0.0001	26	0.134	0.004	23	−0.064	0.140	Y	0.259	0.0001	**3**	0.171	0.0001	**1**	−0.086	0.028	Y
100	0.253	0.0001	9	0.150	0.003	10	−0.061	0.184	Y	0.262	0.0001	**1**	0.170	0.0002	**3**	−0.074	0.088	Y
Mean (±SE)	**0.2374 (±0.0138)**			**0.1150 (±0.0381)**						**0.2557 (±0.0060)**			**0.1640 (±0.0104)**					
Land_G *Forest + forestry plantations + scrublands + agroforestry mosaics + pastures + rocky areas + croplands*										**Land_Gb**								
5	0.217	0.0001	55	0.041	0.027	57	−0.018	0.351	Y	0.249	0.0001	16	0.157	0.0001	7	−0.089	0.007	Y
25	0.223	0.0001	48	0.074	0.002	48	−0.039	0.125	Y	0.249	0.0001	15	0.157	0.0001	8	−0.087	0.009	Y
50	0.229	0.0001	43	0.096	0.000	42	−0.051	0.077	Y	0.248	0.0001	18	0.152	0.0001	9	−0.081	0.017	Y
100	0.237	0.0001	36	0.120	0.000	30	−0.063	0.043	Y	0.246	0.0001	21	0.142	0.0001	16	−0.071	0.036	Y
Mean (±SE)	**0.2263 (±0.0082)**			**0.0827 (±0.0338)**						**0.2481 (±0.0014)**			**0.1519 (±0.0069)**					
ENnb										**EN**								
	0.237	0.0001	35	0.103	0.032	39	−0.003	0.947	Y	0.256	0.000	5	0.146	0.0018	14	−0.030	0.519	Y

Model definitions according to Table 1. There are 3 Mantel tests comprising causal modeling: (1) G*L—simple Mantel test between the candidate model and genetic distance; (2) G*L|Dis—partial Mantel test between the candidate model and genetic distance, partialling out Euclidean distance; (3) G*D|L—partial Mantel test between the Euclidean and genetic distance, partialling out the candidate model. For a candidate model to be supported tests (1) and (2) must be significant, while test (3) must be negative or non-significant. Mantel tests meeting each criterion are italicized. Ranking of each model according to Mantel and partial Mantel r values is included. CMS? Indicates if the model is supported within the causal modeling framework (Y) or not (N).

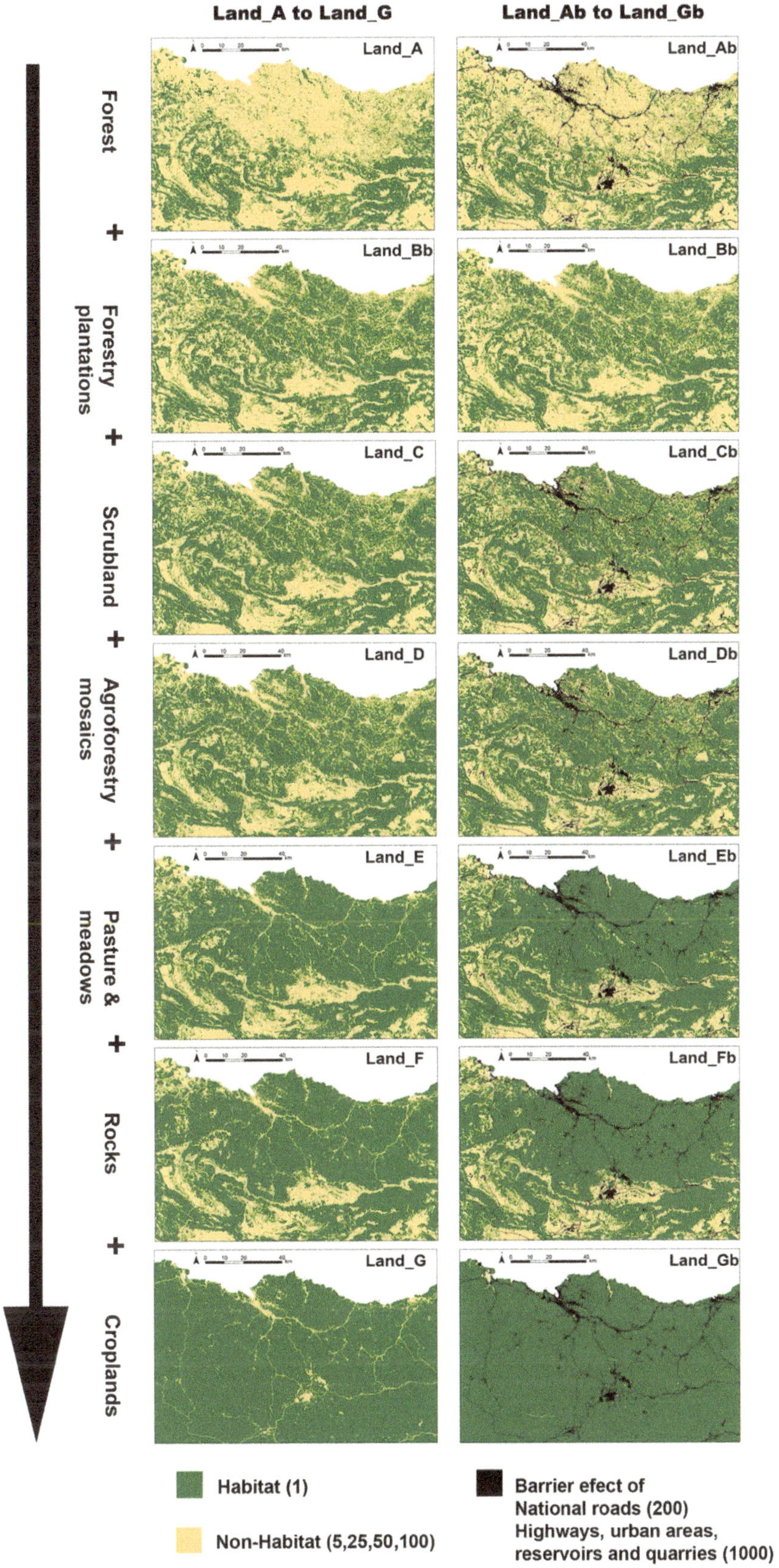

Figure 2. Binary landscape resistance maps on a gradient from greater to lesser preference of the focal species in relation to forest environment. Binary landscape resistance maps, on a gradient from greater to lesser preference of the focal species in relation to forest environment (Land_A to Land_G). Green-coloured cells represent "Habitat" (resistance value 1) and yellow-coloured cells "Non-Habitat" (resistance

values evaluated, 5, 25, 50,100). Models Land_Ab to Land_Gb additionally include black-coloured cells representing the barrier effect of national roads (resistance value 200), highways, urban areas, reservoirs and quarries (resistance value 1000).

assignment of resistance values to non-habitat, we explored 4 different resistance values (5, 25, 50, 100) of non-habitat relative to habitat to verify if it could affect the detectability of landscape genetic relationships, as it has been previously shown in both empirical [24] and simulation [84] studies. In this way, preferential land uses for dispersal were assigned a value of 1 (i.e. Habitat), while non-favourable to dispersal habitat (i.e. non-habitat) sites were assigned a value of 5, 25, 50 or 100, depending on the scenario.

For each of the resistance maps described above, we tested the effect of potential anthropogenic barriers, including national roads (resistance 200), highways, urban areas, reservoirs and quarries (resistance 1000) (Land_Ab to Land_Gb resistance maps; Table 1), using the same resistance values of the ecological network resistance map (see below), for comparative purposes [41]. As each resistance map (Land_A to Land_G; Land_Ab to Land_Gb) was explored for 4 different resistance values, we evaluated 56 different binary resistance maps (i.e. without barrier effect: Land_A_x to Land_G_x; With barrier effect: Land_Ab_x to Land_Gb_x; where X corresponds to the 4 evaluated resistances values per model: 5, 25, 50, and 100) (Table 1).

3) Ecological network resistance map: We evaluated the resistance map previously utilized in the design of ecological corridors linking forest Natura 2000 areas of the Basque Country [41]. The resistance map was based on the assignment of different resistance levels to each land use and parameterized through bibliographical review and expert opinion [41]. The resistance surface outlined in [41] was updated with the new available spatial data regarding land uses in the study area [59,60] (Table 1). Raster breaks in linear barrier elements were avoided by the reinforcement of the size of national roads and highways [14]. Sections of highways which run through viaducts or tunnels were assigned the resistance value corresponding to the land use of the surrounding area. Additionally, a second resistance map was used, with a view to testing the effect of noticeably decreasing the resistance value attributed to the potential barrier effects of national roads, highways, urban areas, reservoirs and quarries (ENnb map) (Table 1).

The raw ascii file of the EN is included in File S1. Following the resistance values outlined in Table 1, all the 59 resistance maps evaluated can be produced from the raw ascii EN resistance map (File S1).

The effective and Euclidean distances between each pair of individuals were calculated with PATHMATRIX 1.1 [15]. Pair-wise effective distances between individuals were calculated as the accumulated cost through the least cost paths (LCP) throughout each resistance surface [14,15] (Fig. 1).

We proposed 59 alternative landscape models: 1) 56 binary landscape resistance maps; 2) two complex resistance maps based on the resistance surfaces used to develop the regional ecological network (EN and ENnb) (Table 1), and 3) the null model of Isolation by Distance.

Relationship between genetic and geographical distances within a reciprocal causal modeling framework

Mantel correlations between genetic distance and alternative resistance hypotheses. The pairwise genetic distances matrix (Rousset's a_r) was correlated with different matrices of geographical and (cost) distances encompassing a total number of 5151 pairwise comparisons, including: i) Euclidean distance, to determine whether the patterns of differentiation follow an isolation by distance pattern (null hypothesis) and ii) the effective distances calculated for each of the 58 resistance maps, to infer landscape structure effects on gene flow. The correlation between distance matrixes was calculated by means of the Mantel test [85] and partial Mantel tests [86] as implemented in the ECODIST package [86] in R version 2.7 (R Development Core Team 2008) with 10,000 permutations. Given the potential sensitivity of Mantel tests to non-linear relationships between genetic and cost-distances [87], we compared results between two sets of analyses, one log transforming the effective and Euclidean distances, and one using the original untransformed cost-distance matrices.

Factorial hypothesis cube randomization: Evaluation of the unimodality of support across landscape models. When hypotheses are constructed across a quantitative range of values for a parameter, it is possible to evaluate the degree to which the analysis indicates a unimodal peak of support for a global best model [30]. The degree of unimodality of model support in a factorial hypothesis cube is one measure of the reliability of model results [13]. This is done by computing the differences in support (in our case partial Mantel r values) among all neighbouring cells (i.e. different models) in the hypothesis cube and comparing the sum of those differences to the distribution of the sum of differences from a large number of randomizations of the hypothesis cube (e.g. [13]). We evaluated the unimodality of support across the 56 binary resistance hypotheses (i.e. without barrier effect: Land_Ax to Land_Gx; With barrier effect: Land_-Abx to Land_Gbx; where X corresponds to the 4 evaluated resistances values per model: 5, 25, 50, and 100) for the transformed and untransformed analyses using the randomization procedure introduced by Cushman et al. [13], in which the order of hypotheses in the hypothesis cube is randomized a large number of times and each time the difference in partial Mantel r (partialling out distance) is calculated between neighboring hypotheses in the cube. The sum of squared neighbor distances from the actual hypothesis cube (Actual Sum Differences, ASD) is then compared to the distribution of squared neighbor distances in the randomized hypothesis cubes (Mean Sum Randomized Differences, MSRD). We conducted this analysis with 1,000,000 randomizations of the hypothesis cube for both the untransformed and transformed analysis. If no randomizations produce a sum of squared neighbor distances as small as observed, it is strong evidence that the analysis has shown a strong peak of support (i.e. unimodal support).

Original causal modeling. In addition to the reciprocal causal modeling approach [84] (see below), we conducted the original causal modeling [30] as a comparative framework, in which the 58 alternative landscape resistance models are tested against the null model of isolation by distance (IBD) as described in Cushman et al. [30]. There were 3 sets of diagnostic Mantel and partial Mantel tests to complete the causal modeling. These included: (i) simple Mantel tests between genetic distance and landscape resistances; (ii) partial Mantel tests between genetic distance and landscape cost distances, partialling out the effects of Euclidean distance; (iii) partial Mantel tests between genetic distances and Euclidean distance, partialling out the effects of landscape resistance. To infer an effect of a landscape resistance scenario on dispersal, we expected (i) and (ii) to be significant, and we expected (iii) to be negative or non-significant if that scenario

a)

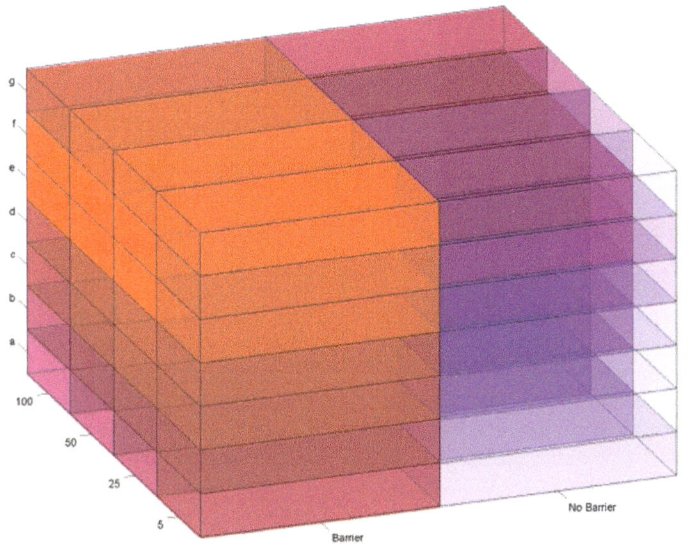

b)

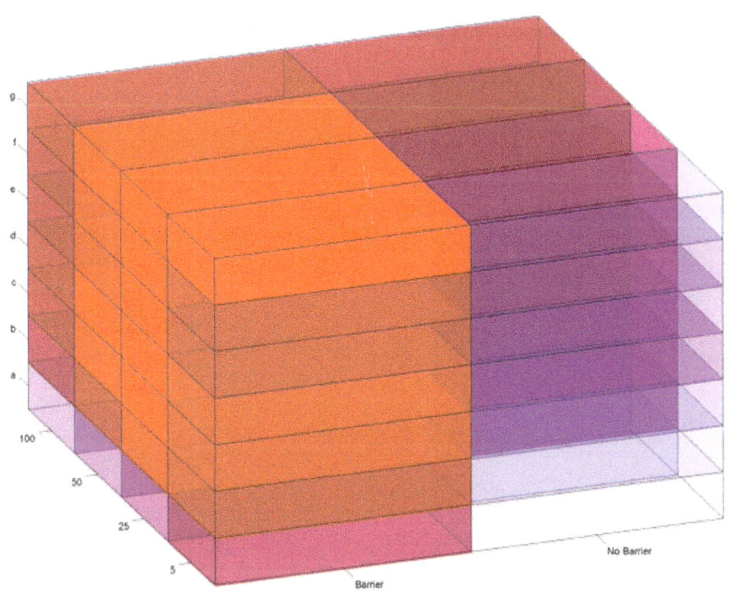

Figure 3. Factorial hypothesis cube randomization. Visualization of the 56 binary landscape-resistance hypotheses after the effects of geographical distance are partialed out on the a) log transformed and b) untransformed cost distances. The cubes each represent one of the 56 binary landscape-resistance models. The cubes are colored in a gradient from blue to red, with red being the most supported models based on the partial Mantel r value. The Mantel r values corresponding to each cube are found in Table 2 and Table S3 for the log transformed and the untransformed matrices, respectively.

'correctly' explained population connectivity in our study population [30].

Reciprocal causal modeling. We used (partial) Mantel tests in a reciprocal causal modeling framework [84] to analyse the influence of landscape structure on gene flow and to determine the extent to which possible landscape resistance models (i.e. resistance maps) explained the spatial pattern of genetic distance between individuals. Cushman and Languth [88] found that the inherent

Table 3. Factorial randomization of the hypothesis cube.

	Untransformed	Transformed
Rank	1	1
Actual Sum Differences (ASD)	19.6325	17.81
Mean Sum Randomized Differences (MSRD)	26.4738	24.17
SD error from MSRD	6.59E+04	6.97E+04

high correlation among alternative resistance models results in a high risk of spurious correlations using simple Mantel tests. Several refinements, including causal modeling [30], have been developed to reduce the risk of affirming spurious correlations and to assist model selection. However, Cushman et al. [84] showed these still suffer from elevated Type I error rates. Consequently, Cushman et al. [84] proposed "reciprocal causal modeling" which they showed greatly lessens Type I error rates in landscape genetic analysis [89]. In reciprocal causal modeling, each alternative resistance hypothesis is tested against all others with partial Mantel tests. A matrix of relative support is calculated by taking the difference between a) the partial Mantel r of each candidate model partialling out each alternative model, and b) the partial Mantel test of the alternative model partialling out the candidate model [84]. A fully supported hypothesis will have positive values of this difference with all alternative models, and no alternative models will have positive values compared to the supported model.

Results

Non-invasive sample collection and species identification

Out of 733 faecal samples collected from the entire study area, 141 were discarded because they were not fresh or because they presumably belong to the same individual (samples separated by < 1km). 494 out of 592 analyzed samples were classified as *Martes* sp. (*M. martes* and *M. foina*) based on genetic species identification results. Thus, unequivocal species identification was possible in 83.45% of the samples. We effectively identified 232 faecal samples as stone marten and 262 as pine marten. Additionally, we obtained 57 tissue samples from road-killed pine martens.

Out of 262 faecal samples identified as pine marten, 108 were not included to the microsatellite genotyping procedure. These samples correspond to the sampling period from 2004–2005, which was used for a first distribution assessment of sympatric martens in the study area and were not potentially fresh enough for microsatellite analysis. Thus, 213 pine marten samples (154 faecal samples and 59 tissue samples) were used for microsatellite genotyping.

Individual identification, genotype checking and genetic diversity

The first quality-screening test, based on 4 replicates of four loci, was not passed by 73 non-invasive samples (47.40%), which were immediately discarded. The remaining 81 samples (52.59%) were amplified at the other 11 loci. After multiple-tubes genotyping, 27 samples from this sub-set (17.53% from the total analyzed samples) were then discarded because they showed <50% PCR success, or because of high failure rates. Full multilocus microsatellite genotypes were obtained for the remaining 54 samples (66.67% from the samples that passed the screening and 35.06% from the total samples analyzed) all showing reliability score> 0.95 [67].

The observed average error rates across loci were: ADO = 0.188 and FA = 0.017. PID analysis showed that the set of 15 loci would produce an identical genotype with a probability of 1.69×10^{-10}, and with a probability of 4.45×10^{-5} for a full-sib, suggesting no "shadow effect" (i.e. all the genotypes identify distinct individuals; [90], and that matching genotypes were recaptures of the same individual).

After a regrouping procedure, we identified 42 individual genotypes from faecal samples. All of the 59 tissue samples were correctly genotyped at 15 loci and all provided new individuals. In total we identified 113 genotypes that corresponded with 101 different individuals. The number of times each individual was detected in the survey varied from 1 to 3, with a total number of 12 re-samplings. Complete genetic profiles and the geographic coordinates for the 101 pine marten individuals are included in Table S2.

The average observed (HO) and expected (HE) heterozygosity values were 0.53 and 0.58, respectively (Table S1). All 15 loci were variable with total numbers of alleles ranging between 3 and 8 per locus. The overall pine marten dataset showed a significant deficit of heterozygotes as compared to Hardy-Weinberg expectations ($p < 0.001$). Despite the broad scale of sampling, the majority of loci were in Hardy-Weinberg proportions (13 out of 15). Only loci Mp0188 and Lut-435 were out of Hardy-Weinberg proportions (Table S1). These results suggest signs of a Wahlund effect, due to the existence of an isolation by distance (Euclidean or effective) pattern in the study area. Linkage disequilibrium was not apparent for any pair of loci after performing Bonferroni corrections.

Correlation between genetic and effective distances

Factorial hypothesis cube randomization. We evaluated the unimodality of support across the 56 binary resistance maps for the log transformed and untransformed data to determine which form of the data should be used for subsequent analyses. After the effects of distance are partialled out, ranking the models by partial Mantel r value provides a means to determine which hypotheses have the greatest support and to identify the most related model to the genetic structure (Table 2, Fig. 3). According to the results outlined in Figure 3, there is a more coherent, unimodal pattern of support in the transformed analysis than the untransformed analysis. Additionally, factorial randomization of the hypothesis cube, in both the transformed and untransformed analyses, no instance of 1,000,000 randomizations produced a sum of squared differences between neighboring hypotheses (MSRD) as small as the actual sum of squared differences (ASD) in partial Mantel r values (Table 3), indicating very high unimodality in both forms of analysis. However, the transformed analysis had higher total support for optimal unimodal support of the best hypothesis as indicated by the larger number of standard errors of MSRD between neighboring hypotheses across the 1,000,000 randomizations (Table 3). Accordingly, all subsequent analyses are restricted to the log transformed resistance distances. As indicated

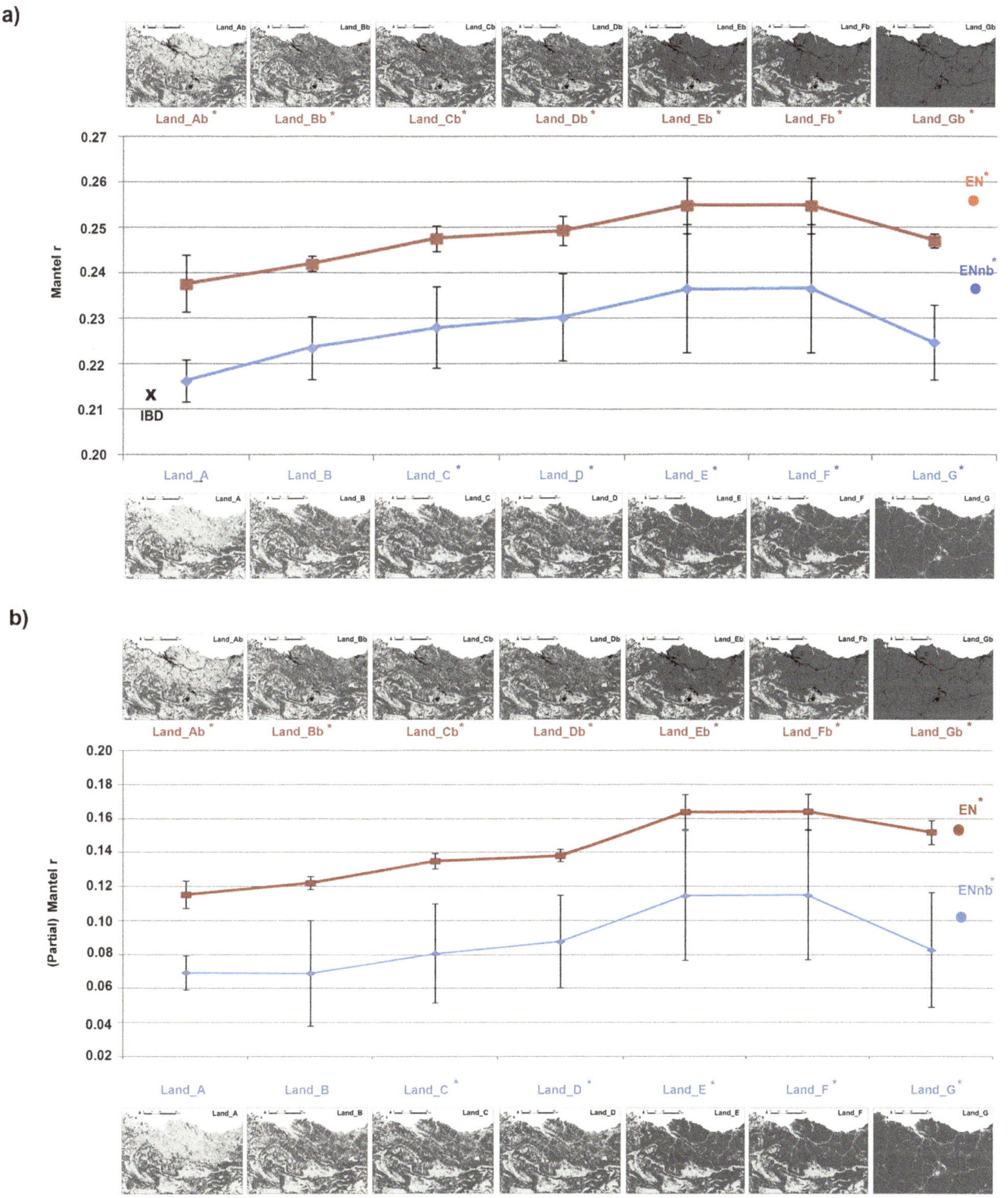

Figure 4. Mantel r results for the different landscape resistance maps evaluated (log transformed). a) Pearson correlation coefficients (Mantel r) between genetic distance and effective distance (log transformed) and **b**) Pearson correlation coefficients (Partial Mantel r) between genetic distance and effective distance (log transformed) after factoring out the effect of the Euclidean distance in the different landscape resistance maps examined. Models marked with an asterisk correspond to the models supported within the causal modeling framework [30].

by the hypothesis cube (Fig. 3), the different resistance values evaluated (5, 25, 50, 100) slightly modified the (partial) Mantel correlation results obtained for each model for both the log transformed (Table 2; Fig. S1 and Fig. S2) and the untransformed distances (Table S3; Fig. S3 and Fig. S4), but overall a consistent pattern was obtained.

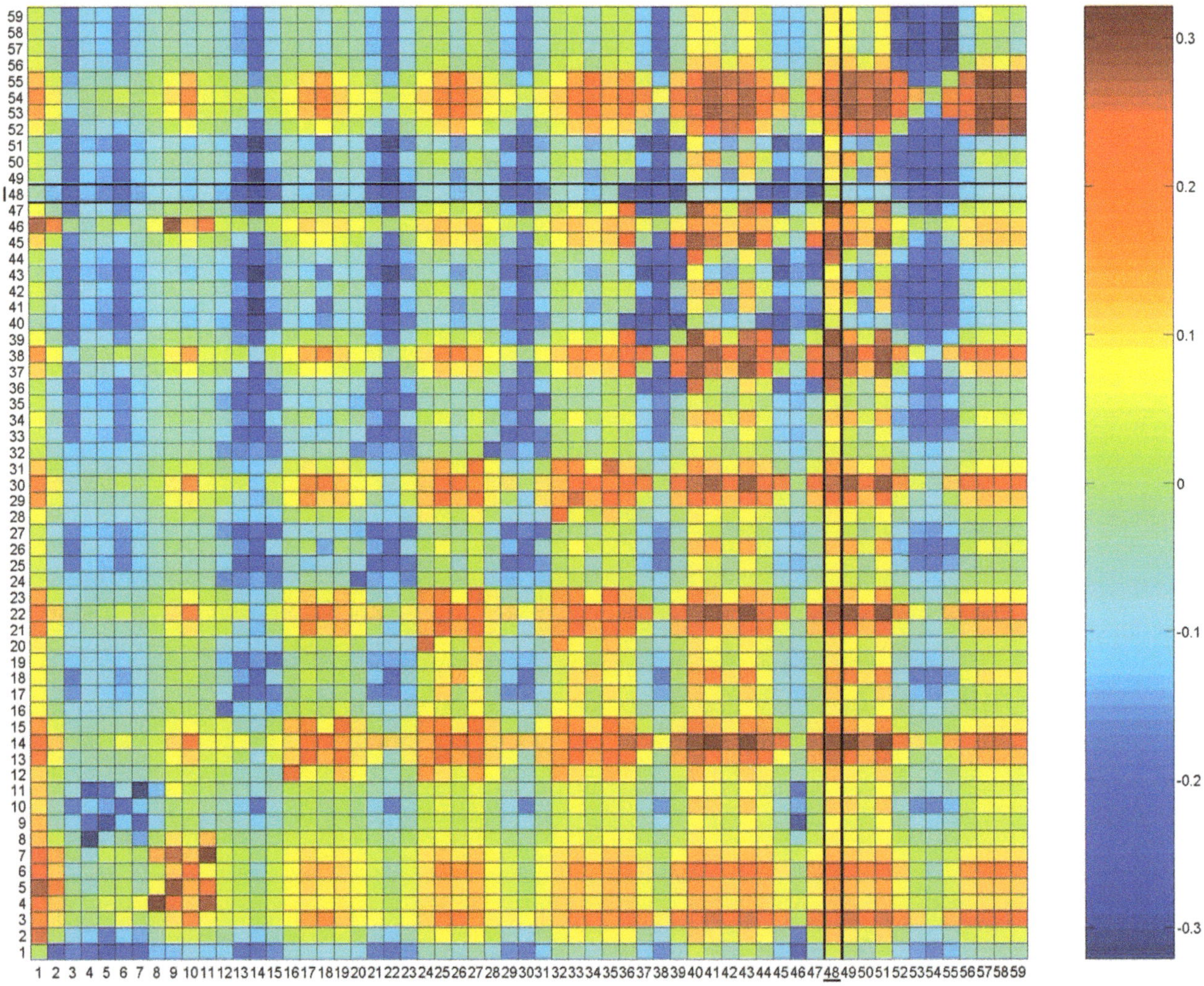

Figure 5. Reciprocal causal modeling results. Results of reciprocal causal modeling on the log transformed cost distances. A single resistance model (Model 48-Land_Fb100) is supported in analysis of the transformed cost distances. Columns indicate focal models, and rows indicate alternative models. The color gradient from blue to red indicates support for the focal model independent of the alternative model (e.g. focal model | alternative model – alternative model | focal model is positive). A fully supported model would have all positive values in the vertical dimension (e.g. that model is supported independently of all other models), and all negative values in the horizontal dimension (no other model is supported independently of the focal model). Model number and associated resistance map: 1 - EN, 2 - ENnb, 3 - Geo_dist, 4 - Land_A100, 5 - Land_A25. 6- Land_A5. 7 - Land_A50, 8 - Land_Ab100, 9 - Land_Ab25, 10 - Land_Ab5. 11 - Land_Ab50. 12 - Land_B100, 13 - Land_B25, 14 - Land_B5, 15 - Land_B50, 16 - Land_Bb100, 17 - Land_Bb25, 18 - Land_Bb5, 19 - Land_Bb50, 20 - Land_C100, 21 - Land_C25, 22- Land_C5, 23 - Land_C50, 24- Land_Cb100, 25 - Land_Cb25, 26 - Land_Cb5, 27 - Land_Cb50, 28 - Land_D100, 29 - Land_D25, 30 - Land_D5, 31 - Land_D50, 32 - Land_Db100, 33 - Land_Db25, 34 - Land_Db5, 35 - Land_Db50, 36 - Land_E100, 37 - Land_E25, 38 - Land_E5, 39 - Land_E50, 40 - Land_Eb100, 41 - Land_Eb25, 42 - Land_Eb5, 43 - Land_Eb50, 44 - Land_F100, 45 - Land_F25, 46 - Land_F5, 47 - Land_F50, 48 - Land_Fb100, 49 - Land_Fb25, 50 - Land_Fb5, 51 - Land_Fb50, 52 - Land_G100, 53 - Land_G25, 54 - Land_G5, 55 - Land_G50, 56 - Land_Gb100, 57 - Land_Gb25, 58 - Land_Gb5, 59 - Land_Gb50.

Simple Mantel correlations between genetic differentiation and alternative landscape models. *1) Isolation by distance:* A significant positive correlation was obtained between the genetic distances and Euclidean distances ($r = 0.214$; $p < 0.0001$), bearing clear witness to the existence of a pattern of isolation by distance (IBD) (Table 2, Fig. 3). However, when the models were ranked based on Mantel r, all the landscape models performed better than the null model (Table 2, Fig. 4).

2) Binary resistance maps: All of the simple Mantel tests were significant when analyzed in log-transformed form (Table 2). The correlation between genetic and effective distance gradually increased on including, in addition to natural forest (Land_A), forestry plantations (Land_B), scrublands (Land_C), agroforestry mosaics (Land_D), and pastures and meadows (Land_E) as environments favouring dispersal (Table 2, Fig. 4). This correlation did not change on including rocky areas as dispersal environments (Land_F), while it decreased on including cultivated land (Land_G). The same pattern was obtained with models which specifically increase the cost value of the main barrier features (national roads, highways, urban areas, reservoirs and quarries; Land_Ab to Land_Gb resistance maps), but with an increase of Mantel r values with respect to Land_A to Land_G models (Table 2, Fig. 4), indicating that including barrier effects due to linear features improves the resistance model. The correlation reached its maximum value on including barrier effects in resistance maps Land_E and Land_F (i.e. resistance maps

Land_Eb and Land_Fb; r = 0.256±0.0060; p <0.0001) (Table 2, Fig. 4). The different resistance values (5, 25, 50, 100) slightly modified the correlation results obtained for each model (see Figure S2 and S4 for further details), with the highest correlations for Land_Fb100 (Table 2, Table S3; Fig. 4).

3) Ecological network resistance map: The effective distances calculated on the basis of the EN map were positively correlated with genetic distances and explained a slightly higher proportion of the observed genetic variance than the Euclidean distances (EN, r = 0.256; p <0.0001; Table 2, Fig. 4). The degree of correlation when using the ENnb map was less than that obtained with EN, though still greater than that obtained using Euclidean distance (r = 0.237; p <0.0001; Table 2, Fig. 4). However, the original model used in the design of the ecological network (EN), which included a higher barrier effect for national roads, highways, urban areas, reservoirs and quarries was better supported than the alternative model (ENnb).

Partial Mantel correlations between genetic differentiation and alternative resistance hypotheses. We found significant effects of nearly all of the landscape resistance models (46 out of 52), as the relationship between genetic distance and effective distance was always significant when Euclidean distance was factored out of the relationship (p <0.05) (Table 2; Table S3).

1) Binary landscape resistance maps: The correlation values after factoring out the effect of Euclidean distance showed the same pattern of increase of that obtained by means of a simple Mantel test, with Land_Fb50 showing the highest partial Mantel r correlation (Table 2; Figure 4 and Fig. S2). However, Land_A_x, Land_B_x, Land_C5 and Land_D5 were not significant when the Euclidean distance was partialled out. Factorial support cubes indicate a clear unimodal peak of support in models Land_Fb50, Land_Eb50, Land_Fb100, Land_Eb100 (Figure 3) that are ranked from first to fourth according to partial r values. Similarly, these models have the highest simple Mantel r values of all of the evaluated models (Table 2; Figure 4). These results suggest that there is a strong peak of support for Land_Fb50 with a clear similarity with Land_Eb50, Land_Fb100 and Land_Eb100. Thus, the best supported models were associated with minimum resistance to movement on forest, forestry plantations, scrublands, agroforestry mosaics and pastures habitats and clear support for the barrier effect.

2) Ecological network resistance map: Both EN and ENnb models appeared better supported than the null model of IBD as this latter retained a significant positive relationship with a_r-based genetic distance after factoring out the effects of Euclidean distance, but less supported than the top resistance models (Land_Eb50, Land_Fb100, Land_Eb100; Table 2).

Original and reciprocal causal modeling. Using the original form of causal modeling approach [30] we found that all binary resistance hypotheses except Land_Ax, Land_Bx, Land_C5 and Land_D5 were supported (Table 2).

Using the novel Cushman et al. [84] method of "reciprocal causal modeling", only one resistance model out of the 59 candidate models was fully supported. The single supported model is model number 48, Land_Fb100. The reciprocal causal modeling method shows that the indexes of relative support of this model [i.e. calculated by taking the difference between 1) the partial Mantel r value of the candidate model partialling out each alternative model and 2) the partial Mantel r of the alternative model partialling out the candidate model (reciprocal partial Mantel test)] are all positive (Fig. 5). In addition to model Land_Fb100, several others are nearly perfectly supported. Land_Eb100 is supported independently of all but model Land_Fb100. Similarly, Land_Fb50 is supported independently of all except Land_Fb100 and Land_Eb100. Model Land_Eb50 is supported independently of all but Land_Fb100, Land_Fb50, and Land_Eb100. This verifies the peak of support seen in the hypothesis cube, with highest support for Land_Fb100, followed by Land_Eb100, Land_Fb50, and Land_Eb50 [i.e. models that included in addition to natural forest, forestry plantations, scrublands, agroforestry mosaics, pastures and meadows (Land Eb) and rocky areas (Land_Fb) as environments favouring dispersal and importance effect of roads as potential barriers to gene flow]. Importantly, the Ecological Network model EN was supported independently of all other models except Land_Fb100, Land_Eb100, Land_Fb50 and Land_Eb50, indicating that it is a highly effective surrogate for landscape resistance to pine marten gene flow.

Discussion

Recent studies suggested that individual-based landscape genetic analysis using partial Mantel tests in a causal modeling framework have high power to correctly identify landscape resistance as a driving process and reject spurious correlations with isolation by distance [84,88]. Cushman et al. [84] showed that the reciprocal causal modeling method we employed here substantially reduces the frequency of Type I errors. In this regards, Castillo et al. [91] recently showed that simulations support the reciprocal causal modeling with partial Mantel tests approach as an effective means to identifying the relationship between gene flow and landscape variables. Although there has been recent controversy over the use of Mantel tests in landscape genetics [92–94], a preferable alternative has yet to be identified that does not also suffer drawbacks [84]. There is no one-size- fits-all approach, and the most appropriate methodology will depend on the research question and landscape under investigation [92]. We use here the more robust modeling framework, proposed by Cushman et al. [84], that is based on the relative support of each candidate model and includes a reciprocal causal modeling step in the model optimization process. Moreover, we used the original causal modeling approach [30], factorial hypothesis cube randomization and ranking by simple Mantel test values as a comparative framework to further explore the performance of these several approaches.

Our results clearly indicate that a standard isolation-by-distance model is not sufficient to explain the observed genetic pattern, and including landscape variables through different resistance maps significantly improves the prediction of the target species gene flow. One model was supported in the transformed analysis using reciprocal causal modeling. This model, Land_Fb100, indicates that pine marten gene flow in northern Spain is facilitated by forests, forestry plantations, scrubland, agroforestry mosaics and pastures and meadows, and that crops have roughly 100 times higher resistance than optimal habitat. Further, this uniquely supported model indicates that anthropogenic barriers, such as national roads, highways, urban areas, reservoirs and quarries and wetlands likely pose much greater resistance to marten gene flow. This suggests that the population connectivity of pine martens in the study area may be vulnerable to habitat loss and fragmentation processes, due to the presence of anthropogenic barriers as has been previously suggested in other forest-dependant species [25,32,95].

The reciprocal causal modeling approach [84] clearly improved discrimination among the competing models, from only one fully supported model (i.e. Land_Fb100) versus more than 80% of models supported equivocally by the original form of causal modeling [30]. Indeed, recent studies showed that the novel reciprocal causal modeling approach is a strong improvement over other methods [91,96]. Even thought reciprocal causal modeling has a greater discrimination to detect the top model, the Cushman et al. [30] method (i.e. causal modeling + model rank + hypothesis cube) reached nearly identical conclusions. We found that ranking models based on simple or partial Mantel r gave consistent support. Likewise, our evaluation of unimodality of support in the hypothesis cube (e.g. [13,24,30]) verified the same peak of support. Reciprocal causal modeling resolved the Type I error problem that we saw in our results from the original form of causal modeling (evaluating support relative to isolation by distance), and was consistent with the peak of support seen in the factorial hypothesis cube. Thus, both forms of causal modeling are complementary and provided independent support to the obtained results, as did evaluating unimodality of support in the hypothesis cube.

In landscape genetics a large number of pairwise genetic relatedness measures have been applied to infer effects of landscape structure on gene flow [12,97], with Rousset's a_r or $\hat{a}$ [75] being one of the most widely used measures (e.g. [79,80,98,99]). Watts et al. [100] proposed a differentiation statistic ($\hat{e}$) that seems to improve Rousset's $\hat{a}$ performance for populations with large neighborhood-size ($D\sigma 2$) values (i.e. weak IBD pattern). However, differences in statistical performance usually highly depend on the data set used and the sampling scheme as well as how well the data set meets the underlying model assumptions [77]. Thus, and even if the results among different genetic distance measures have generally agreed [24,29,30,101] further studies based on empirical analyses of genetic patterns and simulation modeling are needed to properly evaluate the potential effect of different genetic distance estimator on disentangling landscape effects on gene flow.

Influence of land uses on gene flow: insights into pine marten ecology

The most consistent marten-habitat relation appears to be a general association with forest habitats, and avoidance of open, non-forested habitats [48,51,52]. Thus, the marten's unwillingness to cross open habitats may restrict the species' ability to disperse and colonise new forested areas [51,55]. Ruiz-González et al. [52] found that pine marten occurrence in the study area is highly dependent on the presence of forest and consequently sensitive to forest fragmentation as has been previously suggested in other studies across Europe [49,51]. Nevertheless, the presence of forest habitats is not the only factor which explains pine marten gene flow in the study area, indicating that the habitat selection and gene flow of pine martens may be driven by different factors [17,25,31].This may be because gene flow is driven by mating and dispersal events and habitat selection reflects the behaviour of individual organisms to maximize fitness within home ranges (e.g. [102]).

Our results suggest that it is not only forest masses which serve as favourable environments for dispersal. Scrubland, agroforestry mosaics and grassland habitats also potentially favour dispersal, since the correlation increases as, step by step, these environments are included as predictor variables of pine marten gene flow [Land_B(b) <Land_C(b) <Land_D(b) <Land_E(b)<Land_F(b)]. Original causal modeling identified the same pattern and suggested that Land_Eb and Land_Fb for resistance values of 50 and 100 are the most supported models. Likewise, novel reciprocal causal modeling highlighted similar results, identifying Land_Fb100 as the uniquely supported model, due to its greater discriminatory power [84].

These results are in consonance with recent ecological studies of European pine martens, based on radio tracking, which provide new data substantially differing from traditional descriptions in the scientific literature as strictly forest dependant species [48,51,55]. These studies show that martens are not exclusively confined to extensive forest patches but that they also use other patches including scrubland and agroforestry mosaics [48,51,54,55]. Indeed, the inclusion of scrub habitat in marten home ranges is likely to be related to its role in the connectivity of forest habitats [48,55]. In the same way, the improvement in correlation obtained by including pastures and meadows indicates that the species does not always avoid crossing these open spaces areas when there is forest habitat in the immediate vicinity as has been previously suggested by radiotracking data [48]. This is precisely the case in the area under study, where pastures and meadows are typically found in the immediate vicinity of forest. However, the inclusion of homogeneous croplands reduces the correlation between genetic distance and effective distance, suggesting that zones with intensive agriculture potentially impede species dispersal. This could be due to the scarcity of natural vegetation in these zones and the distance separating them from forest in the study area [41].

Additionally, models that increase the barrier effect of major roads and urban areas leads to a substantial improvement in the correlation between genetic distances and cost distances. The correlation with models Land_Ab-Land_Fb was greater than that obtained with models Land_A-Land_F, and the uniquely supported model in reciprocal causal modeling included the barrier effect (Land_Fb100). This suggests that the potential barrier effect of these land uses could have a synergic effect within a fragmented landscape, decreasing the gene flow due to road avoidance behaviour and/or road mortality [103–106].

Similar landscape genetics studies have also been conducted on other forest dependant *Martes* sp., providing contrasting results regarding landscape effects on gene flow [25,32,107–110]. Similar to the results found in this study, Broquet et al. [32], found that American marten (*Martes americana*) dispersal in Ontario is impeded by the loss and fragmentation of suitable habitat. Wasserman et al. [25] showed that gene flow in the Northern Idaho American marten population is driven by a gradient function of elevation, which was a proxy for snowpack, with marten avoiding lower elevations and dispersing in mid to high elevation montane forests. In contrast, Koen et al. [110] found that marten dispersal across Ontario can best be described as neighbour-mating with no directional bias caused by forest-management induced landscape structure, resulting in a pattern of isolation by distance, suggesting that Ontario landscape is well connected with respect to suitable marten habitat. These contrasting landscape hypotheses governing gene flow could be explained by the different limiting factors that could be acting in each of landscape under study [22].

Even though no previous individual-based landscape genetics data was available for the pine marten, Mergey et al. [58] found that genetic diversity is not associated with habitat fragmentation metrics in France, in spite of the existence of a high degree of forest fragmentation in the studied marten populations. However, this result does not demonstrate that the pine marten gene flow is not affected by forest fragmentation processes. Thus, a more detailed individual-based landscape genetics analysis (Larroque et al. Unpublished data), could provide better insights into the landscape processes governing gene flow and an interesting comparative framework with Spanish pine marten populations.

Empirical evaluation of ecological network resistance maps through landscape genetics

Maps of ecological corridors are commonly used in land use planning, but unfortunately are more often the product of expert opinion rather than empirical data [9,10]. Thus, using landscape genetic analysis, we could partially solve this limitation by studying the gene flow of a target species with regards to the resistance maps used to design the ecological networks [11,12]. Here, the parameterization found in the resistance map which was used to design the regional corridors linking forest protected areas of the Basque Country (north Spain) [41] was adequate to explain pine marten gene flow, with one of the highest partial Mantel r value ($r = 0.145$) of all the evaluated models. Based on reciprocal causal modeling only Land_Fb100, Land_Eb100, Land_Fb50 and Land_Eb50 were supported independently of EN. Even though Land_Fb100 better explains pine marten gene flow, the high mantel correlation value between the cost distances for Land_Fb100 and EN (Mantel $r = 0.9362$ $p<0.001$) suggests EN is a good proxy. This indicates that the EN model used to develop regional connectivity networks among protected areas [41] likely performs very well as a surrogate for landscape resistance for pine marten.

Thus, the resistance map with which the regional ecological network was originally designed in the Basque Country (EN), appears to have high congruence with one of its official target species at regional scale [41]. This is a welcome finding, given that most past evaluations of expert-derived resistance values found that they performed poorly in comparison to empirically optimized models [18,24]. Given the importance of pine marten as a bio-indicator of species associated with natural vegetation [41], our results emphasize the importance of incorporating regional corridors into land use planning and management to preserve landscape connectivity for forest dwelling species.

The influence of resistance values and logarithmic transformations to detect landscape genetic relationships

Since our ability to detect the effects of landscape structure on genetic differentiation depends on both the landscape features used and the relative costs of each feature, different resistance values could provide different results [17,19,83]. Previous studies have found that the degree of contrast in resistance to gene flow in habitat as compared to non-habitat could affect whether or not a given landscape configuration will significantly affect genetic differentiation [22]. We found an increasing reliability of predictions as resistance contrast increased, with several models only supported by 25, 50 and 100 resistance values and unimodal peak of support for 50 and 100 resistance values. The hypothesis that was uniquely supported by reciprocal causal modeling indicated that non-habitat was 100 times more resistant than habitat, and anthropogenic barriers may impart additional resistance as high as 1000 times that of optimal habitat.

Some landscape genetic studies have found that the untransformed geographic distances perform better than logarithmic transformation (e.g. [24,30]), while most previously studies used transformed distances without any evaluation. However, the relationship between cost distances and genetic distances is highly dependant on the study area and the focal species. For example, at small extents the relationship between cost and genetic distances is nearly linear and the untrasformed correlations may fit the data better [24,30]. However, when the study area is large in extent relative to the dispersal ability of the species, as in the present study, the relationship between cost distance and genetic distance will be nonlinear and the logarithm transform will improve fit. Thus, taking into account the potential bias due to an incorrect use of transformations, we propose that future landscape genetics should evaluate the unimodality of support among the hypotheses as a means to determine the degree to which the transformation improves the analysis.

Conclusions

This paper presents a comprehensive individual-based landscape genetic analysis of the European pine marten, and the first formal use of landscape genetics to evaluate the effectiveness of regional ecological networks. We compared results from several methods of model selection and found that ranking based on Mantel r or partial Mantel r, the unimodality of support in the hypothesis cube, causal modeling and reciprocal causal modeling all identified the same best models of landscape resistance for European pine marten in northern Spain. Reciprocal causal modeling appeared to provide the strongest differentiation among hypotheses and enabled the identification of a single, independently supported model. Gene flow of European pine marten is facilitated by natural land cover, such as forest, scrublands and pastures and meadows, and is resisted by anthropogenic land uses and linear barriers such as major roads. We confirm that the resistance map used to develop the regional ecological network in the Basque Country is a close surrogate to the empirically optimized resistance model for marten.

Supporting Information

Figure S1 Mantel Correlation values for a) Land_A-Land_G and b) Land_Ab-Land_Gb models for the 4 different cost values evaluated on the log transformed cost distances.

Figure S2 Partial Mantel Correlation values for a) Land_A-Land_G and b) Land_Ab-Land_G models for the 4 different cost values evaluated on the log transformed cost distances.

Figure S3 Mantel Correlation values for a) Land_A-Land_G and b) Land_Ab-Land_G models for the 4 different cost values evaluated on the untransformed cost distances.

Figure S4 Partial Mantel Correlation values for a) Land_A-Land_G and b) Land_Ab-Land_G models for the 4 different cost values evaluated on the untransformed cost distances.

Table S1 Summary of the genetic variability. Genetic variability of the 15 microsatellite loci multiplexed used in this study. Number of alleles (N_A), Observed (H_O) and expected (H_E) heterozygosities for each locus and for whole data set.. Loci marked with an asterisk deviated from Hardy-Weinberg proportions.

Table S2 Sample locations and microsatellite data. Complete genetic profiles (15 microsatellite loci) and the geographic coordinates for the 101 pine marten individuals

Table S3 Results of causal modeling of landscape resistance on genetic distance in European pine marten according to Mantel and partial mantel tests for the untransformed distances.

File S1 Ecological network resistance map. Raw '.asc' file of the EN resistance map from which all the resistance maps evaluated can be produced following the resistance values outlined in Table 1.

Acknowledgments

The authors wish to thank the following persons and institutions for supplying some of the tissue and faecal samples used in this study: Patricia Lizarraga, Ricardo Gutierrez and Laura Elorza (CRF Martioda—Alava Provincial Council); Txema Fernandez and Javier Sesma (IKT S:A); the technical staff and rangers of the Natural Parks of Alava (Juan Carlos Ortíz, Ricardo Ortíz, María Elena García, Kepa García, Sonia Benítez, Elisabeth Cabanillas, Iker Ayala, Arantza Puente, Jesús Gómez, Iñaki Martínez, Lidia Lacha, Jokin Sáez de Camara); Rangers from Alava (Asier Martínez, Tomás Landa, Javier Pinedo, Mario Corral) and Bizkaia (Igor Aginako, Julio Ruiz Guijarro, Ignacio Martínez, Eneko Díaz, Juan Manuel Pérez de Ana, David Vado, Ander Eguia) Provincial Councils; Joseba Carreras (Alava Provincial Council); Jokin Larumbe (Navarre Government); Dr. Fermín Urra (GAVRN); Asun Gómez (TRAGSATEC); Dr. Madis Podra (Asoc. Visón Europeo); Nerea Ruiz de Azua (EKOS, S.L.); Gorka Belamendia (CEA); Felipe Canales and Miguel Ángel Campos (Consultora de Recursos Naturales, S.L); Enrique Arberas; Oskar Berdión; Javier López de Luzuriaga; Dr. Iñigo Zuberogoitia (ICARUS); Dr. Marta Barral (NEIKER); Rebeca Pérez; Haizea Aguirre; Pablo Pérez; Amaia García de Albéniz; Alvaro Osés; and Gerardo Domínguez. Especial thanks to Maria Vergara, Oihana Razkin, Luis Javier Chueca and Peio Lozano (UPV/EHU), the staff of IKT S.A and colleagues of the Conservation genetics laboratory (ISPRA) for their valuable help at different stages of the study.

Author Contributions

Conceived and designed the experiments: ARG MG SC ER BJGM. Performed the experiments: ARG MG MJM. Analyzed the data: ARG MG SC. Contributed reagents/materials/analysis tools: ER BJGM. Wrote the paper: ARG MG SC. Revised the manuscript: ARG MG SC MJM ER BJGM.

References

1. Opdam P, Wascher D (2004) Climate change meets habitat fragmentation: linking landscape and biogeographical scale levels in research and conservation. Biological Conservation 117: 285–297.
2. Fahrig L (2007) Non-optimal animal movement in human-altered landscapes. Functional Ecology 21: 1003–1015.
3. Cushman SA, McKelvey KS, Schwartz MK (2009) Use of Empirically Derived Source-Destination Models to Map Regional Conservation Corridors. Conservation Biology 23: 368–376.
4. Taylor PD, Fahrig L, Henein K, Merriam G (1993) Connectivity is a vital element of landscape structure. Oikos 68: 571–573.
5. Brooks CP (2003) A scalar analysis of landscape connectivity. Oikos 102: 433–439.
6. Cushman SA, Landguth EL (2012) Multi-taxa population connectivity in the Northern Rocky Mountains. Ecological Modelling 231: 101–112.
7. Bennett G, Wit P (2001) The development and application of ecological networks: a review of proposals, plans and programmes. Amsterdam: AIDEnvironment.
8. Jongman R, Pungetti G (2004) Ecological networks and greenways: concept, design, implementation. Cambridge: Cambridge University Press. pp. 368
9. Cushman S, McRae B, Adriansen F, Beier P, Shirley M, et al. (2013) Biological corridors and connectivity. In: MacDonald D, editor. Conservation in Theory and Practice. New York: Wiley. pp. 32.
10. Boitani L, Falcucci A, Maiorano L, Rondinini C (2007) Ecological networks as conceptual frameworks or operational tools in conservation. Conservation Biology 21: 1414–1422.
11. Luque S, Saura S, Fortin M-J (2012) Landscape connectivity analysis for conservation: insights from combining new methods with ecological and genetic data. Landscape Ecology 27: 153–157.
12. Segelbacher G, Cushman SA, Epperson BK, Fortin MJ, Francois O, et al. (2010) Applications of landscape genetics in conservation biology: concepts and challenges. Conservation Genetics 11: 375–385.
13. Cushman SA, Chase M, Griffin C (2010) Mapping landscape resistance to identify corridors and barriers for elephant movement in Southern Africa. In: Cushman SA, Huettmann F, editors. Spatial Complexity, Informatics and Wildlife Conservation. Tokyo: Springer. pp. 349–367.
14. Adriaensen F, Chardon JP, De Blust G, Swinnen E, Villalba S, et al. (2003) The application of 'least-cost' modelling as a functional landscape model. Landscape and Urban Planning 64: 233–247.
15. Ray N (2005) PATHMATRIX: a geographical information system tool to compute effective distances among samples. Molecular Ecology Notes 5: 177–180.
16. Beier P, Majka DR, Spencer WD (2008) Forks in the road: Choices in procedures for designing wildland linkages. Conservation Biology 22: 836–851.
17. Spear SF, Balkenhol N, Fortin MJ, McRae BH, Scribner K (2010) Use of resistance surfaces for landscape genetic studies: considerations for parameterization and analysis. Molecular Ecology 19: 3576–3591.
18. Zeller KA, McGarigal K, Whiteley AR (2012) Estimating landscape resistance to movement: a review. Landscape Ecology 27: 777–797.
19. Rayfield B, Fortin M-J, Fall A (2010) The sensitivity of least-cost habitat graphs to relative cost surface values. Landscape Ecology 25: 519–532.
20. Manel S, Schwartz MK, Luikart G, Taberlet P (2003) Landscape genetics: combining landscape ecology and population genetics. Trends in Ecology & Evolution 18: 189–197.
21. Manel S, Holderegger R (2013) Ten years of landscape genetics. Trends in Ecology & Evolution 28: 614–621.
22. Cushman SA, Shirk AJ, Landguth EL (2013) Landscape genetics and limiting factors. Conservation Genetics 14: 263–274.
23. Epps CW, Wehausen JD, Bleich VC, Torres SG, Brashares JS (2007) Optimizing dispersal and corridor models using landscape genetics. Journal of Applied Ecology 44: 714–724.
24. Shirk AJ, Wallin DO, Cushman SA, Rice CG, Warheit KI (2010) Inferring landscape effects on gene flow: a new model selection framework. Molecular Ecology 19: 3603–3619.
25. Wasserman TN, Cushman SA, Schwartz MK, Wallin DO (2010) Spatial scaling and multi-model inference in landscape genetics: *Martes americana* in northern Idaho. Landscape Ecology 25: 1601–1612.
26. Bull RAS, Cushman SA, Mace R, Chilton T, Kendall KC, et al. (2011) Why replication is important in landscape genetics: American black bear in the Rocky Mountains. Molecular Ecology 20: 1092–1107.
27. Zalewski A, Piertney SB, Zalewska H, Lambin X (2009) Landscape barriers reduce gene flow in an invasive carnivore: geographical and local genetic structure of American mink in Scotland. Molecular Ecology 18: 1601–1615.
28. Perez-Espona S, McLeod JE, Franks NR (2012) Landscape genetics of a top neotropical predator. Molecular Ecology 21: 5969–5985.
29. Blair C, Weigel DE, Balazik M, Keeley ATH, Walker FM, et al. (2012) A simulation-based evaluation of methods for inferring linear barriers to gene flow. Molecular Ecology Resources 12: 822–833.
30. Cushman SA, McKelvey KS, Hayden J, Schwartz MK (2006) Gene flow in complex landscapes: Testing multiple hypotheses with causal modeling. American Naturalist 168: 486–499.
31. Centeno-Cuadros A, Roman J, Delibes M, Antonio Godoy J (2011) Prisoners in Their Habitat? Generalist Dispersal by Habitat Specialists: A Case Study in Southern Water Vole (*Arvicola sapidus*). Plos One 6.
32. Broquet T, Ray N, Petit E, Fryxell JM, Burel F (2006) Genetic isolation by distance and landscape connectivity in the American marten (*Martes americana*). Landscape Ecology 21: 877–889.
33. Wang YH, Yang KC, Bridgman CL, Lin LK (2008) Habitat suitability modelling to correlate gene flow with landscape connectivity. Landscape Ecology 23: 989–1000.
34. Schwartz MK, Copeland JP, Anderson NJ, Squires JR, Inman RM, et al. (2009) Wolverine gene flow across a narrow climatic niche. Ecology 90: 3222–3232.
35. Blair C, Jimenez Arcos VH, Mendez de la Cruz FR, Murphy RW (2013) Landscape Genetics of Leaf-Toed Geckos in the Tropical Dry Forest of Northern Mexico. Plos One 8.
36. Beja-Pereira A, Oliveira R, Alves PC, Schwartz MK, Luikart G (2009) Advancing ecological understandings through technological transformations in noninvasive genetics. Molecular Ecology Resources 9: 1279–1301.
37. Taberlet P, Waits LP, Luikart G (1999) Noninvasive genetic sampling: look before you leap. Trends in Ecology & Evolution 14: 323–327.
38. Piggott MP, Taylor AC (2003) Remote collection of animal DNA and its applications in conservation management and understanding the population biology of rare and cryptic species. Wildlife Research 30: 1–13.
39. Waits LP, Paetkau D (2005) Noninvasive genetic sampling tools for wildlife biologists: A review of applications and recommendations for accurate data collection. Journal of Wildlife Management 69: 1419–1433.

40. Schwartz MK, Monfort SL (2008) Genetic and Endocrine Tools for Carnivore Surveys. In: Long R, MacKay P, Ray J, Zielinski W, editors. Noninvasive survey methods for North American carnivores. Washington D.C.: Island Press. pp. 228–250.
41. Gurrutxaga M, Lozano PJ, del Barrio G (2010) GIS-based approach for incorporating the connectivity of ecological networks into regional planning. Journal for Nature Conservation 18: 318–326.
42. Jongman RHG (2002) Homogenisation and fragmentation of the European landscape: ecological consequences and solutions. Landscape and Urban Planning 58: 211–221.
43. Jongman R, Bouwma I, Van Doorn A (2006) Indicative map of the Pan-European ecological network in Western Europe. Wageningen: Alterra. 104 p.
44. Gurrutxaga M, Rubio L, Saura S (2011) Key connectors in protected forest area networks and the impact of highways: A transnational case study from the Cantabrian Range to the Western Alps (SW Europe). Landscape and Urban Planning 101: 310–320.
45. Mallarach J, Rafa M, Sargatal J (2010) Cantabrian Mountains-Pyrénées-Massif Central-Western Alps great mountain corridor. In: Worboys GL, Francis WL, Lockwood M, editors. Connectivity conservation management A global guide London: Earthscan. pp. 269–279.
46. Proulx G, Aubry K, Birks J, Buskirk S, Fortin C, et al. (2004) World distribution and status of the genus Martes in 2000. In: Harrison D, Fuller A, Proulx G, editors. Martens and fishers (*Martes*) in human- altered environments: an international perspective. New York: Springer-Verlag. pp. 77–98.
47. Zalewski A, Jedrzejewski W (2006) Spatial organisation and dynamics of the pine marten *Martes martes* population in Bialowieza Forest (E Poland) compared with other European woodlands. Ecography 29: 31–43.
48. Pereboom V, Mergey M, Villerette N, Helder R, Gerard JF, et al. (2008) Movement patterns, habitat selection, and corridor use of a typical woodland-dweller species, the European pine marten (*Martes martes*), in fragmented landscape. Canadian Journal of Zoology-Revue Canadienne De Zoologie 86: 983–991.
49. Brainerd SM, Rolstad J (2002) Habitat selection by Eurasian pine martens *Martes martes* in managed forests of southern boreal Scandinavia. Wildlife Biology 8: 289–297.
50. Kurki S, Nikula A, Helle P, Linden H (1998) Abundances of red fox and pine marten in relation to the composition of boreal forest landscapes. Journal of Animal Ecology 67: 874–886.
51. Mergey M, Helder R, Roeder J-J (2011) Effect of forest fragmentation on space-use patterns in the European pine marten (*Martes martes*). Journal of Mammalogy 92: 328–335.
52. Ruiz-Gonzalez A, Rubines J, Berdion O, Gomez-Moliner BJ (2008) A non-invasive genetic method to identify the sympatric mustelids pine marten (*Martes martes*) and stone marten (*Martes foina*): preliminary distribution survey on the northern Iberian Peninsula. European Journal of Wildlife Research 54: 253–261.
53. Bright PW (2000) Lessons from lean beasts: conservation biology of the mustelids. Mammal Review 30: 217–226.
54. Balestrieri A, Remonti L, Ruiz-Gonzalez A, Gomez-Moliner BJ, Vergara M, et al. (2010) Range expansion of the pine marten (*Martes martes*) in an agricultural landscape matrix (NW Italy). Mammalian Biology 75: 412–419.
55. Caryl FM, Quine CP, Park KJ (2012) Martens in the matrix: the importance of nonforested habitats for forest carnivores in fragmented landscapes. Journal of Mammalogy 93.
56. Schwartz M, Ruiz-González A, Pertoldi C, Masuda R (2012) *Martes* conservation genetics: assessing within species movements, units to conserve and connectivity cross ecological and evolutionary time. In: Aubry K, Zielinski W, Raphael M, Proulx G, Buskirk S, editors. *Biology and conservation of marten, sables, and fisher: a new synthesis*. New York: Cornell University Press. pp. 398–428.
57. Wright S (1943) Isolation by distance. Genetics 28: 114–138.
58. Mergey M, Larroque J, Ruette S, Vandel J-M, Helder R, et al. (2012) Linking habitat characteristics with genetic diversity of the European pine marten (*Martes martes*) in France. European Journal of Wildlife Research 58: 909–922.
59. Environment SMft (2006) Mapa forestal de España 1:50.000. Madrid: Ministerio de Medio Ambiente.
60. Institut SNG (2008) Base cartográfica numérica BCN200. Ministerio de Fomento.
61. Ruiz-Gonzalez A, Jose Madeira M, Randi E, Urra F, Gomez-Moliner BJ (2013) Non-invasive genetic sampling of sympatric marten species (*Martes martes* and *Martes foina*): assessing species and individual identification success rates on faecal DNA genotyping. European Journal of Wildlife Research 59: 371–386.
62. Davison A, Birks JDS, Brookes RC, Braithwaite TC, Messenger JE (2002) On the origin of faeces: morphological versus molecular methods for surveying rare carnivores from their scats. Journal of Zoology 257: 141–143.
63. Taberlet P, Griffin S, Goossens B, Questiau S, Manceau V, et al. (1996) Reliable genotyping of samples with very low DNA quantities using PCR. Nucleic Acids Research 24: 3189–3194.
64. Frantz AC, Pope LC, Carpenter PJ, Roper TJ, Wilson GJ, et al. (2003) Reliable microsatellite genotyping of the Eurasian badger (*Meles meles*) using faecal DNA. Molecular Ecology 12: 1649–1661.
65. Stenglein JL, Waits LP, Ausband DE, Zager P, Mack CM (2010) Efficient, Noninvasive Genetic Sampling for Monitoring Reintroduced Wolves. Journal of Wildlife Management 74: 1050–1058.
66. Brzeski KE, Gunther MS, Black JM (2013) Evaluating River Otter Demography Using Noninvasive Genetic Methods. Journal of Wildlife Management 77: 1523–1531.
67. Miller CR, Joyce P, Waits LP (2002) Assessing allelic dropout and genotype reliability using maximum likelihood. Genetics 160: 357–366.
68. Valiere N (2002) GIMLET: a computer program for analysing genetic individual identification data. Molecular Ecology Notes 2: 377–379.
69. Pompanon F, Bonin A, Bellemain E, Taberlet P (2005) Genotyping errors: Causes, consequences and solutions. Nature Reviews Genetics 6: 847–859.
70. Belkhir K, Borsa P, Chikhi L, Raufaste N, Bonhomme F (2004) Genetix 4.02, Logiciel sous windows pour la génétique des populations. Université de Montpellier II, Montpellier, France.
71. Raymond M, Rousset F (1995) Genepop (Version-1.2) - Population-Genetics Software for Exact Tests and Ecumenicism. Journal of Heredity 86: 248–249.
72. Guo SW, Thompson EA (1992) Performing the Exact Test of Hardy-Weinberg Proportion for Multiple Alleles. Biometrics 48: 361–372.
73. Rice WR (1989) Analyzing Tables of Statistical Tests. Evolution 43: 223–225.
74. Van Oosterhout C, Hutchinson WF, Wills DPM, Shipley P (2004) MICRO-CHECKER: software for identifying and correcting genotyping errors in microsatellite data. Molecular Ecology Notes 4: 535–538.
75. Rousset F (2000) Genetic differentiation between individuals. Journal of Evolutionary Biology 13: 58–62.
76. Hardy OJ, Vekemans X (2002) SPAGEDi: a versatile computer program to analyse spatial genetic structure at the individual or population levels. Molecular Ecology Notes 2: 618–620.
77. Vekemans X, Hardy OJ (2004) New insights from fine-scale spatial genetic structure analyses in plant populations. Molecular Ecology 13: 921–935.
78. Coulon A, Cosson JF, Angibault JM, Cargnelutti B, Galan M, et al. (2004) Landscape connectivity influences gene flow in a roe deer population inhabiting a fragmented landscape: an individual-based approach. Molecular Ecology 13: 2841–2850.
79. Blair ME, Melnick DJ (2012) Scale-Dependent Effects of a Heterogeneous Landscape on Genetic Differentiation in the Central American Squirrel Monkey (*Saimiri oerstedii*). Plos One 7.
80. Dudaniec RY, Rhodes JR, Wilmer JW, Lyons M, Lee KE, et al. (2013) Using multilevel models to identify drivers of landscape-genetic structure among management areas. Molecular Ecology 22: 3752–3765.
81. ESRI (2009) ArcMap version 9.3. Redlands, CA, USA.: Environmental Systems Research Institute.
82. Anderson CD, Epperson BK, Fortin MJ, Holderegger R, James PMA, et al. (2010) Considering spatial and temporal scale in landscape-genetic studies of gene flow. Molecular Ecology 19: 3565–3575.
83. Jenkins DG, Carey M, Czerniewska J, Fletcher J, Hether T, et al. (2010) A meta-analysis of isolation by distance: relic or reference standard for landscape genetics? Ecography 33: 315–320.
84. Cushman SA, Wasserman TN, Landguth EL, Shirk AJ (2013) Re-Evaluating Causal Modeling with Mantel Tests in Landscape Genetics. Diversity 5: 51–72.
85. Mantel N (1967) Detection of disease clustering and a generalized regression approach. Cancer Research 27: 209–220.
86. Smouse PE, Long JC, Sokal RR (1986) Multiple-regression and correlation extensions of the mantel test of matrix correspondence. Systematic Zoology 35: 627–632.
87. Rousset F (1997) Genetic differentiation and estimation of gene flow from F-statistics under isolation by distance. Genetics 145: 1219–1228.
88. Cushman SA, Landguth EL (2010) Spurious correlations and inference in landscape genetics. Molecular Ecology 19: 3592–3602.
89. Castillo JA, Epps CW, Davis AR, Cushman SA (2014) Landscape effects on gene flow for a climate-sensitive montane species, the American pika. Molecular Ecology 23: 843–856.
90. Mills LS, Citta JJ, Lair KP, Schwartz MK, Tallmon DA (2000) Estimating animal abundance using noninvasive DNA sampling: Promise and pitfalls. Ecological Applications 10: 283–294.
91. Castillo JA, Epps CW, Davis AR, Cushman S (2014) Landscape effects on gene flow for a climate-sensitive montane species, the American pika. Molecular Ecology 23: 843–856.
92. Balkenhol N, Gugerli F, Cushman SA, Waits LP, Coulon A, et al. (2009) Identifying future research needs in landscape genetics: where to from here? Landscape Ecology 24: 455–463.
93. Guillot G, Rousset F (2013) Dismantling the Mantel tests. Methods in Ecology and Evolution 4: 336–344.
94. Graves TA, Beier P, Royle JA (2013) Current approaches using genetic distances produce poor estimates of landscape resistance to interindividual dispersal. Molecular Ecology 22: 3888–3903.
95. Coulon A, Guillot G, Cosson JF, Angibault JMA, Aulagnier S, et al. (2006) Genetic structure is influenced by landscape features: empirical evidence from a roe deer population. Molecular Ecology 15: 1669–1679.
96. Cushman SA, Max TL, Whitham TG, Allan GJ (2014) River network connectivity and climate gradients drive genetic differentiation in a riparian foundation tree. Ecological Applications 24:1000–1014.
97. Guillot G, Leblois R, Coulon A, Frantz AC (2009) Statistical methods in spatial genetics. Molecular Ecology 18: 4734–4756.

98. Latch EK, Boarman WI, Walde A, Fleischer RC (2011) Fine-Scale Analysis Reveals Cryptic Landscape Genetic Structure in Desert Tortoises. Plos One 6.
99. Quemere E, Crouau-Roy B, Rabarivola C, Louis EE, Chikhi L (2010) Landscape genetics of an endangered lemur (*Propithecus tattersalli*) within its entire fragmented range. Molecular Ecology 19: 1606–1621.
100. Watts PC, Rousset F, Saccheri IJ, Leblois R, Kemp SJ, et al. (2007) Compatible genetic and ecological estimates of dispersal rates in insect (*Coenagrion mercuriale*: Odonata: Zygoptera) populations: analysis of 'neighbourhood size' using a more precise estimator. Molecular Ecology 16: 737–751.
101. Row JR, Blouin-Demers G, Lougheed SC (2010) Habitat distribution influences dispersal and fine-scale genetic population structure of eastern foxsnakes (*Mintonius gloydi*) across a fragmented landscape. Molecular Ecology 19: 5157–5171.
102. Cushman SA, Lewis JS (2010) Movement behavior explains genetic differentiation in American black bears. Landscape Ecology 25: 1613–1625.
103. Balkenhol N, Waits LP (2009) Molecular road ecology: exploring the potential of genetics for investigating transportation impacts on wildlife. Molecular Ecology 18: 4151–4164.
104. Cushman SA, Compton BW, McGarigal K (2010) Habitat fragmentation effects depend on complex interactions between population size and dispersal ability: Modeling influences of roads, agriculture and residential development across a range of life-history characteristics. In: Cushman SA, Huettman F, editors. Spatial complexity, informatics and wildlife conservation. Tokyo: Springer. pp. 369–387.
105. Jackson ND, Fahrig L (2011) Relative effects of road mortality and decreased connectivity on population genetic diversity. Biological Conservation 144: 3143–3148.
106. Cushman SA, Lewis JS, Landguth EL (2013) Evaluating the intersection of a regional wildlife connectivity network with highways. Movement Ecology 1: 1–12.
107. Cushman SA, Raphael MG, Ruggiero LF, Shirk AS, Wasserman TN, et al. (2011) Limiting factors and landscape connectivity: the American marten in the Rocky Mountains. Landscape Ecology 26: 1137–1149.
108. Wasserman TN, Cushman SA, Littell JS, Shirk AJ, Landguth EL (2013) Population connectivity and genetic diversity of American marten (*Martes americana*) in the United States northern Rocky Mountains in a climate change context. Conservation Genetics 14: 529–541.
109. Garroway CJ, Bowman J, Wilson PJ (2011) Using a genetic network to parameterize a landscape resistance surface for fishers, *Martes pennanti*. Molecular Ecology 20: 3978–3988.
110. Koen EL, Bowman J, Garroway CJ, Mills SC, Wilson PJ (2012) Landscape resistance and American marten gene flow. Landscape Ecology 27: 29–43.

11

Nonlinear Effect of Dispersal Rate on Spatial Synchrony of Predator-Prey Cycles

Jeremy W. Fox[1]*, Geoff Legault[1], David A. Vasseur[2], Jodie A. Einarson[1]

1 Department of Biological Sciences, University of Calgary, Calgary, Alberta, Canada, **2** Department of Ecology and Evolutionary Biology, Yale University, New Haven, Connecticut, United States of America

Abstract

Spatially-separated populations often exhibit positively correlated fluctuations in abundance and other population variables, a phenomenon known as spatial synchrony. Generation and maintenance of synchrony requires forces that rapidly restore synchrony in the face of desynchronizing forces such as demographic and environmental stochasticity. One such force is dispersal, which couples local populations together, thereby synchronizing them. Theory predicts that average spatial synchrony can be a nonlinear function of dispersal rate, but the form of the dispersal rate-synchrony relationship has never been quantified for any system. Theory also predicts that in the presence of demographic and environmental stochasticity, realized levels of synchrony can exhibit high variability around the average, so that ecologically-identical metapopulations might exhibit very different levels of synchrony. We quantified the dispersal rate-synchrony relationship using a model system of protist predator-prey cycles in pairs of laboratory microcosms linked by different rates of dispersal. Paired predator-prey cycles initially were anti-synchronous, and were subject to demographic stochasticity and spatially-uncorrelated temperature fluctuations, challenging the ability of dispersal to rapidly synchronize them. Mean synchrony of prey cycles was a nonlinear, saturating function of dispersal rate. Even extremely low rates of dispersal (<0.4% per prey generation) were capable of rapidly bringing initially anti-synchronous cycles into synchrony. Consistent with theory, ecologically-identical replicates exhibited very different levels of prey synchrony, especially at low to intermediate dispersal rates. Our results suggest that even the very low rates of dispersal observed in many natural systems are sufficient to generate and maintain synchrony of cyclic population dynamics, at least when environments are not too spatially heterogeneous.

Editor: Frédéric Guichard, McGill University, Canada

Funding: This work was supported by National Sciences Foundation (NSF) DEB grant 2050803 to David A. Vasseur. (www.nsf.gov). Natural Sciences and Engineering Research Council (NSERC) Discovery Grant RGPIN/311728-2010 to Jeremy W. Fox. (www.nserc.ca). The funders had no role in study design, data collection and analysis, decision to publish, or preparation of the manuscript.

Competing Interests: The authors have declared that no competing interests exist.

* E-mail: jefox@ucalgary.ca

Introduction

Spatially-separated populations in nature often exhibit correlated fluctuations in abundance, population growth rate, and other properties [1]. Even populations separated by hundreds or thousands of kilometers can exhibit spatially-synchronized fluctuations. Although climate synchronization (the Moran effect) can account for some of this phenomenon [2], it cannot account for all of it [3], [4], [5], [6], [7], [8]. Dispersal is a key mechanism generating spatial synchrony in populations experiencing similar density-dependent processes [4], [5], [9], [10]. Dispersal partially mixes spatially-separated populations and couples them together, thereby synchronizing them.

In nature, dispersal rates often are very low, especially at longer distances; most dispersing individuals or propagules don't move very far (reviewed in [11]). It is unclear how low dispersal rates can be while still producing appreciable spatial synchrony. Comparative work suggests that species dispersing longer distances are synchronized over longer distances [3], [12]. This suggests that observed cases of long-distance synchrony represent situations in which even long-distance dispersal rates are high (e.g., [13]). However, theory predicts that the dispersal rate-synchrony relationship can take on various forms, depending on biological details (reviewed in [14]). The spatial scale of synchrony need not align with the spatial scale of the typical dispersal distance [15]. Dispersal rates also can vary over time and space, including due to human activities. For instance, anthropogenic habitat fragmentation can restrict animal movement; conversely, air travel can spread human disease outbreaks around the world in days. Only a few controlled experiments manipulate dispersal rate and examine the effect on spatial synchrony, and these studies consider only one or two non-zero dispersal rates [15], [16], [17], [18]. Ecology currently lacks a quantitative understanding of the expected relationship between dispersal rate and spatial synchrony. For instance, how low can dispersal rates be while still producing appreciable spatial synchrony?

Ecology also lacks a complete understanding of the variation around the expected dispersal rate-synchrony relationship. In general, the dynamics of a single realization of any stochastic process can deviate considerably from the average dynamics. For instance, Figure 1 shows prey dynamics in four simulations of a two-patch predator-prey model with density-independent dispersal of both species and demographic stochasticity, and where local dynamics are cyclic. All four simulations began with the same initial conditions, and all four used the same parameter values, yet the realized spatial synchrony of those dynamics varied widely. Fig. 1a shows a slow transition from initially anti-synchronous, anti-phase cycles to synchronous, in-phase cycles, Fig. 1b shows a rapid transition to synchronous, in-phase cycles, Fig. 1c shows cycles that never transition to synchrony over the timeframe shown, and Fig. 1d shows cycles that go into phase and then drift back out of phase. Figure 1 also illustrates variation in cycle period and amplitude between patches within a single simulation, between simulations, and over time. Only extremely high dispersal rates would be expected to produce a reliable, rapid transition to synchronous, in-phase cycles, with little variability among replicate realizations. Spatially-independent environmental stochasticity also would be expected to generate substantial variation in realized spatial synchrony. Variation in the realized level of synchrony also can arise because the probability distribution of synchrony for any given dispersal rate might be multimodal [19]. That is, there might not be one single expected level of synchrony around which realized levels of synchrony vary, but rather several different expected levels, each of which represents an alternative attractor.

We conducted what is to our knowledge the first quantification of the dispersal rate-synchrony relationship for any system by assembling two-patch predator-prey metapopulations in aquatic laboratory microcosms. Each metapopulation comprised a pair of semi-continuous cultures in which the ciliate protist predator *Euplotes patella* fed on the ciliate protist prey *Tetrahymena pyriformis*. Our previous work established this system as a model system for the study of spatial synchrony of predator-prey cycles [15], [18]. Here we subjected replicate metapopulations to one of 11 different rates of dispersal to quantify the dispersal rate-synchrony relationship, and to quantify among-replicate variation in synchrony arising from demographic and environmental stochasticity. This deliberately-simplified microcosm system allowed us to conduct a controlled experiment that would've been impossible to conduct in any natural system. Our results complement observational comparative data from natural systems on the dispersal rate-synchrony relationship (e.g., [3]). Our results provide a baseline aiding the interpretation of future experiments incorporating additional biological complexities.

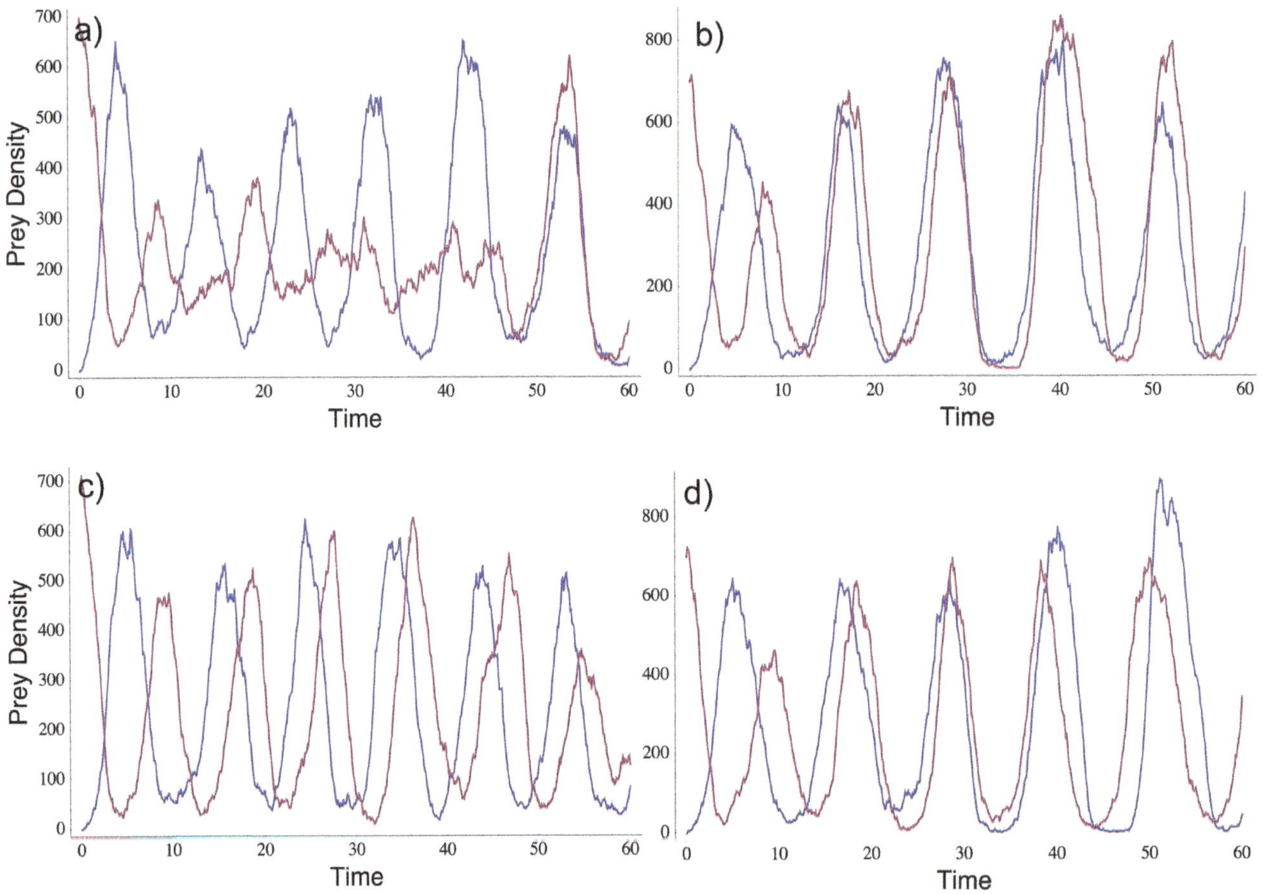

Figure 1. Variation in realized synchrony due to demographic stochasticity. Simulated prey population dynamics in a two-patch Rosenzweig-MacArthur predator-model with demographic stochasticity [33]. In each panel, red and blue lines show prey dynamics in two patches linked by dispersal of prey and predators at the same per-capita rate. The four panels show four different realizations of the model, using the same parameter values and starting from the same, initially-antisynchronous state. Because the model is stochastic, different realizations can have very different behavior, including (a) slow achievement of synchrony (after ~50 time units in this example), (b) rapid achievement of synchrony (after ~12 time units) which is subsequently maintained, (c) failure to achieve synchrony during the simulated time period, and (d) achievement of synchrony (after ~25 time units) which is subsequently lost. Dynamics were simulated using the SSA algorithm of [33], using the following parameter values: attack rate 0.01, handling time 3.0, predator per-capita mortality rate 0.5, predator conversion efficiency 0.4, prey intrinsic rate of increase 2.0 (= per-capita birth rate 3.0 - per-capita mortality rate 1.0), prey carrying capacity 1000, prey and predator per-capita dispersal rate 0.05.

Materials and Methods

Experimental units were pairs of microcosms containing the ciliate protist prey species *Tetrahymena pyriformis* and the ciliate protist predator *Euplotes patella*. Culture methods and sampling techniques closely followed previous work [15], [18]. Culture vessels were 100 ml screw-capped glass bottles containing 80 ml of nutrient medium and one wheat seed. Nutrient medium comprised 0.15 g/L of "Protozoan Pellets" (standardized pellets of crushed, dried plant matter; Carolina Biological Supply, Burlington, NC, USA), and one wheat seed per bottle. Bottles were loosely capped to allow gas exchange while preventing contamination. All materials were sterilized (autoclaved) before use.

Forty-eight hours before use, we inoculated the medium with a mixture of unidentified bacteria isolated from the stock cultures of the protists used in the experiment. On day 0 of the experiment, we inoculated all bottles with a small volume of medium from a *Tetrahymena* stock culture, and inoculated one randomly-chosen bottle in each pair with eight *Euplotes* cells drawn from a stock culture of *Euplotes* growing on *Tetrahymena*. We followed the population dynamics (see below for sampling methods) but conducted no dispersal events, until *Euplotes* reached high density and drove *Tetrahymena* to low density. At that point (day 20), we inoculated the other bottle in each pair with eight *Euplotes* cells drawn from the *Euplotes* stock culture. This created pairs of patches with initially anti-phase (maximally out-of-sync) predator-prey cycles. We then began dispersal treatments. The experiment ran until day 92, with our analyses of synchrony considering only the final 72 d, during which dispersal occurred.

We conducted dispersal events daily on weekdays, after sampling. To conduct a dispersal event, bottles were agitated, and a sample of medium and the organisms in it was withdrawn from each bottle and exchanged with the sample from the other bottle in the pair. We used 11 dispersal rate treatments: exchange of 0, 0.125, 0.5, 1, 2.5, 3, 4, 5, 6, 9, or 12.5% of medium per dispersal event. Our previous work used a dispersal rate of 10% of medium per event and three events per week, approximately equivalent to our 6% dispersal rate treatment since we conducted five dispersal events per week in this experiment [15], [18]. Our dispersal rates thus ranged from no dispersal up to dispersal rates substantially higher than those shown to maintain synchrony in previous work [15], [18]. We replicated each dispersal rate treatment five times, for a total of 55 pairs of bottles. This was the largest experiment feasible, given the equipment and personnel available.

For comparability with previous experiments [15], [18], and to provide an additional challenge to the synchronizing effects of dispersal, we subjected all bottles to daily, spatially-asynchronous temperature fluctuations. Temperature fluctuations started at the same time as dispersal; before that, all bottles were maintained in a 20°C incubator. We imposed temperature fluctuations by swapping bottles between two incubators, set at 20°C and 30°C respectively. Each bottle experienced a predetermined sequence of temperatures, generated using the method of [20]. All bottles experienced the same number of days at each temperature, and experienced temperature fluctuations with the same autocorrelation (a $1/f^{0.5}$ power spectrum). The two bottles in a pair experienced spatially uncorrelated temperature fluctuations (zero cross-correlation). While spatially-uncorrelated temperature fluctuations should tend to inhibit synchronization in all treatments, temperature fluctuations could not create differences in synchrony among treatments because all bottle pairs experienced equally-uncorrelated temperature fluctuations. Each of the five pairs of bottles within a dispersal rate treatment experienced a unique pair of temperature time series, and we used the same five pairs of temperature time series for all dispersal rate treatments. This procedure ensured that our results did not reflect unusual properties of any particular pair of temperature time series, and also ensured that differences among treatments did not reflect differences among the temperature time series used in different treatments.

Once per week beginning on day 7, we agitated the bottles, withdrew 8 ml of medium, and replaced it with 10 ml of fresh sterile medium to prevent exhaustion of the resource base, accumulation of waste products, and to replace medium lost to sampling and evaporation. Medium replacement occurred after sampling and dispersal. Medium replacement did not generate a detectable signal in the population dynamics, and could not create differences among treatments because all bottles experienced the same medium replacement regime.

Bottles were sampled on weekdays starting on day 1, using established methods [18]. Briefly, sampling involved counting the protists in small (*c.* 0.3 ml) samples under a binocular microscope, with dilution and subsampling as necessary to count dense populations. For various technical reasons, it was not possible to automate sampling using a particle counter or image analysis software. Sampling and maintaining the experiment was a full-time job for two people.

Data processing and statistical analysis

In 13 of the 55 metapopulations, either the prey, the predator, or both went extinct in at least one (usually both) of the two bottles early in the experiment, with the last non-zero density being observed before day 68. Apparent extinctions were confirmed at the end of the experiment by pouring bottle contents into petri dishes and scanning under low magnification. We excluded these metapopulations from the analyses. Extinctions occurred randomly with respect to dispersal rate and the particular temperature sequence experienced, occurred too early in the experiment to be related to realized levels of pre-extinction synchrony, and could not be related to the results in any obvious way. Extinctions likely reflected demographic stochasticity, given that predator-prey cycles in this system are characterized by long periods of low abundance for both prey and predators (see RESULTS). We also excluded from the analyses one metapopulation in which both prey populations were driven to very low but non-zero density immediately after predator addition and remained at very low density for the entire experiment. Prey dynamics in this metapopulation were non-cyclic and dominated by sampling error, so that synchrony could not be meaningfully analyzed or compared to that of the cycling populations.

As in our previous experiment using these culture conditions [15], predator densities were low on average. Sampling error prevents meaningful analysis of predator synchrony.

Our time series were fairly short relative to the period of the predator-prey cycle (>20 d; see RESULTS). Further, inspection of the time series indicated that cycle period varied both over time and among bottles, a feature of the data we discuss below. For this reason, we could not use spectral analysis or wavelets to quantify predator-prey cycle periods or the phase difference between paired cycles as in [15]. Instead we summarized prey synchrony in each metapopulation by transforming prey densities as $\log(n/\text{ml}+1)$, and then calculating the cross-correlation between transformed prey densities in the two bottles in each pair. This approach was used in the previous theoretical and experimental work on which the present study builds [15], [18], [21]. Square root transforming prey densities before calculating cross-correlations led to similar results. We calculated cross-correlations over the entire 72 d following initiation of dispersal, rather than trying to quantify temporal changes in synchrony during this period of time, because cross-correlations calculated over short time windows necessarily are based on only a few data points and so are dominated by sampling error. Linearly detrending the transformed prey densities before calculation of cross-correlation coefficients did not qualitatively affect the results and produced only minor quantitative changes. We applied Fisher's z transformation to the cross-correlation coefficients to normalize them before analysis, but note that this made no qualitative difference, as most of the cross-correlations were non-extreme.

We fit several different linear and nonlinear regression models to the relationship between dispersal rate (% medium exchanged per event) and prey synchrony (z-transformed cross-correlation) in order to identify the best model. The most complicated continuous model was a flexible, asymmetric sigmoid model $y = [ax/(b+x)][1+(1-cx)e^{-cx}]$. In this model, the parameter c governs the location of the inflection point of the sigmoid curve on the x--axis; the inflection point is located at $x = 1/c$. This choice of sigmoid function necessarily is somewhat arbitrary, but various other asymmetric sigmoid functions gave similar results (not shown). We included this model among the candidates because recent theory predicts an asymmetrical sigmoid relationship between dispersal rate and spatial synchrony in a system similar to our experimental system [21]. We also considered the simpler saturating model $y = ax/(b+x)$, the linear model $y = ax$, and the null model $y = a$, which lacks any effect of dispersal rate on synchrony. In the limit as $c \rightarrow \infty$, the asymmetrical sigmoid model approaches the saturating model. Finally, we also fit a discontinuous nonlinear model, a piecewise linear regression with a single discontinuity. While there is no biological rationale for a piecewise linear relationship between synchrony and dispersal rate, a piecewise linear function can approximate a relationship with a relatively sharp transition between two phases. Inspection of the residuals for the best-fitting model indicated conformity with statistical assumptions. We compared the models using AIC and (where appropriate) likelihood ratio tests. All statistical analyses were performed in R 2.15.1, with function nls and optimization algorithm "port" used to fit the continuous nonlinear models and function segmented.lm from the segmented package used to fit the discontinuous nonlinear model.

Results

Prey synchrony was near-zero on average in the absence of dispersal, as intended and confirming that weekly medium replacement was too small and infrequent a perturbation to affect synchrony (Fig. 2). Dispersal rates as low as 0.5% per event were capable of producing levels of synchrony substantially higher than those observed in the no-dispersal treatment (Fig. 2). Dispersal rates as low as 2.5% per event were capable of producing quite high synchrony (cross-correlation >0.7).

The saturating model $y = ax/(b+x)$ had the lowest AIC. The saturating model was estimated to be somewhat closer to the unknown true model than the asymmetrical sigmoid model (ΔAIC = 1.9). Further, the maximum likelihood parameter estimates for the asymmetrical sigmoid model produced a curve nearly identical to the saturating model (Fig. 2). The saturating model was estimated to be somewhat closer to the unknown true model than the piecewise linear model (ΔAIC = 3.7), which required more parameters to produce a similar fitted relationship (Fig. 2). The saturating model was estimated to be much closer to the unknown true model than either the linear model or the null model (ΔAIC = 10.8 and 13.4, respectively). The saturating model also fit the data significantly better than the null model $y = a$ in a likelihood ratio test, indicating that synchrony did vary significantly with dispersal rate ($F = 17.9$, $P<0.001$ with 1 df).

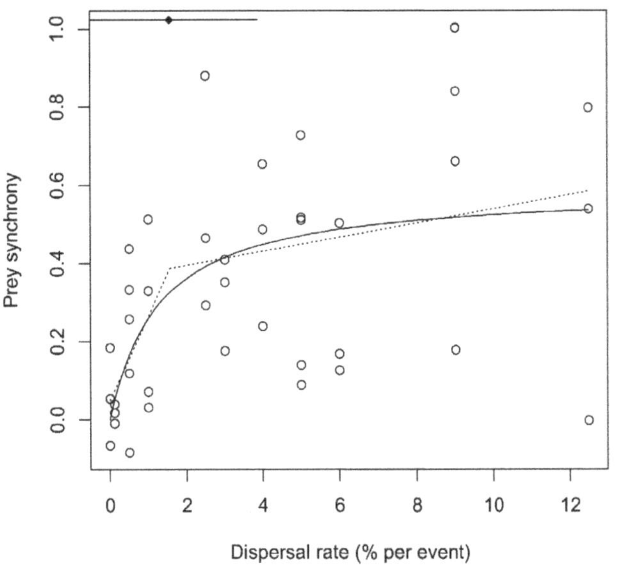

Figure 2. Prey synchrony vs. dispersal rate. Prey synchrony (z-transformed cross-correlation of $\log_{10}$-transformed prey abundances) as a function of dispersal rate. Each open point gives results from one replicate pair of bottles. The solid curve is $y = ax/(b+x)$ with estimated parameters (95% likelihood profile c.i.) of $a = 0.59$ (0.39, 1.25), $b = 1.27$ (0.14,8.50). The curve $y = [ax/(b+x)][1+(1-cx)e^{-cx}]$ with estimated parameters (95% likelihood profile c.i.) of $a = 0.59$ (0.45, 1.26), $b = 1.22$ (0.45, 8.62), $c = 14.66$ (14.66, 44.64) is hidden by the solid curve. The dotted curve is a piecewise linear regression. The black diamond indicates the estimated location and 95% confidence interval for the discontinuity of the dotted curve. See [34] for review of the concept of profile confidence intervals.

While prey synchrony increased on average with increasing dispersal rate, there was substantial variation around this trend (Fig. 2). Replicates with the same dispersal rate often exhibited very different levels of synchrony. Figure 3 illustrates the population dynamics underpinning this variability. As intended, paired bottles invariably were anti-synchronous when dispersal began on day 20, with one prey population at high density and the other at low density. Subsequent dynamics clearly were cyclic in all bottles included in the analysis. However, the realized level of synchrony varied widely. Metapopulations that experienced very low dispersal rates (≤1% per event) rarely went into phase quickly and so never exhibited high prey synchrony (Fig. 3a,b). Conversely, metapopulations that experienced the highest dispersal rates (≥9% per event) typically went into phase quickly and so typically exhibited relatively high prey synchrony (Fig. 3g,h). At dispersal rates in between these extremes, prey cycles sometimes remained out of phase until late in the experiment (Fig. 3c), sometimes went into phase quickly (Fig. 3d), sometimes never went into phase at all (Fig. 3e), and sometimes went into phase quickly only to subsequently drift out of phase (Fig. 3f). This range of behavior closely resembles that produced by a spatial predator-prey model incorporating demographic stochasticity (compare Fig. 3c-f to Fig. 2).

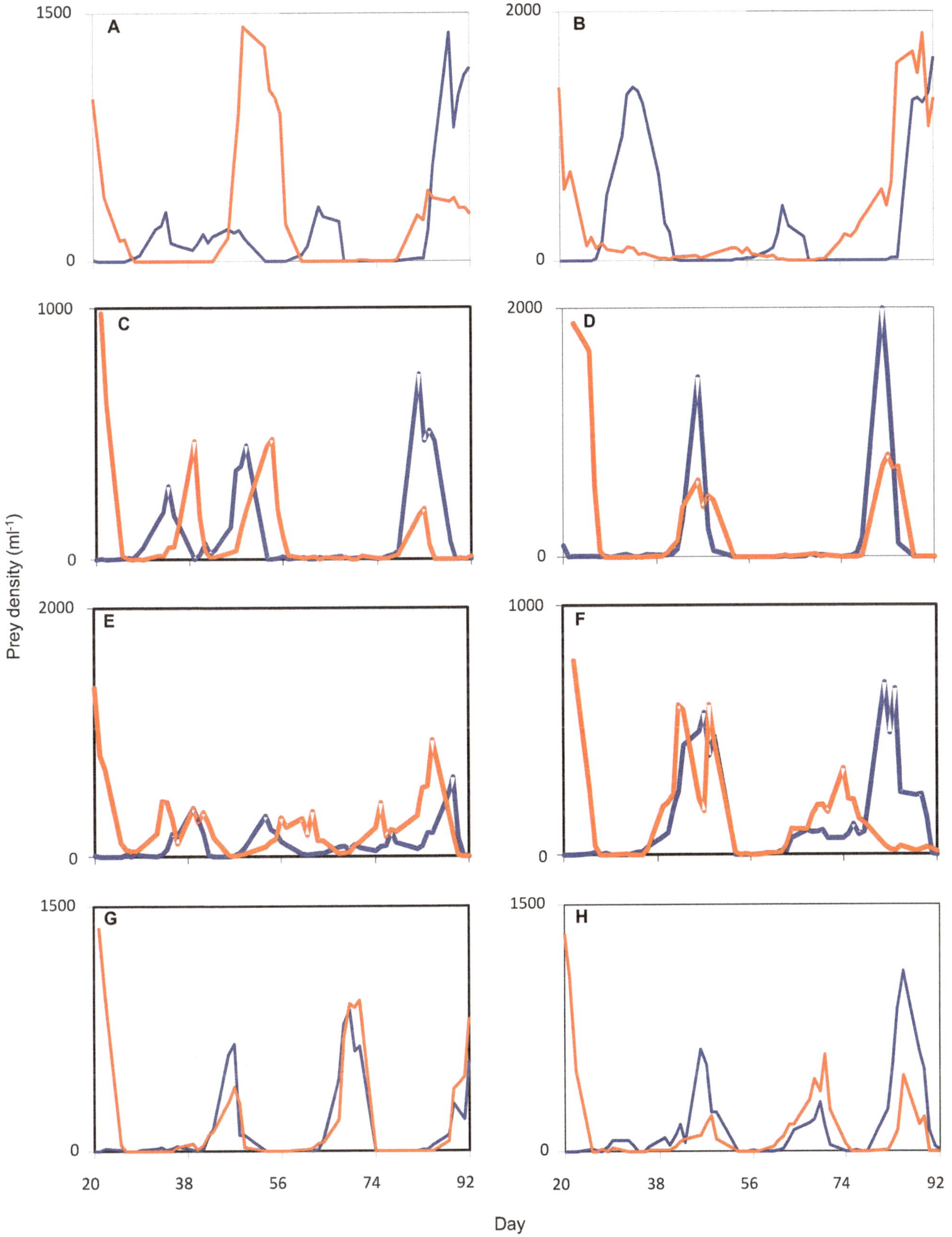

Figure 3. Representative prey population dynamics. Red and blue lines in each panel give prey dynamics in two patches linked by dispersal, starting from day 20 when dispersal was initiated. (a-d) Failure to achieve synchrony with a dispersal rate of 0.125% per event, (c) slow achievement of synchrony with a dispersal rate of 5% per event, (d) rapid achievement of synchrony with a dispersal rate of 5% per event, (e) failure to achieve

synchrony with dispersal rate of 2.5% per event, (f) rapid achievement of synchrony which was subsequently lost with dispersal rate of 2.5% per event, (g-h) rapid achievement of synchrony with a dispersal rate of (g) 9% or (h) 12.5% per event. Compare c-f to Figure 1.

Discussion

Our results provide the first quantitative description of the dispersal rate-synchrony relationship for any ecological system. We find that even extremely low rates of dispersal, here around 0.5-1% per dispersal event (equivalent to ~0.1–0.4% per prey generation), can produce ecologically-substantial levels of synchrony within just 2–3 cycle periods, despite starting from an initially anti-synchronous state and despite spatially-asynchronous environmental fluctuations and demographic stochasticity. It is remarkable that such low rates of dispersal can have such a strong effect on spatial population dynamics. To our knowledge, this is the first experiment in ecology to test the effects of such low rates of dispersal on spatial synchrony. Dispersal rates as low as 2.5% per event were capable of producing high synchrony. Given that interpatch movement rates in nature average approximately 15% per generation [22], our results suggest that natural dispersal rates typically will be sufficient to generate spatial synchrony, so long as natural systems are sufficiently similar to ours in other relevant respects (a caveat discussed further below).

Our results also confirm the theoretical prediction that demographic and environmental stochasticity produce substantial variation around the average behavior. Even for the highest dispersal rates used here (equivalent to >2% per prey generation), it was possible for prey metapopulations experiencing identical dispersal rates and starting from nearly-identical initial conditions to exhibit anything from near-zero synchrony to near-perfect synchrony. Inspection of the population dynamics indicates that among-replicate variability in realized synchrony was not due to sampling error, but rather reflected real variation among metapopulations and over time in the phase and period of the predator-prey cycles (i.e. in the timing and length of predator and prey "outbreaks"; Fig. 3). Just as predicted by theory (Fig. 1), our paired predator-prey populations varied greatly in how quickly they went into phase, and in whether or not they subsequently stayed in phase or drifted back out of phase (Fig. 3).

The relative roles of environmental and demographic stochasticity in generating variation in realized synchrony are unclear. Theory predicts that asynchronous environmental stochasticity can inhibit phase locking of predator-prey cycles. Fox et al. [15] found that environmental stochasticity in this system did not affect the phase of predator-prey cycles, but instead affected the synchrony of low-amplitude, environmentally-generated stochastic fluctuations in abundance superimposed on the dominant cyclic pattern. However, Fox et al. [15] noted that it was unclear if this would remain true in longer experiments, such as the one we conducted. Realized levels of synchrony were not related in any obvious way to the particular temperature time series used. Demographic stochasticity, and its desynchronizing effects, are strong when demographic events (births and deaths of individuals) are rare, as when population sizes are very low [21]. While protists are capable of achieving very large population sizes under the culture conditions used here, they can be affected by demographic stochasticity when population sizes are low, as at the nadir of a predator-prey cycle [23], [24]. That a number of our predator and prey populations went extinct illustrates that protist population sizes can drop low enough for demographic stochasticity to affect their dynamics. In future work we plan to directly test for the desynchronizing effects of demographic stochasticity by varying culture vessel size, thereby varying absolute population sizes.

Recent theory considers factors governing the rate of convergence to synchrony [25]. Our predator *Euplotes* has much slower dynamics than the prey *Tetrahymena*, leading to prey cycles that take the form of relaxation oscillations: rapid increases and subsequent crashes in prey abundance, separated by long intervals of very low density (Fig. 3). In a deterministic world, dispersal should rapidly synchronize relaxation oscillations [25]. However, long periods of low density should increase the desynchronizing effects of demographic stochasticity. Manipulating culture vessel size would allow a test for rapid convergence to synchrony at sufficiently large absolute population sizes.

Theory identifies conditions under which cycling predator-prey metapopulations will converge towards a multimodal distribution of synchrony, leading to wide variation in realized levels of synchrony among replicate metapopulations subject to both dispersal and spatially-independent demographic and environmental stochasticity [19]. It is unclear whether our system satisfies the conditions for a multimodal distribution of synchrony [19]. Testing this interesting possibility would require a much larger experiment than the one we conducted.

We were unable to detect an accelerating (concave up) relationship between synchrony and dispersal rate at very low dispersal rates, although we cannot rule it out. Possibly, upward concavity to the dispersal rate-synchrony relationship is weak or absent, or occurs only at lower dispersal rates than the ones we considered. Alternatively, among-replicate variability in the dispersal rate-synchrony relationship may have been too large for us to detect significant upward concavity at low dispersal rates.

Mean spatial synchrony reached a moderate asymptote (mean prey cross-correlation ~0.5) at high dispersal rates, in contrast to the theoretically-predicted asymptote at near-perfect synchrony [21]. Sampling error can reduce observed synchrony by creating spatially-independent errors in population density estimates, but there are other processes contributing to this contrast. We imposed spatially-independent temperature fluctuations on our paired bottles. In previous work we showed that this reduces the prey cross-correlation in paired bottles by ~0.3-0.4 compared to imposition of spatially-synchronized environmental fluctuations [15], [18]. In light of sampling error and spatially-independent environmental stochasticity, our demonstration of even a moderate "saturating" level of synchrony illustrates the robustness of synchrony in this experiment and ecological systems in general.

Spatial synchrony might have attained a somewhat higher asymptote had the experiment run longer, thereby allowing more time for metapopulations to achieve their asymptotic dynamics from their initially anti-synchronous state. But comparison with previous work suggests that at least our highest-dispersal metapopulations likely were relatively close to their asymptotic levels of synchrony. In previous experiments in which predator-prey cycles in different patches were started in phase or nearly in phase, and in which an intermediate dispersal rate was used, realized levels of prey synchrony under spatially-independent environmental fluctuations were either no higher than the highest levels observed in this experiment [18], or only somewhat higher [15]. Conversely, a longer experiment might not have changed the results, since given sufficient time, paired bottles can drift out of as well as into phase, a phenomenon we observed in our experiment.

We note in passing that sufficiently high rates of dispersal would produce high synchrony for the trivial reason that they would convert the metapopulation into a single well-mixed population.

Our highest dispersal rate was well below the rate required to create a single well-mixed population.

Our results demonstrate that even quite low rates of dispersal, much less than 1% per generation, can generate surprisingly high spatial synchrony, even in the face of demographic and environmental stochasticity. However, demographic and environmental stochasticity can generate substantial variability in realized levels of synchrony even among ecologically-identical metapopulations. While our microcosms were not designed to mimic any particular natural system, our experiment included key factors thought to affect the synchrony of a wide range of systems. Our results suggest that, for natural species exhibiting endogenously-generated cyclic dynamics (which may be as much as 30% of populations; [9]), spatial synchrony is likely to be common. Synchrony is particularly likely in cycling systems with dispersal rates greater than a few percent per generation, since such systems likely exhibit relatively rapid convergence to synchrony.

Our experimental system can be modified to incorporate various ecological complexities, and so be used to test their effects on spatial synchrony. One major mismatch between our experiment and many natural systems is that our experiment lacked spatial environmental heterogeneity, which in nature is predicted to interfere with synchrony by causing different populations to exhibit different dynamics (e.g., cycles of different periods and amplitudes). The periods and amplitudes of microcosm predator-prey cycles can be manipulated by varying the enrichment and thickness of the culture medium and other aspects of culture conditions [26], [27]. A second potentially-important mismatch between our experiment and many natural systems is our imposition of density-independent dispersal. Under some culture conditions, some protists exhibit adaptive, density dependent dispersal when allowed to disperse themselves among microcosms interconnected by tubes [28], [29], but see [30], [31]. Theory predicts that adaptive dispersal can alter both within-patch population dynamics and their spatial synchrony [32]. Testing this theory would be an interesting direction for future work.

Acknowledgments

Thanks to Stephanie Thibodeau for assistance with data collection, and to the referees for comments.

Author Contributions

Conceived and designed the experiments: JWF DAV. Performed the experiments: JE JWF. Analyzed the data: JWF GL. Wrote the paper: JWF GL DAV JE.

References

1. Liebhold A, Koenig WD, Bjørnstad ON (2004) Spatial synchrony in population dynamics. Annu Rev Ecol Evol Syst 35: 467–490.
2. Koenig WD (2002) Global patterns of environmental synchrony and the Moran effect. Ecography 25: 283–288.
3. Sutcliffe OL, Thomas CD, Moss D (1996) Spatial synchrony and asynchrony in butterfly population dynamics. J Anim Ecol 65: 85–95.
4. Blasius B, Huppert A, Stone L (1999) Complex dynamics and phase syncrhonization in spatially extended ecological systems. Nature 399: 354–359.
5. Bjørnstad ON (2000) Cycles and synchrony: two historical 'experiments' and one experience. J Anim Ecol 69: 869–873.
6. Bunnell D, Adams J, Gorman O, Madenjian C, Riley S et al. (2010) Population synchrony of a native fish across three Laurentian Great Lakes: evaluating the effects of dispersal and climate. Oecologia 162: 641–651.
7. de Valpine P, Scranton K, Ohmart CP (2010) Synchrony of population dynamics of two vineyard arthropods occurs at multiple spatial and temporal scales. Ecol Appl 20: 1926–1935.
8. Crone EE (2012) Desynchronization and re-synchronization of reproduction by Astragalus scaphoides, a plant that flowers in alternate years. Ecol Res in press. DOI: 10.1007/s11284-012-0942-8
9. Kendall BE, Prendergast J, Bjørnstad ON (1998) The macroecology of population dynamics: taxonomic and biogeographic patterns in population cycles. Ecol Lett 1: 160–164.
10. [10] Kaitala V, Ranta E, Lundberg P (2001) Self-organized dynamics in spatially structured populations. Proc Roy Soc B 268: 1655–1660.
11. Kot M., Lewis MA, van den Driessche P (1996) Dispersal data and the spread of invading organisms. Ecology 77: 2027–2042.
12. Paradis E, Baillie SR, Sutherland WJ, Gregory RD (1999) Dispersal and spatial scale affects synchrony in spatial population dynamics. Ecol Lett 2: 114–120.
13. Schwartz MK, Mills LS, McKelvey KS, Ruggiero LF, Allendorf FW (2002) DNA reveals high dispersal synchronizing the population dynamics of Canada lynx. Nature 415: 520–522.
14. Abbott KC (2011) A dispersal-induced paradox: synchrony and stability in stochastic metapopulations. Ecol Lett 14: 1158–1169.
15. Fox JW, Vasseur DA, Hausch S, Roberts J (2011) Phase locking, the Moran effect, and distance decay of synchrony: experimental tests in a model system. Ecol Lett 14: 163–168.
16. Dey S, Joshi A (2006) Stability via asynchrony in *Drosophila* metapopulations with low migration rates. Science 312: 434–436.
17. Vogwill T, Fenton A, Brockhurst MA (2009) Dispersal and natural enemies interact to drive synchrony and decrease stability in patchy populations. Ecol Lett 12: 1194–1200.
18. Vasseur DA, Fox JW (2009) Phase locking and environmental fluctuations generate synchrony in a predator-prey community. Nature 460: 1007–1010.
19. Bressloff PC, Lai YM (2012) Dispersal and noise: various modes of synchrony in ecological oscillators. J Math Biol. doi: 10.1007/s00285-012-0607-9
20. Vasseur DA (2007) Environmental colour intensifies the Moran effect when population dynamics are spatially heterogeneous. Oikos 116: 1726–1736.
21. Simonis JL (2012) Demographic stochasticity reduces the synchronizing effect of dispersal in predator-prey metapopulations. Ecology 93: 1517–1524.
22. Bowne DR, Bowers MA (2004) Interpatch movements in spatially-structured populations: a literature review. Landscape Ecology 19: 1–20.
23. Holyoak M, Lawler SP (1996) Persistence of an extinction-prone predator-prey interaction through metapopulation dynamics. Ecology 77: 1867–1879.
24. Holyoak M, Lawler SP, Crowley PH (2000) Predicting extinction: progress with an individual-based model of protozoan predators and prey. Ecology 81: 3312–3329.
25. Goldwyn EE, Hastings A (2008) When can dispersal synchronize populations? Theor Pop Biol 73:395–402.
26. Luckinbill LS (1973) Coexistence in laboratory populations of *Paramecium aurelia* and its predator *Didinium nasutum*. Ecology 54: 1320–1327.
27. Luckinbill LS (1979) Regulation, stability, and diversity in a model experimental microcosm. Ecology 60: 1098–1102.
28. Hauzy C, Hulot FD, Gins A, Loreau M. (2007) Intra- and interspecific density-dependent dispersal in an aquatic prey-predator system. J Anim Ecol, 76: 552–558.
29. Fellous S, Duncan A, Coulon A, Kaltz O (2012) Quorum sensing and density-dependent dispersal in an aquatic model system. PLoS ONE 7: e48436. DOI:10.1371/journal.pone.0048436
30. Donahue MJ, Holyoak M, Feng C (2003) Patterns of dispersal and dynamics among habitat patches of varying quality. Am Nat 162: 302–317.
31. Fox JW (2007) Testing the mechanisms by which source-sink dynamics alter competitive outcomes in a model system. Am Nat 170: 396–408.
32. Abrams PA, Ruokolainen L (2011) How does adaptive consumer movement affect population dynamics in consumer-resource metacommunities with homogeneous patches? J Theor Biol 277: 99–110.
33. Yaari G, Ben-Zion Y, Shnerb NM, Vasseur DA (2012) Consistent scaling of persistence time in metapopulations. Ecology 93: 1214–1227.
34. Hudson DJ (1971) Interval estimation from the likelihood function. J Roy Statis Soc B 33: 256–262.

12

Defining Landscape Resistance Values in Least-Cost Connectivity Models for the Invasive Grey Squirrel: A Comparison of Approaches Using Expert-Opinion and Habitat Suitability Modelling

Claire D. Stevenson-Holt[1]*, Kevin Watts[2], Chloe C. Bellamy[3], Owen T. Nevin[4], Andrew D. Ramsey[5]

1 Centre for Wildlife Conservation, University of Cumbria, Ambleside, Cumbria, United Kingdom, **2** Centre for Ecosystems, Society and Biosecurity, Forest Research, Farnham, Surrey, United Kingdom, **3** Centre for Ecosystems, Society and Biosecurity, Forest Research, Roslin, Midlothian, United Kingdom, **4** School of Medical and Applied Sciences, Central Queensland University, Gladstone, Queensland, Australia, **5** School of Biological and Forensic Sciences, University of Derby, Derby, Derbyshire, United Kingdom

Abstract

Least-cost models are widely used to study the functional connectivity of habitat within a varied landscape matrix. A critical step in the process is identifying resistance values for each land cover based upon the facilitating or impeding impact on species movement. Ideally resistance values would be parameterised with empirical data, but due to a shortage of such information, expert-opinion is often used. However, the use of expert-opinion is seen as subjective, human-centric and unreliable. This study derived resistance values from grey squirrel habitat suitability models (HSM) in order to compare the utility and validity of this approach with more traditional, expert-led methods. Models were built and tested with MaxEnt, using squirrel presence records and a categorical land cover map for Cumbria, UK. Predictions on the likelihood of squirrel occurrence within each land cover type were inverted, providing resistance values which were used to parameterise a least-cost model. The resulting habitat networks were measured and compared to those derived from a least-cost model built with previously collated information from experts. The expert-derived and HSM-inferred least-cost networks differ in precision. The HSM-informed networks were smaller and more fragmented because of the higher resistance values attributed to most habitats. These results are discussed in relation to the applicability of both approaches for conservation and management objectives, providing guidance to researchers and practitioners attempting to apply and interpret a least-cost approach to mapping ecological networks.

Editor: Benjamin Lee Allen, University of Queensland, Australia

Funding: This project was funded by the Forestry Commission GB and the National School of Forestry at the University of Cumbria. The funders had no role in study design, data collection and analysis, decision to publish, or preparation of the manuscript.

Competing Interests: The authors have the following competing interest: This work was funded by the Forestry Commission GB and National School of Forestry at the University of Cumbria. This does not alter the authors' adherence to all the PLOS ONE policies on sharing data and materials.

* Email: claire.stevenson@cumbria.ac.uk

Introduction

Effective biodiversity conservation within fragmented landscapes often requires the modelling of connectivity to define the extent of the problem, target conservation activities and to evaluate the impacts of landscape change [1]. Connectivity is defined as the degree to which the landscape facilitates or impedes species movement among resource patches [2]. A landscape consists of a complex, often dynamic, heterogeneous mixture of habitats and land uses which may impact on important ecological processes, such as species movement, habitat selection and survival, and influence behavioural and physiological responses [2–5]. The study of the impacts of the matrix on species movement, known as functional connectivity [6], is now the subject of much research within modified and fragmented landscapes [7]. Assessing functional connectivity is commonly used to aid conservation strategies by identifying potential movement pathways across fragmented landscapes for species of conservation concern [8–10]. It has also been used to help predict the potential dispersal and movement of invasive species to aid species management by identifying areas to target resources [11,12].

Geographical Information System (GIS), raster-based least-cost analysis techniques are often used to assess functional connectivity by modelling the impact of permeability of the surrounding landscape matrix on species movement [10]. It has been used in conservation [8–10] and invasive species management contexts [11,12]. For example, the population expansion of the grey squirrel (*Sciurus carolinensis*) in Britain, following its first introduction in 1876 [13], has had negative effects upon the forestry industry and native biodiversity [14–16]. In particular, it has occurred simultaneously with the decline and replacement of native red squirrel (*Sciurus vulgaris*) populations through resource competition and disease [14–16]. Therefore, an understanding of how grey squirrels utilise and move through the landscape is essential for effective red squirrel conservation and grey squirrel management. By using least-cost modelling it is possible to identify the potential dispersal areas, in addition to the most probable dispersal corridors, to assess the extent of spread [11]. Developing these models involves defining a species' 'core' or 'source' habitat and assigning resistance values to the surrounding landscape features, based on the actual or perceived impact to species movement at a particular resolution [17]. A cell with a high resistance value is used to represent an area that an individual is unlikely to traverse under typical conditions because of high energy, mortality, or other ecological costs [18]. Using information on a species' maximum dispersal distance, the area around a core habitat patch that is accessible to a species can be mapped with a simple Euclidean buffer. The permeability buffer zone is then taken into account so that the buffer is compressed or stretched according to the cumulative resistance scores assigned to the underlying landscape features. Overlapping buffers therefore signify connections where the species is assumed to be able to move between core habitat patches, forming a functionally connected habitat network.

It is widely acknowledged [4,18,19] that a critical step in least-cost modelling is defining resistance values for each type of landscape feature. Beier et al. [19] highlighted three ranked choices for estimating landscape resistance values with the first being the most highly ranked option: 1. empirical animal movement data, genetic distance or rates of inter-patch movements; 2. animal occurrence, density or fitness; 3. literature review and expert opinion. Ideally, resistance values should be informed and parameterised with independent field data, such as extensive mark release recapture studies, actual movement data from radio-telemetry or Global Positioning System (GPS) studies [11,20], data from experimental studies to record movement through different land cover types [21], or inferred movement data from landscape genetics [9]. However, as these resistance values are species and landscape specific, there is an understandable shortage of such empirical data [22]. Zeller et al. [23] reviewed the different types of data used to parameterize least-cost models and concluded that expert-opinion and occurrence data are most often used. However, they also suggest that comparative studies on the data used to derive resistance values are needed.

Although the use of expert-opinion to parameterise least-cost models is seen as subjective and out performed by values informed by empirical data [24], many studies utilise this type of information to parameterise models [3,12,25]. The use of expert-opinion may be appropriate in some cases, such as where there is a particular shortage of empirical data, an urgency to act, or a focus on general principles, focal species or particular species traits. However, in an attempt to make the setting of landscape resistance values less biased and more data-driven, some researchers [26–31] are starting to utilise species distribution models, such as MaxEnt [32], to parameterise least-cost connectivity models (defined as option 2 by Beier [19]). This study uses MaxEnt, a species distribution model which utilises maximum entropy principles to predict a species' use of a landscape based upon occurrence data and a selected set of environmental predictors [32]. The habitat suitability indices provided by the models can then be used in calculations [26–31] to create least-cost connectivity models. Given that resistance values informed by empirical data are ranked higher [19] and seen to outperform expert-opinion values [24], it is hypothesised that the HSM-informed values will produce a more accurate least-cost network than expert-opinion data. The aim of this study is to investigate how expert-derived resistance values compare against values informed by habitat suitability modelling (HSM). The results of this study provide guidance to researchers and practitioners on the suitability of these approaches for informing management and research objectives relating to both species of conservation concern and invasive species spread.

Materials and Methods

Ethical statement

Ethical clearance for this study was approved by the University of Cumbria Ethics Committee, ref 09/17. This was a desk based study with no field work required. Therefore, research permits and licences were not required.

Study site

To compare expert-derived resistance values against HSM-informed values, grey squirrel within the county of Cumbria UK (Figure 1), are used as the study species. Whilst six large woodlands in Cumbria are designated red squirrel refuge reserves (Figure 1), the grey squirrel remains throughout the county. A number of previous studies have used expert-derived least-cost models to define habitat connectivity for Britain's native red squirrels and invasive grey squirrels [33–36], providing expert-opinion on land cover resistance. In addition, Cumbria has an extensive collection of grey squirrel distribution records available with which to create HSM-informed data for comparison. Cumbria covers an area of 6,768 km^2 and has a sparse population of 490,000 people. The Lake District National Park is located in the centre of Cumbria and has legislation and planning restrictions to conserve the landscape. The National Park Authority are responsible for implementing legislation and planning decisions aimed at conserving the landscape and its species, which means that little has changed regarding land use during the time frame that the species presence data used within this study were recorded (2000–2009). The topography is varied with the Cumbrian Mountain range (≤978 m a.s.l.) that runs approximately west to east across the middle of the county. The majority of land at these higher elevations is used for grazing with little woodland habitat. However, at lower elevations there are numerous woodlands, and other semi-natural habitats, scattered within an agricultural matrix which may provide greater potential for squirrel movement.

Identifying least-cost networks

Land cover types from a highly accurate and up to date vector land cover map (Ordnance Survey Master Map) were reclassified into 21 broad land cover categories for Cumbria (Table 1). The map was rasterised at 10 m resolution to ensure accurate representation of narrow linear features, such as strips of woodland. All woodland patches were classed as core habitat as squirrels use these areas for nesting and breeding [37,38]. This map was then parameterised with five alternative expert-derived resistance sets from previous studies (Table 1). The resistance values given in the different studies varied substantially. An additional set of values was developed by the authors by refining Stevenson's [35] scores (referred to as new expert-derived), following a review of the literature and the ecological underpinning of the values that had been applied previously, as described below.

Coniferous, mixed and broadleaved woodland were all assigned the lowest resistance value of 1, as core habitat. Scrub, coppice, orchard, and garden were given relatively low resistance values because they often contain tree species and are commonly used by

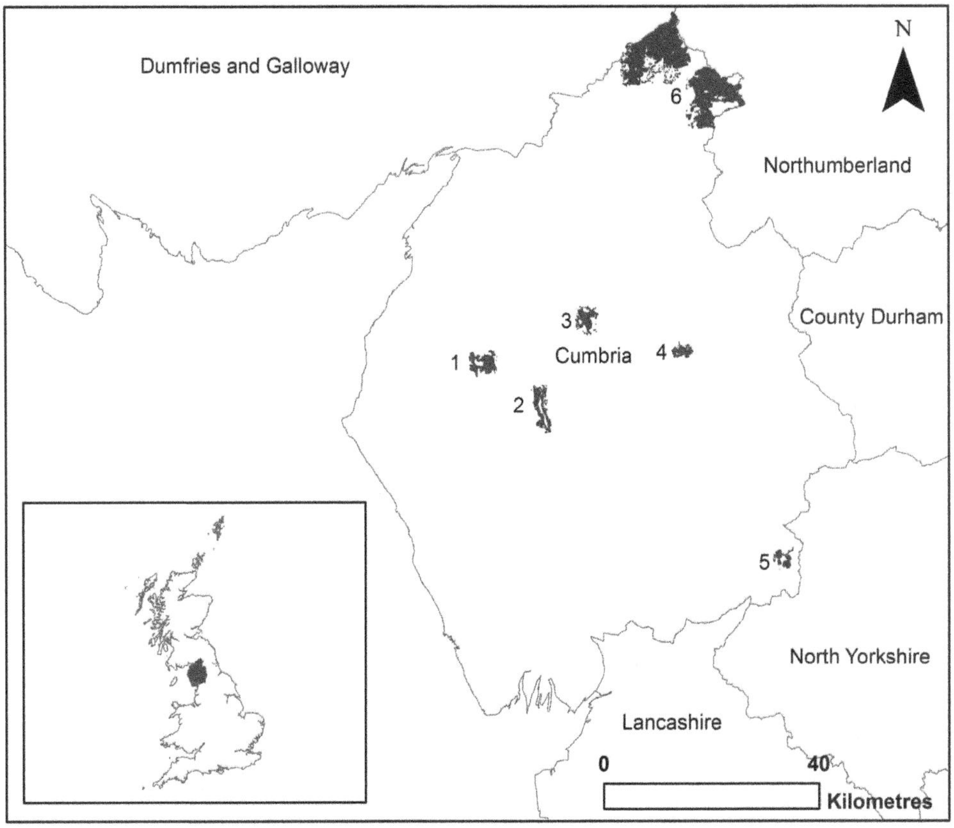

Figure 1. Map of red squirrel reserves in Cumbria and neighbouring counties with reference to its location in the UK. * 1. Whinlatter; 2. Thirlmere; 3. Greystoke; 4. Whinfell; 5. Garsdale/Mallerstang and 6. Kielder (Cumbria proportion of). Boundary lines were obtained through EDINA Digimap Ordnance Survey Service, http://digimap.edina.ac.uk/digimap/home.

grey squirrels for commuting [11,20]. Path, track, road verge, road and railway verge may also be used as commuting corridors, [13], but their use may confer higher mortality risks and therefore they were assigned a relatively high score. Improved/arable/amenity, rough grassland and heath were all attributed higher values still, as squirrel species tend to avoid open habitats [39]. Due to the threat of railways and the difficulty of moving over marsh, water, urban areas, buildings, and rocky areas like cliffs, the high scores assigned in previous studies were maintained.

Least-cost networks were created for each set of resistance values (Table 1) using the least-cost network process outlined in Watts et al. [10]. This network tool analysis utilises ArcView 9.1 and the Spatial Analyst extension (ESRI, Redlands, CA). The first step defined suitable patches of woodland habitat and generated a cost surface raster from the land cover map, by joining the resistance values (Table 1) to the 21 land cover classes. Secondly, the 'cost-distance' function in the Spatial Analyst toolbox was used to create a cost-distance surface between woodland patches. The resulting accumulated cost raster was then reclassified to a standardised maximum dispersal distance of 8 km to ensure comparability between the different resistance sets. The 'region group' function was used to define each discrete network, using an eight-cell rule so that touching cells, either adjacent and diagonally opposite, within the minimum distance of any given patch were considered part of the same network.

Deriving resistance scores from habitat suitability modelling

Records of grey squirrel presence were obtained from Save Our Squirrels (http://www.saveoursquirrels.org.uk/). These consisted of 2,281 verified sightings recorded year-round between 2000 and 2009 given by members of the public from both within woodland habitat (35%) and the wider landscape (65%). The grid references and type of habitat the sightings were recorded and verified by Save Our Squirrels. Sightings that were recorded outside of the grey squirrels known distribution range were also verified by contacting the recorder. The points outside of core woodland habitat are believed to relate to landscape use and movement, rather than indicating suitable foraging, breeding or nesting resources [37,38]. It is these non-woodland records that are used to infer the permeability of the landscape matrix using the habitat suitability modelling software, MaxEnt [40,32]. MaxEnt assigns each raster cell a Habitat Suitability Index (HSI) based on the environmental conditions at locations where a species has been recorded, using the maximum entropy method [41]. There are three output formats given by the MaxEnt programme: raw, cumulative and logistic; the most easily intuitive logistic HSI scores, which indicate the probability of occupancy ranging between 0–1 and assuming that this is 0.5 at an average site [40,32], were used in this study.

Both the species records and environmental data were prepared for modelling with MaxEnt. The squirrel data were filtered to remove locations recorded at a resolution of >100 m. Of the remaining 2,008 points, 842 squirrel presences recorded were

Table 1. Land cover resistance values based on previous least-cost modelling studies and resistance set based on expert opinion.

Land cover classification	Stevenson (2008)	Humphrey et al.(2007)	Williams (2008)	Verbeylen et al (2003) *	Gonzales (2000)	New expert-derived	Land cover ranking
Broadleaf	1	0	0	1	1	1	1
Mixed	1	0	0	1	1	1	2
Coniferous	1	0	0	1	1	1	3
Orchard	30	1	4	300	5	16	4
Scrub	1	2	10	10	5	16	18
Coppice	1	1	4	300	5	16	5
Garden	1	1	20	10	5	11	6
Improved/arable/amenity	30	10	20	800	8	40	20
Rough grassland	30	10	20	800	8	40	15
Heath	22	20	20	300	9	37	19
Path	16	50	20	800	1	27	9
Railway verge	12	10	20	800	4	27	11
Road verge	12	10	20	800	4	27	7
Marsh	76	50	50	800	9	91	14
Water	115	50	100	800	999	130	13
Urban	57	50	50	800	1	72	16
Railway	40	50	50	800	9	55	12
Road	12	50	30	800	9	27	8
Track	12	50	30	800	1	27	10
Building	1000	50	50	1000	999	1000	17
Rock	1000	50	50	1000	999	1000	21

*Note the resistance scores given in Verbeylen et al.'s (2003) study are for red squirrels in an urbanised matrix and are used for comparison.

within the matrix. A categorical land cover raster (gridded data map) was created using the same Ordnance Survey Master Map data and 21 broad land cover categories as previously described (Table 1). However, a coarser resolution of 100 m was used to match the spatial accuracy of the squirrel records. To ensure that linear habitat features were not under-represented, each land cover type was rasterised separately at a 10 m resolution and then aggregated to 100 m using the 'maximum' rule. These rasters were mosaicked using ranks that prioritised the classification of each 100 m square containing more than one land cover type (Table 1). All areas of woodland (core habitat), rocks and buildings (highly impermeable) were removed from the land cover map to prevent their incorporation in the model. In an effort to account for sampling bias towards accessible areas, a well-known and common characteristic of species data collected in an *ad-hoc* or non-systematic way [42], all areas over 500 m from a road, track or path were also removed from the map. This left a total of 665 squirrel records that fell within the remaining areas of the land cover map which were used to train and test the habitat suitability model. Each point (located in the south west of the 100 m grid square) was adjusted by 50 m east and 50 m north to locate each point in the centre of the grid square. This was to ensure that the points matched the 100 m raster landscape i.e. were within one cell, not potentially boarding four.

All models were run in MaxEnt Version 3.3.3k, using primarily default settings (regularisation multiplier = 1; duplicate occurrences removed; maximum number of background points = 10000, as used in Kramer-Schadt et al. [43]). Five-fold cross validation was used to calculate mean Area Under Roc Curve (AUC) and extrinsic omission rates (the average proportion of test points that fall outside the area predicted to be suitable), following use of the occupancy threshold rule that maximises the sum of test sensitivity and specificity (as recommended by Liu et al., [44]). Residual spatial autocorrelation (rSAC) can inflate measures of model performance [45–47] therefore Moran's correlograms were created (1 – predicted HSI for each species record; [48]) using the Spatial Analysis in Macroecology software program (SAM; [49]). Significance of Moran's I was calculated using a randomisation test with 9,999 Monte Carlo permutations, correcting for multiple testing.

The response curves, which showed the mean predicted probability of a species' presence (p; 0–1 scale) within each land cover type, were used to derive the resistance values for each land cover type. For both the new expert-derived and the HSM-informed values, woodland was given a value of 1, as permeable core habitat, and rock and building given values of 1000, as impermeable land cover types. The remaining land cover type values were inverted and standardised to the same scale as the new expert-derived values, (1–130; using 1-($p \times 130$)). These values were then used to identify least-cost networks using the same approach as applied to the new expert-derived resistance scores.

Comparing resistance scores and resulting habitat networks

An area-minimisation methodology was applied to select for the smallest network that captures the majority (≥90%) of the filtered distribution point data (n = 842). This methodology, derived during this study, was based on the principle that when managing invasive species, areas for control must be defined and defensible to provide successful management [50]. As the grey squirrel population continues to expand in the Cumbria study site, it is important that control efforts are targeted to provide effective management. By identifying habitat networks management can be targeted in these specific areas of the landscape. The larger the habitat networks are the more widespread management would have to be. Therefore, the resistance set which produced networks that include a high proportion of distribution points but a small network area are regarded as the better networks as management can be targeted in these focused areas. In addition a chi square test was used to test whether a significant number of distribution points were within the networks when compared to random points.

The HSM-informed resistance scores and the resulting networks were compared to those created with the new expert-derived set selected by the area-minimisation criteria. A Wilcoxon signed ranks test was used to assess the relative difference between scores. The habitat networks produced were also measured and compared visually and using the distribution points. Distribution points that were within the new expert-derived networks but not within the HSM-informed networks were identified along with the land cover type they were in and vice versa.

Results

Habitat Suitability Model performance

The results from five-fold cross-validation test showed that the models performed well (Training sample size = 532; Test sample size = 133; Training AUC = 0.80±0.001; Test AUC = 0.78±0.04; Test gain = 0.70±0.19; Extrinsic omission rate = 0.23, P<0.001), indicating that land cover type provides useful information on the likelihood of grey squirrel presence. No significant residual spatial autocorrelation was found at any distance lag. Moran's I values were <0.05 and statistically insignificant at each distance lag, indicating that the residuals were not spatially autocorrelated.

Selecting an 'optimal model' from expert-derived resistance sets

There was considerable variation between the previous studies and new expert-derived resistance sets, with network area ranging from 78% to 15% of the total landscape area, containing between 99% and 32% of the squirrel point data (Figure 2). However, when the networks were tested against the occurrence data within the matrix all resulting networks contained significantly more distribution points than expected by chance (n = 842, Stevenson 2008, $\chi 2 = 623$, df. = 1, p<0.001; Humphreys et al. 2007, $\chi 2 = 238$, df. = 1, p<0.001; Williams 2008, $\chi 2 = 357$, df. = 1, p< 0.001; Verbeylen et al. 2003, $\chi 2 = 169$, df. = 1, p<0.001; Gonzales 2000, $\chi 2 = 213$, df. = 1, p<0.001; new expert-derived, $\chi 2 = 623$, df. = 1, p<0.001). Using the area-minimisation methodology, the new expert-derived resistance set was shown to have above 90% of sightings within the networks and the lowest networks area of 49% of the total landscape (Figure 2). This was therefore selected and used for further comparison with the HSM-informed networks.

Comparing expert-derived and HSM-informed networks

The HSM-informed network had significantly more grey squirrel distribution points within it than expected by chance (n = 842, $\chi 2 = 836$, df. = 1, p<0.001). However, the new expert-derived network contained significantly more points than the HSM-informed network (n = 842, $\chi 2 = 185$, df. = 1, p<0.001). The majority of land cover types were given higher resistance values using the HSM approach compared to those derived from the new expert-derived values, with relative differences ranging from 7–86% (Table 2). These differences were found to be statistically significant (n = 16, Wilcoxon signed ranks test, $p = 0.002$); water and coppice were the only habitats to be assigned an HSM-informed lower resistance value compared to those derived from the new expert-derived resistance set. The

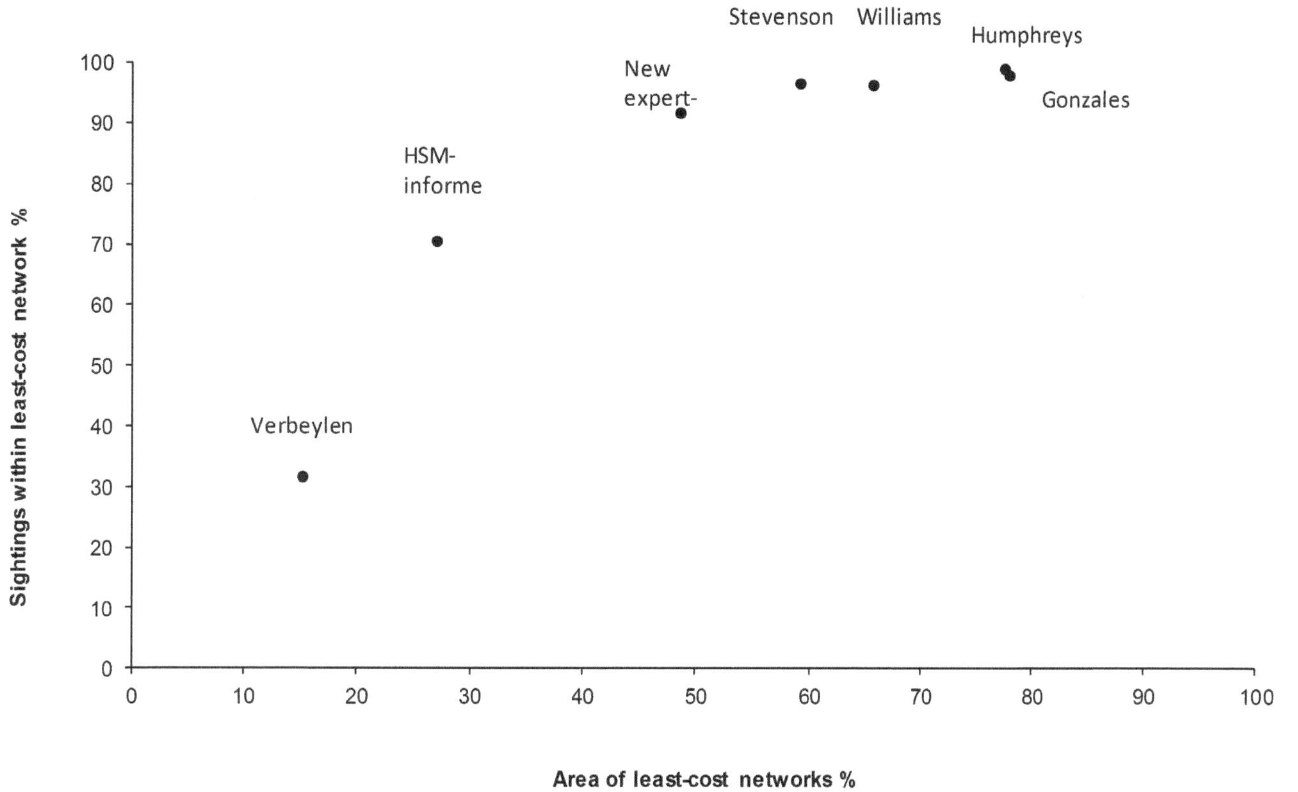

Figure 2. Comparison between expert-derived and Habitat Suitability Model-derived resistance values. Note: values that produce a network with>90% sightings points and the lowest network area is considered the best model for management.

largest differences in the resistance values assigned to habitats by the two approaches were between scrub, tracks, railways and railway verges (Table 2).

The least-cost model parameterised with the new expert-derived resistance values identified 738 discrete networks, although two of these cover substantial areas; habitat network 1 in the north and habitat network 2 in the south (Figure 3). The mean network size was 4.7 km^2 ($\pm$84.4). These networks accounted for 42% of the Cumbrian land cover (3,518 km^2) and appear to be separated by the land cover types within the Cumbrian Mountains. The HSM-informed resistance values generated comparatively smaller and more fragmented networks, owing to the higher resistance scores attributed to most habitat types. This network was 55% the size of the new expert-derived network (1,953 km^2; 34% of land cover) and sat almost entirely inside it, with only 0.2% extending beyond the expert-derived network, over areas of water. The mean network size was 0.3 km^2 ($\pm$5.0) and 5,840 separate networks were identified in Cumbria (Figure 3). Ten of these were relatively large (>20 km^2). The HSM-informed networks also indicated that networks in the north and south of the county were separated by the Cumbrian Mountains range. Both identified Grizdale Forest and surrounding woodlands as a large, well connected grey squirrel habitat network (Figure 3).

The smaller HSM-informed least-cost networks contained 592 (70%) of 842 species records within the habitat network (compared to 772 (92%) using new expert-derived) (Figure 4). As the HSM-informed scores were based upon the actual distribution data it was expected that the resulting networks would include a substantial amount of distribution points. The number of points outside of the HSM-informed least-cost networks was 250; of these points missed by the HSM-informed network 180 were included within the new expert-derived networks. These 180 points were located in improved/arable/amenity land (77%), gardens (8%), rough grassland (6%), urban (3%), road (2%), road verge (1%), tracks (1%), marshland (1%), scrub (1%) or water (1%). The number of points outside of the new expert-derived networks was 70; of these points none were included within the HSM-informed least-cost networks.

Discussion

When estimating resistance values Beier [19] highlighted three ranked choices. Although using animal movement data, genetic distance or rates of inter-patch movements (option 1) is the preferable option to define resistance values, animal occurrence data (option 2) and/or literature review and expert opinion (option 3) may be the only information available to many researchers and conservationists trying to model functional connectivity in fragmented landscapes. In this study resistance values derived from expert-opinion have been compared to HSM-informed values. Both techniques identified least-cost networks that contained significantly more distribution points than would be expected by chance. However, differences occur between the degree of model assumptions and biases (based on the different types of data), resistance values for certain land cover types and the least-cost networks identified. This has implications for the reliability of using such data in meeting conservation and management objectives.

To derive a set of expert-opinion resistance values it is useful to compare previous resistance values from multiple sources, particularly if the studies have similar species and environmental

Table 2. Average probability of grey squirrel presence according to land cover type.

Habitat type	HSM *p* score	HSM-resistance score	New expert-derived resistance score	Difference between HSM and Expert-derived resistance scores
Scrub	0.10	117	16	0.86
Track	0.16	109	27	0.75
Railway	0.17	108	27	0.75
Railway verge	0.17	108	27	0.75
Path	0.34	86	27	0.69
Heath	0.17	108	37	0.66
Garden	0.73	32	11	0.66
Road	0.41	77	27	0.65
Improved/arable/ amenity	0.13	113	40	0.65
Rough grassland	0.13	113	40	0.65
Road verge	0.43	74	27	0.64
Orchard	0.77	30	16	0.47
Urban	0.29	92	72	0.22
Marsh	0.17	108	91	0.16
Broadleaf	N/A	1	1	0.00
Coniferous	N/A	1	1	0.00
Mixed	N/A	1	1	0.00
Building	N/A	1000	1000	0.00
Rock	N/A	1000	1000	0.00
Coppice	0.86	15	16	−0.07
Water	0.18	107	130	−0.21

p = mean predicted probability of presence according to habitat type.

conditions. The resistance values given in previous studies were highly variable, resulting in varied least-cost habitat network areas and number of distribution points within networks. Although the land cover resistance values given in these studies were for red or grey squirrels, the studies took place in different countries with different regional environmental conditions and large scale and inevitable differences in landscape composition and structure. This may account for the differences in values given and resulting networks. Verbeylen et al [3] in particular was focused on red squirrels and based in an urban area which is very different to the largely-rural and sparsely populated Cumbria. However by assessing the range of different resistance values given in these studies and additional literature on land cover use, the new expert-derived resistance set was created. The area-minimisation method suggests that these values appear to be the best set for management purposes in this area, capturing a high percentage of distribution points within the smallest network area.

The resistance values for the new expert derived and HSM informed least-cost models in this study were significantly different from one another. The HSM-informed model provided higher resistance values for most land cover types. The validity of HSM-informed least-cost models may be limited as the probability of occurrence in a particular land cover type does not always equate to the resistance of that land cover type during species movement [19]. In using distribution/occurrence data, certain land cover types may be undervalued when in reality they are used by the species. Conversely there will be land cover types that are overvalued. A key assumption of presence only modelling is that the data has come from random sampling or is representative of the whole landscape [51]. It is questionable whether the degree of bias in presence data can be truly known [51]. Squirrels are well known to use scrub habitat and will use this and linear features to aid dispersal [13,52–54], yet scrub and railway verge (a linear feature) were given high HSM-informed resistance values due to a low number of distribution points. Of the distribution points missed by the HSM-informed networks but included within the new expert-derived networks, 77% were within improved/arable/ amenity land cover type. This suggests that the inverted HSM values for this land cover may be too high, and squirrels may be able to cross these hostile areas quickly and undetected. The dispersal distance used for both expert-derived model and the HSM-informed model were set at 8 km. Therefore, it is the higher resistance values given to certain land cover types using the inverted-HSM that led to the identification of smaller and more fragmented networks.

The HSM-informed networks were 45% smaller than the expert derived networks and were spatially nested inside these networks. The smaller mean size of HSM-informed networks suggests that grey squirrel occurs in a highly fragmented and functionally unconnected landscape. Both models highlight the land cover types of the Cumbrian Mountains as a barrier to movement; the combination of relatively high elevation and intense grazing result in a lack of woodland in the area. Although, some individuals may attempt to cross the barrier, the lack of available habitat will impede dispersal subjecting individuals to high levels of predation and starvation. There are no recorded introductions of the grey squirrel into Cumbria [55,56] and therefore these animals have been able to spread to their present

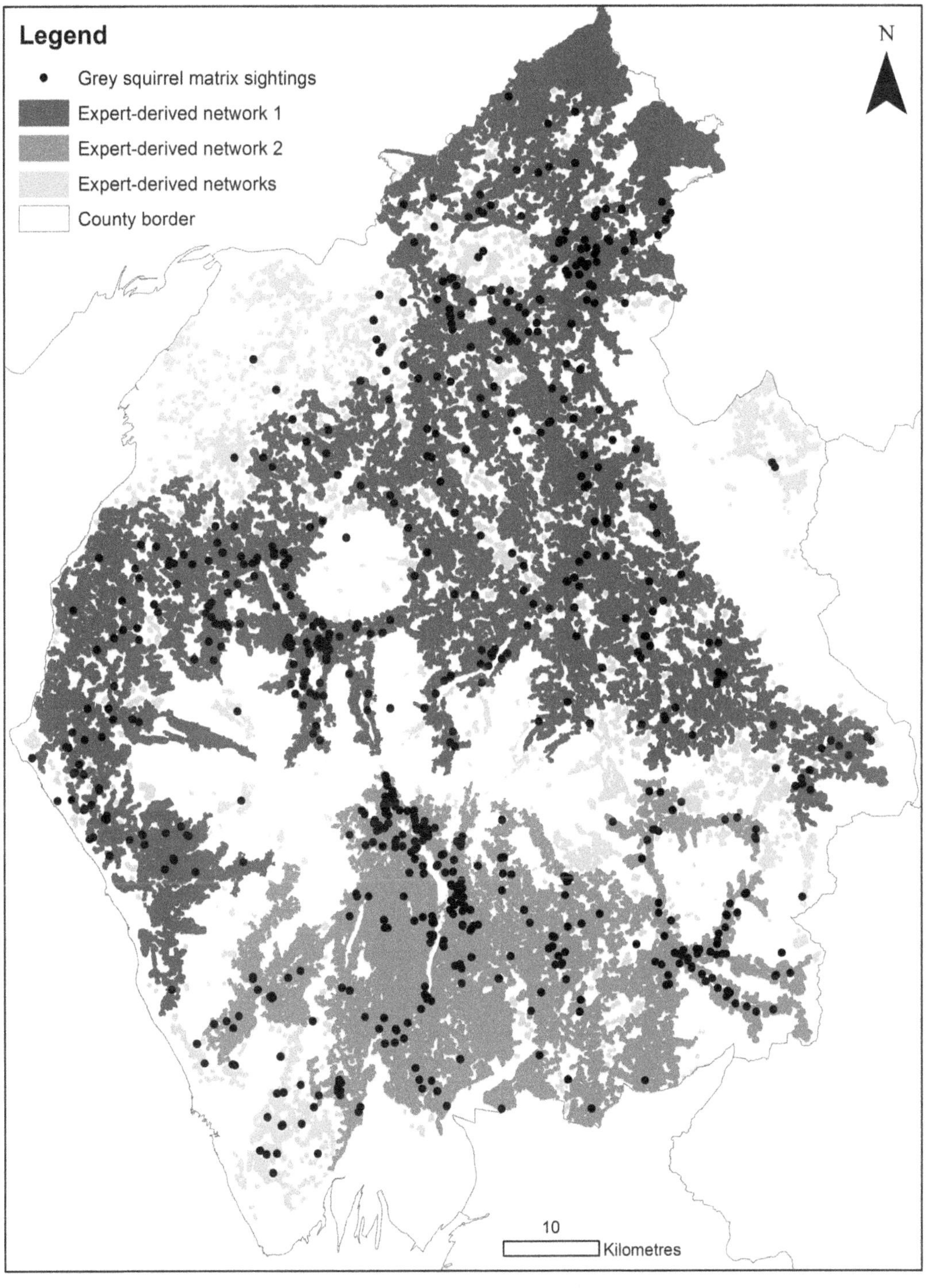

Figure 3. Grey squirrel least-cost habitat networks identified from expert-derived resistance values. Boundary lines were obtained through EDINA Digimap Ordnance Survey Service, http://digimap.edina.ac.uk/digimap/home.

distribution in the north and south of the county by natural means. The expert-derived model identified two large networks, one in the north and one in the south, suggesting a much more connected landscape.

Studies have suggested that expert-opinion based models perform less accurately than models informed by empirical data [24,57,58]. Given that HSM-informed networks are derived from known distribution data, these models could be interpreted as identifying more precise areas in the landscape that are connected for a species. In comparison, the expert-derived networks include those areas where sighting have not been recorded but are judged by experts as permeable to the species during dispersal. Experts may overestimate the importance of certain land cover types erring on the side of caution and therefore rendering the model less accurate [24]. Where actions might require a more precise approach, such as identifying possible protected areas or sites for

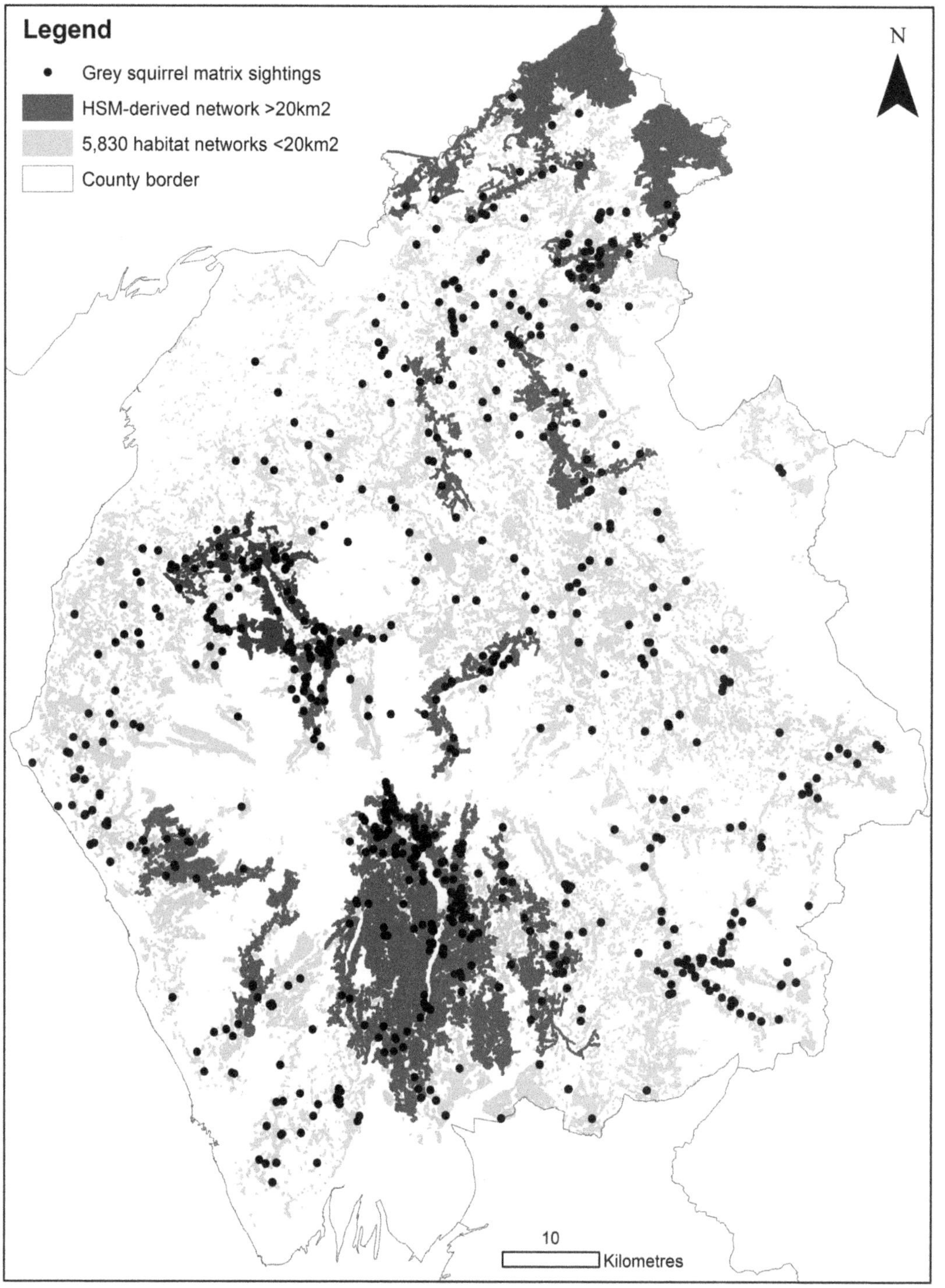

Figure 4. Highly fragmented grey squirrel least-cost habitat networks identified with Habitat Suitability Model derived resistance values. Boundary lines were obtained through EDINA Digimap Ordnance Survey Service, http://digimap.edina.ac.uk/digimap/home.

an efficient and intensive control program, a HSM modelling approach would be appropriate. However, when assessing invasive species it is not just the most likely areas that a species will disperse to, but the entire possible range that needs identified. In an invasive species context, it may be more appropriate to apply a conservative less precise model, such as the expert-derived model, to enable all possible areas of dispersal to be included within the network.

In the case of invasive species the assessment of potential movement and impact is needed as soon as possible to aid management planning. This method is not dependent upon extensive species distribution data and can therefore be produced relatively quickly. Clevenger et al. [24] found that expert only derived resistance values had a weaker correlation with empirical-derived values than literature-derived values. Systematically collecting expert opinion, as promoted by Eycott et al [59], in combination with published data on land cover usage will enable resistance values to be assigned in the initial stages to give an indication of species movements whilst other empirical data is collected where possible. Adriensen et al [18] suggested that once a 'starter kit' of resistance values has been identified, sensitivity studies can be initiated and multiple alternative resistance sets can be tested [60]. Once species distribution data is collected, HSM-informed least-cost networks can be identified and used to aid the selection of most likely used sites to focus monitoring or eradication programs. It should not be assumed that using distribution data (option 2 in Beier et al. [19]) to identify resistance values is better or worse than using well developed expert-opinion (option 3 in Beier et al. [19]) as the choice of which method to use may depend upon the aims and objectives of the user and the appropriate precision of the approach.

This paper describes the first step towards developing least-cost habitat networks using *ad hoc* species records and a simple, land cover-based habitat suitability model. It is acknowledged, however, that species respond to their surrounding environment over a range of spatial scales and that both local and landscape features will affect both the suitability of the core habitat and the permeability of the surrounding matrix [5,61]. More complex models incorporating multiscale information on the terrain, built environment, and the composition, structure and arrangement of habitat patches are likely to provide more accurate and useful models [45], providing predictions at each location, rather than assuming consistent levels of permeability for a particular land cover type. This spatially explicit technique would enable landscape level decision making, improving our ability to identify important networks of habitat and enabling a targeted and informed approach to both conservation and infrastructural development.

Conclusion

Even though approaches to gather expert opinion are becoming more systematic and robust, it should not be seen as a blanket substitute for empirical data. Empirical data will continue to be important for studies on single species, where there is considerable uncertainty or where there is significant investment in time and money on conservation activities. Conservation planners must be aware of the subjectivity and pitfalls of the different types of data used in least-cost models, without any further validation or sensitivity testing of model values. If expert opinion is the only option available it should be used as a first step by systematically combining multiple expert opinions and published data, but with the knowledge that further assessment of resistance values through sensitivity analysis and empirical data will be needed. Where distribution data is already available, the type of data collection and the subjective translation issues of over and under valuing land cover types must be assessed with expert knowledge or empirical data and explicitly stated in methodologies [51,62].

This study successfully compared expert-derived and HSM-informed resistance values used in least-cost modelling. Although the results of the models differed, both identified equally useful least-cost networks. For the grey squirrel in Cumbria, both expert-derived and HSM-informed networks have shown that there is a separation between north and south of Cumbria due to the land cover types and lack of habitat of the Cumbrian Mountain range. The expert-derived networks indicate a conservative less precise least-cost network that indicates the potential dispersal range of the grey squirrel and suggests that there may be multiple infiltration routes into the county from the north and south. This conservative expert-derived approach is useful when dealing with invasive or generalist species to identify the potential extend of spread. When assessing endangered or specialist species, or areas that are highly likely to contain target species, the HSM-informed network provides smaller precise networks. These precise networks should be used to inform targeted conservation to increase connectivity for species of conservation concern, or to inform targeted management to prevent the incursion of invasive species. The variable but acceptable precision of both expert-derived and HSM-informed least-cost networks highlights the need to consider data reliability and environmental context when deciding on the most appropriate management of invasive species.

Acknowledgments

The Authors would like to thank Dr Sallie Bailey for advice and comments, Phillip Handley for GIS advice and Simon O'Hare for sightings data. Country and county outlines in Figures 1, 3 and 4 and OSMM data were obtained through EDINA Digimap Ordnance Survey Service, http://digimap.edina.ac.uk/digimap/home.

Author Contributions

Conceived and designed the experiments: CDSH KW OTN ADR. Performed the experiments: CDSH KW CB. Analyzed the data: CDSH KW CB. Contributed reagents/materials/analysis tools: CDSH KW CB OTN ADR. Wrote the paper: CDSH KW CB OTN ADR.

References

1. Worboys G, Francis WL, Lockwood M (2009) Connectivity conservation management: A global guide (with particular reference to mountain connectivity conservation). London: Earthscan/James & James.
2. Taylor PD, Fahrig L, Henein K, Merriam G (1993) Connectivity is a vital element of landscape structure. Oikos 68: 571–573.
3. Verbeylen G, De Bruyn L, Adriaensen F, Matthysen E (2003) Does matrix resistance influence red squirrel (*Sciurus vulgaris* L. 1978) distribution in an urban landscape? Landscape Ecology 18: 791–805.
4. Spear SF, Balkenhol N, Fortin MJ, McRae BH, Scribner K (2010) Use of resistance surfaces for landscape genetic studies: Considerations for parameterization and analysis. Mol Ecol 19: 3576–3591.
5. Ricketts TH (2001) The matrix matters: Effective isolation in fragmented landscapes. The American Naturalist 158: 87–99.
6. Tischendorf L, Fahrig L (2000) On the usage and measurement of landscape connectivity. Oikos 90: 7–19.
7. Crooks KR (2006) Connectivity conservation. Cambridge: Cambridge Univ Pr.
8. Ferreras P (2001) Landscape structure and asymmetrical inter-patch connectivity in a metapopulation of the endangered Iberian lynx. Biol Conserv 100: 125–136.
9. Epps CW, Wehausen JD, Bleich VC, Torres SG, Brashares JS (2007) Optimizing dispersal and corridor models using landscape genetics. J Appl Ecol 44: 714–724.
10. Watts K, Eycott AE, Handley P, Ray D, Humphrey JW, et al. (2010) Targeting and evaluating biodiversity conservation action within fragmented landscapes: An approach based on generic focal species and least-cost networks. Landscape Ecology 25: 1305–1318.
11. Stevenson CD, Ferryman M, Nevin OT, Ramsey AD, Bailey S, et al. (2013) Using GPS telemetry to validate least-cost modeling of gray squirrel (*Sciurus carolinensis*) movement within a fragmented landscape. Ecology and Evolution 3: 2350–2361.
12. Gonzales EK, Gergel SE (2007) Testing assumptions of cost surface analysis- a tool for invasive species management. Landscape Ecology 22: 1155–1168.
13. Middleton AD (1930) The ecology of the American grey squirrel (*Sciurus carolinensis* gmelin) in the British isles. J Zool, Lond. 100: 809–843.
14. Gurnell J, Wauters LA, Lurz PWW, Tosi G (2004) Alien species and interspecific competition: Effects of introduced eastern grey squirrels on red squirrel population dynamics. J Anim Ecol 73: 26–35.
15. Gurnell J, Mayle B (2003) Ecological impacts of the alien grey squirrel (*Sciurus carolinensis*) in Britain. In Bowen CP, editor. MammAliens – A one day conference on the problems caused by non- native British mammals. London: Peoples Trust for Endangered Species/Mammals Trust UK. pp.40–45.
16. Kenward RE (1983) The causes of damage by red and grey squirrels. Mamm Rev 13: 159–166.
17. Sawyer SC, Epps CW, Brashares JS (2011) Placing linkages among fragmented habitats: Do least-cost models reflect how animals use landscapes? J Appl Ecol 48: 668–678.

18. Adriaensen F, Chardon JP, De Blust G, Swinnen E, Villalba S, et al. (2003) The application of 'least-cost' modelling as a functional landscape model. Landscape and Urban Planning 64: 233–247.
19. Beier P, Majka DR, Spencer WD (2008) Forks in the road: Choices in procedures for designing wildland linkages. Conserv Biol 22: 836–851.
20. Driezen K, Adriaensen F, Rondinini C, Doncaster CP, Matthysen E (2007) Evaluating least-cost model predictions with empirical dispersal data: A case-study using radio tracking data of hedgehogs (*Erinaceus europaeus*). Ecological Modelling 209: 314–322.
21. Stevens VM, Polus E, Wesselingh RA, Schtickzelle N, Baguette M (2005) Quantifying functional connectivity: Experimental evidence for patch-specific resistance in the natterjack toad (*Bufo calamita*). Landscape Ecol 19: 829–842.
22. Eycott AE, Stewart GB, Buyung-Ali LM, Bowler DE, Watts K, et al. (2012) A meta-analysis on the impact of different matrix structures on species movement rates. Landscape Ecol 27: 1263–1278.
23. Zeller KA, McGarigal K, Whiteley AR (2012) Estimating landscape resistance to movement: A review. Landscape Ecol 27: 777–797.
24. Clevenger AP, Wierzchowski J, Chruszcz B, Gunson K (2002) GIS-generated, expert-based models for identifying wildlife habitat linkages and planning mitigation passages. Conserv Biol 16: 503–514.
25. Chardon JP, Adriaensen F, Matthysen E (2003) Incorporating landscape elements into a connectivity measure: A case study for the speckled wood butterfly (*Pararge aegeris* L.). Landscape Ecology 18: 561–573.
26. Wang Y, Yang K, Bridgman CL, Lin L (2008) Habitat suitability modelling to correlate gene flow with landscape connectivity. Landscape Ecol 23: 989–1000.
27. Richards-Zawacki CL (2009) Effects of slope and riparian habitat connectivity on gene flow in an endangered Panamanian frog, *Atelopus varius*. Divers Distrib 15: 796–806.
28. Decout S, Manel S, Miaud C, Luque S (2010) Connectivity loss in human dominated landscape: Operational tools for the identification of suitable habitat patches and corridors on amphibian's population. Landscape International Conference IUFRO, Portugal.
29. Wang IJ, Summers K (2010) Genetic structure is correlated with phenotypic divergence rather than geographic isolation in the highly polymorphic strawberry poison-dart frog. Mol Ecol 19: 447–458.
30. Decout S, Manel S, Miaud C, Luque S (2012) Integrative approach for landscape-based graph connectivity analysis: A case study with the common frog (*Rana temporaria*) in human-dominated landscapes. Landscape Ecol 27: 267–279.
31. Howard A, Bernardes S (2012) A maximum entropy and least cost path model of bearded capuchin monkey movement in Northeastern brazil. Intergraph. Available: https://intergraphgovsolutions.com/assets/white-paper/A_Maximum_Entropy-2_Imagery.sflb.pdf Accessed 2013.
32. Phillips SJ, Anderson RP, Schapire RE (2006) Maximum entropy modeling of species geographic distributions. Ecol Model 190: 231–259.
33. Gonzales EK (2000) Distinguishing between modes of dispersal by introduced eastern grey squirrels (*Sciurus carolinensis*). MSc Thesis, University of Guelph.
34. Humphrey J, Smith M, Shepherd N, Handley P (2007) Developing lowland habitat networks in Scotland: Phase 2. Edinburgh: Forestry Commission.
35. Stevenson CD (2008) Modelling red squirrel population viability under a range of landscape scenarios in fragmented woodland ecosystems on the Solway plain, Cumbria. London: People's Trust for Endangered Species. Available http://insight.cumbria.ac.uk. Accessed 2013.
36. Williams S (2008) Red squirrel strongholds consultation. Edinburgh: Forestry Commission.
37. Lowe VPW (1993) The spread of the grey squirrel (*Sciurus carolinensis*) into Cumbria since 1960 and its present distribution. J Zool, Lond 231: 663–667.
38. Skelcher G (1997) The ecological replacement of red by grey squirrels. In: Gurnell J, Lurz P, editors. The conservation of red squirrels, *Sciurus vulgaris* L. London: People's Trust for Endangered Species. pp.67–78.
39. Nixon CM, McClain MW, Donohoe RW (1980) Effects of clear-cutting on grey squirrels. Journal of Wildlife Management 44: 403–412.
40. Phillips SJ, Dudik M (2008) Modeling of species distributions with maxent: New extensions and a comprehensive evaluation. Ecography 31: 161–175.
41. Phillips SJ, Dudík M, Elith J, Graham CH, Lehmann A, et al. (2009) Sample selection bias and presence-only distribution models: Implications for background and pseudo-absence data. Ecol Appl 19: 181–197.
42. Warton DI, Renner IW, Ramp D (2013) Model-based control of observer bias for the analysis of presence-only data in ecology. Plos One 8: e79168.
43. Schadt S, Knauer F, Kaczensky P, Revilla E, Wiegand T, et al. (2002) Rule-based assessment of suitable habitat and patch connectivity for the Eurasian lynx. Ecol Appl 12: 1469–1483.
44. Liu C, White M, Newell G (2013) Selecting thresholds for the prediction of species occurrence with presence-only data. J Biogeogr 40: 778–789.
45. Bellamy C, Scott C, Altringham J (2013) Multiscale, presence-only habitat suitability models: Fine-resolution maps for eight bat species. J Appl Ecol 50: 892–901.
46. Merckx B, Steyaert M, Vanreusel A, Vincx M, Vanaverbeke J (2011) Null models reveal preferential sampling, spatial autocorrelation and overfitting in habitat suitability modelling. Ecol Model 222: 588–597.
47. Veloz SD (2009) Spatially autocorrelated sampling falsely inflates measures of accuracy for presence-only niche models. J Biogeogr 36: 2290–2299.
48. De Marco P, Diniz-Filho JA, Bini LM (2008) Spatial analysis improves species distribution modelling during range expansion. Biol Lett 4: 577–580.
49. Rangel TF, Diniz-Filho JAF, Bini LM (2010) SAM: A comprehensive application for spatial analysis in macroecology. Ecography 33: 46–50.
50. Zalewski A, Piertney SB, Zalewska H, Lambin X (2009) Landscape barriers reduce gene flow in an invasive carnivore: Geographical and local genetic structure of American mink in Scotland. Mol Ecol 18: 1601–1615.
51. Yackulic CB, Chandler R, Zipkin EF, Royle JA, Nichols JD, et al. (2013) Presence-only modelling using MaxEnt: When can we trust the inferences? Methods in Ecology and Evolution 4: 236–243.
52. Taylor KD, Shorten M, Lloyd HG, Courtier FA (1971) Movements of the grey squirrel as revealed by trapping. J Appl Ecol 8: 123–146.
53. Fitzgibbon CD (1993) The distribution of grey squirrel dreys in farm woodland: The influence of wood area, isolation and management. J Appl Ecol 30: 736–742.
54. Wauters LA, Gurnell J, Currado I, Mazzoglio P (1997) Grey squirrel *Sciurus carolinensis* management in Italy - squirrel distribution in a highly fragmented landscape. Wildlife Biology 3: 117–123.
55. Shorten M (1954) Squirrels. London: Collins.
56. Shorten M (1957) Squirrels in England, Wales and Scotland, 1955. J Animal Ecol 26: 287–294.
57. Pearce J, Cherry K, Whish G (2001) Incorporating expert opinion and fine-scale vegetation mapping into statistical models of faunal distribution. J Appl Ecol 38: 412–424.
58. Seoane J, Bustamante J, Diaz-Delgado R (2005) Effect of expert opinion on the predictive ability of environmental models of bird distribution. Conserv Biol 19: 512–522.
59. Eycott AE, Marzano M, Watts K (2011) Filling evidence gaps with expert opinion: The use of Delphi analysis in least-cost modelling of functional connectivity. Landscape Urban Plann 103: 400–409.
60. Rayfield B, Fortin MJ, Fall A (2010) The sensitivity of least-cost habitat graphs to relative cost surface values. Landscape Ecol 25: 519–532.
61. Wiens JA (1989) Spatial scaling in ecology. Funct Ecol 3: 385–397.
62. Beier P, Majka DR, Newell SL (2009) Uncertainty analysis of least-cost modeling for designing wildlife linkages. Ecological Applications 19: 2067–2077.

13

Understanding Uncertainties in Non-Linear Population Trajectories: A Bayesian Semi-Parametric Hierarchical Approach to Large-Scale Surveys of Coral Cover

Julie Vercelloni[1,2]*, M. Julian Caley[2], Mohsen Kayal[3], Samantha Low-Choy[1], Kerrie Mengersen[1]

1 School of Mathematical Sciences, Queensland University of Technology, Brisbane, Queensland, Australia, **2** Australian Institute of Marine Science, Townsville, Queensland, Australia, **3** Bren School of Environmental Science and Management, University of California Santa Barbara, Santa Barbara, California, United States of America

Abstract

Recently, attempts to improve decision making in species management have focussed on uncertainties associated with modelling temporal fluctuations in populations. Reducing model uncertainty is challenging; while larger samples improve estimation of species trajectories and reduce statistical errors, they typically amplify variability in observed trajectories. In particular, traditional modelling approaches aimed at estimating population trajectories usually do not account well for nonlinearities and uncertainties associated with multi-scale observations characteristic of large spatio-temporal surveys. We present a Bayesian semi-parametric hierarchical model for simultaneously quantifying uncertainties associated with model structure and parameters, and scale-specific variability over time. We estimate uncertainty across a four-tiered spatial hierarchy of coral cover from the Great Barrier Reef. Coral variability is well described; however, our results show that, in the absence of additional model specifications, conclusions regarding coral trajectories become highly uncertain when considering multiple reefs, suggesting that management should focus more at the scale of individual reefs. The approach presented facilitates the description and estimation of population trajectories and associated uncertainties when variability cannot be attributed to specific causes and origins. We argue that our model can unlock value contained in large-scale datasets, provide guidance for understanding sources of uncertainty, and support better informed decision making.

Editor: Yong Deng, Southwest University, China

Funding: The authors received no specific funding for this work.

Competing Interests: The authors have declared that no competing interests exist.

* Email: julie.vercelloni@gmail.com

Introduction

Effective decision making in conservation and management can be hindered by uncertainty surrounding our understanding of population trajectories [1]. Improving our understanding of the variability in temporal fluctuations in species abundances is a particular challenge when using ecological data because it includes both endogenous variation in species dynamics as well as errors of observation and estimation [2–5]. This situation is complicated further because the long-term and broad-scale surveys that are needed to efficiently capture ecological trends typically exhibit variability that increases with the scale of observation [6].

There is an acknowledged lack of statistical tools able to efficiently deal with non-linear trajectories, multi-scale observations and the resulting uncertainties in ecology [7–9]. This lack of tools is particularly relevant for coral reefs where highly variable species dynamics and complex multi-dimensional interaction pathways make it difficult to construct models that adequately describe both overall (linear) trends and (nonlinear) fluctuations or deviations around these trends, and also sufficiently accommodate uncertainties inherent in these species trajectories [10–18]. Ignoring such nonlinearities and uncertainties, however, can lead to imprecise inferences and compromise effective management of ecosystems. For example, differences in model specification have led to disagreements in interpretation of reef trajectory and health [13,18–21].

Globally, coral reefs face diverse threats that are increasing in frequency and intensity, yet the knowledge required to manage them effectively is currently unavailable [22]. These threats are contributing to a global decline of corals, the foundation species in these ecosystems, through direct sources of mortality (e.g. destructive fishing, coastal development, anchorage damage) and degradation of reef environments from different forms of pollution, over-harvesting, and climate change [23]. These pressures can affect corals across spatial scales ranging from a few cm^2 to thousands of km^2, and across temporal scales ranging from seconds to decades and centuries [17,24,25]. Despite general awareness that corals are being seriously impacted by the simultaneous effects of these multiple sources of degradation, little attention has been given to the importance of spatial scale in models of coral trajectories [9,13,26]. Indeed, most models of coral trajectories have focused on a single spatial scale using either the scale of observation, or by aggregating data from multiple sources to achieve some degree of generalization at a larger scale [27–29]. Both of these approaches, however, fail to capture the causal mechanisms leading to the actual trajectory observed in nature, and their variability in time and space [18,30,31].

Recent multi-scale approaches to understanding coral trajectories have used generalized linear mixed models (GLMMs). These trajectories are typically fitted using linear equations [12,16,17] or natural splines [13,14]. In such cases, coral trends at larger spatial scales (e.g. regions and subregions) are usually fitted by taking into account the hierarchical structure of smaller scales (e.g. habitats, reefs and sites) nested within these larger levels [12,13,16,32]. GLMMs typically accommodate this hierarchical structure in the trends, or fixed components, by including random effects in the parameterization of the model [33]. However, when using such methods, care is required to account realistically for the spatial and/or temporal structure affecting random effects [2]. In all of these studies, random effects are typically assumed to be identically, normally distributed and independently at a particular spatial scale, such as reef [12–14,16]. Since random effects are not directly observed in such an analytical framework, the validity of these assumptions is difficult to test, and if violated, can lead to model misspecifications [2]. Of concern is that violation of these assumptions may be particularly important when investigating coral trajectories. Indeed, corals as well as other sessile habitat-forming communities are often studied using long-term surveys over large spatial scales [11,13,27]. Data from such surveys are likely to exhibit temporal and/or spatial structure, with measurements made close together in space and or time being closely correlated. Such correlations are also likely to be reinforced by large-scale disturbances that can affect multiple locations simultaneously and for long periods [12,16,34–36]. Consequently, commonly used statistical approaches described above, specify spatio-temporal structure in the fixed-effects component, but not in the random effects. Such methods, therefore, are limited in their ability to realistically describe the underlying generative processes and the different sources of uncertainty commonly encountered in ecological monitoring data [4,26,37].

An alternative approach to modelling large-scale survey data is to break the overall model into a series of hierarchically organized sub-models and embed them in a Bayesian analytical framework. These sub-models collectively describe a joint probability distribution for parameters of interest and for predictions. By construction, the model also accounts for differing sources of uncertainty emerging at differing stages in the investigation [2,38], including measurement error in data collection (stage 1), process error in representation of the ecological process(es) of interest (stage 2), and statistical error in the estimation of parameters (stage 3). The flexibility of this formulation, coupled with simulation-based computational methods for parameter estimation such as Markov Chain Monte Carlo and related algorithms, can be harnessed to produce more ecologically sensitive nonlinear models. Using this approach, the posterior predictive distribution, which incorporates all three levels and other relevant prior information, can be used to evaluate the adequacy of the model in terms of its ability to describe the observed data as well as predict unobserved data including future or missing observations [3]. These models have the added advantage of being able to be updated when more data become available, thereby facilitating continual refinement of knowledge about ecological processes and considerably increasing their predictive capability [39].

We introduce a Bayesian semi-parametric hierarchical model to simultaneously quantify nonlinearities in the relationships between a response variable, coral cover, and its covariates, in this case, time, uncertainties associated with the model structure and associated parameters, and error associated with multiple scales of observation. We illustrate this approach using 14 years of population estimates for the major reef-building coral genus *Acropora*, from the northern Great Barrier Reef (GBR), Australia. We provide posterior estimates and associated uncertainties for coral population trajectories among sites-, reefs- and habitats within a sub-region, and within reefs and sites across time. Our purpose is not to show a full analysis of the dataset, but rather to illustrate the modelling concept and its broader application. We demonstrate that this modelling approach is well suited to estimation of population trajectories and corresponding uncertainty in cases where population variability cannot be assigned to unique causes and origins. We argue that this approach provides a useful tool for investigating environmental drivers of population trajectories, the scales at which they act, and which is applicable to a wide range of species. In doing so, it facilitates greater comprehension of the uncertainties associates with trajectories of these species and, thereby supports more informed decision making.

Methods

Data

We used estimates of coral cover from the Australian Institute of Marine Science's Long Term Monitoring Program (LTMP) of the GBR [40]. The LTMP sampled benthic communities annually from 1994 to 2004, and then every second year, on 47 reefs throughout the GBR using five permanent 50×1 m^2 photo- (prior to 2006) and video-transects (from 2006) between 6 and 9 m depth [41]. Hard coral cover was estimated at the genus level and expressed as a percentage, based on estimates taken from 200 random points along each transect (see [41] for further details). In this study, we restrict our investigation to the dynamics of acroporid corals which dominate the GBR (i.e. 51% of the coral cover) and are considered responsible for most of the variability observed in its coral community trajectories [12].

To minimize geographical variability in coral dynamics [14], we parameterized the model using data from a single sub-region within the GBR, the Cooktown-Lizard Island section for the period 1994 to 2010 inclusive (Fig. 1). This sub-region is the northern-most section of the GBR sampled by the LTMP, spanning a latitudinal range from 14°S to 15°50'S. Within this area, the LTMP samples three reef habitats defined by their positions on the continental shelf. Inner-reefs are the closest to the coast and are most exposed to terrestrial and human influences [23]. The mid-shelf habitat extends over a large part of the GBR lagoon, with reefs situated at various distances between inner and outer habitats of the barrier reef. The outer-reef habitat extends into more oceanic conditions. In this sub-region, the survey is spatially replicated on two to three reefs per habitat, each reef being itself sampled at three distinct sites (Fig. 1). We investigated variability in coral dynamics at four spatial scales of observation of this hierarchical sampling design ranging from $\sim$250 m^2 for the site scale, 1 km^2 for the reef scale, 5–45 km^2 for the habitat scale, and $\sim$500 km^2 for the sub-regional scale of the Cooktown-Lizard Island section of the GBR. Observations at the transect scale within sites were pooled because there were no significant differences among transects after testing.

Model

A Bayesian longitudinal semi-parametric regression model was used to assess trends in the study region. Longitudinal models are suitable for situations like this where repeated measurements on several individuals (here sites within reefs) have been made to reflect how a variable of interest behaves through time. Temporal dynamics are often too complex to be modelled parametrically,

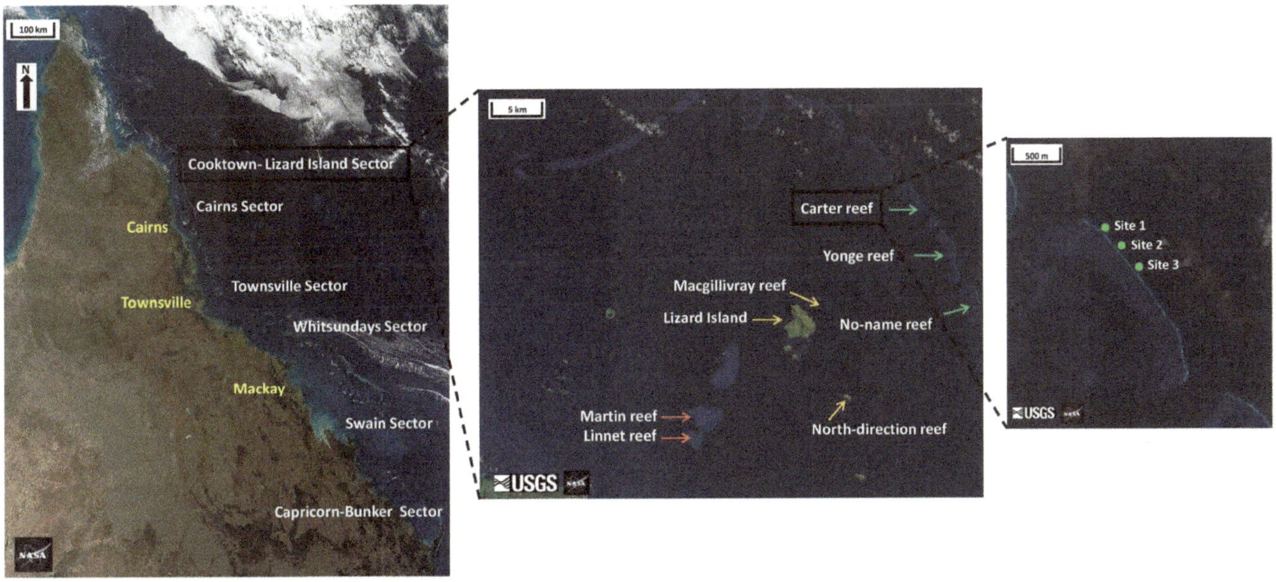

Figure 1. Sampling design of the Long Term Monitoring Program showing the hierarchical levels of observation for the Cooktown-Lizard Island sub-region, one of six sub-regions of Australia's Great Barrier Reef. The sub-region is divided into three shelf-positions or habitats (inner reef (red arrows), mid-shelf reef (orange arrows), and outer reef (green arrows)). Each habitat is sampled on three sites (green dots) in each of two or three reefs. We highlight three sites at Carter reef from the outer reef habitat. Modified from original satellite images © Landsat and MODIS satellite imagery courtesy of NASA Goddard Space Flight Center and US Geological Survey.

when the relationship between time and the response variable cannot be assumed to follow a specific and consistent pattern. Such patterns are typical for ecological data from corals and many other natural populations and communities whose dynamics are driven by myriad interacting biotic and abiotic factors [7]. Semi-parametric regressions offer an attractive approach for modelling such data. They combine a linear model and a smooth non-linear function, and therefore, provide a trade-off between flexibility and interpretability of results [42]. In this paper, penalized splines were used to describe the non-linear components of the model [43]. Within- and between-group variability was accounted for by the inclusions of random effects terms and differing levels of spatial hierarchy in the structure of the model. The preferred model was selected based on Deviance Information Criterion (DIC) diagnostics, which take into account both goodness of fit and model parsimony [44].

Let y_{ij} represent the proportion of *Acropora* spp. observed at time t_{ij} for a site i at the jth time point. There are a number of potential ways for describing the distribution of y_{ij}; for example, it could be represented by a Beta distribution (continuous over the range 0 to 1) or by a normal distribution based on an arcsine or angular transformation [45]. Here, we adopt the latter approach, so that the dynamics of *Acropora* are modelled as:

$$\arcsin\left(\sqrt{y_{ij}}\right) \sim N(\mu_{ij}, \sigma_\varepsilon^2) \qquad \text{(Eq. 1)}$$

where the transformed value of *Acropora* coverage is assumed to be normally distributed with an expected value μ_{ij} and measurement error term σ_ε. We use a very weakly informative conjugate prior for the variance with $\sigma_\varepsilon^{-2} \sim Gamma(10^{-3}, 10^{-3})$.

The expected value μ_{ij} is modelled as:

$$\mu_{ij} = f_s(t_{ij}) + f_{h(s)}(t_{ij}) + f_{r(sh)}(t_{ij}) + f_{i(shr)}(t_{ij}) \qquad \text{(Eq. 2)}$$

where f_s (.) describes the overall mean curve at the scale of the sub-region, and deviations from this overall curve represent hierarchically the habitat-, reef-, and site-specific trajectories. These scales are indexed h, r and i respectively with $1 \leq h \leq N_{habitats}(=3)$, $1 \leq r \leq N_{reefs}(=8)$ and $1 \leq i \leq N_{sites}(=24)$. In the following, for clarity and where it does not cause confusion, the nested subscripts are ignored.

All four contributions to the expected response are modelled as combinations of linear trends and splines:

$$f_s(t) = \beta_0 + \beta_1 t + \sum_{k=1}^{K_1} c_k z_{tk}^s, s = 1, k = 1, \ldots, K_1$$

$$f_h(t) = \gamma_{0h} + \gamma_{1h} t + \sum_{k=1}^{K_2} d_{hk} z_{tk}^h, h = 1, \ldots, N_{habitats}, k = 1, \ldots, K_2$$

$$f_r(t) = \theta_{0r} + \theta_{1r} t + \sum_{k=1}^{K_3} e_{rk} z_{tk}^r, r = 1, \ldots, N_{reefs}, k = 1, \ldots, K_3$$

$$f_i(t) = \delta_{0i} + \delta_{1i} t + \sum_{k=1}^{K_4} g_{ik} z_{tk}^i, i = 1, \ldots, N_{sites}, k = 1, \ldots, K_4 \qquad \text{(Eq. 3)}$$

The matrix z_{tk}^s is the $(t, k)^{\text{th}}$ entry of the design matrix [43], for the penalized spline random coefficients $\{c_k\}$, corresponding to the sub-region mean function f_s $(\cdot)$. Similarly, z_{tk}^h z_{tk}^r, and z_{tk}^i are defined as the $(t, k)^{\text{th}}$ entries of the design matrices for random coefficients corresponding to habitat level f_h $(\cdot)$, the reef level f_r $(\cdot)$, and site scale curves f_i $(\cdot)$. At the habitat scales one set of random coefficients $\{d_{hk}\}$ is allocated to each habitat. Similarly at reef and site scales one set of random coefficients $\{e_{rk}\}$ or $\{g_{ik}\}$ are allocated to each reef or site, respectively. Four knots were used for the splines for each curve (i.e. $K_1 = \ldots = K_4 = 4$). Note that because the smoothing is controlled by a penalty parameter, the number of knots K is not a crucial parameter in the model [46]. For modelling trajectories over time t, the model allows random slopes and intercepts θ_r and δ_s at the reef- and site-specific scales. Trajectories at the sub-regional and habitat scales are considered as fixed effects, modelled via β_s and γ_h. The time regressor (t) is centered on the year 2001 to facilitate model convergence, and to minimize correlation among random effects [17]. Autocorrelated temporal and spatial terms are not explicitly defined in this model, since the correlation exhibited in the data was largely incorporated through the hierarchical structure and the assignment of random effects at each stage of the model [8,17]. This was checked by plotting residuals at each modelled spatial scale and using the autocorrelation function to confirm an absence of temporal structure in the fitted model (results not shown).

The model also assumes that the c, d, e, g, θ and δ parameters are mutually independent, with hierarchical priors defined to model the random effects σ^2 as follows:

$$c_k \sim N(0,\sigma_c^2), k=1,\ldots,K_1$$

$$d_{hk} \sim N(0,\sigma_d^2), h=1,\ldots,N_{habitats}, k=1,\ldots,K_2$$

$$e_{rk} \sim N(0,\sigma_e^2), r=1,\ldots,N_{reefs}, k=1,\ldots,K_3$$

$$g_{ik} \sim N(0,\sigma_g^2), i=1,\ldots,N_{sites}, k=1,\ldots,K_4$$

$$\theta_{0r} \sim N(0,\sigma_{\theta_0}^2), r=1,\ldots,N_{reefs}$$

$$\theta_{1r} \sim N(0,\sigma_{\theta_1}^2), r=1,\ldots,N_{reefs}$$

$$\delta_{0i} \sim N(0,\sigma_{\delta_0}^2), i=1,\ldots,N_{sites}$$

$$\delta_{1i} \sim N(0,\sigma_{\delta_1}^2), i=1,\ldots,N_{sites} \qquad \text{(Eq. 4)}$$

The amount of shrinkage induced by $\sigma_c^2, \sigma_d^2, \sigma_e^2$, σ_g^2 is allowed to differ for each level of the model and $\sigma_{\theta_0}^2, \sigma_{\theta_1}^2, \sigma_{\delta_0}^2$ and $\sigma_{\delta_1}^2$ represent variability of intercepts, and slopes among reefs and sites, respectively.

Temporal trends at sub-regional and habitat scales are considered specific to each area, and hence fitted using fixed effects, with vaguely informative prior distributions:

$$\beta_0, \beta_1, \gamma_{0h}, \gamma_{1h} \sim N(0,10^3), h=1,\ldots,N_{habitats} \qquad \text{(Eq. 5)}$$

We use conjugate Gamma priors on the random effects parameters:

$$\sigma_c^{-2}, \sigma_d^{-2}, \sigma_e^{-2}, \sigma_g^{-2}, \sigma_{\theta_0}^{-2}, \sigma_{\theta_1}^{-2}, \sigma_{\delta_0}^{-2}, \sigma_{\delta_1}^{-2} \sim Gamma(10^{-3},10^{-3})$$

$$\sigma_e^{-2} \sim Gamma(10^{-6},10^{-6}) \qquad \text{(Eq. 6)}$$

Other prior formulations for the precisions were also evaluated, including the Uniform and half-Cauchy distributions for the corresponding standard deviations [47], with no substantive effect on the posterior distributions or inferences (results not shown).

The Directed Acyclic Graphic (DAG) presented in Figure 2 shows connections between the three hierarchical stages of the model. At the first stage, the data level, the arcsine transformed value of the response y_{ij} is distributed around a population mean μ_{ij} with measurement error σ_ε (Eq. 1). The second stage comprises the model for the ecological process and the associated process error (Eq. 2–4). The third stage includes uncertainty in the trend parameters (Eq. 5) and variance components (Eq. 6).

Analysis was performed using the R package R2WinBUGS [48] to call the Bayesian software analysis WinBUGS [49]. We provide R and WinBUGS code in the supporting information file. Posterior distributions of parameters were approximated by Markov Chain Monte Carlo with 200,000 iterations. Convergence diagnostics were assessed by using visual (trace and density-plots of parameters and autocorrelation plots between MCMC draws) and statistical (Gelman and Rubin diagnostic) functions from the R package coda [50]. Convergence was satisfied using a burn-in of 100,000 iterations. Three MCMC chains were simultaneously run to further evaluate and confirm convergence to stationarity. A thinning rate of 50 iterations, mainly used to reduce computer storage space, also improved the independence of the simulated values. Parameter inferences were drawn from posterior distribu-

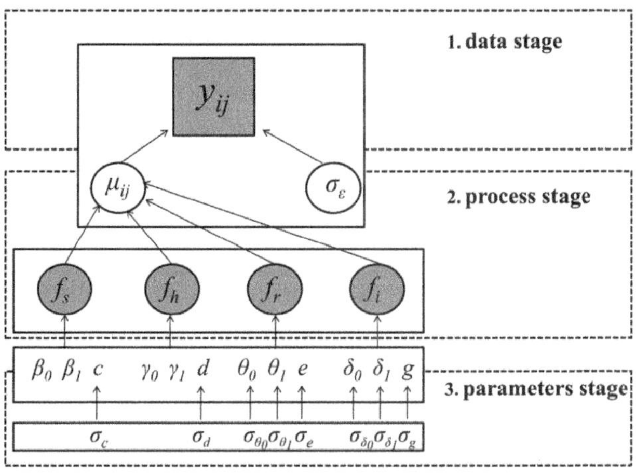

Figure 2. Directed Acyclic Graph showing how *Acropora* cover y_{ij} at a site i and time j is fitted in three stages. Here, the transformed data y_{ij} are modelled using a normal distribution with an expected value μ_{ij} and variance σ_ε. The expected value μ_{ij} is a function of four spatial scales (sub-region f_s, habitat f_h, reef f_r, and site f_i). Linear trends at sub-region and habitat scale are modelled using fixed effects (β, γ). At the site and reef scale, trend parameters (θ, δ) are considered as random. Trend parameters are denoted with the subscript 0 corresponding to the intercept and 1 corresponding to the slope. Random effects require specification of a variance component (σ), specified for each spatial scale of the model. Refer to the methods section for equations.

tions constructed from the retained 2000 iterations from the MCMC chains. Goodness-of-fit was assessed by overall model fit diagnostics (DIC), approximate normal distributions for the root-mean-squared error (RMSE) terms, precision of parameter estimates (width of credible intervals) and posterior predictive fit (whether the observed values were contained in the 95% credible intervals obtained from the respective posterior predictive distributions). Autocorrelations between parameters within a spatial scale were also examined to confirm independence between linear slope and intercept terms. The accuracy of model outputs at each spatial scale was assessed by inspecting the posterior distributions of the trend parameters: γ_{0h}, γ_{1h}, θ_{0r}, θ_{1r}, δ_{0i} and δ_{1i}. The fitted trajectory model was then decomposed into linear and non-linear components by splitting the respective equations (Eq. 3) into two parts (for example, linear: $\delta_{0i}+\delta_{1i}t$ and non-linear: $\sum_{k=1}^{K_4} g_{ik}z^i_{tk}, i=1,\ldots,N_{sites}, k=1,\ldots,K_4$ for $f_i(t)$). This decomposition was visualized for each spatial scale. Note that we examine different components of the expected value μ_{ij} depending on the spatial scale. Coverage of *Acropora* at the habitat scale is a function of the contribution of the sub-region and habitat (f_s+f_h)whereas at the reef level it is a function of the contribution of the sub-region, habitat and reef ($f_s+f_h+f_r$). Finally at the site scale we retrieve the expected value of observations ($\mu_{ij}=f_s+f_h+f_r+f_i$).

Results

Posterior distributions of model parameters and uncertainty

Visualization of posterior predictions compared to observations indicated that the model successfully captured spatial and temporal variability in *Acropora* dynamics at the three spatial scales nested within sub-region (Fig. 3). However, posterior distributions of the slope and intercept in trend (i.e, fixed component) show that their precision decreased with the scale of observation (see top-left inserts on plots in Fig. 3, Fig. S1). At the finer scale of the site, linear components of the trend were relatively tightly estimated: slopes were centred around 0 (range of 95% credible interval, RCI<0.05) and intercept terms had slightly larger variances (RCI≈0.1) with occasional non-zero central values. At the intermediate scale of the reef, the variance of the trend parameters was also relatively small (RCI<0.1). At the broadest scale, however, the estimated slopes and intercepts in trend for all three habitats were poorly estimated (Fig. S1). These results demonstrate that the temporal trends in the model successfully capture patterns at the site and reef scales, but not at the broader scale of habitat. At the finer spatial scales, linear dynamics were different between and within reefs and sites, with a consistent increase in the amount of uncertainty in more recent years (top-right inserts on plots in Fig. 3). The estimated non-linear dynamics were consistent within each spatial scale with spline contributions close to zero, but the degree of uncertainty differed between scales and was smallest at the reef scale, intermediate at the site, and largest at the habitat scale (top-right insert on each plot in Fig. 3). As expected, these uncertainties increased when moving from the centre to the edges of the surveyed period. The relatively narrow range of the measurement error (RCI≈0.01, Fig. 4) illustrates that coral dynamics were well explained by the multi-scale dynamics model given the data, model structure and distributional assumptions.

Modelled coral trajectory patterns

Most of the uncertainty in estimated coral trajectories occurred among habitats (Fig. 4). Indeed, *Acropora* dynamics were substantially different at this scale of observation. Overall coral cover was least on mid-shelf reefs, intermediate in inner habitats, and largest on outer reefs (Fig. 3). Inner- and mid-shelf reefs showed a slow increase in coral cover over the 14 years of survey, with some evidence of previous decline in the mid-shelf habitat. Outer reefs showed more pronounced trajectories, with a rapid increase in coral cover between 1994 and 2000, a sharp decline from 2000 to 2008, and early signs of recovery afterward.

Within the sub-region of Cooktown-Lizard Island, *Acropora* population trajectories were relatively consistent among sites and reefs with narrow uncertainty in model parameters estimation (RCI<0.01), however, the degree of uncertainty was higher within reefs and sites particularly in the estimation of variance components for slope in the trend,σ_{θ_1}andσ_{δ_1}(Fig. 4). These results indicate that estimation of coral cover is affected by diverse sources of uncertainty acting at different stages of the model and levels of observation.

Discussion

Mathematical description of non-linear trends and associated uncertainties for hierarchically structured data over different spatial scales is a very big challenge which is extremely difficult to address using existing statistical approaches. This limitation severely impacts our abilities to understand a diverse range of ecological systems, and other systems more broadly. For example, long-term surveys of ecosystems are typically based on hierarchical observations [3,8]; in epidemiology, spatial trends in health outcomes are monitored among different hierarchies in populations; in finance, temporal patterns in returns are monitored among financial sectors; and so on. In this paper, we introduce a new statistical approach that explicitly accounts for variability and uncertainty present in population trajectory models by coupling hierarchical and semi-parametric methods. Hierarchical modelling by its very nature allows the partitioning of variability into multiple spatial scales and model stages. It has been well demonstrated that ecological processes can be highly variable both within and across different spatio-temporal scales and that the relationship between two variables can change according to the spatial scale considered [8,51]. As a consequence, robust quantitative descriptions of natural patterns need to be able to connect broad-scale patterns to fine-scale processes, and in doing so transfer information across these scales [52]. Complemented by a semi-parametric formulation, our approach here facilitates a more flexible representation of trends over time, free of parametric constraints, and thereby, more effectively "letting the data speak for themselves" [7]. In addition, we modelled multifaceted trajectories of populations across space and time by decomposing dynamics into linear and non-linear components which represent deviations from linearity. Thus, we are able to simultaneously identify long-term trends (linear) as well as more temporally localized variations (often non-linear) in population trajectories [12], contributing to better diagnosis of sources of variability inherent in long-term ecological data.

There is a growing literature on the philosophical, practical and inferential benefits of the Bayesian framework for modelling ecological data [2,3,6,31,38,53]. In the field of coral reef ecology, applications of Bayesian hierarchical regression models are fairly new. Published examples have examined: the effects of temperature anomalies on coral cover declines from a global meta-analysis [17] and spatial variability of reef fish community structure in French Polynesia [9] and Australia [26]. These

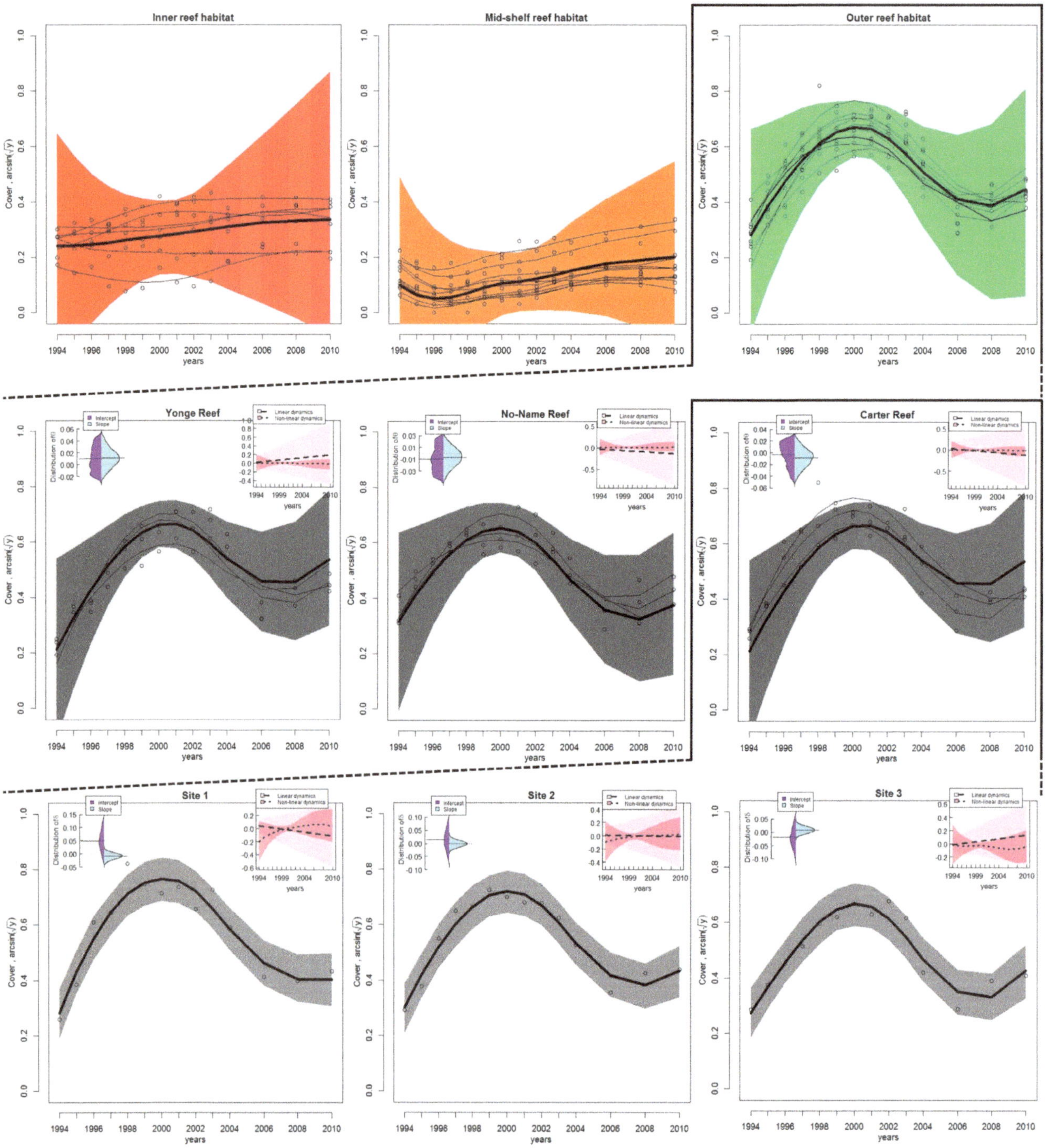

Figure 3. Observed (dots) and fitted (curves) ***Acropora*** **dynamics and associated parameter estimates and uncertainties.** Coral dynamics are modelled at the three spatial scales within the sub-region of Cooktown Lizard Island; with plotted lines for: habitats (top); reefs (middle); and sites (bottom). Shaded areas encompass 95% posterior predictive intervals around estimated coral trajectories and 95% credible intervals around model parameters. Top-right inserts on plots illustrate the linear and non-linear components of coral trajectories extracted from equations for f_h, f_r and f_i. Top-left inserts illustrate posterior distributions of linear parameters (from the top to the bottom, plotted lines are for: θ_0 and θ_1, δ_0 and δ_1 respectively; refer to Fig. 2, see equations in main text). Intercept terms were indexed by 0 and slope terms by 1 and shown with their 95% credible interval. Thin black lines on reef- and habitat-scale plots (mid- and top-line) show the fitted dynamics of nested individual sites. Note different y-axis scales in inserted graphs. Estimates of coral cover trajectories are illustrated at the three sites at Carter Reef from the outer reef habitat depicted in Figure 1.

studies, however, investigated the effects of different covariates without full consideration of the associated uncertainties in their models. As a consequence, these studies have not fully utilized a major benefit of Bayesian computation, which resides in its capacity to simultaneously quantify uncertainties associated with estimated data, model structure and model parameters. Indeed,

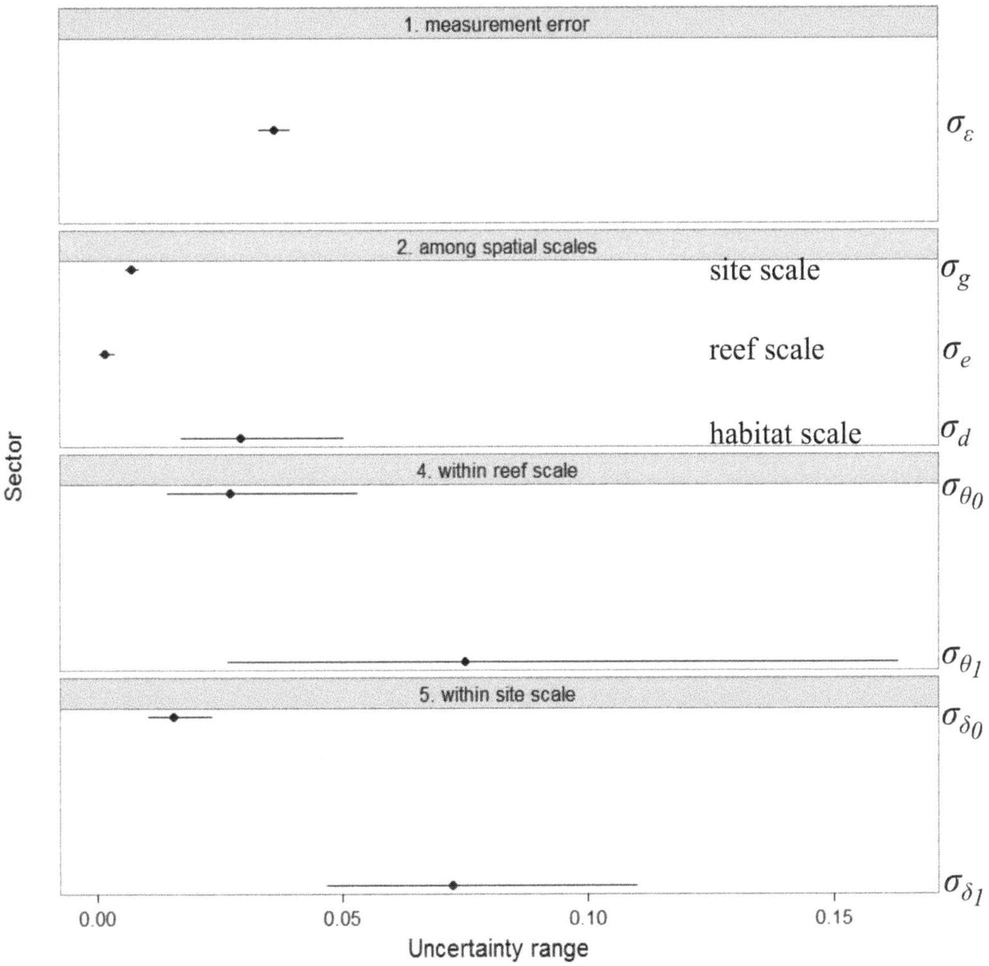

Figure 4. Estimation of the uncertainty in the three hierarchical stages of the model. The measurement error σ_ε corresponds to the uncertainty at the data stage, and all other variance components σ account for uncertainty in the process model, decomposed into the variability in trend (intercept and slope) of coral cover at the three spatial scales of habitat, reefs and sites.

we are no longer restricted to making a choice: whether to account for the spatio-temporal structure in the data whilst only tracking positive or negative effects of covariates on a response variable; or to simplify the spatial structure whilst tracking non-linear effects of the covariates on the response. By accounting for non-linear trends (in the fixed component of the model), we may then focus on the noise at the appropriate spatio-temporal scales (in the random effects component). For coral reef trajectories, we must allow sufficient complexity in both of these fixed and random components. Doing so will help identify and potentially reduce sources of uncertainty associated with modelling and will contribute to improved knowledge of the dynamics of populations studied in this way.

The Bayesian semi-parametric hierarchical model presented here also provides a more ecologically relevant way of modelling population trajectories, in cases where variability is large and cannot be assigned to unique causes and origins. By decomposing this variation into variance components at multiple spatial scales and model stages, and directly assessing the posterior distributions of these components, we have shown that it is not possible to accurately estimate trend parameters at a spatial scale larger than that of a reef in the sub-region considered here. That is, coral trajectories were consistent at the km^2 scale of sites within reefs but diverged at the larger 5 km^2 scale of reefs within habitats. These results suggest that time as a single covariate is insufficient to explain the coral trajectories at the habitat or sub-regional scales, and would benefit from additional explanatory factors, such as processes operating at these larger scales. For example, a historical review of *Acropora* populations within the Cooktown-Lizard Island sub-region modelled here reveals that they were affected by outbreaks of the coral predator crows-of-thorns starfish (*Acanthaster planci*) on mid-shelf reefs between 1995 and 1999, and by white syndrome disease from 2000 to 2003 on outer reefs [12]. Crowns-of-thorns starfish outbreaks are relatively slow and diffusive disturbances whose propagation is driven by prey availability [16]. Similarly, the spread of diseases and impacts from other major disturbances such as cyclones and coral bleaching are not homogenous on reefs and typically attenuate as the result of multiple factors acting at differing scales [12,28]. As a consequence, the effects of disturbances are seldom homogeneous across reefs, particularly at larger spatial scales such as, in the case of the GBR, shelf-position, making it difficult to explain coral trajectories using time as a single covariate. Nonetheless, this sort of approach is sometimes adopted [12–14]. Moreover, the three categories of shelf-position sampled by the LTMP were initially defined for management purposes but in a largely *ad hoc* fashion and were based on limited knowledge of reef ecology at that time [21]. Indeed, the observed reduction in the precision of parameter estimates at larger spatial scales can be partly attributed to a lack of trends in the data at this habitat scale. A remedy for this may be to include informative covariates in the model or use more informed prior information [54], obtained perhaps from similar analyses of other coral reef systems; these options are the subject of ongoing research. Moreover, based on the more extensive information extracted from our statistical approach, more informative spatial sampling programs could be designed to address specific management issues in an adaptive learning framework as advocated for example by [55]. At present, however, our results indicate that in the absence of better prior knowledge, other explanatory covariates and/or other spatial designs, conclusions about GBR coral cover trajectories become more uncertain at a scale larger than individual reefs. Therefore, in this context, management actions and the assessment of their efficacy may be better focused at the reef scale.

In conclusion, the Bayesian semi-parametric hierarchical approach introduced here facilitates flexible and environmentally relevant description of non-linear population trajectories and associated uncertainties. As illustrated, it can be used to identify critical spatial thresholds beyond which ecological data reveal divergence in the trajectories and so hinder model efficiency. Without this decomposition of uncertainties at multiple spatial scales and model stages, patterns remain concealed and conclusions regarding population trajectories can be considerably compromised. In contrast, we argue that our model can unlock information contained in spatially extensive time-series data, facilitate the design of better future surveys, provide guidance for understanding sources of uncertainty, and support better informed decision making.

Supporting Information

Figure S1 Illustration of the poor estimation of slope (γ_{1h}) and intercept (γ_{0h}) parameters based on three MCMC chains from a simulation of 200,000 iterations, a burn-in of 100,000 iterations and a thinning rate of 50 iterations. Differences in estimated values in MCMC chains demonstrate the non-convergence of the algorithm to a unique posterior distribution for each parameter. This non-convergence also affects the variance of the parameters with ranges of the 95% credible intervals (RCIs) varying between 0.1 and 1.

Code S1 R and WinBUGS code to implement the Bayesian semi-parametric hierarchical model.

Acknowledgments

We thank Matt Moores for assistance with satellite imagery and Matthew Spencer for his thoughtful and constructive review.

Author Contributions

Conceived and designed the experiments: JV MJC MK SLC KM. Analyzed the data: JV SLC KM. Wrote the paper: JV MJC MK SLC KM.

References

1. Moritz C, Agudo R (2013) The future of species under climate change: resilience or decline? Science 341: 504–508.
2. Wikle CK (2003) Hierarchical Bayesian models for predicting the spread of ecological processes. Ecology 84: 1382–1394.
3. Cressie N, Calder CA, Clark JS, Ver Hoef JM, Wikle CK (2009) Accounting for uncertainty in ecological analysis: the strengths and limitations of hierarchical statistical modeling. Ecological Applications 19: 553–570.
4. Miao S, Carstenn S, Nungesser M (2009) Unprecedented challenges in Ecological research: past and present. In: Real world in Ecology: Large-Scale and Long-Term Case Studies and Methods. Springer, New York, pp. 1–19.
5. Kujala H, Burgman MA, Moilanen A (2012) Treatment of uncertainty in conservation under climate change. Conservation Letters 6: 73–85.
6. Clark JS (2005) Why environmental scientists are becoming Bayesians. Ecology Letters 8: 2–14.
7. Wood SN (2001) Partially specified ecological models. Ecological Monographs 71: 1–25.
8. McMahon SM, Diez JM (2007) Scales of association: hierarchical linear models and the measurement of ecological systems. Ecology Letters 6: 437–452.
9. MacNeil MA, Graham NAJ, Polunin NVC, Kulbicki M, Galzin R, et al. (2009) Hierarchical drivers of reef-fish metacommunity structure. Ecology 1: 252–264.
10. Caley MJ, St John J (1996) Refuge availability structures assemblages of tropical reef fishes. The Journal of Animal Ecology 65: 414–428.
11. Hughes TP, Connell JH (1999) Multiple stressors on coral reefs: A long-term perspective. Limnology and Oceanography 44: 932–940.
12. Osborne K, Dolman AM, Burgess SC, Johns KA (2011) Disturbance and the dynamics of coral cover on the Great Barrier Reef (1995–2009). PloS one 6: e17516.
13. Sweatman H, Delean S, Syms C (2011) Assessing loss of coral cover on Australia's Great Barrier Reef over two decades, with implications for longer-term trends. Coral Reefs 30: 521–531.
14. De'ath G, Fabricius KE, Sweatman H, Puotinen M (2012) The 27-year decline of coral cover on the Great Barrier Reef and its causes. Proceedings of the National Academy of Sciences of the United States of America 109: 17995–17999.
15. Hughes TP, Baird AH, Dinsdale EA, Moltschaniwskyj NA, Pratchett MS, et al. (2012) Assembly rules of reef corals are flexible along a steep climatic gradient. Current Biology 22: 736–741.
16. Kayal M, Vercelloni J, Lison de Loma T, Bosserelle P, Chancerelle Y, et al. (2012) Predator crown-of-thorns starfish (Acanthaster planci) outbreak, mass mortality of corals, and cascading effects on reef fish and benthic communities. PloS one 7: e47363.
17. Selig ER, Casey KS, Bruno JF (2012) Temperature-driven coral decline: the role of marine protected areas. Global Change Biology 18: 1561–1570.
18. Connolly S, Vercelloni J, Mengersen K, Caley MJ (2014) Meta-analysis of heterogeneous coral reef time series: a critical review. The ANZIAM Journal In press.
19. Hughes TP, Bellwood DR, Baird AH, Brodie J, Bruno JF, et al. (2011) Shifting base-lines, declining coral cover, and the erosion of reef resilience: comment on Sweatman et al. (2011). Coral Reefs 30: 653–360.
20. Sweatman H, Syms C (2011) Assessing loss of coral cover on the Great Barrier Reef: A response to Hughes et al. (2011). Coral Reefs 30: 661–664.
21. Brodie J, Waterhouse J (2012) A critical review of environmental management of the 'not so Great' Barrier Reef. Estuarine, Coastal and Shelf Science 104: 1–22.
22. Fisher R, Knowlton N, Brainard RE, Caley MJ (2011) Differences among major taxa in the extent of ecological knowledge across four major ecosystems. PLoS ONE 6: e26556.
23. Bellwood DR, Hughes TP, Folke C, Nyström M (2004) Confronting the coral reef crisis. Nature 429: 827–833.
24. Nyström M, Folke C (2001) Spatial Resilience of Coral Reefs. Ecosytems 4: 406–417.
25. Riegl B, Berumen M, Bruckner A (2013) Coral population trajectories, increased disturbance and management intervention: a sensitivity analysis. Ecology and Evolution doi: 10.1002/ece3.519
26. Cheal AJ, Emslie M, MacNeil MA, Miller I, Sweatman H (2013) Spatial variation in the functional characteristics of herbivorous fish communities and the resilience of coral reefs. Ecological Applications 23: 174–188.
27. Gardner TA, Côté IM, Gill JA, Grant A, Watkinson AR (2003) Long-term region wide declines in Caribbean corals. Science 301: 958–960.
28. Bruno JF, Selig ER (2007) Regional decline of coral cover in the Indo-Pacific: timing, extent, and subregional comparisons. PLoS ONE 8: e711.
29. Graham NAJ, Nash KL, Kool JT (2011) Coral reef recovery dynamics in a changing world. Coral Reefs 30: 283–294.
30. Austin MP (2002) Spatial prediction of species distribution: an interface between ecological theory and statistical modelling. Ecological Modelling 157: 101–118.
31. Clark JS (2003) Uncertainty and variability in demography and population growth: a hierarchical approach. Ecology 6: 1370–1381.
32. Mellin C, Bradshaw CJA, Meekan MG, Caley MJ (2010) Environmental and spatial predictors of species richness and abundance in coral reef fishes. Global Ecology and Biogeography 19: 212–222.
33. Bolker BM, Brook ME, Clark CJ, Geange SW, Poulsen JR, et al. (2009) Generalized linear mixed models: a practical guide for ecology and evolution. Trends in Ecology and Evolution 24: 127–135.

34. Connell JH (1997) Disturbance and recovery of coral assemblages. Coral Reefs 16: 101–113.
35. Pandolfi JM, Bradbury RH, Sala RE, Hughes TP, Bjorndal KA, et al. (2003) Global trajectories of the long-term decline of coral reef ecosystems. Science 301: 955–958.
36. Halford AR, Caley MJ (2009) Towards an understanding of resilience in isolated coral reefs. Global Change Biology 15: 3031–3045.
37. Hadfield JD, Wilson AJ, Garant D, Sheldon BC, Kruuk LEB (2010) The misuse of BLUP in ecology and evolution. The American Naturalist 175: 116–125.
38. Clark JS, Gelfand AE (2006) A future for models and data in environmental science. Trends in Ecology and Evolution 21: 375–380.
39. Gelman A (2005) Multilevel (hierarchical) modeling: what it can and can't do. Technometrics 48: 432–435.
40. Sweatman H, Cheal A, Coleman G, Emslie M, Johns K, et al. (2008) Long-term monitoring of the Great Barrier Reef. Statut report number 8, Australian Institute of Marine Science, 369 pp.
41. Jonker M, Johns K, Osborne K (2008) Surveys of benthic reef communities using underwater digital photography and counts of juvenile corals. – Long-term Monitoring of the Great Barrier Reef Standard Operation Procedure Number 10.
42. Ruppert D, Wand MP, Carroll RJ (2003) Semiparametric Regression. Cambridge University Press, Cambridge.
43. Crainiceanu CM, Ruppert D, Wand MP (2005) Bayesian Analysis for penalized spline regression using WinBUGS. Journal of Statistical Software 14: 1–24.
44. Spiegelhalter DJ, Best NG, Carlin BP, Van Der Line A (2002) Bayesian measures of model complexity and fit. Journal of the Royal Statistical Society, Series B, 64: 583–639.
45. Rao PV (1998) Statistical Research Methods in the Life Sciences. Brooks/Cole Publishing Company, Pacific Grove, California.
46. Ruppert D (2002) Selecting the Number of Knots for Penalized Splines. Journal of Computational and Graphical Statistics 11: 735–757.
47. Gelman A (2006) Prior distributions for variance parameters in hierarchical models. Bayesian Analysis 3: 515–533.
48. Sturtz S, Ligges U, Gelman A (2005) R2WinBUGS: A Package for Running WinBUGS from R. Journal of Statistical Software 12: 1–16.
49. Spiegelhalter DJ, Thomas A, Best N, Lunn D (2003) WinBUGS Version 1.4 Users manual. MRC Biostatistics Unit, Cambridge.
50. Plummer M, Best N, Cowles K, Vines K, Sarkar D, et al. (2006) CODA: Convergence Diagnosis and Output Analysis for MCMC. R News 6: 7–1.
51. Wiens JA (1989) Spatial scaling in ecology. Functionnal Ecology 3: 385–397.
52. Wu J, David JL (2002) A spatially explicit hierarchical approach to modelling complex ecological systems: theory and applications. Ecological Modelling 153: 7–26.
53. Ellison AM (2004) Bayesian inference in ecology. Ecology Letters 7: 509–520.
54. Low Choy S, O'Leary R, Mengersen K (2009) Elicitation by design in ecology: using expert opinion to inform priors for Bayesian statistical models. Ecology 90: 265–277.
55. Caley MJ, Fisher R, Mengersen K (2014) Global species richness estimates have not converged. Trends in Ecology and Evolution 4: 187–188.

14

Indigenous Burning as Conservation Practice: Neotropical Savanna Recovery amid Agribusiness Deforestation in Central Brazil

James R. Welch[1]*, Eduardo S. Brondízio[2,3], Scott S. Hetrick[3], Carlos E. A. Coimbra Jr[1]

1 Escola Nacional de Saúde Pública, Fundação Oswaldo Cruz, Rio de Janeiro, Rio de Janeiro, Brazil, **2** Department of Anthropology, Indiana University, Bloomington, Indiana, United States of America, **3** Anthropological Center for Training and Research on Global Environmental Change, Indiana University, Bloomington, Indiana, United States of America

Abstract

International efforts to address climate change by reducing tropical deforestation increasingly rely on indigenous reserves as conservation units and indigenous peoples as strategic partners. Considered win-win situations where global conservation measures also contribute to cultural preservation, such alliances also frame indigenous peoples in diverse ecological settings with the responsibility to offset global carbon budgets through fire suppression based on the presumed positive value of non-alteration of tropical landscapes. Anthropogenic fire associated with indigenous ceremonial and collective hunting practices in the Neotropical savannas (cerrado) of Central Brazil is routinely represented in public and scientific conservation discourse as a cause of deforestation and increased CO_2 emissions despite a lack of supporting evidence. We evaluate this claim for the Xavante people of Pimentel Barbosa Indigenous Reserve, Brazil. Building upon 23 years of longitudinal interdisciplinary research in the area, we used multi-temporal spatial analyses to compare land cover change under indigenous and agribusiness management over the last four decades (1973–2010) and quantify the contemporary Xavante burning regime contributing to observed patterns based on a four year sample at the end of this sequence (2007–2010). The overall proportion of deforested land remained stable inside the reserve (0.6%) but increased sharply outside (1.5% to 26.0%). Vegetation recovery occurred where reserve boundary adjustments transferred lands previously deforested by agribusiness to indigenous management. Periodic traditional burning by the Xavante had a large spatial distribution but repeated burning in consecutive years was restricted. Our results suggest a need to reassess overreaching conservation narratives about the purported destructiveness of indigenous anthropogenic fire in the cerrado. The real challenge to conservation in the fire-adapted cerrado biome is the long-term sustainability of indigenous lands and other tropical conservation islands increasingly subsumed by agribusiness expansion rather than the localized subsistence practices of indigenous and other traditional peoples.

Editor: Brock Fenton, University of Western Ontario, Canada

Funding: Financing was provided by the Brazilian National School of Public Health (INOVA-ENSP program), the Fulbright Commission (Fulbright-Hays DDRAF no. P022A040016), and the Brazilian Research Council (CNPq grants 475674/2008-1, 403569/2008-7, and 500072/2010-8). The funders had no role in study design, data collection and analysis, decision to publish, or preparation of the manuscript.

Competing Interests: The authors have declared that no competing interests exist.

* E-mail: welch@ensp.fiocruz.br

Introduction

Efforts to reduce the negative environmental and biodiversity impacts of commercial agriculture and pasture activities are increasingly recognized to benefit from locally based knowledge and practices [1]. The new United Nations Intergovernmental Science-Policy Platform on Biodiversity and Ecosystem Services (IPBES) program, for instance, recognizes explicitly the need to include local and indigenous knowledge as part of its assessment and policy reach mandates [2]. Nevertheless, international conservation discourse generates important misconceptions by presuming the destructiveness of human alteration of tropical landscapes and, specifically, overgeneralizing about the effects of anthropogenic fire in diverse cultural and ecological settings.

Indigenous reserves and other types of protected areas have become the most important policy mechanism for controlling deforestation [3,4] and fire [5] associated with the booming expansion of large-scale agriculture in Northern and Central Brazil. Today these areas represent a mosaic of conservation islands amid expanding monoculture landscapes and are often considered win-win situations where global conservation goals and reduction in carbon emissions also contribute to cultural preservation [3,6,7]. Nevertheless, the generalized association between fire impacts on conservation, CO_2 emissions, and climate change is contributing to important misrepresentations when extrapolated without attention to regional and local contexts.

Increasingly part of the widespread implementation of Reducing Emissions from Deforestation and Forest Degradation (REDD) and REDD+ programs [8], these ideas regarding anthropogenic fire are also widely incorporated by interest groups as political narratives contesting indigenous rights to land in Brazil. They promote the notion that indigenous burning activities represent a destructive mentality ("culturally endorsed pyromania," to borrow an expression applied to indigenous practices in northern Australia

[9]) out of line with the conservation agenda set forward at global, national, and local levels to revert biodiversity losses and climate change. Thus, localized traditional landscape management practices are subjected to the scrutiny of unexpected alliances (e.g., large commercial ranchers using conservation arguments). Increasingly, indigenous and other traditional peoples in Brazil are framed with the responsibility to offset a substantial portion of the national and global carbon budget through conservation stewardship and REDD+ projects financed by high-emission industrialized countries [10]. Nevertheless, industrial agriculture and ranching expansion since the late 1960s has transformed the Central Brazilian cerrado into one of the most threatened biomes in the country [11,12].

Biodiversity and Fire in the Cerrado

The cerrado, included among the 35 most important "hotspots" in the world [13,14], is a diverse but highly threatened tropical savanna-like biome covering over 2 million hectares, approximately 24% of the total area of Brazil, mainly in the Central Brazilian Plateau. According to one estimate, of the approximately 10,000 plant species identified in the cerrado, about 4,400 are endemic [15]. The agricultural potential of the cerrado and its positive economic benefits for Brazil have been widely publicized for decades [16]. At least 35% of the cerrado region has been completely converted to intensive human use since 1960, predominantly for pasture, intensive monoculture and, more recently, eucalyptus cultivation [11]. Recent land conversion to soybean and sugarcane biofuel production in the cerrado results in large carbon debts estimated to require 17 to 37 years to repay [17]. In lieu of buffering the Amazon from further expansion of agricultural commodities and biofuels, in coming years the cerrado region is prone to increasing pressures with associated climatic impacts [18,19] and escalation of conflicts between agribusiness and indigenous interests [20].

Differently from the Brazilian rainforests – Amazonian and Atlantic forests, adaptation to periodic surface fires is an evolutionary characteristic of cerrado ecology independently from human action [21,22]. Lightning, which is common during the rainy season, is considered the principal environmental ignition factor in the cerrado [23] and explains evidence of cerrado wildfires as early as 32,000 BP, long before human presence in the region [24,25]. The recurrence of fire in a determined cerrado landscape is thought to have resulted in over 70% of the total cerrado biomass being subterranean, as well as other anatomical adaptations providing resistance to fire damage and favoring vigorous vegetation regrowth immediately after the fires, even before first rains [21,26,27]. The diversity of structural forms in the cerrado biome is considerable. Depending on the environmental factors that predominate in a given region, such as soil depth and drainage, level of acidity, and aluminum saturation, vegetation can range from dense forests to open grasslands, with many intermediate configurations in close proximity [27].

The presence and potential positive ecological effects of anthropogenic fire in the cerrado has been noted by botanists travelling through the region since the nineteenth century. In 1892, botanist Eugene Warming observed that plant form and vegetation structure in the cerrados of Minas Gerais, Brazil, benefitted from fire, with most plants undergoing vigorous sprouting and flowering shortly after being burned [28]. As he wrote, "the most beautiful, also the most blooming and freshest green field I have ever seen was precisely one that was burned in October" [28]. More recently, based on observations of cerrados in Xavante territory, botanist George Eiten wrote that frequencies of indigenous fires greater than every one or two years had no evident long-term effects on plant physiognomy and recently

Figure 1. Xavante elder during collective hunt with fire, Pimentel Barbosa Indigenous Reserve, Brazil, 2005. Photograph by James R. Welch. The legal representative of the depicted individual provided informed written consent for the publication of his image.

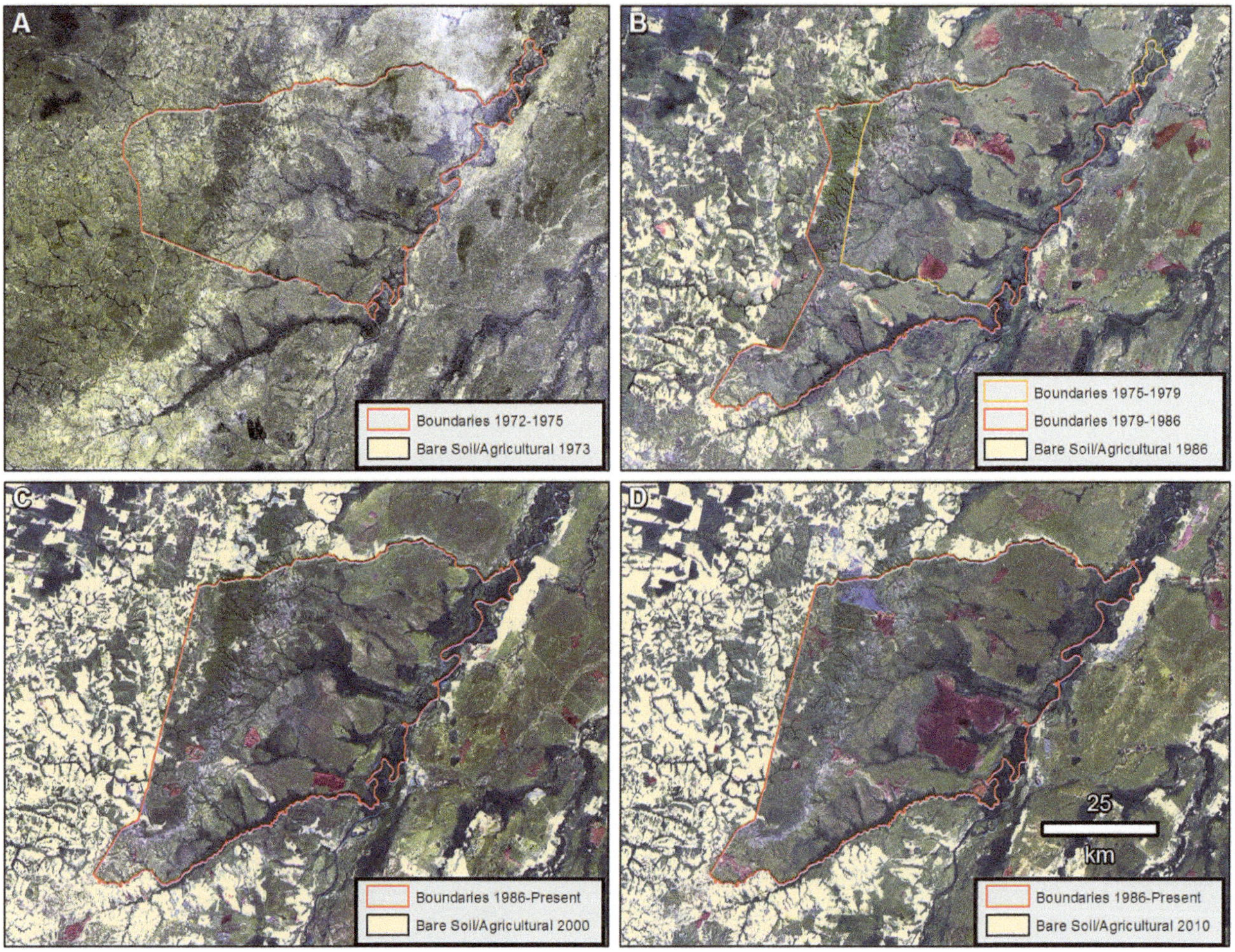

Figure 2. Mosaic of calibrated satellite imagery overlaid with reserve boundaries and deforestation mapped from 1973 to 2010, Pimentel Barbosa Indigenous Reserve, Brazil, 1973–2010. Landsat MSS and TM imagery used in our analyses was obtained courtesy of the United States Geological Survey (USGS) and from the Brazilian National Institute for Space Research (INPE) under Creative Commons Attribution-ShareAlike 3.0 Unported License.

burned areas attracted game animals through the creation of mineral licks and the abundant production of tender plant shoots [27].

Recent ecological studies show that cerrado landscapes may be even more highly resistant to the impacts of fire, including anthropogenic burning, than previously thought [21,22]. Also, fire regimes that vary in intensity, frequency, and seasonality may have significantly different effects on biomass and vegetation structure [26]. Periodic burning has been found to result in less intense and more fragmented, and therefore less destructive, fires than occur in protected cerrado areas where fire suppression causes the accumulation of large amounts of dead aboveground biomass [22,23,29]. Conversely, sporadic fires produce positive effects on cerrado biodiversity [25,30]. According to botanists, fires benefit cerrado plants due to adaptations such as fire-dependent sexual reproduction and dispersal, as well as rapid regeneration and increased flowering and germination [29,30]. These adaptations in turn benefit wildlife through increased availability of such foods as pollens, nectars, shoots, and fruits [29].

Although fire can effect major short term landscape change through its impact on cerrado vegetation, the resilience of fauna populations is indicated by scant evidence of local animal extinctions [22,31]. Additionally, the low temperatures and patchy distribution of cerrado surface fires allow animals ample escape routes and areas of cover. Ecological evidence also shows that moderate intensity fires increase the availability of foods for some taxa, including small vertebrates and arthropods, by stimulating vegetative growth, flowering, and fruiting [32–34]. Ecological studies specifically evaluating the impacts of indigenous burning activities in the cerrado also found no evidence that animal populations, including those of economically important game species, decreased following fires [35–39].

Cultural Aspects of Hunting with Fire

Hunting with fire is a characteristic subsistence activity of many indigenous inhabitants of the Central Brazilian cerrado, including such Gê-speaking indigenous groups as the Kayapó, Krahô, Canela, and Xavante [25]. Contrary to journalistic representations of Xavante hunting with fire as a cause of cerrado degradation and deforestation [40,41] or an impediment to vegetation recovery [42], ethnographic accounts indicate periodic burning by these indigenous peoples has minimal or desirable effects on cerrado vegetation structure and productivity [43,44]. Our previous ethnographic research [45] revealed that the Xavante people,

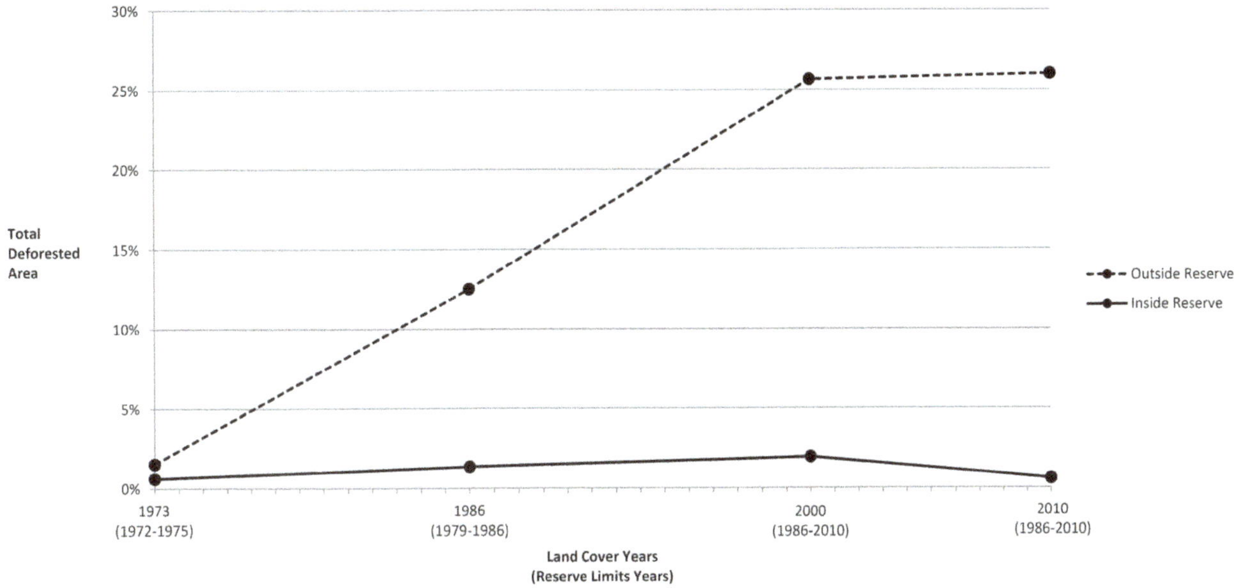

Figure 3. Total deforested area (%) relative to total area inside and outside (20 km buffer) corresponding limits of Pimentel Barbosa Indigenous Reserve, Brazil, 1973–2010.

eastern Mato Grosso State, use fire for hunting annually during the dry season (primarily July to September and less frequently the moister months of May and June) in widely dispersed locations throughout the reserve.

Xavante collective hunting with fire often occurs in conjunction with such ceremonial occasions as weddings and adolescent rites of passage [45]. Hunting excursions follow a ritualized format, whereby representatives from opposite exogamous moieties set fire to the vegetation as they run footraces in two wide arcs towards a distant finishing place. As the fire expands in a shifting mosaic pattern, numerous hunters act individually or in collaboration with other hunters to dispatch fleeing game animals (Figure 1). At the conclusion of hunting excursions, the acquired game meat is often pooled and ceremoniously presented to designated recipients, such

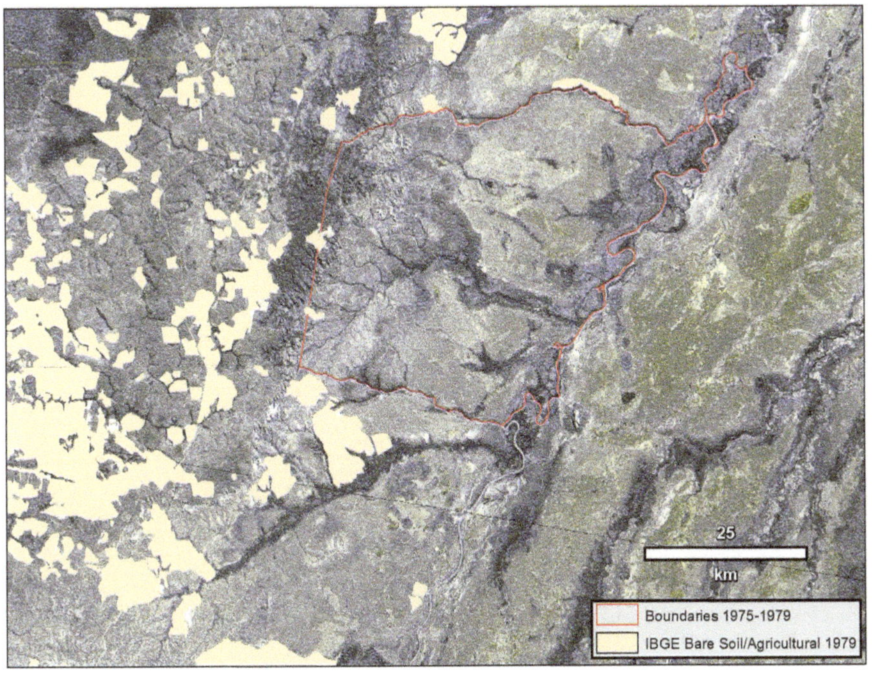

Figure 4. Mosaic of calibrated satellite imagery (1976) overlaid with reserve boundaries (1975–1979) and areas of bare soil used for livestock pasture and agricultural fields (1979), Pimentel Barbosa Indigenous Reserve, Brazil. Landsat MSS imagery used in our analyses was obtained from the Brazilian National Institute for Space Research (INPE) under Creative Commons Attribution-ShareAlike 3.0 Unported License. Cartographic data were digitized with permission from IBGE topographic maps [54].

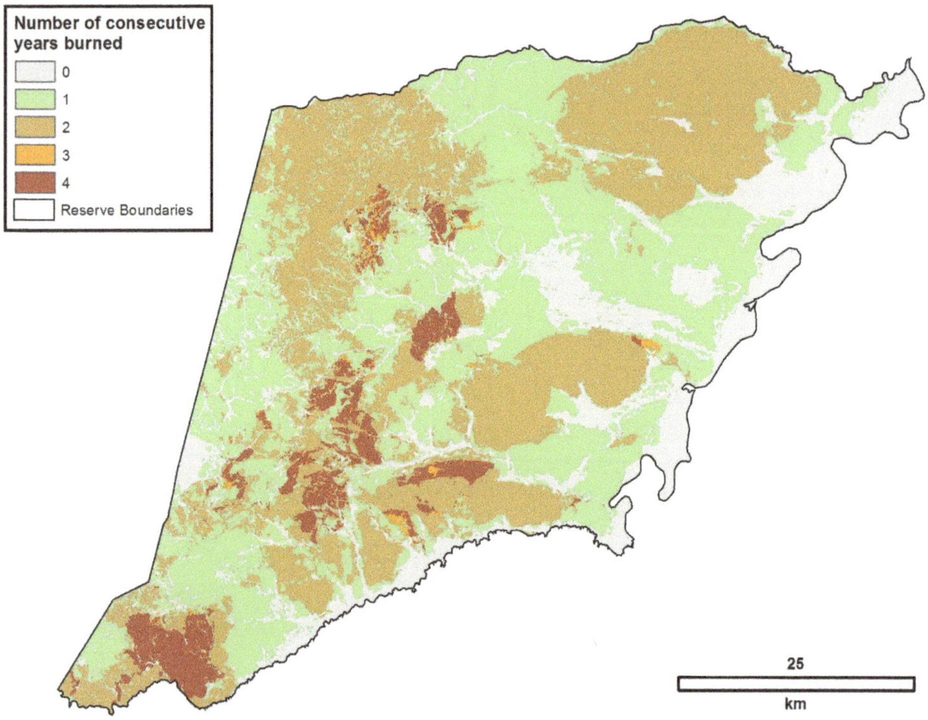

Figure 5. Burn scar transition matrix analysis showing land area by number of sequential years burned, Pimentel Barbosa Indigenous Reserve, Brazil, 2007–2010. Landsat TM and ETM+ imagery used in our analyses were obtained courtesy of the United States Geological Survey (USGS).

as in-laws and initiation leaders. From the Xavante perspective, collective hunting with fire is an essential component of the ceremonial activities that distinguish them ethnically and causes no undesirable ecological changes when practiced according to traditional protocols. To the contrary, Xavante cultural knowledge recognizes that traditional periodic burning accelerates the vegetative growth of some plants, increases the availability of certain animal foods, and reduces the intensity of future fires.

Although the Xavante practice agriculture and in the 1970s sold a small portion of rice yields as part of a government development project, today this activity is largely restricted to domestic production in small rotating patches, primarily in narrow gallery

Figure 6. Agribusiness deforestation (right) along western border of Pimentel Barbosa Indigenous Reserve (left), Brazil, 2009. Photograph by James R. Welch.

Figure 7. Remains of mechanized tree-chaining on private property adjacent to western boundary of Pimentel Barbosa Indigenous Reserve, Brazil, 2012. Photograph by Carlos E. A. Coimbra Jr.

forests along river margins [46]. In contrast to agriculture, periodic hunting with fire has a large spatial distribution and therefore greater potential for causing landscape-scale ecological effects within reserve boundaries. However, Xavante strategies for hunting with fire are highly regulated by specific traditional protocols and knowledge aimed to maximize both short-term hunting productivity and long-term sustainability. Our ethnographic research [45] shows that when planning fires, Xavante hunters discuss such diverse factors as season, weather, winds, soil and foliage moisture, type of vegetation, and natural barriers in order to control their intensity, spatial distribution, and frequency through time (see also [47–49]). Among these considerations ample attention is given to the potential negative ecological effects of burning the same patches of vegetation in consecutive years, especially during extended droughts.

Aims of this study

In this paper, we evaluate conservation narratives regarding the relative long-term ecological impacts of indigenous and agribusiness management of Brazilian neotropical savanna landscapes based on estimates of diachronic environmental changes and indigenous burning patterns in a cerrado landscape in Mato Grosso State. Specifically, we compared land cover change over four decades inside and within a 20 km buffer area around Pimentel Barbosa Indigenous Reserve (Xavante ethnic group) in order to ascertain their respective deforestation patterns. Multitemporal spatial analyses of changes in vegetation cover (1973–2010) were performed utilizing satellite and georeferenced field data, as well as archival research for reconstructing changes in reserve boundaries since 1972. Additionally, in order to evaluate the contemporary Xavante burning regime contributing to observed results at the end of this sequence, we analyzed the spatial and temporal distribution of anthropogenic fire within the reserve utilizing a four year sample (2007–2010). This integrated study expands upon our long-term ethnographic and interdisciplinary research carried out in collaboration with the Xavante people since the 1990s [45,50–53].

Methods

Ethics statement

Permission for the project was provided by the National Indian Foundation (FUNAI) and the National Ethics Committee (CONEP, permission #652-2011).

Location and population

Research was conducted in the Pimentel Barbosa Indigenous Reserve, Mato Grosso State, Brazil. The majority of this territory pertains to the Araguaia River Basin, although a small portion at the western boundary pertains to the Xingu River Basin. The ecological biome is cerrado, with a division occurring between western portions of the reserve, in which cerrado *sensu stricto* and dry forests predominate, and eastern portions, in which inundated grasslands are more prevalent. The boundaries of this reserve underwent a series of alterations during the study period (through decrees in 1972, 1975, and 1979, and ratification in 1986).

The 2010 population of Pimentel Barbosa Indigenous Reserve was approximately 1,400 individuals residing in nine villages [45]. The entire population pertains to the Xavante ethnic group and all individuals speak the Xavante language (Gê language family).

Georeferenced field data

Field data was collected for 117 georeferenced training samples inside and outside the reserve. These included multiple locations for 16 land cover classes (water, bare soil, agricultural field/pasture, fire scar, open grassland, arboreous (*murundu*) grassland, scrub grassland, cerrado *sensu stricto*, dense cerrado, rupicolous cerrado, *cerradão* woodland, dry upland forest, gallery forest, bamboo forest, *buriti* palm forest, and *ipuca* flooded forest). Data collected included vegetation type, land cover, land use history, regrowth status, notable species, and photographs. Elder members of the indigenous community accompanied all fieldwork activities and assisted with the identification of land use histories at all training sample locations.

Multi-temporal remote sensing: classification and deforestation analysis

Using Landsat MSS (WRS I Path/Row –240/69; 240/70; 80 meter spatial resolution) and TM (WRS II Path Row –223/69; 223/70; 224/69; 224/70; 30 meter spatial resolution) imagery from the United States Geological Survey (USGS) and Brazilian National Institute for Space Research (INPE), a multi-temporal data set was established for the image dates 1973, 1986, 2000, and 2010. Each date was created using a mosaic of two (1973) and four (1986, 2000, 2010) contiguous images. All images were subjected to radiometric and atmospheric calibrations, georeferenced, and image subsets were layer stacked to form multitemporal images. Image classification was carried out using ERDAS IMAGINE 11.0.4. Deforestation analysis utilized a land cover classification system consisting of water, bare soil/agricultural, grassland/cerrado, and forest.

Thematic data extraction for reserve boundaries and a 20 km buffer zone around the reserve was carried out using the tabulate area tool in ArcGIS 10.1. Classification procedure for the 1973 image was based on unsupervised procedures only; for 1986, 2000, and 2010 classification was based on a hybrid approach (i.e., involving integration of unsupervised and supervised classifiers). Images were independently classified. Image classification benefited from field collected georeferenced training samples. Field sites representative of natural vegetation were used as references for all dates, while recently altered sites were used as references for 2010. A layer of field sites was created and referenced into the multi-temporal image dataset.

Unsupervised analysis using ISODATA started with high-dimensional unsupervised clustering (50 classes) of the whole image area. Classes were analyzed according to spectral-structural differences (relationship between spectral data and indicators of vegetation structure), spatial distribution, and statistical values (mean, standard deviation, and covariance) for all spectral bands except thermal. Using supervised procedures, training samples with known vegetation and history of land use were created and used to develop spectral signatures for the classes of interest in each data set for 1986, 2000, and 2010. Training samples and unsupervised classes were combined into a single spectral signature file. Analysis of training samples and unsupervised signatures include a combination of Transformed Divergence separability analysis, spectral signature comparison, and correlation analysis of spectral-vegetation data. Based on these procedures, signatures were merged (or eliminated) to produce a final signature set representing a gradient of vegetation structure, from bare soil/agricultural to high forest. Signature sets were submitted to a probability-based Gaussian Maximum-Likelihood classifier. Classification accuracy of the 2010 classified image was assessed for aggregated classes (i.e., bare soil/agricultural, grassland/cerrado, and forests) used in transition matrix analysis. A set of 30 test fields of known features (i.e., training samples not used for supervised classification) was used to assess the accuracy of each class. Classification accuracy ranged from 91% for bare soil/agricultural, 86% for grassland/cerrado, and 84% for forests. Confusion between classes occurred mainly in areas of transitional vegetation structure with mixed elements of soil, herbaceous, arbustive, and arboreal coverage.

Legal descriptions of historical reserve boundaries (dates 1972–1975, 1975–1979, 1979–1986, and 1986-present) were digitally reconstructed from archival maps and descriptive documents of delimiting landmarks and overlaid on the multi-temporal image dataset in ArcGIS. Analyses of deforestation trajectories were developed by calculating transition matrices in ERDAS IMAGINE for each subsequent pair of images starting with the earliest date (1973–1986, 1986–2000, and 2000–2010). For the transition matrix analysis grassland and cerrado classes were aggregated. Deforestation was considered when a shrub or arboreous class converted into bare soil or agricultural fields/pasture) at a subsequent date. Areas of bare soil/agricultural that returned into grassland/cerrado or forest classes were recoded as such to indicate a reforestation process. The final thematic layer produced eight classes, including those that remained unaltered across dates (i.e., water, bare soil/agricultural, grassland/cerrado, and forest) and those involving change, including deforestation before 1973, deforestation between 1973 and 1986, deforestation between 1986 and 2000, and deforestation between 2000 and 2010. The final layer was recoded to show the area deforested for each period. We used cross-tabulation in ArcGIS to allow estimation of deforestation during each period within (respective reserve boundaries) and outside (respective buffer zones) the reserve.

We used ancillary data to verify the expansion of agriculture and pasture within the reserve during the late 1970s, years not represented in the analysis sequence described above but described by Xavante observers and historical sources as the apex of agribusiness encroachment and deforestation in and around Pimentel Barbosa Indigenous Reserve [45,50]. A thematic map of land cover for 1979 was produced based on cartographic maps by the Brazilian Institute for Geography and Statistics (IBGE) printed at a scale of 1:100,000 [54]. These IBGE maps were developed from aerial photographs of the area from 1967 and fieldwork from 1979. The map was digitized to produce a thematic layer, which was overlaid to a Landsat MSS image (1976) for correction. This map discriminates cerrado and areas opened for agriculture and pasture. The resulting thematic layer provides important supporting evidence for the deforestation of cerrado areas within reserve boundaries and the buffer zone towards the end of the 1970s.

Fire matrix analysis

Burned areas within reserve boundaries in the years 2007–2010 were mapped using Landsat TM and ETM+ imagery from the United States Geological Survey (USGS). Images free of clouds were acquired for each month from July-September for each of the four years to capture the primary annual burning season. As with the land-cover classifications, the imagery used for burned area mapping were coregistered to ensure spatial accuracy. Each image was classified using the ISODATA clustering algorithm in ERDAS IMAGINE. Fifty clusters were initially created using a convergence threshold of 0.995. Burned areas were isolated from other land covers through spectral signature analysis and visual interpretation. Mixed pixels indicating burning scars were manually digitized if necessary. Classifications from all months in the same year were then mosaicked and manually cleaned before running a 3×3 majority filter to complete the burned area classification. Each pair of years was subjected to a transition matrix analysis to estimate areas of consecutive burnings: first the transition from 2007 to 2008, second the transition from 2007/2008 to 2009, and third the transition from 2007/2008/2009 to 2010. A final transition matrix image 2007–2010 comprised of 5 classes was created to illustrate land area by number of consecutive years burned (i.e., burning in no more than 0, 1, 2, 3, and 4 consecutive years).

Results

The evolution of reserve boundaries overlaid on deforestation maps for 1973, 1986, 2000, and 2010 show the progression of deforestation and vegetational recovery inside and outside the reserve (Figure 2). Deforestation outside the reserve indicates a pattern of encroachment from the west with its eastern limit shifting through time as reserve boundaries were adjusted. Comparison of intermediate reserve boundaries (Figure 2B:1975–1979 and 1979–1986) indicate that most bare soil and agricultural fields or pasture inside the reserve in the years 1986 and 2000 was in locations previously outside or bordering the reserve limits. Additionally, large deforested patches evident inside the reserve in 1986 and 2000, especially near its western and southern boundaries, were greatly reduced in size by 2010.

Considering the historical reserve boundaries associated with each image year (Figure 3), the deforested area inside the reserve remained stable at 0.6% between 1973 (1,645.9 ha) and 2010 (1,989.0 ha). However, intermediate years showed higher proportions (1.3% in 1986 and 1.9% in 2000) associated with land previously impacted by agricultural operations being incorporated into the reserve and, to a lesser extent, with agricultural projects promoted by the federal government inside the reserve. The deforested area inside the reserve decreased by 68.9% (4,412.5 ha) from its peak in 2000 to the end of the sequence in 2010. The proportion of deforested land in a 20 km buffer area outside historical reserve limits increased continually from 1.5% (9,589.6 ha) in 1973 to 26.0% (175,412.4 ha) in 2010, with the largest increase observed in the period 1973–1986 (718.4%; 76,470.8 ha) and the smallest in 2000–2010 (1.4%; 2,383.4 ha).

Data from Brazilian topographic maps based on 1979 aerial photography (Figure 4) identify the distribution of commercial livestock pasture and agricultural fields when reserve boundaries were most restricted (1975–1979). These boundaries are also shown without temporally corresponding satellite imagery in Figure 2B. Figure 4 shows that many of the locations first appearing as bare soil/agricultural within the reserve in 1986 (Figure 2B) were in fact previously deforested by agribusiness before they were incorporated into the reserve by legal changes to the reserve limits.

Vegetation burning matrix analyses show the distribution of lands inside the reserve by number of consecutive years burned from 2007 to 2010 (Figure 5). The majority of the reserve burned at least once during the four-year period (83.2%; 273,979.26 ha) but only 40.7% (134,043.84) burned in two or more consecutive years. The largest areas were burned one (42.5%; 139,935.42 ha) or two (35.2%; 115,829.01 ha) years sequentially. Only 0.6% (2,019.24 ha) and 4.9% (16,195.59 ha) of the reserve was burned in three and four consecutive years, respectively. Additionally, 16.8% (55,496.52 ha) did not burn in any year.

Implications for cerrado conservation

The proportion of deforested land inside reserve boundaries increased in 1973–1986 and 1986–2000, but decreased in 2000–2010. Considering the entire period 1973–2010, the proportion of deforestation remained constant at an extremely low level (0.6%) inside the reserve. Most deforestation inside the reserve occurred along the western and southern boundaries, which suffered the greatest impacts of regional eastward economic expansion since the mid-1970s (Figure 6).

A notable factor in this progression is the evolution of reserve boundaries through time. In particular, Xavante political action [45] resulted in substantial increases in the reserve size (62.5% between 1975 and 1986), mainly through reclaiming traditional territory from adjacent agribusiness lands to the west and south. By returning these areas to indigenous management, previously deforested lands subsequently underwent vegetational recovery, thereby restoring the proportion of deforested land inside the reserve to the value observed four decades earlier (0.6%). This interpretation is corroborated by evidence from topographic maps generated in 1979, when the reserve was reduced to its smallest size, showing extensive agribusiness deforestation in areas which were subsequently incorporated or reincorporated into the reserve. A substantial portion of these areas were still classified as bare soil/agricultural in 1986 and 2000 but returned to grassland/cerrado or forest by 2010.

In contrast, outside reserve boundaries the proportion of deforested land increased in all periods, reaching 26% in 2010. The comparatively smaller increase observed in 2000–2010 contrasts with a regional pattern of rapid agribusiness expansion in the same period [55]. This difference may be attributable to decreased economic advantage of agricultural expansion along the western boundary of Pimentel Barbosa Indigenous Reserve given the extensive availability of lands nearby that were less intensively deforested before 2000.

The temporal burn scar pattern observed during the annual burning seasons of the last four years of this four decade sequence (2007 to 2010) indicates that almost 60% of the reserve was never burned or not burned in consecutive years. Only a very small portion of the reserve, about 5%, was burned in three or more consecutive years. Comparison of Figures 2D and 5 reveals that vegetation cover was maintained or recovered even in areas of high fire periodicity within the reserve. Thus, the observed Xavante burning pattern involved predominantly low fire periodicity and did not cause deforestation even where fires occurred in multiple consecutive years. Although these data do not allow us to draw direct comparisons with non-anthropogenic cerrado fire regimes, differences are likely to occur in terms of seasonality and frequency. In particular, whereas we observed the Xavante hunt with fire most often during the dry season, most cerrado fires not ignited through human agency are caused by lightning during rainy or transitional seasons [23,29,56]. However, in general terms, both scenarios involve highly varied surface fire periodicity and patchy distributions.

Ecological studies demonstrate that periodic cerrado fires can have positive effects on species diversity, as well as other environmental measures, and ought to be considered a strategic technique for environmental management in this biome [21,22,25,26,29,30]. This practice has yet to be incorporated in environmental management policies by the Brazilian government as it is in some other parts of the world with fire-adapted landscapes [57–59]. This oversight may result from a tendency for the Brazilian rainforests, which are not fire adapted, to receive more attention from conservationists and environmental policies than the cerrado biome and to frame narratives of conservation and climate change mitigation in Brazil.

Additionally, some scientific research on climate change [60] conflates rainforests with cerrados, despite their enormous ecological distinctions, because they are both included in Brazil's geopolitical division "Legal Amazon" (*Amazônia Legal*). However, the cerrado landscape and the indigenous reserves in Central Brazil face different threats than the Brazilian rainforests due to their distinct economic potentials for commercial agricultural development and ecological responses to anthropogenic fire [3,15,22,61]. For the past 50 years, the cerrado has been massively exploited through large scale clearing techniques largely absent in the Brazilian rainforests, involving mechanized tree-chaining that uproots trees in such a manner that regrowth is completely averted

(Figure 7). Conversely, commercial exploitation of Amazonian rainforests (in some cases involving indigenous peoples) involves some deforestation activities that are less frequent in the cerrado of Central Brazil, including illegal timber logging, large-scale industrial burning, and gold mining. As the Brazilian government and the agribusiness sector respond to international pressures to curb carbon emission from deforestation in the Amazonian rainforests, economic pressure on the cerrado and associated forest clearing activities tend to increase.

Much of policy and public discourse on tropical conservation presumes the positive value of non-alteration of tropical landscapes. For example, although international initiatives such as REDD and REDD+ promote the inclusion of indigenous peoples and their interests in adopting policies for reducing carbon emissions, these programs nevertheless often aim at halting traditional subsistence activities presumed to be responsible for CO_2 emissions, irrespective of their particular social and ecological contexts [62]. Similarly, scientific conservation discourse in Brazil often assumes that anthropogenic burning in the cerrado, whether practiced by indigenous peoples or others, is uniformly destructive, contributes to global warming, and threatens the value of indigenous reserves and other conservation areas [63–68]. However, as demonstrated by the results of the present study, indigenous Xavante landscape management practices over four decades, including periodic collective hunting with fire and political advocacy for federal recognition of traditional lands, maintained the integrity of the cerrado landscape and sustained vegetational recovery as compared to adjacent lands under non-indigenous management.

Our findings call into question the widespread assumption, evident in international conservation narratives such as those associated with the implementation of REDD and REDD+ programs, that the non-alteration of tropical landscapes through the suppression of indigenous anthropogenic burning is a beneficial strategy for reducing deforestation in indigenous reserves in the cerrado biome. This point is especially pertinent because rapid plant regrowth in the cerrado quickly re-assimilates carbon emissions from appropriate fire regimes [22,26]. Furthermore, our results suggest a need to reassess the implications of overreaching conservation and reduction in carbon emission narratives, increasingly shared by contrasting interest groups in Brazil and elsewhere, about the purportedly destructive nature of indigenous land use, particularly landscape management with fire [40,41]. In this context, the real challenge of cerrado conservation is to address the long-term sustainability of 'conservation islands,' such as indigenous reserves, and the carbon offset approach to deal with the inter-connected nature of land use systems that now characterizes Brazil and elsewhere, whereby conservation units are increasingly subsumed by the impacts of the larger landscapes to which they pertain.

Acknowledgments

Welch and Coimbra thank the members of the Xavante villages Pimentel Barbosa and Etênhiritipá for their participation and support during research activities. In particular, our field activities benefitted from assistance by Tsuptó Bruprewem Wairi Xavante, José Paulo Seriuwarão Xavante, Marco Aurelio Serenho Ihi Xavante, Tsidowi Wai'adzatse', Wahipó Xavante, and Pari'õwá Xavante. Brondízio and Hetrick express thanks for the support of Indiana University and the Anthropological Center for Training and Research on Global Environmental Change (ACT), the Institut d'Études Avancées (IEA), Paris, and the Institut des Hautes Études de l'Amérique Latine at Université Sorbonne Nouvelle, Paris.

Author Contributions

Conceived and designed the experiments: JRW SSH ESB CEAC. Analyzed the data: SSH ESB. Wrote the paper: JRW SSH ESB CEAC. Conducted the fieldwork: JRW CEAC.

References

1. Turnhout E, Bloomfield B, Hulme M, Vogel J, Wynne B (2012) Conservation policy: Listen to the voices of experience. Nature 488: 454–455.
2. United Nations (2012) Report of the second session of the plenary meeting to determine modalities and institutional arrangements for an intergovernmental science-policy platform on biodiversity and ecosystem services. Nairobi: United Nations Environment Programme.
3. Brondízio ES, Ostrom E, Young OR (2009) Connectivity and the governance of multilevel socio-ecological systems: The role of social capital. Annu Rev Environ Resour 34: 253–278.
4. Soares-Filho B, Moutinho P, Nepstad D, Anderson A, Rodrigues H, et al. (2010) Role of Brazilian Amazon protected areas in climate change mitigation. Proc Natl Acad Sci USA 107: 10821–10826.
5. Nepstad D, Schwartzman S, Bamberger B, Santilli M, Ray D, et al. (2006) Inhibition of Amazon deforestation and fire by parks and indigenous lands. Conserv Biol 20: 65–73.
6. Ricketts TH, Soares-Filho B, Fonseca GAB, Nepstad D, Pfaff A, et al. (2010) Indigenous lands, protected areas, and slowing climate change. PLoS Biol 8: e1000331.
7. Adeney JM, Christensen Jr NL, Pimm SL (2009) Reserves protect against deforestation fires in the Amazon. PLoS ONE 4: e5014.
8. Fairhead J, Leach M, Scoones I (2012) Green grabbing: A new appropriation of nature? J Peasant Stud 39: 237–261.
9. Whitehead PJ, Bowman DMJS, Preece N, Fraser F, Cooke P (2003) Customary use of fire by indigenous peoples in northern Australia: Its contemporary role in savanna management. Int J Wildland Fire 12: 415–425.
10. Birrell K, Godden L, Tehan M (2012) Climate change and REDD+: Property as a prism for conceiving Indigenous peoples' engagement. Journal of Human Rights and the Environment 3: 196–216.
11. Klink CA, Moreira AG (2002) Past and current human occupation, and land use. In: Oliveira PS, Marquis RJ, editors. The Cerrados of Brazil: Ecology and Natural History of a Neotropical Savanna. New York: Columbia University Press. 69–88.
12. Phalan B, Bertzky M, Butchart SHM, Donald PF, Scharlemann JPW, et al. (2013) Crop expansion and conservation priorities in tropical countries. PLoS ONE 8: e51759.
13. Myers N, Mittermeier RA, Mittermeier CG, Fonseca GAB, Kent J (2000) Biodiversity hotspots for conservation priorities. Nature 403: 853–858.
14. Mittermeier RA, Robles Gil P, Hoffman M, Pilgrim J, Brooks T, et al, editors (2004) Hotspots Revisited: Earth's Biologically Richest and Most Endangered Terrestrial Ecoregions. Chicago: Conservation International.
15. Klink CA, Machado RB (2005) Conservation of the Brazilian cerrado. Conserv Biol 19: 707–713.
16. Abelson PH, Rowe JW (1987) A new agricultural frontier. Science 235: 1450–1451.
17. Fargione J, Hill J, Tilman D, Polasky S, Hawthorne P (2008) Land clearing and the biofuel carbon debt. Science 319: 1235–1238.
18. Georgescu M, Lobell DB, Field CB, Mahalov A (2013) Simulated hydroclimatic impacts of projected Brazilian sugarcane expansion. Geophys Res Lett 40: 972–977.
19. Davidson EA, de Araujo AC, Artaxo P, Balch JK, Brown IF, et al. (2012) The Amazon basin in transition. Nature 481: 321–328.
20. Garfield S (2001) Indigenous Struggle at the Heart of Brazil: State Policy, Frontier Expansion, and the Xavante Indians, 1937–1988. Durham, NC: Duke University Press.
21. Simon MF, Grether R, Queiroz LP, Skema C, Pennington RT, et al. (2009) Recent assembly of the cerrado, a neotropical plant diversity hotspot, by in situ evolution of adaptations to fire. Proc Natl Acad Sci USA 106: 20359–20364.
22. Pivello VR (2011) Use of fire in the cerrado and Amazonian rainforests of Brazil: Past and present. Fire Ecol 7: 24–39.
23. Ramos-Neto MB, Pivello VR (2000) Lightning fires in a Brazilian savanna national park: Rethinking management strategies. Environ Manage 26: 675–684.
24. Ledru M-P (2002) Late quaternary history and evolution of the cerrados as revealed by palynological records. In: Oliveira PS, Marquis RJ, editors. The Cerrados of Brazil: Ecology and Natural History of a Neotropical Savanna. New York: Columbia University Press. 33–50.
25. Pivello VR (2006) Fire management for biological conservation in the Brazilian cerrado. In: Mistry J, Berardi A, editors. Savannas and Dry Forests: Linking people with Nature. Aldershot, England: Ashgate Publishing. 129–154.

26. Castro EA, Kauffman JB (1998) Ecosystem structure in the Brazilian Cerrado: A vegetation gradient of aboveground biomass, root mass and consumption by fire. J Trop Ecol 14: 263–283.
27. Eiten G (1972) The cerrado vegetation of Brazil. Bot Rev 38: 201–341.
28. Warming E (1892) Lagoa Santa: Et Bidrag til den biologiske Plantegeografi. Copenhagen: Bianco Luno.
29. Coutinho LM (1990) Fire in the ecology of the Brazilian cerrado. In: Goldammer JG, editor. Fire in the Tropical Biota: Ecosystem Processes and Global Challenges. New York: Columbia University Press. 82–105.
30. Miranda HS, Bustamante MMC, Miranda AC (2002) The fire factor. In: Oliveira PS, Marquis RJ, editors. The Cerrados of Brazil: Ecology and Natural History of a Neotropical Savannah. New York: Columbia University Press. 51–68.
31. Frizzo TLM, Bonizário C, Vasconcelos HL (2011) Revisão dos efeitos do fogo sobre a fauna de formações savânicas do Brasil. Oecologia Australis 15: 365–379.
32. Cavalcanti RB, Alves MAS (1997) Effects of fire on savanna birds in Central Brazil. Ornitologia Neotropical 8: 85–87.
33. Prada M, Marinho-Filho JS, Price PW (1995) Insects in flower heads of *Aspilia foliacea* (Asteraceae) after a fire in a central Brazilian savanna: Evidence for the plant vigor hypothesis. Biotropica 27: 513–518.
34. Vitt LJ, Caldwell JP (1993) Ecological observations on cerrado lizards in Rondônia, Brazil. J Herpetol 27: 46–52.
35. Prada M (2001) Effects of fire on the abundance of large mammalian herbivores in Mato Grosso, Brazil. Mammalia 65: 55–61.
36. Prada M, Marinho-Filho JS (2004) Effects of fire on the abundance of Xenarthrans in Mato Grosso, Brazil. Austral Ecol 29: 568–573.
37. Villalobos MP (2002) Efeito de Fogo e da Caça na Abundância de Mamíferos na Reserva Xavante do Rio das Mortes, MT, Brasil [Ph.D. dissertation]. Brasília: Universidade de Brasília. 81 p.
38. Briani DC, Palma ART (2004) Post-fire succession of small mammals in the Cerrado of Central Brazil. Biodivers Conserv 13: 1023–1037.
39. Leite DLP (2007) Efeitos do Fogo sobre a Taxocenose de Lagartos em Áreas de Cerrado *Sensu Stricto* no Brasil Central [Master thesis]. Brasília: Universidade de Brasília.
40. Oliveira M (2001) Fazendeiros controlam fogo perto de reserva: Área Pimentel Barbosa teve parte de seu território destruída pelo fogo. Diário de Cuiabá. 14 August 2001. Cuiabá. Available: http://www.diariodecuiaba.com.br/detalhe.php?cod = 64004. Accessed 2013 Jan 12.
41. Kassu, Milanez F (2010) Terra Indígena de Areões: Queimadas continuam. Água Boa News. 12 December 2010. Água, MT. Available: http://www.aguaboanews.com.br/portal/index.php?option = com_content&view = article&id = 10599. Accessed 2013 Jan 12.
42. Nascimento LJ (2013) Incêndios atingem 31 mil hectares de terra indígena xavante em MT. G1 Mato Grosso. 20 August 2013. Cuiabá, MT. Available: http://g1.globo.com/mato-grosso/noticia/2013/08/incendios-atingem-31-mil-hectares-de-terra-indigena-xavante-em-mt.html. Accessed 2013 Aug 22.
43. Hecht SB (2009) Kayapó savanna management: Fire, soils, and forest islands in a threatened biome. In: Woods WI, Teixeira WG, Lehmann J, Steiner C, WinklerPrins A, et al., editors. Amazonian Dark Earths: Wim Sombroek's Vision. Berlin: Springer. 143–162.
44. Mistry J, Berardi A, Andrade V, Krahô T, Krahô P, et al. (2005) Indigenous fire management in the cerrado of Brazil: The case of the Krahô of Tocantíns. Hum Ecol 33: 365–386.
45. Welch JR, Santos RV, Flowers NM, Coimbra CEA Jr. (2013) Na Primeira Margem do Rio: Território e Ecologia do Povo Xavante de Wedezé. Rio de Janeiro: Museu do Índio/FUNAI.
46. Gross DR, Eiten G, Flowers NM, Leoi FM, Ritter ML, et al. (1979) Ecology and acculturation among native peoples of Central Brazil. Science 206: 1043–1050.
47. Leeuwenberg FJ, Robinson JG (2000) Traditional management of hunting by a Xavante community in Central Brazil: The search for sustainability. In: Robinson JG, Bennett EL, editors. Hunting for Sustainability in Tropical Forests. New York: Columbia University Press. 375–394.
48. Melo MM, Saito CH (2011) Regime de queima das caçadas com uso do fogo realizadas pelos Xavante no cerrado. Biodiversidade Brasileira 1: 97–109.
49. Melo MM (2013) The practice of burning savannas for hunting by the Xavante Indians based on the stars and constellations. Soc Natur Resour 26: 478–487.
50. Coimbra CEA Jr., Flowers NM, Salzano FM, Santos RV (2002) The Xavánte in Transition: Health, Ecology, and Bioanthropology in Central Brazil. Ann Arbor: University of Michigan Press.
51. Santos RV, Coimbra CEA Jr., Welch JR (2013) A half-century portrait: Health transition in the Xavante Indians from Central Brazil. In: Brondízio ES, Moran EF, editors. Human-Environment Interactions: Current and Future Directions. Dordrecht: Springer. 29–52.
52. Flowers NM (1983) Seasonal factors in subsistence, nutrition, and child growth in a Central Brazilian community. In: Hames RB, Vickers WT, editors. Adaptive Responses of Native Amazonians. New York: Academic Press. 357–390.
53. Welch JR, Ferreira AA, Santos RV, Gugelmin SA, Werneck G, et al. (2009) Nutrition transition, socioeconomic differentiation, and gender among adult Xavante Indians, Brazilian Amazon. Hum Ecol 37: 13–26.
54. Instituto Brasileiro de Geografia e Estatística (IBGE) (1981) Topographic Maps (1:100,000) Corixão da Mata Azul (SD-22-X-A-IV), Bandeirantes (SD-22-X-C-IV), Canarana (SD-22-V-D-IV), Cascalheira (SD-22-V-B-V), Luiz Alves (SD-22-X-C-I), Matinha (SD-22-V-D-V), Riberão Água Preta (SD-22-V-D-VI), Rio Darro ou Feio (SD-22-V-B-IV), Rio das Mortes (SD-22-V-D-III), São João Grande (SD-22-V-B-VI), Rio Tanguro (SD-22-V-D-I), Rio Turvo (SD-22-V-D-II). Brasília: IBGE.
55. VanWey LK, Spera S, de Sa R, Mahr D, Mustard JF (2013) Socioeconomic development and agricultural intensification in Mato Grosso. Philos Trans R Soc Lond B Biol Sci 368: 20120168.
56. Mistry J (1998) Fire in the cerrado (savannas) of Brazil: An ecological review. Prog Phys Geog 22: 425–448.
57. McGregor S, Lawson V, Christophersen P, Kennett R, Boyden J, et al. (2010) Indigenous wetland burning: Conserving natural and cultural resources in Australia's World Heritage-listed Kakadu National Park. Hum Ecol 38: 721–729.
58. Boer MM, Sadler RJ, Wittkuhn RS, McCawa L, Grierson PF (2009) Long-term impacts of prescribed burning on regional extent and incidence of wildfires: Evidence from 50 years of active fire management in SW Australian forests. For Ecol Manage 259: 132–142.
59. van Wilgen BW, Forsyth GG, Klerk Hd, Das S, Khuluse S, et al. (2010) Fire management in Mediterranean-climate shrublands: A case study from the Cape Fynbos, South Africa. J Appl Ecol 47: 631–638.
60. Aragão LEOC, Shimabukuro YE (2010) The incidence of fire in Amazonian forests with implications for REDD. Science 328: 1275–1278.
61. Ratter JA, Ribeiro JF, Bridgewater S (1997) The Brazilian cerrado vegetation and threats to its biodiversity. Ann Bot 80: 223–230.
62. van Dam C (2011) Indigenous territories and REDD in Latin America: Opportunity or threat? Forests 2: 394–414.
63. Leonel MM (2000) O uso do fogo: O manejo indígena e a piromania da monocultura. Estudos Avançados 14: 231–250.
64. Diaz MCV, Nepstad D, Mendonça MJC, Motta RS, Alencar A, et al. (2002) O Preço Oculto do Fogo na Amazônia: Os Custos Econômicos Associados às Queimadas e Incêndios Florestais. Belém: Instituto de Pesquisa Ambiental da Amazônia.
65. Pereira CA, Fiedler NC, Medeiros MB (2004) Análise de ações de prevenção e combate aos incêndios florestais em unidades de conservação do cerrado. Revista Floresta 34: 95–100.
66. Soares RV, Santos JF (2002) Perfil dos incêndios florestais no Brasil de 1994 a 1997. Revista Floresta 32: 219–232.
67. Falleiro RM (2011) Resgate do manejo tradicional do cerrado com fogo para proteção das terras indígenas do oeste do Mato Grosso: um estudo de caso. Biodiversidade Brasileira 1: 86–96.
68. Nascimento DTF, Araújo FM, Ferreira Jr LG (2011) Análise dos padrões de distribuição espacial e temporal dos focos de calor no bioma cerrado. Revista Brasileira de Cartografia 63: 577–589.

15

Comparative Landscape Genetics of Three Closely Related Sympatric Hesperid Butterflies with Diverging Ecological Traits

Jan O. Engler[1,2]*, Niko Balkenhol[2], Katharina J. Filz[3,4], Jan C. Habel[5], Dennis Rödder[1]

1 Zoological Research Museum Alexander Koenig, Bonn, Germany, **2** Department of Wildlife Sciences, University of Göttingen, Göttingen, Germany, **3** Department of Biogeography, Trier University, Trier, Germany, **4** Museum of Natural History Dortmund, Dortmund, Germany, **5** Department of Ecology and Ecosystemmanagement, Technical University Munich, Freising-Weihenstephan, Germany

Abstract

To understand how landscape characteristics affect gene flow in species with diverging ecological traits, it is important to analyze taxonomically related sympatric species in the same landscape using identical methods. Here, we present such a comparative landscape genetic study involving three closely related Hesperid butterflies of the genus *Thymelicus* that represent a gradient of diverging ecological traits. We analyzed landscape effects on their gene flow by deriving inter-population connectivity estimates based on different species distribution models (SDMs), which were calculated from multiple landscape parameters. We then used SDM output maps to calculate circuit-theoretic connectivity estimates and statistically compared these estimates to actual genetic differentiation in each species. We based our inferences on two different analytical methods and two metrics of genetic differentiation. Results indicate that land use patterns influence population connectivity in the least mobile specialist *T. acteon*. In contrast, populations of the highly mobile generalist *T. lineola* were panmictic, lacking any landscape related effect on genetic differentiation. In the species with ecological traits in between those of the congeners, *T. sylvestris*, climate has a strong impact on inter-population connectivity. However, the relative importance of different landscape factors for connectivity varies when using different metrics of genetic differentiation in this species. Our results show that closely related species representing a gradient of ecological traits also show genetic structures and landscape genetic relationships that gradually change from a geographical macro- to micro-scale. Thus, the type and magnitude of landscape effects on gene flow can differ strongly even among closely related species inhabiting the same landscape, and depend on their relative degree of specialization. In addition, the use of different genetic differentiation metrics makes it possible to detect recent changes in the relative importance of landscape factors affecting gene flow, which likely change as a result of contemporary habitat alterations.

Editor: Daniele Canestrelli, Tuscia University, Italy

Funding: JOE received financial support by the German Federal Environmental Foundation (DBU) fellowship programme. KJF was funded by the Friedrich-Ebert-Stiftung. The funders had no role in study design, data collection and analysis, decision to publish, or preparation of the manuscript.

Competing Interests: The authors have declared that no competing interests exist.

* Email: j.engler.zfmk@uni-bonn.de

Introduction

In the theory of island biogeography, McArthur & Wilson [1] predicted the evolution of biodiversity on islands based on two key factors: habitat size and isolation. Later, this island based model was adopted to explain population structure of organisms in mainland ecosystems consisting of habitat patches surrounded by a semi- or non-permeable matrix. This mainland transformation of the theory of island biogeography inspired the fundamental paradigm of the metapopulation concept [2–3] and also of the neutral theory in both macroecology and population genetics [4–5]. Ultimately, island biogeography theory revolutionizes our thinking on habitat fragmentation and conservation biology (summarized in [6]). Apart from habitat size and isolation, spatial biodiversity patterns are also influenced by additional factors such as habitat quality [7], intrinsic characteristics of species-specific dispersal behaviour [8–9] and ecological tolerance [10] of species. Importantly, population responses are highly species-specific, when the quality of the landscape matrix in between suitable habitat patches is reduced [11]. This would also have consequences for global biodiversity [12–13] and large scale conservation efforts [14].

Understanding the effects of the landscape matrix on realized dispersal and functional population connectivity is also a major focus of landscape genetics [15–17]. Incorporating spatial landscape information with population genetic data goes far beyond the classical analysis of isolation-by-distance (IBD) [18]. Species respond differently to the landscape, in terms of their dispersal, which ultimately affects the rates of gene flow among local populations [19–20]. While the classical isolation-by-distance approach introduced by Wright [18] accounts for the geographic (Euclidean) distance between sampled populations only, other approaches such as the recently proposed isolation-by-resistance (IBR) concept [21] accounts for these species-specific responses to different landscape components that impede or favor gene-flow across a given landscape matrix.

Many studies assess landscape effects on gene flow in only a single species. However, to understand how landscape effects on gene flow differ between species, and to take effective conservation actions, it is important to analyze multiple species in the same landscape using identical methods [22]. However, past studies comparing different species mostly focused on species that inhabited comparable habitats, but were taxonomically independent [19–20,23]. A different comparative approach is to analyze landscape genetic relationships in closely related taxa inhabiting the same landscape. Such a focus on taxonomically related sympatric species (i.e. congeneric species which have the same or overlapping geographic ranges, regardless of whether or not they co-occur at the same locality) allows the assessment of traits that gradually change between the congeners independently from confounding effects that may arise in relation to different evolutionary histories or environments, respectively [24–25]. Next to dispersal propensity, niche breadth (i.e. the degree of specialization on specific habitat traits) is a very important trait in this respect, as it is directly associated with the available habitat within a landscape.

Generalist species can be found in a broader variety of ecosystems, showing higher abundances and broader spatial distributions. In contrast, specialist species demanding certain habitat conditions are often geographically restricted to specific habitats and usually occur in lower local abundances [26]. Apart from ecological demands, connectivity among local populations is further influenced by the dispersal propensity of species. Typically, sedentary species are mostly characterized by rather limited individual exchange compared to species with strong dispersal behavior. These ecological and behavioral traits also affect the genetic structure of generalist versus specialist species [10,26–27]. Organisms with specific habitat demands and restricted dispersal behavior should generally be characterized by low gene flow resulting in strong genetic differentiation. In contrast, species with weaker habitat specificity and higher dispersal propensity should show increased levels of gene flow, leading to a lack of genetic differentiation. Importantly, landscape influences on gene flow and resulting genetic patterns could also differ between generalist and specialist species inhabiting the same landscape.

In this study, we present a comparative landscape genetic analysis involving three closely-related butterfly species, to assess the impact of landscape parameters (i.e. land use, topography and climatic conditions) on the genetic structure of sympatric species with different ecological traits. We re-analyzed a molecular dataset taken from a previous study [28], where landscape effects were previously ignored, involving three congeneric, but ecologically divergent skipper species of the genus *Thymelicus* (Hubner 1890). The three species include the generalist *T. lineola*, which occurs in high abundances and shows strong dispersal propensity; the specialist *T. acteon* which is sedentary and occurs restricted to specific habitats; and *T. sylvestris*, which lies in between these two extremes in terms of habitat specificity and dispersal abilities. Using these three species, we (i) investigate the impact of ecological traits on species-specific functional landscape connectivity and (ii) determine the overall relevance of landscape characteristics for connectivity in each species, as well as the individual importance of topography, climatic conditions and land-use parameters. We hypothesized that species-specific landscape effects on gene flow should follow the cline of specialization in the three Hesperid butterflies, with strongest landscape effects on genetic differentiation in the most specialized *T. acteon* and weakest landscape effects in the generalist *T. lineola*.

Material and Methods

Ethics statement

The research was conducted under permission, to collect butterflies and to work in several protected areas, by the local authorities of Saarbrücken (Germany, Saarland), Koblenz (Germany, Rhineland-Palatinate), Luxembourg, and Metz (Loraine, France). Imagoes of the respective species were stored in liquid nitrogen until genetic analysis.

Study area and species

Our study area is located in the south-west of Germany and includes adjacent parts of France and Luxembourg (Fig. 1, Table S1). The sampling sites covered an area of approximately 120 km in north-south direction and 100 km in east-west direction. The landscape is characterised by a mosaic of residential areas, agricultural land, meadows, forests and semi-natural calcareous grasslands. Especially grasslands, but also some meadows and forest skirts provide suitable habitats for the three selected *Thymelicus* species, acting as valuable retreats and stepping stones [29].

The three selected model species *T. sylvestris*, *T. lineola* and *T. acteon* are closely related to each other with *T. lineola* and *T. acteon* being most distant related and where *T. sylvestris* clusters to a monophylum with *T. acteon* (Material S1). The three species show different habitat demands, even if they are co-occurring at suitable grassland patches: *T. lineola* occupies a broad ecological niche [30] and exhibits strong dispersal behaviour [31]. This combination of wide occurrence and strong dispersal behaviour results in a widespread, spatially continuous distribution. In contrast, *T. acteon* demands specific habitat characteristics like xerothermic climatic conditions and consequently occurs only in highly restricted, geographically disjunct calcareous grasslands. The third, intermediate species, *T. sylvestris* stands in-between both extremes, showing a broad ecological tolerance [30], similar to the generalist *T. lineola*, but shows a rather restricted dispersal behaviour [31].

Molecular data and genetic cluster analysis

For our comparisons, we used a population genetic dataset based on 15 polymorphic allozymes published previously by [28] who did not account for landscape effects. Several studies have shown that the implications as drawn from allozymes and, where available, microsatellites loci were highly congruent in butterflies [32–34]. Here, the use of allozymes instead of other marker systems such as microsatellites has two advantages. 1) In Lepidopterans, locus-specific microsatellites are difficult to find and suitable polymorphic loci are consequently rare [35–38]. This is most likely due to almost identical flanking regions in the Lepidopteran microsatellite DNA [36,39]. However, specificity of these regions is a crucial prerequisite for successful primer annealing [39]. 2) From a landscape genetic perspective, the use of potentially adaptive marker systems might be beneficial in the detection of spatial genetic differentiation in contrast to neutral marker systems, because spatial signals in markers under selection would appear more rapidly [40].

The data set comprised in total 1,063 individuals (417 *T. sylvestris*, 380 *T. lineola*, 160 *T. acteon*) sampled at 12 locations which were distributed across the same study area. Sample sizes ranged from 17 to 44 individuals per species and location. *Thymelicus sylvestris* and *T. lineola* were sampled at identical locations, while *T. acteon* was not found at four of the sampled locations and the data set was supplemented by one additional location (Fig. 1). The 15 enzyme systems provide the following 18 loci: MDH (2 loci), G6PDH, ACON, MPI, AAT (2 loci), FUM, PGI, ME, HBDH, APK, PGM, 6PGDH, IDH (2 loci), GPDH and $PEP_{Phe\text{-}Pro}$. Most of these enzymes showed polymorphisms, except enzyme ME in *T. lineola* and GPDH in *T. sylvestris*. Details about the analytical procedure and the specific running conditions are given in [28]. We used the resulting dataset to

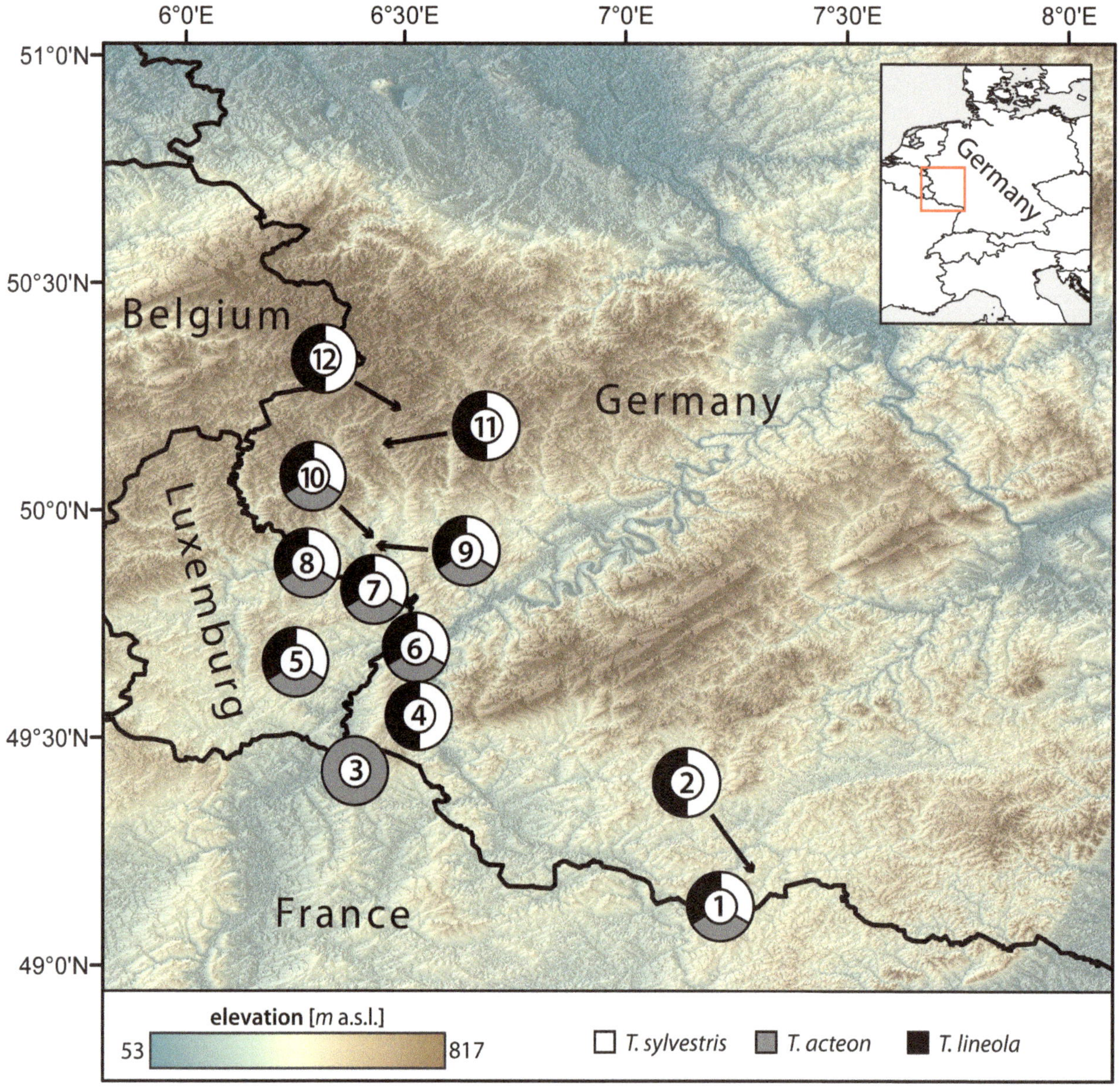

Figure 1. Locations of populations studied for all three *Thymelicus* species in southwestern Germany and adjoining areas in France and Luxemburg.

estimate pairwise F_{ST} and D_{est} for each species in programmes ARLEQUIN 3.1 [41] and SMOGD [42], respectively. The use of these two different measures of inter-population differentiation was recently recommended [43], because of the different underlying assumptions of either measure so that their combination might provide a more detailed impression into the underlying evolutionary processes of differentiation (see [43] and discussion in this study for further details). Tests for Hardy-Weinberg equilibrium and summary statistics for genetic diversity and differentiation were also calculated in ARLEQUIN 3.1.

Prior to inferring landscape effects on genetic differentiation, the number of genetic groups (K) as well as their spatial delineation was evaluated for each species separately using the genetic clustering method implemented in the software GENELAND [44]. This was necessary because (*i*) genetic differences can occur without any obvious landscape pattern (e.g. along secondary contact zones after postglacial expansion from distinct refugia or through anthropogenic introductions from another source population), which in turn would lead to (*ii*) erroneous conclusions on isolation-by-distance IBD/isolation-by-resistance IBR analyses on spatially independent structured data. GENELAND assigns geo-referenced individuals to genetics clusters (K) that maximize Hardy-Weinberg-and Linkage-Equilibrium. K was treated as unknown to allow GENELAND to vary K within a given range between 1 and the maximum number of populations depending on the species (i.e. 7 in *T. acteon* and 11 in both *T. sylvestris* and *T. lineola*). Markov Chains were run for 3,000,000 generations and sampled every 1000th generation, after an initial burn-in of 300 samples after thinning (10%). Markov Chains with these settings were run 10 times independently and the iteration with the highest log posterior probability was chosen for inferring the most likely K and individual assignments.

Modelling landscape effects on genetic differentiation

To test for landscape influences on genetic differentiation in each species, we modeled multiple species distribution models (SDM) incorporating topographic, bioclimatic and/or land use features. We then used resulting SDMs as resistance surfaces to derive inter-population connectivity estimates based on electrical circuit theory, and statistically compared these connectivity estimates to actual genetic differentiation. SDMs are increasingly applied for resistance surface parameterization in landscape genetic studies [23,45] even under longer evolutionary time scales [46,47], since they avoid the subjective parameterization of resistance surfaces which was criticized in the past [48].

Species records

To model SDMs for the three *Thymelicus* species in the study area, presence data were taken from personal observations of JCH, D. Louy and T. Schmitt (Germany) covering the years 2003–2012. Further presence data were added from high resolution records downloaded from the GBIF database (www.gbif.org). The final datasets comprised 67 records for *T. sylvestris*, 62 for *T. lineola* and 28 records for *T. acteon*. Given their specific habitat demands and the sampling effort that was performed across the study area for either species (Fig. 1), we are confident to have compiled a representative sample that covered the realized distribution of the species in our study area.

Environmental layers

For construction of the SDMs, we used freely available GIS based environmental layers. Bioclimatic data based on monthly averaged temperature and precipitation data with 30 arc seconds spatial resolution was obtained from the Worldclim Database (Vers. 1.4; www.worldclim.org; [49]). The comprehensive set of 19 bioclimatic variables are thought to be highly relevant for shaping species' Grinnellian (abiotic) niches [50]. In order to minimize the degree of inter-correlation among the variables (i.e. to keep pair-wise Pearson's $R^2<0.75$), we selected a subset of variables (bio3, 7, 8, 9, 10, 11, 12, 15, 18, see Tab. 3/Appendix S3 for definitions) which were assumed to be most relevant for the study species. Topography-related data were derived from the SRTM Shuttle mission in 90 meters resolution (available through USGS seamless server; Table S2). Based on the altitude layers, we calculated slope and aspect using ArcGIS 9.3 (ESRI Redlands, California, USA). Finally, CORINE land use related data was obtained from the European Environmental Agency (www.eea.europa.eu). We either used CORINE 2006 data to assess current habitat availability as well as CORINE 1990 data for assessing recent land use changes. All environmental layers were re-sampled to uniform grid resolution of 90 m.

Calculating the Potential Connectivity Model

We defined a set of hypotheses based on the available environmental data and generated five variable sets for comparing landscape effects on species-specific gene flow (therein called scenarios, Table S2). These scenarios represent various habitat characteristics (i.e. climate, topography and land use) that were found to be important for butterfly distributions at different spatial scales in previous studies [51–57]. Based on these variable sets and the respective species records, we computed species distribution models (SDMs) with the software Maxent 3.3.3e [58] to generate maps displaying habitat suitability for each species under a given scenario. As many other presence-pseudoabsence SDM algorithms, Maxent links environmental conditions at presence records of a given taxon to those environmental conditions available within a specific geographic area (background) to predict spatial patterns of environmental suitability. The SDM output represents the likelihood of species potential occurrence across a geographic area of interest (projection; for a detailed description see: [59]). We used Maxent instead of other available algorithms because it frequently outperforms other approaches [60–61], even if the number of presence locations is rather limited [62–63]. We ran Maxent with the default settings but used a bootstrap approach, which allows random selection of 70% of presence locations for model training and the remaining 30% for model testing. This procedure was repeated for 100 times and an averaged map of suitable habitats was generated across all repetitions. As output we selected the logistic format which ranges linearly from 0 (not suitable) to 1 (fully suitable). For model evaluation, the area under the receiver operating characteristic curve (AUC) was used [64]. In particular, the AUC as internally computed in Maxent is a measure for the ability of the model to distinguish the given presence records from the background data accounting for the proportion of the study area which is predicted to be suitable for the target species [58]. The AUC ranges between 0.5 (random prediction) to 1.0 (perfect discrimination between presence and pseudo-absence).

For the land use change scenario, we used land use data from CORINE 2006 as a categorical environmental layer - just as we had done for the land use scenario. However, we subsequently projected the model fit onto the CORINE 1990 layer to assess habitat change in terms of a stability surface. The stability surface is the average of both CORINE layers, with high values indicating suitable habitat patches that remain stable over the 16 years time span, whereas low values represent low habitat suitability, a strong habitat change in time, or both. This approach for calculating stability surfaces is commonly used to estimate land use change and habitat suitability across time (see [46–47] for examples).

The resulting SDMs were used as conductance surfaces (i.e. high values indicate good conductivity between two sites, whereas low values indicate poor conductivity [65]) in Circuitscape v.3.4.1 to calculate resistances to movement and gene flow among sampling locations [65]. Circuitscape is based on electrical circuit theory, which was recently adapted from electrical engineering for the assessment of landscape ecological questions [65]. In particular, Circuitscape defines nodes (grid cells) and associated unit resistors (the resistance value) that connecting two nodes and calculates resistance distances between focal locations based on a nodal analysis algorithm as described in [21]. As the habitat matrix had a very high extent (i.e. ~7.6 Mio. cells), we chose a four-neighbor-connection scheme in order to meet the available computational capacities. It has been previously shown that four and eight-neighbor-connection scheme lead to highly similar outcomes [66].

Comparing connectivity estimates with genetic data

Resulting resistance values among locations were statistically compared to estimates of genetic differentiation (i.e. F_{ST} and D_{est}) using linear regression models as well as multiple regressions on distance matrices (MRDM) [67] in R v.2.14.1 [68]. For the linear regression models, the Akaike Information Criterion corrected for small sample sizes (AICc) was used for model comparisons within each species [69]. Despite their sensitivity for non-independence in pair-wise comparisons, multi-model inference based on information theory has been frequently applied in landscape genetic analyses [19–20] as the error entering the comparison was assumed to be equal for each model, which did not affects model ranking and thus still allows for assessing the relative model performance. To ascertain results obtained with the AIC model selection, we also estimated significance of MRDM models using 1,000 permutations. For MRDMs, the *ecodist* package for R was used [70].

Results

Genetic structures

No significant deviation from Hardy-Weinberg equilibrium was detected for any population in the respective species. Genetic diversity was comparatively low in *T. lineola* (mean ± SE; $AR = 1.78 \pm 0.17$ $H_E = 9.6 \pm 2.1$, $H_O = 9.2 \pm 2.1$), while *T. acteon* showed highest genetic diversities ($AR = 1.88 \pm 0.18$, $H_E = 14.9 \pm 2.9$, $H_O = 12.5 \pm 2.6$). *Thymelicus sylvestris* showed an intermediate level of genetic diversity, as compared to its congeners ($AR = 1.80 \pm 0.10$, $H_E = 11.9 \pm 1.5$, $H_O = 11.0 \pm 1.4$). The genetic differentiation was low in *T. lineola* ($F_{ST} = 0.0081$; $D_{est} = 0.0012$; p = n.s.), while we detected highest genetic differentiation for *T. acteon* ($F_{ST} = 0.0718$; $D_{est} = 0.0143$; $p < 0.0001$). Again, *Thymelicus sylvestris* showed an intermediate level of genetic differentiation, with a rather low among-population variance ($F_{ST} = 0.0179$; $D_{est} = 0.0039$; $p < 0.0001$) (Table 1).

Genetic clustering results

The posterior density and log-likelihood levels of all GENELAND runs stabilized long before the end of the Markov Chains, indicating that convergence was reached (Figure S1). For each of the species, all 10 replicate MCMC runs converged on K = 1 panmictic cluster (Appendix S4), indicating no absolute barriers affecting IBD or IBR assumptions.

Species Distribution Models

AUC values derived from the SDMs ranged from 'poor' (AUC = 0.66, scenarios 'land use' and 'land use change' in *T. sylvestris*, Table 2) to 'good' (AUC = 0.86, scenario 'all' in *T. lineola*, Table 2) according to the classification scheme for model quality from [71] adapted from [64]. Variable contributions in multi-factorial SDMs (scenarios 'climate', 'topography' and 'all') differed between species (Table 3). For the topography scenario, *slope* contributed most to the SDM in all three species, followed by *aspect* and *altitude* (Table 3). In *T. acteon* a different set of variables had higher explanative power with respect to the climate scenario. Here, *precipitation of the warmest quarter* (bio18) was most important, followed by a set of temperature related variables (bio3, 7, 8, 9, 11; Table 3). In contrast, *Thymelicus lineola* and *T. sylvestris* had very similar variable contributions as a result of the highly similar distribution of occurrence records. In these species, the *mean temperature of the coldest quarter* followed by the *temperature annual range* contributed to more than half of the total model (Table 3). Finally, considering the entire predictor set, a combination of *slope* and *land use* contributed most in all species, but where *T. lineola* and *T. sylvestris* had again more similar variable contributions rather than *T. acteon* (Table 3). In accordance, *T. lineola* and *T. sylvestris* showed similar potential distributions containing large continuous areas of high suitability, whereas *T. acteon* shows a highly patchy distribution with large unsuitable areas surrounding potential habitat patches (Fig. 2).

Landscape effects of genetic differentiation

Results obtained with the various SDM-based connectivity estimates differed strongly among the three model species (Table 2). The generalist species *T. lineola* showed neither IBD nor any form of IBR using F_{ST} (max $\Delta AICc = 0.86$). Using D_{est}, the IBD scenario produced the best model ($AICc = -806.68$, $\omega = 0.48$) however with a weak relationship ($R^2 = 0.045$, $p = 0.064$). Furthermore, MRDM showed no landscape related signals for either estimate of genetic differentiation in *T. lineola*, suggesting that gene flow in this species is not affected by any spatial or landscape features at this scale. The most specialized species, *T. acteon* showed no significant IBD, but significant IBR for two scenarios (land use & land use change) with both F_{ST} and D_{est} under multi-model inference. These signals become also prominent using MRDM for inference, even though models were slightly insignificant at $p = 0.05$ (land use change F_{ST}: $R^2 = 0.232$, $p = 0.051/D_{est}$: $R^2 = 0.190$, $p = 0.102$). The combined results from AIC and MRDM suggest that land use and land use change both affect genetic differentiation among *T. acteon* populations. Genetic differentiation in *Thymelicus sylvestris* corresponded most strongly to the connectivity estimates derived from the SDM incorporating all variables ($AICc = -271.89$, $\omega = 0.67$) using F_{ST} and the information-theoretic approach. The climate related scenario was also within the most reliable models under AICc ($\Delta AICc = 1.65$, $\omega = 0.29$). However, MRDM suggested that land use and land use change were also important for explaining genetic differentiation in this species. The opposite becomes obvious using D_{est} as differentiation metric. Here, the information theoretic approach reveals climate, land use and land use change as highly informative, with climate being most important ($AIC = -723.08$, $\omega = 0.45$). Surprisingly, the scenario covering the entire variable set contributed nearly no information ($\Delta AICc = 4.44$, $\omega = 0.05$). In addition, MRDM highlighted only climate as significantly related to genetic differentiation. In summary, the combined results of different differentiation metrics and inference methods suggest that the climatic conditions across the study site deliver the most important and stable relationship for adjusting gene flow in the intermediate species, with additional effects of land use. Classical IBD received less support against IBR models

Table 1. Summary statistics for genetic diversity and differentiation for the three *Tymelicus* buttlerflies.

	T. lineola	*T. acteon*	*T. sylvestris*	*source*
AR	1.78±0.17	1.88±0.18	1.80±0.10	Louy *et al.* 2007
H_E	9.6±2.1	14.9±2.9	11.9±1.5	Louy *et al.* 2007
H_O	9.2±2.1	12.5±2.6	11.0±1.4	Louy *et al.* 2007
P_{tot}	52.0±9.7	66.0±9.1	42.9±7.9	Louy *et al.* 2007
P_{95}	36.4±9.4	49.3±13.4	32.3±4.2	Louy *et al.* 2007
F_{ST}	0.0081	0.0755	0.0179	Louy *et al.* 2007
D_{est}	0.0012	0.0143	0.0039	Habel *et al.* 2013

Table 2. Comparison of the genetic structure in three *Thymelicus* butterflies with different landscape parameter sets.

		SDM	Linear regression model					MRDM	
Model F_{ST}		**AUC**	**AICc**	**ΔAICc**	**ω**	**R^2**	**p**	**R^2**	**p**
T. lineola									
	Fst~distance	-	−321.65		0.21	−0.003	0.359	0.016	0.569
	Fst~topography	0.76	−321.44	0.21	0.19	−0.007	0.424	0.012	0.603
	Fst~climate	0.81	−321.41	0.23	0.19	−0.007	0.431	0.012	0.602
	Fst~all	0.86	−320.87	0.77	0.14	−0.017	0.741	0.002	0.828
	Fst~landusechange	0.68	−320.84	0.81	0.14	−0.018	0.795	0.001	0.854
	Fst~landuse	0.67	−320.78	0.86	0.14	−0.019	0.893	0.000	0.926
T. acteon									
	Fst~landusechange	**0.69**	**−94.70**		**0.56**	**0.202**	**0.009**	0.232	0.051
	Fst~landuse	**0.71**	**−93.90**	**0.80**	**0.37**	**0.179**	**0.014**	0.209	0.069
	Fst~distance	-	−88.12	6.58	0.02	−0.009	0.393	0.028	0.433
	Fst~climate	0.79	−87.87	6.83	0.02	−0.018	0.476	0.020	0.748
	Fst~topography	0.79	−87.44	7.26	0.01	−0.034	0.737	0.004	0.821
	Fst~all	0.84	−87.41	7.29	0.01	−0.035	0.771	0.003	0.772
T. sylvestris									
	Fst~all	**0.85**	**−273.53**		**0.67**	**0.252**	**<0.0001**	**0.266**	**0.002**
	Fst~climate	**0.78**	**−271.89**	**1.65**	**0.29**	**0.229**	**<0.0001**	**0.244**	**0.010**
	Fst~land use	0.66	−266.28	7.26	0.02	0.147	0.002	**0.162**	**0.024**
	Fst~land use change	0.66	−265.52	8.02	0.01	0.135	0.003	**0.151**	**0.035**
	Fst~distance	-	−263.78	9.75	0.01	0.107	0.009	0.123	0.068
	Fst~topography	0.78	−262.73	10.81	0.00	0.09	0.015	0.106	0.102
Model D_{est}									
T. lineola									
	Dest~distance	-	**−806.68**		**0.48**	0.045	0.064	0.063	0.176
	Dest~topography	0.76	−803.93	2.75	0.12	−0.004	0.373	0.015	0.559
	Dest~climate	0.81	−804.10	2.58	0.13	−0.001	0.329	0.018	0.535
	Dest~landusechange	0.68	−803.59	3.09	0.10	−0.010	0.493	0.009	0.647
	Dest~all	0.86	−803.31	3.37	0.09	−0.015	0.652	0.004	0.780
	Dest~landuse	0.67	−803.10	3.58	0.08	−0.019	0.951	0.000	0.968
T. action									
	Dest~landusechange	**0.69**	**−274.21**		**0.45**	**0.159**	**0.021**	0.190	0.102
	Dest~landuse	**0.71**	**−274.09**	**0.12**	**0.42**	**0.155**	**0.022**	0.186	0.090
	Dest~climate	0.79	−269.81	4.40	0.05	0.015	0.244	0.052	0.608

Table 2. Cont.

Model F_{ST}	SDM	Linear regression model					MRDM	
	AUC	AICc	ΔAICc	ω	R^2	p	R^2	p
Dest~distance	-	−268.92	5.29	0.03	−0.016	0.460	0.021	0.614
Dest~all	0.84	−268.41	5.79	0.02	−0.035	0.765	0.004	0.784
Dest~topography	0.79	−268.36	5.85	0.02	−0.037	0.845	0.001	0.893
T. sylvestris								
Dest~climate	**0.78**	**−723.08**		**0.45**	**0.099**	**0.011**	**0.115**	**0.049**
Dest~land use	**0.66**	**−721.12**	**1.96**	**0.17**	**0.066**	**0.033**	0.083	0.086
Dest~land use change	**0.66**	**−721.18**	**1.89**	**0.18**	**0.067**	**0.032**	0.084	0.085
Dest~distance	-	−720.50	2.57	0.12	0.055	0.046	0.073	0.118
Dest~all	0.85	−718.64	4.44	0.05	0.023	0.139	0.041	0.259
Dest~topography	0.78	−717.61	5.47	0.03	0.004	0.272	0.023	0.410

Genetic differentiation was inferred by F_{ST} (upper half) and D_{est} (lower half) respectively. SDM AUC values for each scenario (excepting classical IBD) showing the model quality are given as well as parameters for both, linear regression models and multiple regression based on distance matrices (MRDM). Bold values highlight models with highest support (ΔAICc<2 in combination with a significant R^2 in linear regression models; significant R^2 in MRDMs).

(Table 2, Figure S2) in all species. Interestingly, topography seems to play no role at all for any of the species.

Discussion

Studying taxonomically related species inhabiting the same environment makes it possible to infer how species-specific ecological traits affect population genetic structuring without confounding effects of different landscapes or phylogenetic history [24]. By conducting a comparative landscape genetic study involving ecologically diverging Hesperid butterflies, we found different impacts of landscape parameters on the genetic structure of the three study species.

The obtained results show strong genetic differentiation and high genetic diversities in the specialist species *T. acteon*, and low genetic differentiation with accompanying low genetic diversities in the generalist species *T. lineola* with *T. sylvestris* standing in-between the two congeners. The amount of genetic diversity is typical for butterflies in this region (reviewed in [26]). Our analyses indicate that climate has a strong impact on the connectivity of *T. sylvestris* but that other variables (such as land use) might have become more influential in the most recent times. Land use as well as changes in land use patterns (i.e. assessed over a 16yr period) influences the connectivity of *T. acteon* populations. In contrast, *T. lineola* populations were panmictic, lacking any landscape related effects on genetic differentiation at this spatial scale.

Diverging responses to identical landscape conditions

Our data illustrate that closely related species representing a gradient of ecological traits (i.e. from generalist to specialist/from highly mobile to rather philopatric) also show a gradient of changing genetic structures and even more interesting of changing landscape genetic associations (Fig. 3). This highlights that ecological traits determine the species-specific resistance of the landscape matrix, so that its effect on population connectivity can differ strongly among closely-related species inhabiting the same landscape.

The strong genetic differentiation in *T. acteon* is concordant with its patchy occurrence predicted in our SDMs (Fig. 2b), which were best explained by the land-use parameters derived from the CORINE dataset. Furthermore, land-use related scenarios were the only ones that host an IBR-related signal among all competing scenarios in this species (Table 2). Here, the two scenarios 'land-use' and 'land-use-change' fit equally well, irrespective of the genetic differentiation metric or statistical inference method used. Thus, the landscape genetic signal in this specialist species is highly consistent among different analyses, leading to high certainty of inferences.

The slight differences between these two scenarios might be stochastic. However, since there is also consistence about the ranking across all approaches (i.e. land use change steadily explains slightly more variance under each situation than land use), land-use-change might be even more important, when addressing land-use-change over an even larger time period than the 16 years used here. Unfortunately, there is no information available to assess past land-use-changes covering this large geographical extent further into the past. Keeping time-lags between fragmentation and genetic responses accompanying these fragmentations in mind (e.g. as reviewed in [72]) there is some evidence that 16 years are not adequate to detect genetic impacts of altered habitats in this time period in a species with an annual generation time. Changes over this period result just in slightly different resistance surfaces between the scenarios 'land-use' and 'land-use-change'. Nevertheless, *T. acteon* is becoming increas-

Table 3. Averaged variable contributions for the scenarios 'topography', 'climate' and 'all'.

Scenario	Variable	*T. acteon*	*T. lineola*	*T. sylvestris*
Topography	alt	7.1	12.1	10.6
	aspect	21.1	29.7	32.3
	slope	71.8	58.2	57.1
Climate	bio3 (isothermality)	12.0	9.7	8.7
	bio7 (temperature annual range)	10.3	23.8	23.5
	bio8 (mean temperature of wettest quarter)	12.0	3.4	4.0
	bio9 (mean temperature of driest quarter)	11.3	10.1	10.0
	bio10 (mean temperature of warmest quarter)	3.6	4.6	5.7
	bio11 (mean temperature of coldest quarter)	16.1	32.0	31.4
	bio12 (annual precipitation)	5.2	10.0	11.3
	bio15 (precipitation seasonality)	5.4	3.3	3.0
	bio18 (precipitation of warmest quarter)	24.3	3.1	2.4
all	land use	37.7	24.9	23.2
	alt	1.3	8.1	7.7
	aspect	9.3	12.4	14.4
	slope	31.2	24.0	29.7
	bio3 (isothermality)	2.8	3.3	3.3
	bio7 (temperature annual range)	2.3	10.0	8.0
	bio8 (mean temperature of wettest quarter)	4.2	1.9	2.0
	bio9 (mean temperature of driest quarter)	1.0	0.1	0.3
	bio10 (mean temperature of warmest quarter)	0.1	0.7	0.4
	bio11 (mean temperature of coldest quarter)	2.1	6.5	4.7
	bio12 (annual precipitation)	2.1	5.8	4.6
	bio15 (precipitation seasonality)	1.5	1.1	0.8
	bio18 (precipitation of warmest quarter)	4.5	1.2	0.8

Note that land use dependent scenarios are not shown herein as they contain one single variable.

ingly vulnerable in large parts of Europe [73] and has likely declined during the past 30 years within the study area due to habitat loss [29]. Thus, the slightly stronger signal of the land-use-change scenario in comparison to the land-use scenario might become even more prominent when extrapolating these changes further decades into the past, highlighting habitat loss as serve danger for this species.

The genetic diversities (such as heterozygosity or mean number of alleles) are highest in *T. acteon* compared to the other two species. This result is somewhat surprising, as the consequence of restricted gene flow and strong geographic restriction of local populations usually leads to rising genetic differentiation and declining genetic diversity, as frequently observed for species demanding specific habitat qualities and/or sedentary dispersal behaviour [10,74–76]. However, there are also examples where genetic diversities in rare species exceed those of their common congeners [77–79]. This contrasting pattern to neutral genetic theory might be a result from hybridization ([80], but see [78]) or because of time-lags that display the past genetic diversity, when connectivity between populations was much higher than today [79,81]. Indeed, genetic differentiation responds to habitat changes quicker than genetic diversity [82–83] so that the high genetic diversity observed for *T. aceton* may not yet reflect the negative consequences of on-going habitat alterations for this species.

In contrast to the specialist *T. acteon*, the generalist *T. lineola* represents opposing genetic features: the species shows a broad ecological amplitude and a much higher mobility [31]. This combination led to higher abundance pattern in combination with increased inter-population migration rates. These species traits lead to a rather panmictic genetic structure in our study area that appears to prevent landscape genetic relationships or IBD. This coherence between wide ecological amplitudes, high rates of individual exchanges (e.g. gene flow) and thus low genetic differentiation were frequently observed in other studies [84–85]. However, it needs to be considered that on a larger study extent, barriers such as oceans, large lakes, mountain ranges might become important for gene flow acting on a macro-scale [86–87]. The landscape matrix in our study area did not enable the assessment of such macro-scale effects, since the landscape matrix is rather continuous at this scale and large barriers are lacking, as indicated by the GENELAND results.

Finally, the species standing in-between these two extremes, *T. sylvestris*, has an abundance like *T. lineola* but shows a sedentary dispersal behavior comparable to that of *T. acteon* [31]. The reduced dispersal propensity of this species coupled with its wide occurrence makes the colonization of a habitat nearby much more likely than of far distant habitats. Consequently, we obtain IBD and IBR signals for many sets of variables in this species (Table 2). However, when combining the information from the different assessment methods (F_{ST} vs. D_{est}/multimodel inference vs.

a) *Thymelicus lineola*

b) *Thymelicus acteon*

c) *Thymelicus sylvestris*

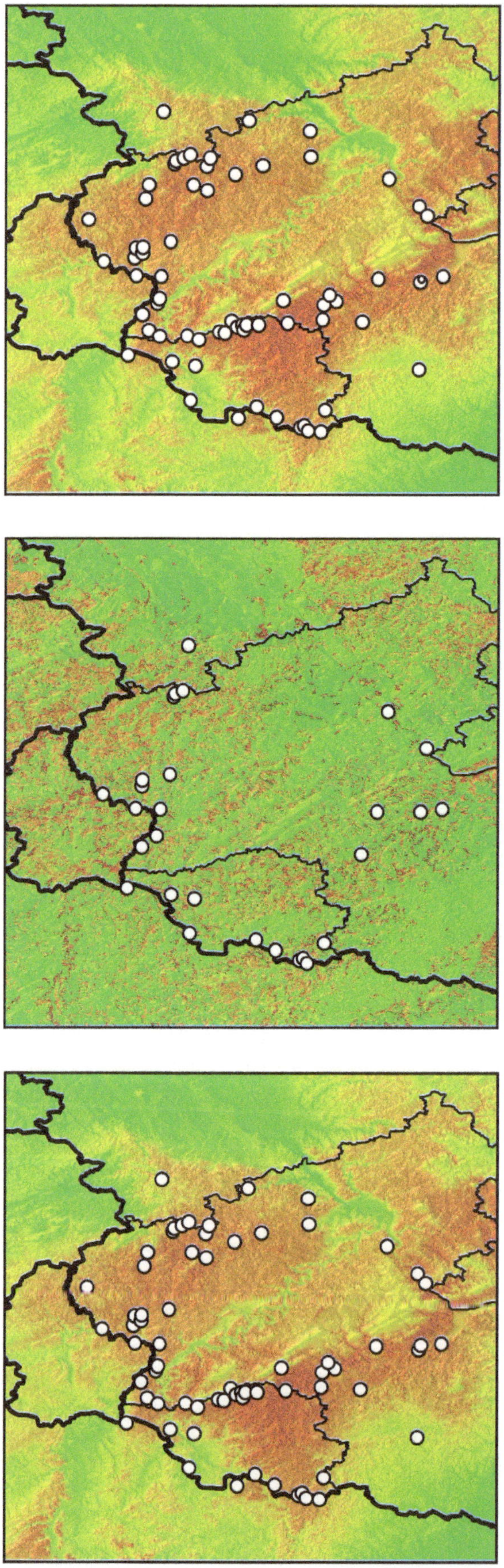

Figure 2. SDM output for *Thymelicus lineola* (A) *T. acteon* (B) and *T. sylvestris* (C) respectively. White circles on SDMs are presence locations used for modeling; Warmer colors (red) indicate higher suitability depending on the best model as presented in Table 2 (climate for *T. sylvestris*; land use change for *T. acteon*; note that *T. lineola* does not have a best model because of its panmictic state. Therefore, also climate is represented here).

MRDM), landscape resistance based on the climate scenario was most important, delivering a consistent strong signal across the different inference methods used (Table 2, see also below). This contrasts to the IBR of *T. acteon*, where climate plays no role at all. In contrast to land-use, climate acts on a meso-scale at our study area (i.e. masking larger areas of the study extent rather than small habitat patches). In *T. sylvestris* the climate related SDM revealed high resistances along river valleys as well as on the higher elevations of the low mountain ranges (Fig 2c). These potential barriers act at a much larger scale and extent compared to the small and patchy habitat islands enclosed by more or less unfavourable habitats in *T. acteon*. Consequently, the different landscape features contributing to the IBR signals in these two species highlight the importance of scale and shape of the connective elements (or their respective barriers) in the landscape matrix where methodological shortcomings can be excluded (Engler, unpublished). However, the obtained IBR models explain only up to 24% of the variance in our dataset. That in turn indicates that the remaining variance of our data can only be explained by additional factors such as ecological traits and habitat requirements. These can be even more relevant for butterfly species than habitat size and habitat isolation, e.g. as shown for the Heath butterfly *Coenonympha tullia* [7]. Nevertheless, the extent of the relationships in our IBD/IBR comparisons are in concert with other studies [88] indicating that gene flow can be interpreted

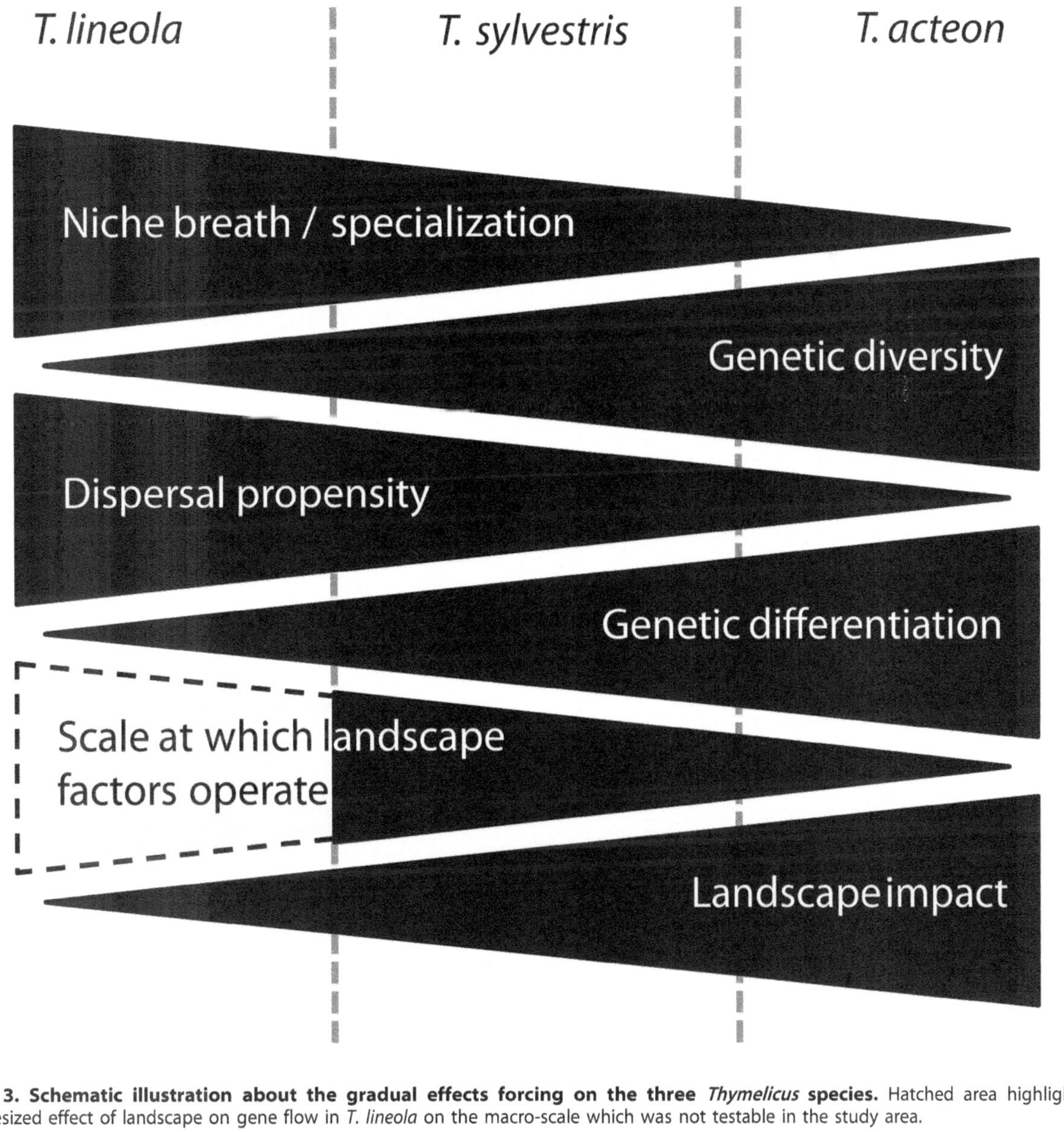

Figure 3. Schematic illustration about the gradual effects forcing on the three *Thymelicus* species. Hatched area highlights the hypothesized effect of landscape on gene flow in *T. lineola* on the macro-scale which was not testable in the study area.

as an important component out of a variety of mechanisms influencing population genetic structure.

Accounting for F_{ST} and D_{est} in landscape genetic studies

Interestingly, in the case of *Thymelicus sylvestris*, the prominent signal under F_{ST} arising from the SDM using all landscape variables becomes completely eliminated when using D_{est} as a differentiation metric. The fact that different metrics can lead to different conclusions is also evident in the ongoing debate about the utility of different genetic differentiation measures [89–94]. For example, traditional F_{ST} -like metrics are more sensitive to recent demographic changes (which depends e.g. on effective population size) than metrics which are independent of effective population size, such as D_{est} [43,90,93–94]. This makes F_{ST} more sensible to effects of gene flow or drift in comparison to D_{est}. Thus, from a landscape genetic perspective, using different types of differentiation metrics allows to test for the contribution of landscape effects in contemporary versus past times. If landscape composition change over time (and consequently the amount of gene flow mediated by the landscape), F_{ST} would respond much quicker to those changes while D_{est} remains rather stable over time. In the case of *Thymelicus sylvestris,* this means that D_{est} may highlight the landscape effect (here climate) of highest importance for gene flow in this species in former times, whereas F_{ST} highlights more recent landscape effects on genetic structure that involves also other landscape elements beside climate such as land use and topographical elements.

In contrast to the climate-only scenario, connectivity estimates involving all variables did not give highest importance to climatic factors. In particular, land use and slope contribute almost 54% of the total importance of this scenario, whereas the best performing variables from the climate scenario, bio11 (mean temp of coldest quarter) and bio7 (temp annual range) that contribute together 54.9%, contributing under the full model just 12.7% of the total importance. This might highlight the change of landscape factors important to gene flow in this species. As *T. sylvestris* is indeed common but not very mobile, anthropogenic land transformations of the past decades might now lead to a stronger fragmentation of populations which ultimately lead to changes in the contributions of landscape factors shaping gene flow as shown elsewhere [95]. Consequently, this might mean that this species is just at the tipping point of being of conservation concern (*sensu* [27]) where population trends swapping from stable to decreasing. Its congeners *T. acteon* and *T. lineola* showing both consistent results across the different metrics underpinning their stable state in terms of their abundance (insentinent and widespread vs. sensible and endangered) and specialization (generalist vs. specialist).

Conclusions

Taxonomically close relatives serve as ideal model systems to study interspecific characteristics in ecological traits without confounding effects derived from different evolutionary histories. Yet, studies investigating the role of landscape on gene flow of closely related taxa inhabiting the same environment are still scarce. Our results reveal that even between sibling species, gene flow is affected by the landscape in very different ways. Thus, it is challenging to predict landscape genetic relationships in one species from a study involving another species, even if the two species are taxonomically closely related. Nevertheless, some generalizations are possible for specialist versus generalist species. In our study, the genetic structure of the generalist species with high dispersal propensities remained unaffected by the current landscape matrix, whereas specialist species were highly sensitive to fine scale habitat features. Changes of these features might therefore affect specialists more readily than generalist species with the negative consequences for their genetic setup. Species with an intermediate degree of specialization (here *T. sylvestris*) also interact with the landscape but at coarser scales in comparison to specialist species (here *T. acteon*). However, in light of global change such species might be on the highest risk due to negative genetic effects such as inbreeding depression, because changes in the habitat matrix can push former meta-population into isolated remnants [27]. This becomes also evident in *T. sylvestris* comparing the genetic structure under either F_{ST} or D_{est}. Further studies focusing on the degree of habitat specialization in addition to dispersal capabilities are needed, ideally conducted with closely related taxa in other areas. Such comparative studies will greatly expand our current understanding of landscape genetic relationships and ultimately lead to more effective conservation and management of biodiversity.

Supporting Information

Figure S1 Estimation of the number of panmictic clusters for each species. A) Convergence of the MCMC after thinning (see methods for details). Values prior to burn-in (indicated as red dashed line) were not considered as chain does not reached convergence. B) Frequency of the estimated number of populations along the chain after burn-in.

Figure S2 Scatterplots showing the differences of isolation by distance patterns with isolation by resistance patterns in the two species that show a spatial genetic structure (*Thymelicus sylvestris* is shown at the upper half, *T. acteon* at the lower half). Note that just the most prominent isolation by resistance pattern is shown (i.e. climate in *T. sylvestris* and land use change in *T. acteon*).

Table S1 Geographic coordinates of the sampling locations. ID numbers correspond to those stated in Fig. 1.

Table S2 Description of the landscape data used for resistance surface building depending on the scenario assumed. Note that the scenarios 'land use' and 'land use change' used the same data source. SDM refers to species distribution model.

Material S1 Evolutionary history of the three *Thymelicus* butterflies.

Acknowledgments

We are grateful to Deborah A. Dawson, Alain Frantz, Joseph Chipperfield, and Lisette P. Waits for valuable comments on an earlier draft of this article.

Author Contributions

Conceived and designed the experiments: JOE NB DR JCH. Performed the experiments: JCH JOE DR NB. Analyzed the data: JOE NB DR JCH KJF. Contributed reagents/materials/analysis tools: JCH JOE NB DR. Wrote the paper: JOE JCH NB DR KJF.

References

1. MacArthur R-H, Wilson E-O (1967) The theory of island biogeography. Princeton: Princeton University Press. 205 p.
2. Levins R (1969) Some demographic and genetic consequences of environmental heterogeneity for biological control. Bull Entomol Soc America 15: 237–240.
3. Hanski I (1999) Metapopulation ecology. Oxford: Oxford University Press. 319 p.
4. Hu XS, He F, Hubbel SP (2006) Neutral theory in macroecology and population genetics. Oikos 113: 548–556.
5. Lomolino MV, Brown JH (2009) The reticulating phylogeny of island biogeography theory. Quart Rev Biol 84: 357–390.
6. Laurance WF (2009) Beyond island biogeography theory: understanding habitat fragmentation in the real world. In: Losos JB, Ricklefs RE, editors. The theory of island biogeography revisited. Princeton: Princeton University Press. 476 p.
7. Dennis RLH, Eales H (1997) Patch occupancy in *Coenonympha tullia* (Müller 1764) (Lepidoptera: Satyrinae): Habitat quality matters as much as patch size and isolation. J Insect Conserv 1: 167–176.
8. Conradt L, Roper TJ, Thomas CD (2001) Dispersal behaviour of individuals in metapopulations of two British butterflies. Oikos 95: 416–424.
9. van Dyck H, Baguette M (2005) Dispersal behaviour in fragmented landscapes: routine or special movements? Basic App Ecol 6: 535–545.
10. Habel JC, Meyer M, Schmitt T (2009a) The genetic consequence of differing ecological demands of a generalist and a specialist butterfly species. Biodiv Conserv 18: 1895–1908.
11. Goodwin BJ (2003) Is landscape connectivity a dependent or independent variable? Landscape Ecol 18: 687–699.
12. Fahrig L (2003) Effects of habitat fragmentation on biodiversity. Ann Rev Ecol Evol Sys 34: 487–515.
13. Hof C, Levinsky I, Araújo MB, Rahbek C (2011) Rethinking species' ability to cope with rapid climate change. Glob Chan Biol 17: 2987–299.
14. Seiferling IS, Proulx R, Peres-Neto PR, Fahrig L, Messier C (2012) Measuring protected-area isolation and correlations on isolation with land-use intensity and protection status. Conserv Biol 26: 610–618.
15. Holderegger R, Wagner HH (2008) Landscape genetics. Bioscience 58: 199–208.
16. Storfer A, Murphy MA, Spear SF, Holderegger R, Waits LP (2010) Landscape genetics: where are we now? Mol Ecol 19: 3496–3514.
17. Manel S, Holderegger R (2013) Ten years of landscape genetics. Trends Ecol Evol 28: 614–621.
18. Wright S (1943) Isolation by distance. Genetics 28: 114–138.
19. Goldberg CS, Waits LP (2010) Comparative landscape genetics of two pond-breeding amphibian species in a highly modified agricultural landscape. Mol Ecol 19: 3650–3663.
20. Richardson JL (2012) Divergent landscape effects on population connectivity in two co-occurring amphibian species. Mol Ecol 21: 4437–4451.
21. McRae BH (2006) Isolation by resistance. Evolution 60: 1551–1561.
22. Schwenk WS, Donovan TM (2011) A multispecies framework for landscape conservation planning. Conserv Biol 25: 1010–1021.
23. Poelchau MF, Hamrick JL (2012) Differential effects of landscape-level environmental features on genetic structure in three codistributed tree species in Central America. Mol Ecol 21: 4970–4982.
24. Steele CA, Baumsteiger J, Storfer A (2009) Influence of life-history variation on the genetic structure of two sympatric salamander taxa. Mol Ecol 18: 1629–1639.
25. Dawson MN (2012) Parallel phylogeographic structure in ecologically similar sympatric sister taxa. Mol Ecol 21: 987–1004.
26. Habel JC, Rödder D, Lens L, Schmitt T (2013) The genetic signature of ecologically different grassland Lepidopterans. Biodiv Conserv 22: 2401–2411.
27. Habel JC, Schmitt T (2012) The burden of genetic diversity. Biol Conserv 147: 270–274.
28. Louy D, Habel JC, Schmitt T, Assmann T, Meyer M, et al. (2007) Strongly diverging population genetic patterns of three skipper species: the role of habitat fragmentation and dispersal ability. Conserv Gen 8: 671–681.
29. Wenzel M, Schmitt T, Weitzel M, Seitz A (2006) The serve decline of butterflies on western German calcareous grasslands during the last 30 years: a conservation problem. Biol Conserv 128: 542–552.
30. Asher J, Warren M, Fox R, Harding P, Jeffcoate G, et al. (2001) The millennium atlas of butterflies in Britain and Ireland. Oxford: Oxford University Press. 433 p.
31. Bink FA (1992) Ecologische Atlas van de Dagvlinders van Noordwest-Europa. Haarlem: Schuyt & Co. Uitgevers en Importeurs. 512 p.
32. Meglécz E, Pecsenye K, Varga Z, Solignac M (1998) Comparison of differentiation pattern at allozyme and microsatellite loci in *Parnassius mnemosyne* (Lepidoptera) populations. Hereditas 128: 95–103.
33. Habel JC, Zachos FE, Finger A, Meyer M, Louy D, et al. (2009b) Unprecedented long term genetic monomorphism in an endangered relict butterfly species. Conserv Gen 10: 1659–1665.
34. Habel JC, Rödder D, Schmitt T, Nève G (2011) Global warming will affect the genetic diversity and uniqueness of *Lycaena helle* populations. Glob Chan Biol 17: 194–205.
35. Meglécz E, Solignac M (1998) Microsatellite loci for *Parnassius mnemosyne* (Lepidoptera). Hereditas 128: 179–180.
36. Meglécz E, Petenian F, Danchin E, Coer D'Acier A, Rasplus J-Y, et al. (2004) High similarity between flanking regions of different microsatellites detected within each of two species of Lepidoptera: *Parnassius Apollo* and *Euphydrias aurinia*. Mol Ecol 13: 1693–1700.
37. Habel JC, Meyer M, Schmitt T, Assmann T (2008) Polymorphic microsatellite loci in the endangered butterfly *Lycaena helle* (Lepidoptera: Lycaenidae). Eur J Enthomol 105: 361–362.
38. Finger A, Zachos FE, Schmitt T, Meyer M, Assmann T, et al. (2008) The genetic status of the Violet Copper *Lycaena helle* – a relict of the cold past in times of global warming. Ecography 32: 382–390.
39. Zang D-X (2004) Lepidopteran microsatellite DNA: redundant but promising. Trends Ecol Evol 19: 507–509.
40. Landguth E, Balkenhol N (2012) Relative sensitivity of neutral versus adaptive genetic data for assessing population differentiation. Conserv Gen 13: 1421–1426.
41. Excoffier L, Larval G, Schneider S (2005) Arlequin ver. 3.0: an integrated software package for population genetics data analysis. Evol Bioinform Onl 1: 47–50.
42. Crawford NG (2009) SMOGD: software for the measurement of genetic diversity. Mol Ecol Res 10: 556–557.
43. Leng L, Zhang D (2011) Measuring population differentiation using Gst or D? A simulation study with microsatellite DNA markers under a finite island model and nonequilibrium conditions. Mol Ecol 20: 2494–2509.
44. Guillot G, Santos F (2009) A computer program to simulate multilocus genotype data with spatially auto-correlated allele frequencies. Mol Ecol Res 9: 1112–1120.
45. Ortego J, Riordan EC, Gugger PF, Sork VL (2012) Influence of environmental heterogeneity on genetic diversity and structure in an endemic southern Californian oak. Mol Ecol 21: 3210–3223.
46. Bell RC, Parra JL, Tonione M, Hoskin CJ, Mackenzie JB, et al. (2010) Patterns of persistence and isolation indicate resilience to climate change in montane rainforest lizards. Mol Ecol 19: 2531–2544.
47. Devitt TJ, Cameron Devitt, SE, Hollingworth BD, McGuire JA, Moritz C (2013) Montane refugia predict population genetic structure in the Large-blotched Ensatina salamander. Mol Ecol 22: 1650–1665.
48. Spear SF, Balkenhol N, Fortin M-J, McRae BH, Scribner K (2010) Use of resistance surfaces for landscape genetic studies: considerations for parameterization and analysis. Mol Ecol 19: 3576–3591.
49. Hijmans RJ, Cameron SE, Parra JL, Jones PG, Jarvis A (2005) Very high resolution interpolated climate surfaces for global land areas. Int J Climat 25: 1965–1978.
50. Beaumont LJ, Hughes L, Poulsen M (2005) Predicting species distributions: use of climatic parameters in BIOCLIM and its impact on predictions of species' current and future distributions. Ecol Mod 186: 250–269.
51. Weiss SB, Murphy DD, White RR (1988). Sun, slope, and butterflies: topographic determinants of habitat quality for *Euphydryas editha*. Ecology 69: 1486–1496.
52. Warren MS, Hill JK, Thomas JA, Asher J, Fox R, et al. (2001) Rapid responses of British butterflies to opposing forces of climate and habitat change. Nature 414: 65–69.
53. Pe'er G, Saltz D, Thulke HH, Motro U (2004) Response to topography in a hilltopping butterfly and implications for modeling nonrandom dispersal. Anim Behav 68: 825–839.
54. Pe'er G, Heinz SK, Frank K (2006) Connectivity in heterogeneous landscapes: analyzing the effect of topography. Landscape Ecol 21: 47–61.
55. Pin Koh L (2007) Impacts of land use change on South-east Asian forest butterflies: a review. J App Ecol 44: 703–713.
56. Marini L, Fontana P, Kimek S, Battisti A, Gaston KJ (2009) Impact of farm size and topography on plant and insect diversity of manages grasslands in the Alps. Biol Conserv 142: 394–403.
57. Filz KJ, Engler JO, Stoffels J, Weitzel M, Schmitt T (2013) Missing the target? A critical view on butterfly conservation efforts on calcareous grasslands. Biodiv Conserv 22: 2223–2241.
58. Phillips SJ, Anderson RP, Schapire RE (2006) Maximum entropy modeling of species geographic distributions. Ecol Mod 190: 231–259.
59. Elith J, Phillips SJ, Hastie T, Dudík M, Chee YE, et al. (2011) A statistical explanation of MaxEnt for ecologists. Divers Distrib 17: 43–47.
60. Elith J, Graham CH, Anderson RP, Dudik M, Ferrier S, et al. (2006) Novel methods improve prediction of species' distribution from occurrence data. Ecography 29: 129–151.
61. Heikkinen RK, Luoto M, Araújo MB, Virkkala R, Thuiller W, et al. (2006) Methods and uncertainties in bioclimatic envelope modelling under climate change. Prog Phys Geogr 30: 751–777.
62. Hernandez PH, Graham CH, Master LL, Albert DL (2006) The effect of sample size and species characteristics on performance of different species distribution methods. Ecography 29: 773–785.
63. Wisz MS, Hijmans RJ, Li J, Peterson AT, Graham CH, Guisan et al. (2008) Effects of sample size on the performance of species distribution models. Divers Distrib 14: 763–773.
64. Swets JA (1988) Measuring the accuracy of diagnostic systems. Science 240: 1285–1293.

65. McRae BH, Dickson BG, Keitt TH, Shah VB (2008) Using circuit theory to model connectivity in ecology, evolution, and conservation. Ecology 89: 2712–2724.
66. McRae BH, Beier P (2007) Circuit theory predicts gene flow in plant and animal populations. PNAS 104: 19885–19890.
67. Lichstein JW (2007) Multiple regression on distance matrices: a multivariate spatial analysis tool. Plant Ecol 188: 117–131.
68. R development core team (2012) R: a language and environment for statistical computing. Available: http://www.R-project.org.
69. Burnham KP, Anderson DR (2002) Model selection and multimodel inference: a practical information-theoretic approach. New York: Springer. 488 p.
70. Goslee SC, Urban DL (2007) The *ecodist* package for dissimilarity-based analysis of ecological data. J Stat Softw 22: 1–19.
71. Araújo MB, Pearson RG, Thuiller W, Erhard M (2005) Validation of species-climate impact models under climate change. Glob Chan Biol 11: 1504–1513.
72. Keyghobadi N (2007) The genetic implications of habitat fragmentation for animals. Can J Zool 85: 1049–1064.
73. van Swaay CAM, Warren M (1999) Red data book of European butterflies (Rhopalocera). Strasbourg: Council of Europe Publishing. 269 p.
74. Kassen R (2002) The experimental evolution of specialists, generalists, and the maintenance of diversity. J Evol Biol 15: 173–190.
75. Packer L, Zayed A, Grixti C, Ruz L, Owen RE, et al. (2005) Conservation genetics of potentially endangered mutualisms: reduced levels of genetic variation in specialist versus generalist bees. Conserv Biol 19: 195–202.
76. Zachos FE, Althoff C, v Steynitz Y, Eckert I, Hartl GB (2007) Genetic analysis of an isolated red deer (*Cervus elaphus*) population showing signs of inbreeding depression. Eur J Wildl Res 53: 61–67.
77. Gitzendanner MA, Soltis PS (2000) Patterns of genetic variation in rare and widespread plant congeners. Am J Bot 87: 783–792.
78. Ellis JR, Pashley CH, Burke JM, McCauley DE (2006) High genetic diversity in a rare and endangered sunflower as compared to a common congener. Mol Ecol 15: 2345–2355.
79. Kadlec T, Vrba P, Kepka P, Schmitt T, Konvicka M (2010) Tracking the decline of the once-common butterfly: delayed oviposition, demography and population genetics in ther hermit *Chazara briseis*. Anim Conserv 13: 172–183.
80. Richards ZT, van Oppen MJH (2012) Rarity and genetic diversity in Indo-Pacific *Acropora* corals. Ecol Evol 2: 1867–1888.
81. Severns PM, Liston A, Wilson MV (2011) Habitat fragmentation, genetic diversity, and inbreeding depression in a threatened grassland legume: is genetic rescue necessary? Conserv Gen 12: 881–893.
82. Balkenhol N, Pardini R, Cornelius C, Fernandes F, Sommer S (2013) Landscape-level comparison of genetic diversity and differentiation in a small mammal inhabiting different fragmented landscapes of the Brazilian Atlantic Forest. Conserv Gen 14: 355–367.
83. Keyghobadi N, Roland J, Matter SF, Strobeck C (2005) Among- and within-patch components of genetic diversity respond at different rates to habitat fragmentation: an empirical demonstration. Proc R Soc London B 272: 553–560.
84. Brouat C, Chevallier H, Meusnier S, Noblecourt T, Rasplus J-Y (2004) Specialization and habitat: spatial and environmental effects on abundance and genetic diversity of forest generalist and specialist *Carabus* species. Mol Ecol 13: 1815–1826.
85. Habel JC, Schmitt T (2009) The genetic consequences of different dispersal behaviours in Lycaenid butterfly species. Bull Entomol Res 99: 513–523.
86. Lee-Yaw JA, Davidson A, McRae BH, Green DM (2009) Do landscape processes predict phylogeographic patterns in the wood frog? Mol Ecol 18: 1863–1874.
87. Kekkonen J, Seppä P, Hanski IK, Jensen H, Väisänen RA, et al. (2011) Low genetic differentiation in a sedentary bird: house sparrow population genetics in a contiguous landscape. Heredity 106: 183–190.
88. Groot AT, Classen A, Inglis O, Blanco CA, López Jr J, et al. (2011) Genetic differentiation across North America in the generalist moth *Heliothis virescens* and the specialist *H. subflexa*. Mol Ecol 20: 2676–2692.
89. Heller R, Siegismund HR (2009) Relationship between three measures of genetic differentiation G_{ST}, D_{EST} and G'_{ST}: how wrong have we been? Mol Ecol 18: 2080–2083.
90. Ryman N, Leimar O (2009) G_{ST} is still a useful measure of genetic differentiation – a comment on Jost's *D. Mol Ecol* 18: 2084–2087.
91. Jost L (2009) D vs. G_{ST}: Response to Heller and Siegismund (2009) and Ryman and Leimar (2009). Mol Ecol 18: 2088–2091.
92. Gerlach G, Jueterbock A, Kraemer P, Deppermann J, Harmand P (2010) Calculations of population differentiation based on G_{ST} and D: forget G_{ST} but not all of statistics! Mol Ecol 19: 3845–3852.
93. Meirmans PG, Hedrick PW (2011) Assessing population structure F_{ST} and related measures. Mol Ecol Res 11: 5–18.
94. Raeymaekers JAM, Lens L, Van den Broeck F, Van Dongen S, Volckaert FAM (2012) Quantifying population structure on short timescales. Mol Ecol 21: 3458–3473.
95. Pavlacky Jr DC, Goldizen AW, Prentis PJ, Nicholls JA, Lowe AJ (2009) A landscape genetics approach for quantifying the relative influence of historic and contemporary habitat heterogeneity on the genetic connectivity of a rainforest bird. Mol Ecol 18: 2945–2960.

16

Rocky Intertidal Zonation Pattern in Antofagasta, Chile: Invasive Species and Shellfish Gathering

Juan Carlos Castilla[1]*, Patricio H. Manríquez[2], Alejandro Delgado[1], Verónica Ortiz[1], María Elisa Jara[2], Manuel Varas[1]

1 Centro de Conservación Marina, Estación Costera de Investigaciones Marinas, Las Cruces, Facultad de Ciencias Biológicas, Pontificia Universidad Católica de Chile, Santiago, Chile, **2** Laboratorio de Ecología y Conducta de la Ontogenia Temprana (LECOT), Centro de Estudios Avanzados en Zonas Áridas (CEAZA), Coquimbo, Chile

Abstract

Background: Biological invasions affecting rocky intertidal zonation patterns, yield information on species interactions. In the Bay of Antofagasta, northern Chile, the non-indigenous tunicate *Pyura praeputialis*, originally from Australia, has invaded (in the past century or so) and monopolized a major portion of the mid-intertidal rocky shore, displacing upshore the native mussel *Perumytilus purpuratus*. In Antofagasta the tunicate is subjected to intensive exploitation. Monitoring protocols show that in the past 10 years Antofagasta's tunicate population has experienced a drastic decline, affecting the intertidal zonation pattern.

Methodology/Principal Findings: A 12.5 km of coastline, on the southern eastern shore of the Bay of Antofagasta, was studied. Eight sites were systematically (1993–1994) or sporadically (2003–2014) monitored for the seaward-shoreward expansion or reduction of the tunicate *Pyura praeputialis,* and native mussel and barnacle bands. A notable reduction in the mid-intertidal band of *P. praeputialis* and a seaward expansion of the mussel, *Perumytilus purpuratus*, and barnacle bands was observed. We suggest that the major cause for the decline in the tunicate is due to its intensive exploitation by rocky shore *Pyura*-gathers. The rate of extraction of tunicates by professional *Pyura*-gathers ranged between 256–740 tunicates hour^{-1}. Between 2009–2014 the density of professional *Pyura*-gather ranged between 0.5–4.5 km^{-1} per low tide. Hence, 10 professional *Pyura*-gathers working 1 h for 10 low tides per month, during 6 months, will remove between 307–888 m^2 of tunicates. A drastic decline in tunicate recruitment was observed and several *P. praeputialis* ecosystems services have been lost.

Conclusion and Significance: In Antofagasta, the continuous and intensive intertidal gathering of the invasive tunicate *Pyura praeputialis,* has caused a drastic reduction of its population modifying the zonation pattern. Thereby, native mussel *Perumytilus purpuratus* has regained its ecological center in the intertidal zone. We recorded a *Pyura* recruitment failure and loss of ecosystem services.

Editor: Erik V. Thuesen, The Evergreen State College, United States of America

Funding: This long-term study was initially funded by a Chilean Presidential Chair in Science to JCC (1997–1999). Additional financial support was received from projects FONDAP-FONDECYT N° 1501-0001, "Center for Advanced Studies in Ecology and Biodiversity" (2000–2011); projects FONDECYT N° 3020035 (2002–2005), FONDECYT N° 1050841 (2005–2007), FONDECYT N° 1080023 (2008–2011) to PHM, the Núcleo Milenio "Centro de Conservación Marina", Ministerio de Economía, Fomento y Turismo, Chile (2010–2014). MV received an undergraduate thesis grant from Minera Escondida Limitada (MEL) (1993–1995). Also MEL has supported biological research in the Bay of Antofagasta through a CREA project, Universidad Católica de la Santísima Concepción, Chile to JCC and PHM (2012–2014). During the final edition of this manuscript PHM was under the tenure of project FONDECYT N° 1130839 (2013–2016). The funders had no role in the study design, data collection, data analysis, or decision to plan, prepare or publish the manuscript.

Competing Interests: The authors have declared that no competing interests exist.

* Email: jcastilla@bio.puc.cl

Introduction

One of the main goals of ecology is to understand the mechanisms and processes responsible for the abundance and distribution patterns of species in their environment, which includes humans. Rocky intertidal food-gathering by artisan or subsistence fishers affecting populations of carnivore keystone species, herbivores and primary producers, leading to the modification of population, community, trophic web and alternative ecological states, have been reported for central and southern Chile [1–7], South Africa [8–10] and Australia [11]. In these reports the ecological role played by humans, as resource extractors, has been characterized as extremely efficient and corresponding to a super-keystone species [12]. The study of non-indigenous invasive highly competitive species, in new environments, particularly if the invasive species is consumed, offers a unique opportunity for analyzing key ecological processes. In natural systems, rigorous ecologically designed experiments are not always possible due to the spatial scales involved, prohibitive logistics or ethical constraints. Instead, long-term monitoring and large-scale empirical observations can be used to determine the

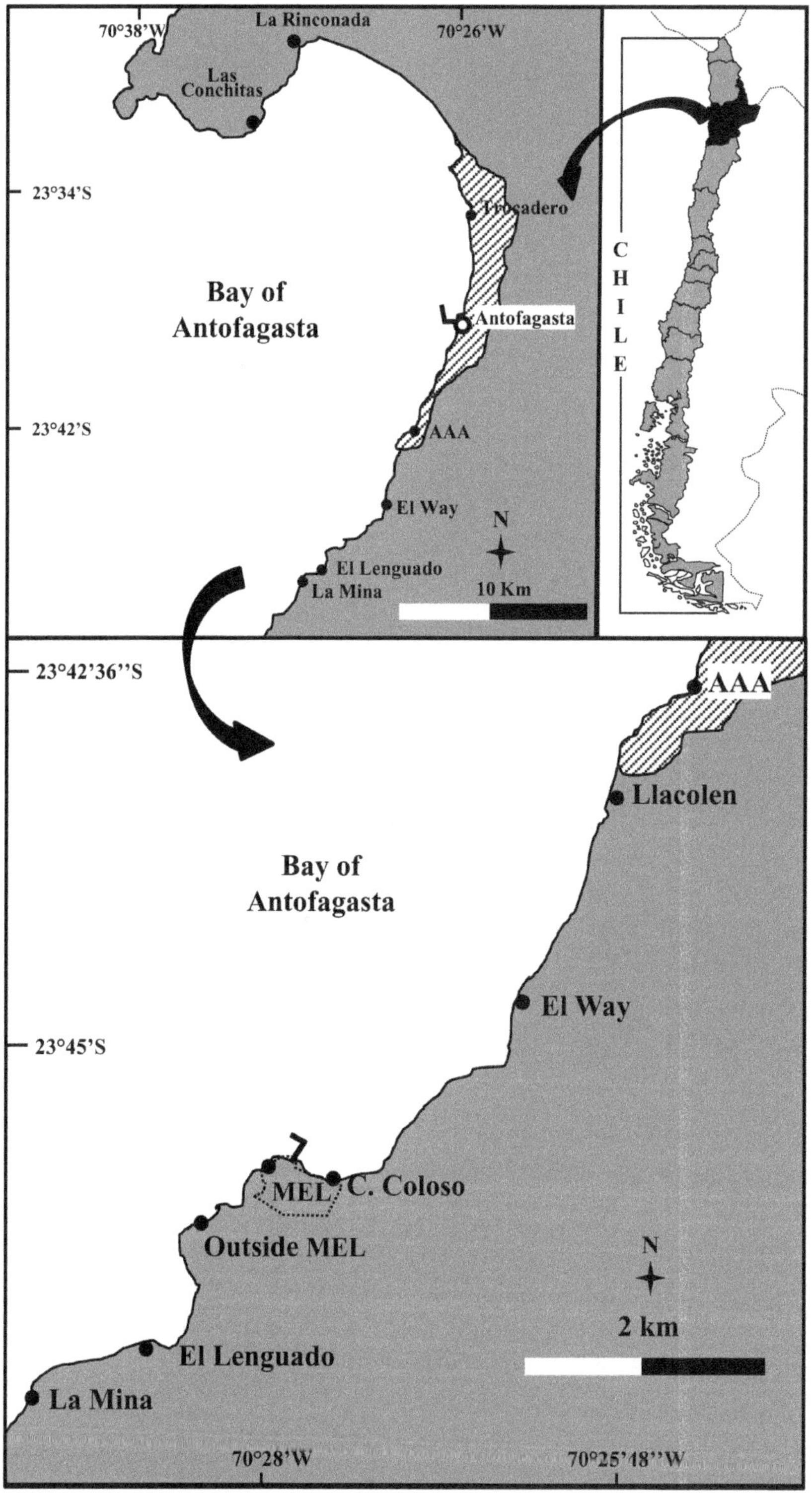

Figure 1. Map of the Bay of Antofagasta depicting the working area and main locations (see further details in reference 37).

possible mechanistic (cause-effect) interactions among such impacts along with complexes of species.

Rocky intertidal non-indigenous invasive highly competitive sedentary species cause profound effects on shore biota, altering ecological balances [13–21]. Biological invasions can be seen as press bio-perturbations [22], resulting in a new ecological equilibrium. One of the best-documented examples is the rocky intertidal invasion of the mussel *Mytilus galloprovincialis* in South Africa [23–26]. Following the invasion, around 1979, the species rapidly dominated thousands of kilometers of exposed rocky shores

Figure 2. ***Pyura praeputialis.*** Photographic record of two extracting tools: (a) knife. (b) hand saw-blade. After the tough tunic is cut the top tunic is opened by hand to remove the edible parts or siphons (S). (c). The basal tunic (Bt) including the remains of the internal visceral tissues (V) and the top tunic (Tt) still connected with the basal tunic remains attached to the rocky substrate. (d) Only the upper parts of the soft tissues, mainly the siphons (S) are retrieved for consumption. (e) After a few hours local birds, fishes and wave action will remove the flesh and only the remains of contiguous tunics are observed. The green coloration (white arrow) on top of the matrices are the green algae *Ulva* spp. that grow on the siphonal zone of the tunicates. The scale bars = 5 cm (c), = 2 cm (d) and = 10 cm (e).

along the west coast of South Africa, resulting in a modification of the intertidal zonation pattern, local biodiversity and causing a marked upshore displacement of the ecological center of gravity of the indigenous mussels *Choromytilus meridionalis* and *Aulacomya ater* [17].

Similarly, in the Bay of Antofagasta, Chile, it has been proposed that the intertidal and shallow subtidal barrel-shaped tunicate *Pyura praeputialis*, which has invaded and monopolized a major portion of the intertidal rocky shore, has changed the original intertidal zonation pattern [18]. *P. praeputialis*, originally from Australia, exhibits a conspicuous disjointed geographical distribution [27], in Chile it is known as "piure" and in Australia as "convejoy" [28,29]. It has been argued that upon its arrival, probably more than 100 years ago [29,30], the tunicate took over a large portion of the mid-intertidal zone (several meters of flat rocky platforms) and restricted the indigenous inferior competitor mussel *Perumytilus purpuratus* to the mid-upper intertidal fringe [18,30]. This zonation pattern in the bay greatly contrasts with that observed in mid-intertidal fringes on other parts of the Chilean coast, where in the absence of the tunicates, the mussel dominates the mid-intertidal primary substratum [1,31,32]. Environmental and ethical constraints [33,34] have prevented us from undertaking large-scale removals (dozens or hundreds of meters) of tunicate or mussel inside the bay to test at a large-scale the inter-specific competitive hypothesis. Instead, we carried out small-scale manipulative and transplant field experiments, in caging units of less than 400 cm^2, to test the competition strength and survival between *P. praeputialis* and *P. purpuratus* [18,30]. The results show that in the mid-low intertidal fringe of the bay, where the tunicate fully dominates, it exhibits greater competitive strength than the native mussel. While further upshore, where the mussel dominates (mid-upper intertidal fringe), the survival of the tunicate is significantly reduced by the presence of the mussel.

Can the results of these small-scale experiments explain the overall intertidal zonation pattern inside the Bay of Antofagasta? In this paper we address this question, focusing on an area of 12.5 km of coast. We have taken advantage of monitoring information (decadal data series) on the intertidal zonation pattern of the tunicate, the mussel and barnacles and have combined that information with studies on the impact of intertidal extraction on *P. praeputialis* by food-gathers. Over the last few decades, probably as result of continuous and accumulative intertidal *Pyura*-gathering, a very substantial reduction in their matrices has occurred. The hypothesis to be tested is that under this large-scale decline (absence) of the invasive superior competitive dominant tunicate, a reinstatement of the original intertidal zonation pattern (as hypothesized as existing prior to the arrival of the tunicate) will occur. Thereby, beds of the inferior competitive native mussel *P. purpuratus* or stands the barnacles *Jehlius cirratus, Chthamalus scabrosus*, will expand down-shore (seaward), into the mid-low fringe, monopolizing rocky zones previously dominated by the superiorly competitive tunicate. Moreover, since in *P. praeputialis* settlement takes place mainly on the border of contiguous conspecifics or on the surface of the siphonal zone [35], we test the hypothesis that the severe reduction of *P. praeputialis* populations will be followed by a reduction in its recruitment in the study area.

Materials and Methods

Ethics statement

This long term research consisted solely of natural observations in fully public coastal areas, and data were collected in a manner that could not cause personal identification or harm. All reported results were obtained with informed oral consent prior to conducting observations and intertidal food-gathers and occasional gathers were given the opportunity to ask questions about our research, mainly regarding descriptions of hypotheses, goals and/or purposes of the study. If consent was not obtained, the observation was not made. We certify that when monitoring started in 1993 the Chilean research/academic system did not have in place any Institutional Review Board to approve the oral consent method described above.

The Bay of Antofagasta and study sites

The Bay of Antofagasta (23°36′S; 70°24′W) is a semi enclosed upwelling-shadow [36] embayment of about 70 km in length, with a 35 km wide southern facing mouth, between Punta Tetas in the north and Punta Coloso in the south (see oceanographic characteristics in [37]). The southeastern shore of the bay shows extended exposed intertidal flat rock platforms. The majority of the platforms in the area are reddish-brown continental sedimentary rocks, belonging to the Coloso formation, from the upper Jurassic or lower Cretaceous and to a lesser extent granitic rocks [38]. At Antofagasta there is a homogeneous semidiurnal tidal regime with a maximum extension of 1.68 to 1.78 m and all intertidal monitoring was done during diurnal low tides ranging between 0.20–0.40 m [39]. Our study area of 12.5 km is located along the southeastern shore of the bay: i) Asociación de Automovilistas de Antofagasta (AAA, 23°42′50″S; 70°25′43″W), ii) Llacolen (23°42′49″S; 70°26′00″W), iii) El Way (23°44″12″S, 70°26′ 28″W), iv) Caleta Coloso (23°45′23″S; 70°27′31″W), v) Minera Escondida Limitada at Punta Coloso (MEL, 23°45′17″S; 70°28′02″W), vi) Outside MEL (23°45′39″S; 70°28′21″W), vii) El Lenguado (23°46′12″S; 70°28′28″W), and viii) La Mina (23°46″42″S; 70°29′03″W) (Fig. 1). Sites were monitored monthly (1993–1994) and sporadically, 3–9 times per year (1997–2014) by a research team (same leader, JCC). Between 1989–1997 the rocky shore of MEL, approximately 2.5 km in length, was protected and fenced by the mining company and only occasional unauthorized visits by food-gathers were reported [40].

Intertidal zonation in the bay and long-term monitoring

The intertidal zonation pattern inside the Bay of Antofagasta, described for the first time in 1959 [41] is unique in Chile [42]. The upper intertidal band (covered only during high tide) consists of 90–95% bare rock and is characterized by abundant populations of the littorinids *Nodilittorina araucana* and *Nodilittorina peruviana*. Moving down-shore there is an intertidal mid-upper band dominated by dense tridimensional matrices of the mussel *P. purpuratus* [43], with a marked and constant landward limit (long-term observations using stainless hex head screw-bolts), likely set by exposure tolerance during low tides. Patches of the barnacles *Jehlius cirratus* and *Notochthamalus scabrosus* and algae such as *Ulva* spp., *Centroceras* spp., *Ceramium* spp. and *Colpomenia* spp. are also present. Lower down the mid-low band is dominated by dense matrices of the tunicate *Pyura praeputialis* that are frequently covered by an abundant growth of *Ulva* spp. (Fig. 2e). The height above sea level of the mid-intertidal *P. praeputialis* band in the different sampling sites ranged between 0.68 and 0.86 m above zero datum. The seaward limit is defined by the primary substratum cover of lithothamnioid/melobesioid algae (*Sinarthrophyton* spp., *Lithophyllum* spp.) and the barnacle *Australomegabalanus psittacus* [44]. The *P. praeputialis* population structure, biomass and physical characteristics of matrices have been described [45,46].

Long-term changes in tunicate band width

To measure changes in the mid-low band width of *Pyura praeputialis* we monitored, between October 1993–October 1994, the distance between the landward and seaward edges of the band at 6 sites, from Llacolen to La Mina (Fig. 1; [47]. Sites were geo-referenced (hand-held Magellan Global Positioning System), photographed and mapped. Each site was 150–200 m long, separated by *ca.* 0.5–3 km and the monitored flat platforms had slopes ranging from 2° to 4°. Measurements were done monthly, during seven low tides, using 3–6 random replicates with a tape measure (0.5 cm accuracy). In January 2014 the same measurements were repeated. At this time, the *P. praeputialis* band was substantially reduced and the primary substrate was dominated by mussels (*P. purpuratus*) with the presence of barnacles (*Jehlius cirratus* and *Chthamamalus scabrosus*). Thus, this monitoring included measurements of the remaining tunicate band width plus data on the mussels and barnacles now installed covering the mid-low intertidal fringe. Additionally, in December 2003 and January 2014, we monitored (see above) changes in band width of the tunicate at three platforms with slopes <3° at AAA (Fig. 1). Platforms and intertidal fringes were marked with stainless hex head screw-bolts, photographed and mapped.

Changes in mid-intertidal zonation patterns

From February 1999 to February 2014 (10 low tides) we monitored a single flat rock platform with a slope of <5° at El Lenguado (Fig. 1). This monitoring was done directly in the field or via photographs to measure the width of the mid-intertidal belt of *Perumytilus purpuratus* and that of *Pyura praeputials*. Additionally, from December 2003 to January-February 2014 (48 low tides), we monitored the same biotic fringe widths on three flat rock platforms at AAA and two flat rock platforms at El Way (slopes <3°) (Fig. 1). For photograph analysis we used Image Pro-Plus image analysis software.

Recruitment of *P. praeputialis* monitoring

Censuses of new recruits of *P. praeputialis* were conducted during low tides in January and February of 2003 to 2008 and 2011 to 2014 on rocky intertidal platforms at El Way and AAA. *P. praeputialis* recruits were defined as individuals between 0.5–1.5 cm in maximum diameter. The count of recruits was done with the naked eye using a square quadrat of 0.04 m^2 (0.20×0.20 m, n = 6 to 20) placed randomly on the matrices of adult individuals (>60% cover). In compact matrices (100% cover) the recruits were counted around the siphons; but when the cover was below 100% the counts also included recruits present on the lateral surfaces of adult individuals. A 1-way ANOVA followed by Tukey post hoc tests determined whether total number of recruits in each sampling locality varied among sampled years.

Rocky intertidal food-gathering

Rocky shore food-gathering in Antofagasta is common and is carried out by professional and occasional (holidaymakers) gathers. Extractive activities take place during 7–10 low tides per month. From October 1993 to October 1994, we systematically monitored rocky shore food-gathers, along the 12.5 km study area [47]. At monthly intervals and over 7 days of low tides, the study area was monitored from a vehicle using binoculars every 0.5 km for 10–15 min, and the number of gathers recorded. Additionally, during 2009–2011 we occasionally (30 low tides) monitored the number of food-gathers present in the study area and during 2013–2014 we occasionally (20 low tides) monitored the number of food-gathers at Trocadero, north of the study area (Fig. 1). Monitoring was made particularly during January and February (austral summer), when the number of intertidal food-gathers in Antofagasta increases substantially due to the arrival of tourists [47]. The category of the extractors was recorded as: i) *Pyura praeputialis*-gather, targeting exclusively tunicates, ii) shellfish-gather, mainly targeting the muricid mollusk *Concholepas concholepas*, iii) *Octopus mimus*-gather, targeting exclusively octopuses, iv) skin diver operating from the rocky shore, v) anglers on the rocky shore. The activities were easily identified as the gathers use different extracting tools (i.e., knife, blade hand saw, curved iron rod or "chope", or a long stick with a fake crab on the end to catch octopuses). For *P. praeputialis*-gathers we distinguish 3 major categories: i) professional *Pyura*-gathers, who use a sharp kitchen knife (Fig. 2a) to open tunicates one by one, including those extracted for consumption or for bait; ii) professional *Pyura*-gathers, who use a small blade hand-saw (Fig. 2b) to cut several *Pyura* at once (up to ¼ m^2), including those extracting tunicates for consumption or for bait (Fig. 2d); iii) occasional *Pyura*-gathers, such as holidaymakers, that are less efficient extractors and can use knife, blade hand-saw or other tools. All *P. praeputialis* gathers

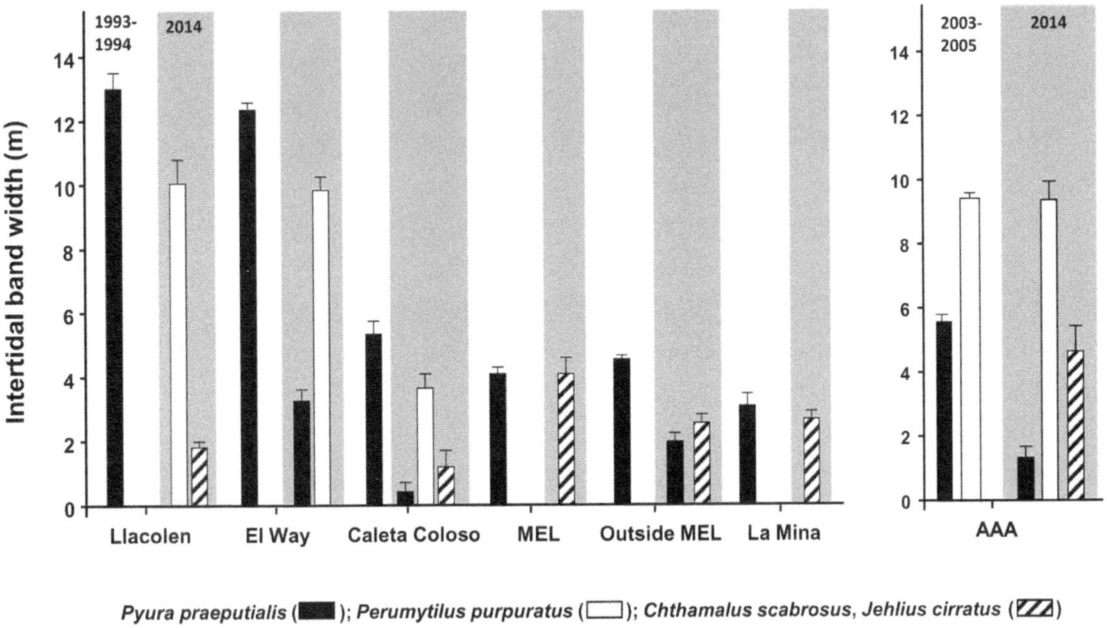

Figure 3. Mean intertidal band width (m; SE) of the tunicate *Pyura praeputialis* (black bars), the mussel *Perumytilus purpuratus* (white bars) and barnacles (*Chthamalus scabrosus, Jehlius cirratus*) (striped bars) at 7 sites in the bay of Antofagasta. For the first 6 sites the comparison is between 1993–1994 and 2014. For the seventh site the comparison is between 2003 and 2014 (b).

make a transverse cut across the upper portion of the fibrous tunic of the tunicate to gain access to the edible siphon and upper portion of the soft body parts (Fig. 2c,d). The observations lasted for 20–160 minutes each and we recorded the time spent by the gathers on extractions and the number and size of tunicates extracted (caliper 1 mm accuracy). The tunicate extraction rate was standardized as the number of tunicates extracted per gather and per hour. The numbers of tunicates extracted per hour by professional and occasional gathers and the diameter of the extracted tunicates over the entire observation period and regardless of the sampling locality were compared by 1-way ANOVA followed by Tukey post hoc tests.

Results

Long term changes in tunicate band width

In the mid-intertidal zone, over time, a sustained and notable reduction of the intertidal band of *Pyura preaputialis* was observed (Fig. 3). Considering the 1993–1994 and 2014 sampling at Llacolen, MEL and La Mina, the tunicate intertidal band disappeared and the few tunicates that remained did not form continuous bands. At El Way, C. Coloso and Outside MEL the band width of *P. praeputialis* diminished between 60 and 90% (Fig. 3). In those sites, the characteristic mid-low intertidal band of tunicates has been totally or partially replaced by the mussel *P. purpuratus* and/or mixed stands of barnacles *J. cirratus* and *C. scabrosus* (Fig. 3). A similar situation was observed on three extra intertidal platforms at AAA and on two platforms at El Way. The sequential photographs (2003–2014) at these two sampling sites and at El Lenguado (1999–2014) show a dramatic modification of the zonation pattern in the mid-intertidal (Fig. 4a–c). At AAA the original 2003 tunicate band width of *ca.* 7 m (average) diminished, from 2003 onwards, reaching a width of 2 m in 2008 and had disappeared by 2014 (Fig. 4a). At El Way the original 2003 tunicate band width of *ca.* 8 m (average) diminished to around 2 m by 2014 (Fig. 4b). At El Lenguado, the original 1999 tunicate band width of *ca.* 8 m (average), started to diminish by 2006 and completely disappeared by 2014 (Fig. 4c). Conversely, at AAA and El Way the initial mussel band width increased to around 12 and 8 m respectively by 2014 (Fig. 4a–c). Mid-intertidal band monitoring at these three sites, shows rates of band width reduction or expansions that started around 2004–2006. This coincides with the observations at the intertidal sites monitored during 1993–1994 and during 2014 along the coast line analyzed (Fig. 3).

Recruitment of Pyura praeputialis

At El Way and AAA, during 2003–2014, two distinctive phases of recruitment abundances of *P. praeputialis* were distinguished (Fig. 5a–b). At El Way, during the first phase (2003–2008), the mean abundance of recruits was relatively constant, with 51.3 recruits m^{-2}, this was followed (2011–2014) by a drastic drop, to practically no recruits (Fig. 5). At AAA, during 2003–2006, the mean abundance was *ca.* 48 recruits m^{-2}, followed (2008–2014) by a drop to mean abundances of *ca.* 10 recruits m^{-2}. The analyses detected that the mean recruit abundances were significantly higher during the first phase than during the second phase at both sampling sites (1-way ANOVA $F_{8,171}$; 8.32; $p<0.0001$ and $F_{8,133}$; 2.90; $p<0.005$ respectively; Fig. 5).

Rocky shore shellfish and *Pyura praeputialis* extraction in Antofagasta

The most frequent rocky intertidal extractors, during 1993–1994, were shellfish-gathers, with an overall mean per low tide of 1 individual km^{-1}, followed by *P. praeputialis*-gathers with a mean of 0.86 km^{-1}. They were followed by anglers at 0.56 km^{-1} and *Octopus*-gathers at 0.14 km^{-1} (-recalculated from [47] to fit the extension of our study site-). *P. praeputialis*-gathers exhibited a seasonal summer behavior concentrated during January and February. During those months, with the exception of MEL (about 2.5 km of fenced coastline, where no gatherers were recorded) the abundance of *Pyura*-gathers per low tide ranged from 6.8 km^{-1} in

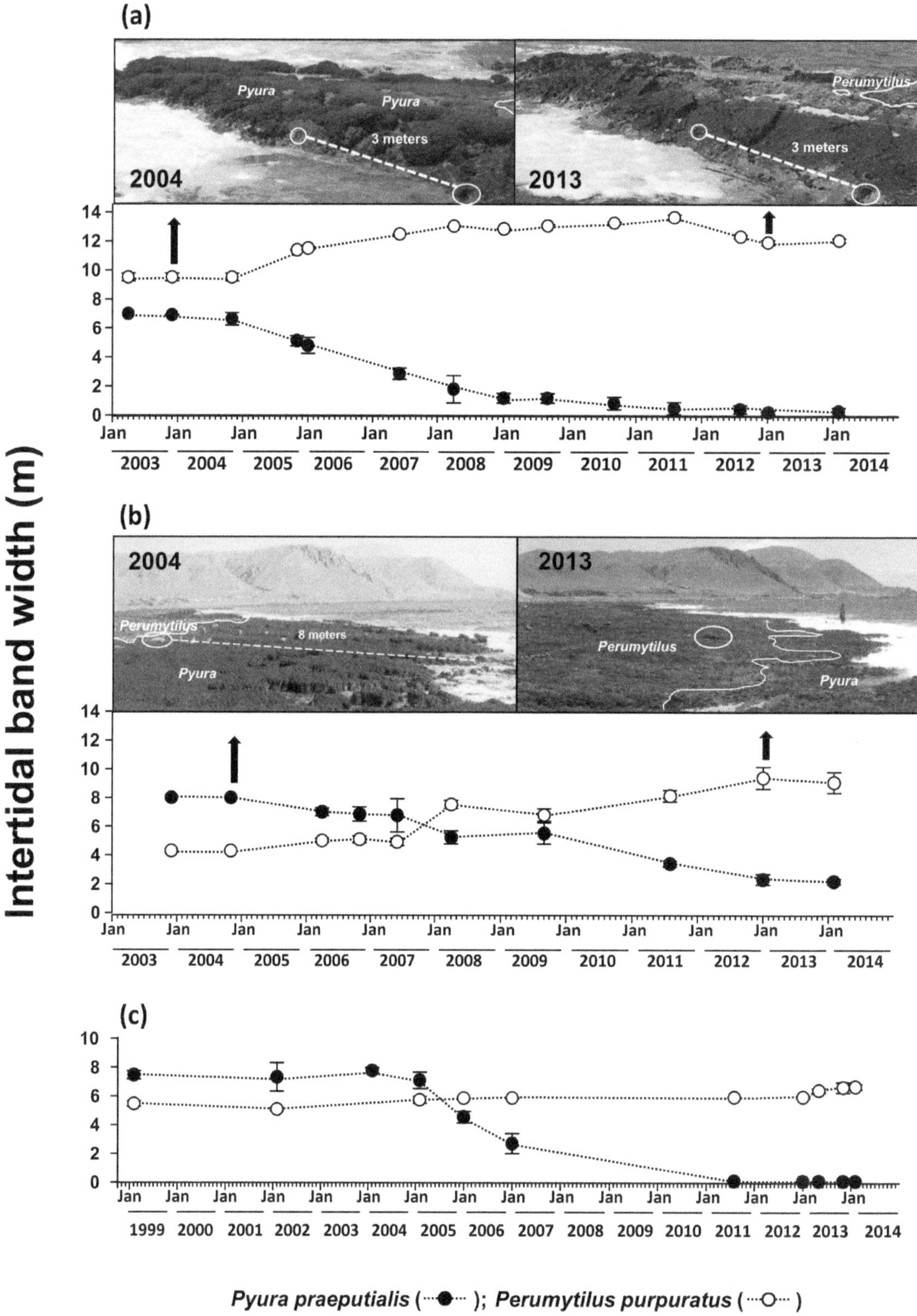

Figure 4. Temporal series of mean intertidal band width (m; SE) of the tunicate *Pyura praeputialis* (black circle) and the mussel *Perumytilus purpuratus* (white circle) in the bay of Antofagasta between April 2003 and February 2014 for a site at AAA (a) and a site at El Way (b) and between February 1999 and February 2014 for El Lenguado (c). For further details see text.

summer to 0.38 km^{-1} in winter (Fig. 6). At four intertidal sites (C. Coloso, El Way, AAA and Trocadero; Fig. 1), monitored during summer months (2009–2014), the mean abundance of professional *Pyura*-gathers ranged between 0.5–4.5 km^{-1} per low tide and the gathering excursion lasted from 6 to 120 min (Table 1). At these sites *Pyura*-gathers, using knives and blade hand-saws, extracted

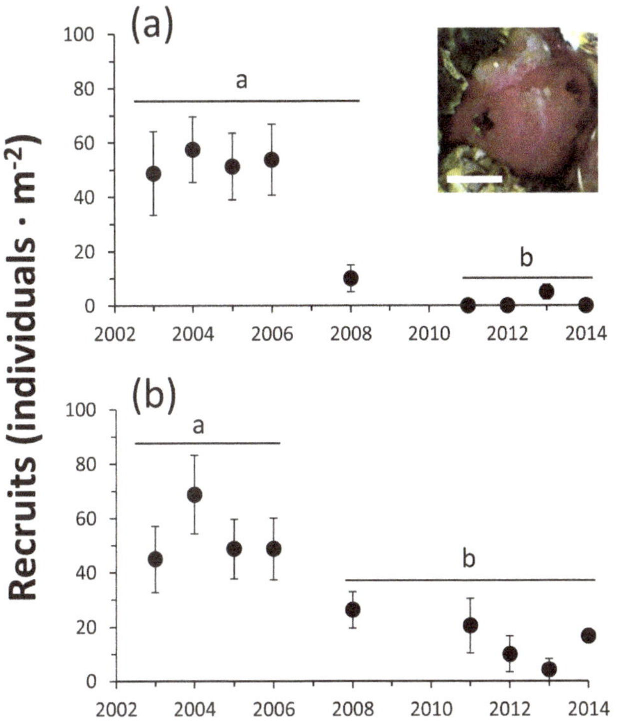

Figure 5. *Pyura praeputialis.* Temporal changes in the number (mean ±SE) of recruits (individuals m^{-2}) recorded around the siphonal and basal area of conspecifics on rocky intertidal platforms at AAA (a) and El Way (b). Values with different letters depict mean values with significant differences. The inserted picture shows a small recruit in the siphonal area of an adult conspecific (scale bar = 0.3 cm).

256–740 tunicates $hour^{-1}$ (Table 2). Regardless of the gathering tool and considering the entire study period the mean (±SD) extractions made by professional *Pyura*-gathers was 417.95 (±61.79) tunicates $hour^{-1}$. This value was significantly higher than the 111.73 (±27.10) tunicates $hour^{-1}$ extracted by occasional gathers over the same temporal period (1-way ANOVA $F_{1,46} = 13,07$; $p<0.001$). Moreover, the average size of the tunicates extracted by professional *Pyura*-gathers was significantly higher than for occasional gathers (1-way ANOVA; $F_{1,1646} = 86,35$; $p<0.0001$) (Table 2). Our conservative estimate of the removal of tunicates by professional *Pyura*-gathers (Tables 1, 2) indicates that 10 gathers working 1 h for 10 low tides per month, over 6 months, will remove between 153.360–444.000 tunicates; which would be equivalent to 307–888 m^2 of compact tunicate matrices (Tables 1, 2; considering a mean density of 500 tunicates per m^2 [46]. This estimate does not consider extractions made by holidaymakers or other less specialized food-gathers; information that is difficult to obtain.

Discussion

Our monitoring along 12.5 km of shore on the southeastern border of Antofagasta Bay (Fig. 1), recording the width of the main intertidal bands, has shown that following the decline in the tunicate matrices, the intertidal zonation pattern inside the bay has drastically changed (Figs. 3, 4). Showing that changes in the zonation pattern reported in the present study had taken place across the entire sampling area. The native mussel and barnacles populations have relocated down-shore their ecological centers of abundance. The conspicuous mid-intertidal seascape dominated by the tunicate *Pyura praeputialis* has substantially decreased or disappeared and been replaced by mussel and/or barnacle bands. This shift is in line with predictions made by made by experimental small-scale caging and manipulative experiments which suggested that in the bay the invasive tunicate *P. praeputialis* outcompeted the native mussel in the mid-low fringe and thereby substantially modified the zonation pattern [18,30]. We suggest that the major cause for the decline in the tunicate matrices in this area is the intensive and long-lasting exploitation, particularly by highly specialized and efficient intertidal rocky shore *Pyura*-gathers. This is the first large-scale demonstration that intertidal predatory human impacts can reverse the zonation pattern due to the invasion of a non-indigenous aggressive sedentary competitor. Of the various types of *Pyura* harvesters, professional gathers likely have the greatest impact. The use of the small blade hand saw is the extractive technique most detrimental for *P. praeputialis* matrices, because the removal is massive and patchy (up to 0.25 m^2 per sawing; Fig. 2e). The extractive technique produces large openings in the matrices that weaken

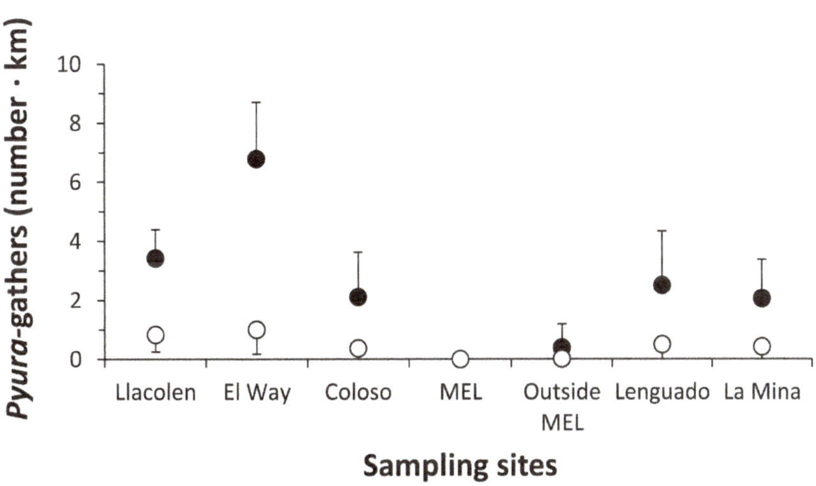

Figure 6. Mean number (±SD) of *Pyura*-gathers per km recorded over 13 months (1993–1994), in January and February (black circles) and the rest of the year (white circles) (taken and modified from 47).

Table 1. Standardized abundance of food-gathers and time spent on extraction activities in Antofagasta Bay (C. Coloso, El Way, AAA and Trocadero, Fig. 1) between 2009 and 2014.

Category	Abundance of gathers (number km^{-1} low tide)	Food gathering time (min)
Pyura-gathers		
Professional (knife)	0.5–2.5 (n = 10)	20–120 (n = 23)
Professional (saw)	0.5–2.5 (n = 4)	6–120 (n = 7)
Occasional	0.5–4.5 (n = 7)	30–80 (n = 14)
Shellfish-gathers		
Professional	1–4 (n = 3)	-
Occasional	3 (n = 1)	-

For further details see text.

and expose matrices to further removal by mechanical forces associated with waves and storms, which are capable of removing large sections of the matrix [35,48]. The knife extracting technique (Fig. 2a) is done individual by individual and therefore does less damage to the matrices. In Australia, [48] also concluded that substantial changes in density, for the same species, resulted from bait harvesting. Moreover, intertidal trampling may be another source of damage to the *Pyura* matrices. For instance, during the weekends of January and February 2014 in the study area we estimated 128–165 holidaymakers km^{-1} of coast, of which *ca.* 20–30% were camping and remained at the beaches beyond the weekends. During low tides (6 low tides, 20 h of observation over study area) holidaymakers were in the intertidal zone at numbers between 20–25 people km^{-1} of coast. At Antofagasta, there are social and economic incentives for the exploitation of tunicates. A professional *Pyura*-gather, working 7 days per month (total of 10–14 h) and selling *ca.* 150 kg of tunicates directly to restaurants can obtain *ca.* US$ 400–600. This is approximately the current official minimum monthly wage in Chile. In Antofagasta's fish markets raw *P. praeputialis* is sold at US$ 15–18 kg^{-1} and is considered a delicacy. Antofagasta is the fastest-growing city in Chile and tourism has flourished over the past 10 years, with increasing demands for marine products. Hence, over time, the tunicate intertidal matrix patchiness generated by intensive human gathering, waves, storms or predation [18,35,46,49], may act additively or synergistically leading to severe matrix reductions. Nevertheless, other additive or synergistic causes for the reduction of this species, such as diseases [50], recruitment failure [48], natural predation [18], pollution or a reduction in the genetic variability in Antofagasta's isolated tunicate population cannot be ruled out.

Our results show a clear reduction in the abundance of recruits of *P. praeputialis*, which began around 2008, when the adult population was notably reduced (see Fig. 5). Given that settlement of *P. praeputialis* takes place mainly on the border of contiguous conspecifics or on the surface of the siphonal zone ([35], present study), it is likely that the severe reduction of intertidal populations of *P. praeputialis* has had negative consequences on the recruitment of the species. The drastic reduction in populations of *P. praeputialis* at Antofagasta has had also impacts on local biodiversity, since it has been shown that macro-invertebrate species richness is significantly higher in *P. praeputialis* matrices compared to *P. purpuratus* matrices [33]. Moreover, the existence of a dense *Pyura* band increases the abundance of intertidal

Table 2. Average (±SD; N) descriptors of Pyura-gathers in Antofagasta Bay (C. Coloso, El Way, AAA, Trocadero, Fig. 1) divided into three temporal periods.

	Extraction (unit · h^{-1})	Mean diameter of the extracted *Pyura* (cm)
(1993–1994)		
Pyura-gathers		
Professional (knife & saw)	330.36 (8.78; 3)	8.57 (0.85; 752)
Occasional	220.76 (125.09; 5)	8.24 (1.03; 207)
(2009–2011)		
Pyura-gathers		
Professional (knife & saw)	474.22 (252.03; 5)	6.50 (1.89; 48)
Occasional	149.51 (49.29; 3)	5.81 (1.85; 31)
(2013–2014)		
Pyura-gathers		
Professional (saw)	740.51 (548.98; 7)	6.69 (2.20; 338)
Professional (knife)	256.31 (106.11; 13)	7.09 (1.67; 55)
Occasional	71.69 (50.57; 11)	6.16 (1.83; 196)

For further details see text.

populations of *Octopus mimus*, which inhabits the crevices between tunicates. *Octopus* gathering in the intertidal zone in the study area has almost completely ceased ([47], present study). A comprehensive analysis of the biological and ecological state of *P. praeputialis* populations inside the Bay of Antofagasta is urgently needed. This will help to determine whether or not the reported drastic reduction of intertidal populations of the tunicate in the analyzed area of the bay is generalized or not. Future research should aim to identify in the bay the existence of pristine *P. praeputialis* populations, which may serve to repopulate the rest of the bay, with the aim of conserving this unique non-indigenous species in Chile. It may sound controversial to advocate the conservation of a marine non-indigenous invasive species; nevertheless, in this case we think it makes sense since the invasive tunicate has increased the biodiversity of Antofagasta Bay [33] and provided numerous positive ecosystem services [20,21,47].

Acknowledgments

We sincerely acknowledge financial support by the funders. We thank Dr. Matthew Lee for improving the English and critical reviews made by Professor Charles Griffiths and two anonymous referees for their comments, which substantially improved the manuscript.

Author Contributions

Conceived and designed the experiments: JCC PHM. Performed the experiments: JCC PHM VO MEJ MV AD. Analyzed the data: JCC PHM MV VO. Contributed reagents/materials/analysis tools: JCC PHM. Wrote the paper: JCC PHM VO AD MV.

References

1. Castilla JC, Durán LR (1985) Human exclusion from the rocky intertidal zone of Central Chile: The effects on *Concholepas concholepas* (Gastropoda). Oikos 45: 391–399.
2. Moreno CA, Lunecke K, Lépez M (1986) The response of an intertidal *Concholepas concholepas* population (Gastropoda; Muricidae) to protection from man in southern Chile and its effects on benthic sessile assemblages. Oikos 46: 359–364.
3. Castilla JC, Bustamante R (1989) Human exclusion from intertidal rocky shore of Las Cruces, Central Chile: Effects on *Durvillaea antarctica* (Phaeophyta, Durvilleales). Mar Ecol Prog Ser 50: 203–214.
4. Godoy C, Moreno CA (1989) Indirect effects of human exclusion from the rocky intertidal in southern Chile: a case of cross-linkage between herbivores. Oikos. 54: 101–106.
5. Moreno CA (1993)Efecto de la intervención humana en la mantención de la Asociación Gymnogongrus–Choromytilus en comunidades intermareales del Sur de Chile. Ph.D. Thesis. Pontificia Universidad Católica de Chile. Facultad de Ciencias Biológicas. Santiago. 136 p.
6. Branch GM, Moreno CA (1994) Intertidal and subtidal grazers. In: Siegfried WR, editor.Rocky shores: Exploitation in Chile and South Africa. Ecological Studies Edition, Springer-Verlag. pp.75–100.
7. Castilla JC (1999) Coastal marine communities: trends and perspectives from human-exclusion experiments. Trends Ecol Evol 14: 280–283.
8. Barkai A, Branch GM (1988) Contrast between the benthic communities of subtidal hard substrata at Marcus and Malgas Islands: A case of alternative stable states? S Afr J Mar Sci 7: 117–137.
9. Barkai A, McQuaid CD (1988) Predator-prey role reversal in a marine benthic ecosystem. Science 242: 672–675.
10. Siegfried WR (1994) Rocky sores: Exploitation in Chile and South Africa Springer-Verlag Berlin Heidelberg, Germany. 177 p.
11. Kinsford MJ, Underwood AJ, Kennelly SJ (1991) Humans as predators on rocky reefs in New South Wales, Australia, Mar Ecol Prog Ser 72: 1–14.
12. Castilla JC (1993) Humans: Capstone strong actors in the past and present coastal ecological play. In: McDonnell MJ, Pickett STA, editors. Humans as Components of Ecological Systems., Springer-Verlag, pp.158–162.
13. Pastorino G, Darrigran G, Martin S, Lunaschi L (1993) *Limnoperna fortunei* (Dunker, 1857) (Mytilidae), nuevo bivalvo invasor en aguas del Río de la Plata. Neotropica 39(101–102): 34.
14. Vallarino EA, Elias R (1997) The dynamics of an introduced *Balanus glandula* population in the southwestern Atlantic rocky shores: the consequence on the intertidal community. PSZN I: Mar Ecol 18: 319–335.
15. Robinson TB, Griffiths CL, McQuaid CD, Rius M (2005) Marine alien species of South Africa — status and impacts. Afr J Mar Sci 27: 297–306.
16. Rilov G, Crooks J (2009) Marine Bioinvasions: conservation hazards and vehicles for ecological understanding. In: Rilov G, Crooks JA, editors.Biological Invasions in Marine Ecosystems: Ecological, Management, and Geographic Perspectives. Ecological Studies Series. Springer, Heidelberg, Germany.
17. Griffiths CL, Robinson TB, Mead A (2009) The status and distribution of marine alien species in South Africa. In: Biological Invasions in marine ecosystems: ecological, management, and geographic perspectives (eds.) Rilov G, Crooks JA. Springer-Verlag Berlin Heidelberg, pp: 393–408.
18. Castilla JC, Guiñez R, Caro AU, Ortiz V (2004) Invasion of a rocky intertidal shore by the tunicate *Pyura praeputialis* in the Bay of Antofagasta, Chile. Proc Natl Acad Sci USA 101: 8517–8524.
19. Castilla JC, Uribe M, Bahamonde N, Clarke M, Desqueyroux-Faúndez R, et al. (2005) Down under the Southeastern Pacific: marine non-indigenous species in Chile. Biol Invas 7: 213–232.
20. Castilla JC, Neill PE (2009) Marine Bioinvasions in the Southeastern Pacific: Status, Ecology, Economic Impacts, Conservation and Management (Chapter 26). In: Rilov G, Crooks JA, editors.Biological Invasions in Marine Ecosystems. Ecological, Management and Geographic Perspectives Series: Ecological Studies, vol 204. Springer-Verlag, Berlin Heidelberg. pp.439–457.
21. Castilla JC, Camaño A (2001) El Piure de Antofagasta, *Pyura praeputialis* (Heller, 1878): un competidor dominante e ingeniero de ecosistemas. En: Alveal K & A Antezana (eds). Sustentabilidad de la biodiversidad, pp. 719–729. Editorial Universidad de Concepción, Concepción.
22. Bender EA, Case TJ, Gilpin ME (1984) Perturbation experiments in community ecology: theory and practice. Ecology 65: 1–13.
23. Hockey PAR, van Erkom-Schurink C (1992) The invasive biology of the mussel *Mytilus galloprovincialis* on the southern African coast. Trans R Soc S Afr 48: 123–140.
24. Griffiths CL, Hockey PAR, Van Erkom Schurink C, Le Roux PJ (1992) Marine invasive aliens on South African shores: implications for community structure and trophic functioning. S Afr J Mar Sci 12: 713–722.
25. Steffani CN, Branch GM (2003) Temporal changes in an interaction between an indigenous limpet *Scutellastra argenvillei* and an alien mussel *Mytilus galloprovincialis*: effects of wave exposure. Afr J Mar Sci 25: 213–229.
26. Robinson TB, Branch GM, Griffiths CL, Govender A, Hockey PAR (2007) Changes in South African rocky intertidal invertebrate community structure associated with the invasion of the mussel *Mytilus galloprovincialis*. Mar Ecol Prog Ser 340: 163–171.
27. Castilla JC, Guiñez R (2000) Disjoint geographical distribution of intertidal and nearshore benthic invertebrates in the southern Hemisphere. Rev Chil Hist Nat 73: 586–603.
28. Castilla JC, Collin AG, Meyer CP, Guiñez R, Lindberg DR (2002) Recent introduction of the dominant tunicate, *Pyura praeputialis* (Urochordata, Pyuridae) to Antofagasta, Chile. Mol Ecol 11: 1579–1584.
29. Rius M, Teske PR (2011) A revision of the *Pyura stolonifera* species complex (Tunicata, Ascidiacea), with a description of a new species from Australia. Zootaxa 2754: 27–40.
30. Caro AU, Guiñez R, Ortiz V, Castilla JC (2011) Competition between a native mussel and a non-indigenous invader for primary space on intertidal rocky shores in Chile. Mar Ecol Prog Ser 428: 177–185.
31. Navarrete S, Castilla JCC (1990) Barnacles walls as mediators of intertidal mussel recruitment: Effects of patch size on the utilization of space. Mar Ecol Prog Ser 68: 113–119.
32. Caro AU, Navarrete SA, Castilla JC (2010) Ecological convergence in a rocky intertidal shore metacommunity despite high spatial variability in recruitment regimes. Proc Natl Acad Sci USA 107: 18528–18532.
33. Castilla JC, Lagos N, Cerda M (2004) Marine ecosystem engineering by the alien ascidian *Pyura praeputialis* on a mid-intertidal rocky shore. Mar Ecol Prog Ser 268: 119–130.
34. Bunting D, Coleman RA (2014) Ethical consideration in invasion ecology: A marine perspective. Ecol Manag Res 15: 64–70.
35. Alvarado JL, Pinto R, Marquet P, Pacheco C, Guiñez R, et al. (2001) Patch recolonization by the tunicate *Pyura praeputialis* in the rocky intertidal of the Bay of Antofagasta, Chile: evidence for self-facilitation mechanisms. Mar Ecol Prog Ser 224: 93–101.
36. Graham WM (1993) Spatio-temporal scale assessment of an "upwelling shadow" in northern Monterey Bay, California. Estuaries 16: 83–91.
37. Castilla JC, Lagos NA, Guiñez R, Largier JL (2002) Embayments and nearshore retention of plankton: the Antofagasta Bay and other examples. In: Castilla JC, Largier JL, editors. The Oceanography and Ecology of the Nearshore and Bays in Chile, Ediciones Universidad Católica de Chile, Santiago, Chile, pp. 179–203.
38. Ferraris F, Di Biase F (1978) Hoja Antofagasta, región de Antofagasta. Instituto de Investigaciones Geológicas, Carta Geológica de Chile N°30, 48p.
39. Tablas de marea de la costa de Chile (2014) Servicio hidrográfico y oceanográfico de la Armada de Chile. Publicación SHOA N° 300940.
40. Castilla JC, Rho E (1997) Marine littoral conservation and resource restocking in Northern Chile: The case of Minera Escondida at Punta Coloso, Antofagasta, Chile. Estudios Oceanológicos 16: 115–14041.
41. Guiler ER (1959) Intertidal belt-forming species on the rocky coasts of northern Chile. Pap proc Roy Soc Tasmania 93: 35–58.

42. Paine RT, Suchanek TH (1983) Convergence of ecological processes between independently evolved competitive dominants, a tunicate-mussel comparison. Evolution 37: 821–831.
43. Guiñez R, Castilla JC (1999) A tridimensional self-thinning model for multilayered intertidal mussels. Am Nat 154(3): 341–35744.
44. Castilla JC (1998) Las comunidades intermareales de la Bahía San Jorge: estudios de línea base y el programa ambiental de Minera Escondida Ltda. en Punta Coloso. In: Minería del cobre, ecología y medio ambiente costero. Arcos D (Ed), Editorial Anibal Pinto, pp.221–244.
45. Guiñez R, Castilla JC (2001) An allometric tridimensional model of self-thinning for a gregarious tunicate. Ecology 82: 2331–2341.
46. Castilla JC, Guiñez R, Alvarado JL, Pacheco C, Varas M (2000) Distribution, population structure, population biomass and morphological characteristics of the tunicate *Pyura stolonifera* in the bay of Antofagasta, Chile. PSZNI: Mar Ecol 121: 161–174
47. Varas M (1996) Pesca de subsistencia en el sector rocoso costero al sur de la ciudad de Antofagasta, con énfasis en el recurso "loco" *Concholepas concholepas* (Bruguière, 1789) y la concesión de Minera Escondida en Punta Coloso. Tesis Universidad Arturo Prat, Iquique, Chile 80 p.
48. Fairweather PG (1991) A conceptual framework for ecological studies of coastal resources: An example of a tunicate collected for bait on Australian seashores. Ocean Shore Manage 15: 125–142.
49. Pacheco CJ, Castilla JC (2001) Foraging behavior of the American Oystercatcher *Haematopus palliatus pitanay* (Murphy 1925) on the intertidal ascidian *Pyura praeputialis* (Heller 1878) in the Bay of Antofagasta, Chile. J Ethol 19: 23–26.
50. Hanekom N (2013) Environmental conditions during mass mortalities of the Ascidian *Pyura stolonifera* (Heller) in the Tsitsikamma Marine Protected Area. Afr Zool 48: 167–172.

17

A Multi-Scale Distribution Model for Non-Equilibrium Populations Suggests Resource Limitation in an Endangered Rodent

William T. Bean[1]*, Robert Stafford[2], H. Scott Butterfield[3], Justin S. Brashares[4]

1 Humboldt State University, Arcata, California, United States of America, **2** California Department of Fish and Game, Los Osos, California, United States of America, **3** The Nature Conservancy, San Francisco, California, United States of America, **4** Department of Environmental Science, Policy and Management, University of California, Berkeley, California, United States of America

Abstract

Species distributions are known to be limited by biotic and abiotic factors at multiple temporal and spatial scales. Species distribution models, however, frequently assume a population at equilibrium in both time and space. Studies of habitat selection have repeatedly shown the difficulty of estimating resource selection if the scale or extent of analysis is incorrect. Here, we present a multi-step approach to estimate the realized and potential distribution of the endangered giant kangaroo rat. First, we estimate the potential distribution by modeling suitability at a range-wide scale using static bioclimatic variables. We then examine annual changes in extent at a population-level. We define "available" habitat based on the total suitable potential distribution at the range-wide scale. Then, within the available habitat, model changes in population extent driven by multiple measures of resource availability. By modeling distributions for a population with robust estimates of population extent through time, and ecologically relevant predictor variables, we improved the predictive ability of SDMs, as well as revealed an unanticipated relationship between population extent and precipitation at multiple scales. At a range-wide scale, the best model indicated the giant kangaroo rat was limited to areas that received little to no precipitation in the summer months. In contrast, the best model for shorter time scales showed a positive relation with resource abundance, driven by precipitation, in the current and previous year. These results suggest that the distribution of the giant kangaroo rat was limited to the wettest parts of the drier areas within the study region. This multi-step approach reinforces the differing relationship species may have with environmental variables at different scales, provides a novel method for defining "available" habitat in habitat selection studies, and suggests a way to create distribution models at spatial and temporal scales relevant to theoretical and applied ecologists.

Editor: Stephanie S. Romanach, U.S. Geological Survey, United States of America

Funding: Funding was provided by The Nature Conservancy, U.S. Department of Agriculture, Bureau of Land Management, California Department of Fish and Game, and the A. Starker Leopold Wildlife Graduate Student Fund. The funders had no role in study design, data collection and analysis, decision to publish, or preparation of the manuscript.

Competing Interests: The authors have declared that no competing interests exist.

* Email: bean@humboldt.edu

Introduction

Species distribution models (SDMs) have become a cornerstone of theoretical (e.g., [1]) and applied (e.g., [2]) ecological research [3,4]. In these models, species occurrence data and environmental correlates are used to define the limits of a species distribution ([4]). However, understanding and predicting the relationship between environmental resources and species distributions is complicated by the temporal and spatial scale of analysis, with most SDMs aimed at mapping range-wide associations using abiotic climatic factors. Perhaps because of the broad temporal and spatial scale at which these analyses are conducted, most recommendations suggest that SDMs operate best for populations at equilibrium (e.g.,[5]). By contrast, most species, especially those of conservation concern, are rarely, if ever, at equilibrium [6].

Guisan and Thuiller [7] provide a framework for modeling species distributions at disparate scales. At broad (*e.g.*, biogeographic) spatial and temporal scales, species' distributions tend to be limited primarily by abiotic factors [8]. At finer spatial and temporal scales, species are limited by local community interactions such as resource factors, dispersal, predation, and competition. Guisan and Thuiller's work suggests a multi-step approach to modeling. That is, they encourage practitioners to first define a species' range-wide distribution, and then model limiting factors within that area to better understand relationships with environmental factors at finer spatial or temporal scales. Echoing Hutchinson [9], Guisan and Thuiller [7] refer to the broad-scale, bioclimatic range as the "potential distribution", and they define the "realized distribution" as the bioclimatic range filtered through dispersal, disturbance, and biotic interactions.

Guisan and Thuiller's [7] research closely parallels work in the field of habitat selection. Johnson [10] defined habitat selection as a strictly hierarchical process, with first-order selection occurring at the level of the physical or geographical range, second-order selection determining the home range, and so on. While Johnson [10] described habitat selection at each scale as a decision-based process by the individual animal, and Guisan and Thuiller [7] formulate it as an environmental filtering process, both clearly suggest that local occurrences are separately constrained within a higher hierarchical biogeographic distribution. As Wiens et al. [11] demonstrated in their landmark study of shrubsteppe birds, not only are these hierarchical levels of habitat selection distinct, animals may select habitat in contrasting directions at different spatial or temporal scales. Environmental factors that predict habitat selection at macro scales (*e.g.* vegetation, cover, temperature, rainfall) may have little predictive value at finer scales, or may even be correlated with selection in opposite directions.

Habitat selection studies have long recognized this problem of temporal or spatial scale incompatibility [12], and resource selection studies frequently examine habitat selection at multiple scales [13]. Despite the long history of research on habitat selection, the problem of defining "available" habitat has been a common and recurring one [14]. Typically, researchers use some measure of a home range, a buffer around used points, or some meaningful political or biological boundary [12]. We suggest that a more appropriate definition of available habitat would follow the well-understood construction of hierarchical habitat selection. That is, a study of habitat selection at multiple scales should follow the theory of Guisan and Thuiller [7] and Johnson [10] by explicitly modeling habitat selection at each hierarchical stage.

Guisan and Thuiller's [7] multi-step approach has been used to model distribution limited by dispersal [15], and habitat type [16], but to our knowledge has not been used to examine the role of resource availability. Resource availability has long been hypothesized as a key factor limiting species' distributions [17], and recent work has supported this (e.g., [18], [19], [20]). In particular, the temporal dynamics of resource availability can be critical to fine-scale distribution modeling in either space or time. While species at broad spatial and temporal scales may be considered at equilibrium, managers are frequently tasked with understanding shifts in distribution at much finer time intervals, such as between years or even seasons [21,22]. At such temporal scales, variability of resources can greatly impact species distributions, particularly where the presence of a species is positively or negatively related with resource availability [23].

Recent advances in remote sensing techniques have allowed for estimates of resource abundance at fine temporal scales [24]. In particular, the Normalized Difference Vegetation Index (hereafter "NDVI") has been used as a reliable estimate of biomass in grassland systems [25], and population dynamics in herbivores have been shown to be correlated with NDVI (e.g., [26], [27], [28]). Recent work has shown that NDVI can be a useful predictor of distribution in large herbivores [29].

In this study, we created a multi-step species distribution model for the giant kangaroo rat (*Dipodomys ingens*, hereafter "GKR"). The GKR is an endangered rodent endemic to southern-central California [30]. GKRs are believed to be limited to areas with loamy soils, flat or gently rolling hills, and to areas with mean annual precipitation no greater than approximately 30 cm [31,32]. First, we estimated the potential distribution (or first-order habitat selection) of the GKR using population-wide occurrence data and static environmental predictor variables (slope, soil particle size, and six climatic variables relating to temperature and precipitation) using the machine-learning method Maxent [33]. Maxent represents an ideal method for modeling the "potential distribution" because it assumes the most uniform distribution of a species' occurrence across the study area, minimally constrained by the provided environmental correlates. Maxent is a presence-background model, and in fact its authors suggest the results may represent the species' potential distribution [33]. Over broader spatial scales and longer time scales, we predicted a negative relationship between GKR presence and precipitation.

We used this model of potential distribution to define available habitat in order to understand finer scale temporal dynamics in GKR population extent (i.e., annual changes in the "realized" distribution). These temporal models incorporated a suite of primary productivity estimates based on the NDVI. In particular, we predicted, based on previous research [30,32,34], that GKR presence would show a positive correlation with resource abundance within their potential distribution, possibly with a time lag reflecting a delayed demographic response of GKR to resource availability. Due to the GKR's strong association between population demographics and precipitation, other factors that may also limit population extent (e.g., predation and competition) were not considered in these models.

Methods

Study site and focal species

The GKR is a state and federally endangered, burrowing, granivorous rodent endemic to deserts grasslands of California, USA [32]. Once widespread in the western San Joaquin Valley, habitat loss from agriculture and other development have severely restricted its range to a half-dozen populations in and around the California Coast Range [30]. The GKR is considered both a keystone species and an ecosystem engineer [35,36]. As grasses begin to senesce in April, GKRs remove all herbaceous vegetation from the top of their burrows [31,37]. This behavior results in clear circles of bare soil, 2–7 m in diameter, where GKRs are present. Aerial surveys have therefore been a useful tool in mapping GKR population extent in years of high primary productivity [38].

This study is primarily focused on the Carrizo Plain National Monument (hereafter "Carrizo"), an area that contains the largest remaining population of GKRs. Carrizo represents the largest representative landscape of San Joaquin Valley annual grassland [39]. Carrizo experiences variable precipitation (mean = 20 cm, sd = 10 cm) that contributes directly to variability in primary productivity, which in turn may drive dramatic annual changes in GKR distribution [32]. Based on aerial surveys, GKR population extent in Carrizo was estimated to expand more than 50% between 2001 and 2006 [38]. Understanding the role of primary productivity in driving these changes is crucial to biodiversity management for this endangered ecosystem. Both the size of the GKR population and its management and monitoring history make Carrizo an ideal study site for examining the role of resource availability on species distributions.

Other factors that often limit a species' realized distribution – predation, parasitism, competition and dispersal – were not believed to be limiting factors for GKR in Carrizo. Within the study area, the open and flat topography, coupled with GKR reproductive habits allow for rapid dispersal. Within Carrizo, GKR appear to be competitively dominant [31,36]. Because of these features of their ecology, GKR distribution was less likely to be affected by dispersal or competition and, thus, the GKR was a good species for testing models of realized distribution based solely on resource abundance.

This study was carried out in strict accordance with the recommendations in the Guidelines of the American Society of Mammalogists for the Use of Wild Mammals in Research. The protocol was approved by the Animal Care and Use Committee at the University of Californa, Berkeley (R304).

GKR Distribution

We obtained estimates of GKR distribution from three sources: (1) historical occurrence records from the Global Biodiversity Information Facility (GBIF); (2) contemporary trapping sites throughout GKR range; and (3) aerial surveys of GKR population extent within Carrizo.

Occurrence records were downloaded from the GBIF using the *dismo* package [40] in R [41], and limited to points collected since 1950 (N = 38). We obtained an additional 185 records of GKR presence or absence from trapping conducted in 2010 and 2011. 157 points were selected randomly throughout GKR range, and trapped for three nights with five traps [42]. Eight additional sites were stratified across a range of habitat suitability values from a preliminary distribution model constructed in 2008, and a final 20 presence points were obtained from ongoing trapping in the center of Carrizo [36]. Of the 185 sites trapped, 120 were occupied in either 2010 or 2011, and thus included in the range-wide potential distribution model. Additional details on trapping methodology are provided in Bean et al. [42].

In 2001, 2006, 2010 and 2011, we conducted Carrizo-wide aerial flight surveys in late summer to estimate GKR extent. Using 800 m wide transects with two observers (i.e., monitoring 400 m on each side) and a global positioning system (GPS), we mapped the total extent of active burrows. GPS points were recorded whenever the observers entered or left areas of observable GKR activity. These points were then connected as lines and buffered 400 m on each side to create an estimate of total extent. These surveys were shown to be a reliable estimate of GKR population extent in a given year [38].

Potential Distribution Modeling

We created a multi-step model to estimate GKR distribution. We first used Maxent to estimate the potential GKR distribution with range-wide occurrence data (museum records and our trapping data] and static environmental variables. Second, we used logistic regression to estimate limits to the potential distribution based on local resource abundance (Fig. 1).

We used the software package Maxent to estimate GKR potential distribution [33]. Maxent uses a maximum entropy approach to estimate the most uniform distribution of a species' occurrence across the study area, minimally constrained by the provided environmental correlates. Maxent is a presence-background model, and therefore may better model the species' potential distribution [33].

To estimate the potential distribution for GKR, we selected a suite of environmental variables believed to limit GKR distribution range-wide. We obtained 19 climate layers [43] frequently used in distribution modeling as independent variables [44]. Bioclim layers are estimated as mean conditions from 1950 to 2000. We limited the variables to six we believed sufficient in describing GKR distribution, and that had limited correlation with each other. These included annual mean temperature (BIO1); annual precipitation (BIO12); minimum temperature of the coldest month (BIO6); precipitation of the driest month (BIO14); and precipitation of the driest quarter (BIO17). In addition, we used slope [45], and soil particle size derived from the SSURGO database [46]. Soil particle size was converted to raster format using ArcGIS 9.2, and all inputs were analyzed at 30 s resolution (the coarsest resolution of all predictor variables). Soil particle size was classified as categorical, with the rest classified continuous.

The output of this initial Maxent distribution model was a map of habitat suitability (Fig. 1), with each 30 s cell representing an index of suitability. To convert the map from a continuous suitability distribution to a binary map of potential distribution, we selected a threshold, above which cells were classified as potential GKR distribution and below which cells were classified as outside potential GKR distribution. A number of methods have been proposed for selecting thresholds [47–49]. However, the optimal thresholds recommended in previous work focused on best predicting overall presence or absence for a species. In this case, we were interested in defining the maximum potential distribution for the species. Therefore, in order to err on the side of inclusiveness, we selected a threshold (0.059) that included 99% of presence points from the modeled potential distribution.

Realized Distribution Modeling

Having produced an estimate of the potential distribution for GKR, we then examined the effects of resource availability on GKR realized distribution for four study years. Because the GKR relies on grass seeds as a food resource, we expected a positive correlation between primary productivity and GKR presence. GKRs dry and cache most of the seeds they collect in underground chambers [37], so GKR presence in a given area may lag primary productivity for a year or more.

To create a spatially explicit measure of primary productivity in Carrizo we acquired 16-day composites (250 m×250 m) of NDVI measured by the Moderate Resolution Imaging Spectroradiometer ("MODIS") platform [50]. The NDVI is calculated as

$$(NIR - R) / (NIR + R) \qquad \text{(Equation 1)}$$

where NIR represents spectral reflectance within the near infrared band (841–876 nm), and R represents the visible red band (620–670 nm). Values approaching −1.0 tend to represent areas with water, while areas greater than 0 and approaching 1.0 tend to represent areas of photosynthetic activity [51]. Pre-processed 16-day composites of NDVI measured from MODIS have been shown to better measure primary productivity than single measures. These composites correlate well with biomass in grassland systems [25].

We created a suite of generalized linear models (GLM) to predict GKR presence using the NDVI for each year (2001, 2006, 2010 and 2011) [13]. We examined two drivers of GKR presence: first, and of primary interest, we tested the effect of primary productivity (*i.e.*, resource abundance) on GKR presence. Second, we tested if GKR presence in the previous year would also be a significant predictor of GKR presence in the current year. The independent variables included in the model to evaluate these predictions represented resource abundance in the current or previous year, or were proxies of GKR presence in the previous year (Table 1). These hypotheses were first tested independently before being included in the suite of models (Fig. 2).

First, to estimate resource abundance, we used the highest recorded NDVI value for a given growing season (November through May, the typical rainy season in the Mediterranean climate of coastal California) as an estimate of primary productivity for that 250 m×250 m cell. In estimating distribution for GKR in 2006, for example, we estimated primary productivity in the previous year as the peak NDVI from November to May, 2004–2005; and primary productivity in the current year as peak NDVI from November to May, 2005–2006. NDVI can be inflated by soil moisture if the soil is visible [51] and NDVI appeared to peak approximately 1–2 weeks before the typical peak growth in Carrizo, suggesting that soil moisture was influencing NDVI measurement. However, precipitation and aboveground biomass are correlated and, despite the lag in measurements, peak NDVI has repeatedly been shown to correlate strongly with peak aboveground biomass in grasslands [24,52].

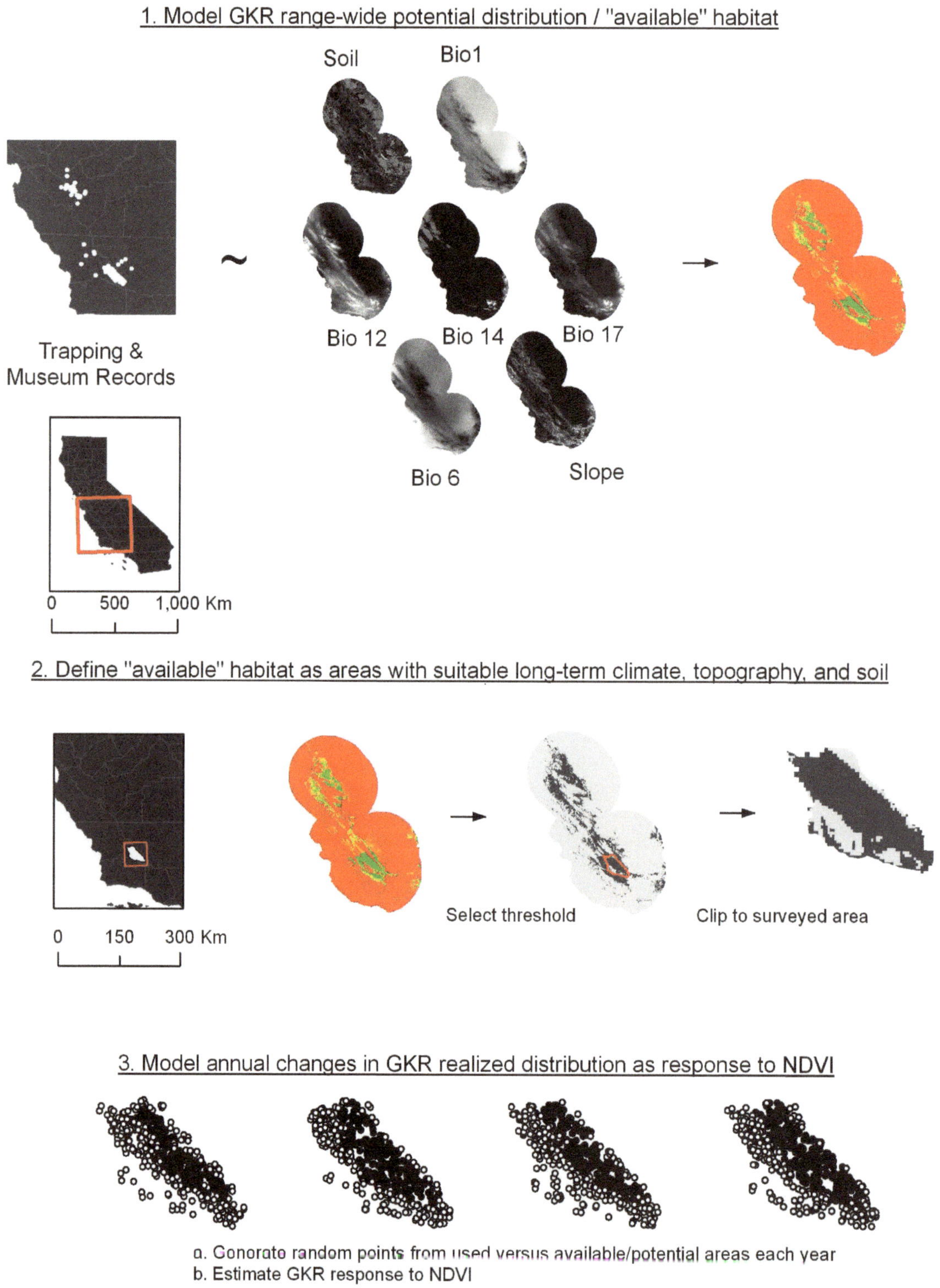

Figure 1. Flow chart of multi-step modeling approach. Here we present a multi-method, multi-scale approach to estimating species distributions. In the first step, Maxent is used to relate contemporary trapping and historical museum records with static environmental variables. The result is a model of potential distribution at a range-wide scale. Predictor variables included soil particle size ("Soil"), annual mean temperature ("Bio1"), minimum temperature of coldest month ("Bio 6"), annual precipitation ("Bio 12"), precipitation of driest month ("Bio 14"), precipitation of driest quarter ("Bio 17"), and slope. We then selected a threshold to define all available habitat for GKR, with the 99% Maxent value for training data used as the threshold. Finally, within the potential habitat in the Carrizo Plain National Monument, we examined annual changes in population extent based on aerial surveys and driven by measures of resource availability (NDVI).

Table 1. Hypothesized relationships between estimates of primary productivity (NDVI) and the local presence of the giant kangaroo rat.

Candidate Predictors	Hypothesized Mechanism
Maximum $NDVI_{T1,T0}$	Estimate of primary productivity, a bottom-up limitation on GKR presence (with potential one year lag)
Minimum $NDVI_{T0}$	Proxy for GKR presence in previous year
NDVI slope during plant $senescence_{T0}$	Proxy for GKR presence in previous year (GKR remove vegetation more quickly than it senesces)

T1 represents the Normalized Difference Vegetation Index data from the same year as the rat distribution was estimated; T0 is data from the previous year.

For three of the four years of surveys, no estimate was available for GKR presence in the previous year. Instead of a direct estimate from aerial surveys, it was therefore necessary to create proxies of GKR presence in the previous year. Because GKR clear their burrow mounds of vegetation, we assumed that GKR would have a direct effect on the NDVI after peak green up. First, we assumed that later in the summer, the areas with GKR would have lower plant biomass than areas without GKR. We therefore included the lowest measured NDVI value from later in the year (April to December) as a proxy for GKR presence, assuming a negative correlation between the two (i.e., areas with GKR would have lower minimum NDVI). Second, we assumed that GKR removed vegetation from around their burrows faster than vegetation naturally senesced. To estimate vegetation removal by GKR, we measured the slope of NDVI decline from its peak. We subtracted the NDVI value from one time step (i.e. 16 days) after peak from

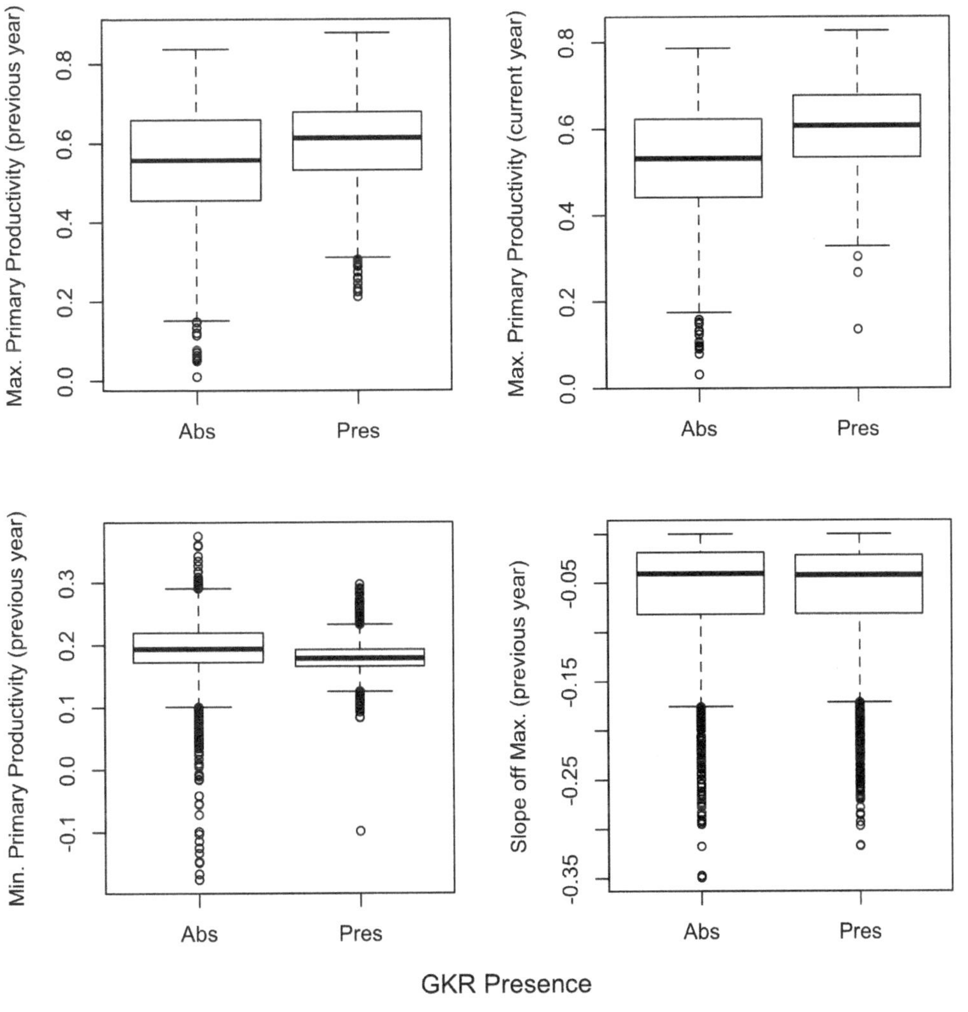

Figure 2. Relationships between primary productivity (measured as NDVI) and GKR presence. GKR were expected to have a positive relationship with maximum primary productivity in the previous and current year; a negative relationship with the minimum primary productivity measured in the previous year; and a negative relationship with the rate of decrease of primary productivity in the previous year. Relationships are shown from 500 random points estimated from aerial surveys in 2011. All differences were significant (t-test, $p<0.05$).

the peak NDVI value, and again hypothesized that a larger difference would suggest GKR activity. These two measurements (minimum NDVI and NDVI slope) were used as proxies for GKR presence in an area, and in effect represent a null model of GKR distribution: if current GKR distribution could be predicted solely from the prior year's presence, plant biomass would not be considered a factor limiting the realized GKR distribution.

Although individual GKR burrows (~27–36 m^2) represent a small fraction of a single MODIS pixel (250 m×250 m), the heterogeneity of the landscape supports analyses at this scale. The density of the GKR burrows, and the strong difference in signal between the perturbed bare soil on burrow and dried grass off burrow, suggest that a mixed pixel with GKR activity has a significantly different signal than one without GKR activity.

GKR distribution models were ranked using Akaike Information Criteria (AIC) [53]. Models were created for all of the presence points, with year included as a fixed effect. For each year of the model, we used 500 random points from the potential distribution, 250 within GKR realized distribution and 250 outside active areas.

The accuracy of the best model (as identified using AIC) was assessed with the PresenceAbsence package in R [54]. For each model, we calculated a threshold to test predicted presence and absence points for each model. Each threshold was set to the observed prevalence [48]. We then calculated the percent correctly classified (PCC), Cohen's kappa, sensitivity, specificity, and the true skill statistic (TSS, sensitivity + specificity -1), a prevalence-independent measure of accuracy [55]. Testing data was obtained in two ways: for all four years, we randomly selected 500 new points from the aerial surveys in each year, in the same manner as the training data. We also used the set of 105 GKR trapping points in Carrizo collected in 2010 and 2011 to test the models in those years.

Results

Potential Distribution Modeling

As expected, GKR potential distribution is limited to a narrow band of habitat on the western San Joaquin Valley and nearby Coast Ranges (Fig. 1). The most important variables in predicting GKR distribution included precipitation of the driest quarter, precipitation of the driest month, and minimum temperature of the coldest month (Table 2). Surprisingly, annual precipitation was not an important predictor of GKR distribution. Instead, precipitation in the driest month and driest quarter were more important predictors. Probability of GKR presence was highest in areas where the driest month received a mean of 0 mm precipitation. Similarly, probability of GKR presence was highest in areas where the driest quarter received a mean of 4 mm precipitation. GKRs inhabited areas that have a narrow band of mean annual temperatures between 14° and 16°C.

The area classified in the Maxent model as the potential distribution of GKR closely resembled the combined distribution from 2001, 2006, 2010 and 2011 (Fig. 3). However, there were portions of Carrizo in the northwest and southeast classified as suitable that were not part of the realized distribution in any of the years monitored. AUC for the potential distribution model was 0.98.

Realized Distribution Modeling

In the best model of GKR realized distribution, population extent was positively related to primary productivity in both the previous and current year, suggesting a strong influence of bottom-up regulation on GKR distribution (Table 3). GKR presence in the previous year also was an important predictor of GKR presence in the current year. Both proxies of prior GKR presence performed as expected: GKR distribution was negatively correlated to both minimum NDVI and the slope from the peak NDVI from the previous year. Realized distribution model accuracy from the testing data was "useful" (AUC = 0.74). The threshold was set at 0.50 (as expected, due to the prevalence of the model data [48]). Using the aerial surveys as testing data, model sensitivity was 0.70 and specificity 0.71. The model Kappa score and the true skill statistic (TSS) were 0.40. The best model correctly classified 70.1% of all test points as inside or outside the GKR's estimated realized distribution. Using the trapping data (obtained independently of the training data), sensitivity = 0.65; specificity = 0.66; Kappa = 0.29 and TSS = 0.29, while 65.8% of all test points were correctly classified.

Discussion

This study joins a growing body of literature that attempts to use ecological theory on limits to population extent and species ranges to inform, interpret and advance species distribution models (e.g. [1,16,56,57]). Specifically, we presented a technique of multi-step modeling to define a species' potential and realized distribution, and in doing so explored the relationship between primary productivity and animal distribution.

Consistent with theory on potential and realized distributions [4,7], our results showed that the potential distribution of GKR was larger than any of the distributions observed in the four years of aerial surveys. In other words, there were areas within Carrizo that should have been suitable for GKR, but monitoring documented them as uninhabited. This result supports conclusions of Guisan and Thuiller [7], and Grinnell [17] and Hutchinson [9] before them, who suggested that species' distributions are limited by more than fixed environmental conditions, a fundamentally important concept for distribution modelers and ecologists.

The fact that distribution models built only on static bioclimatic factors may poorly estimate realized distributions has several important implications for how these models are applied to questions in biodiversity conservation. For example, distribution models are often relied upon to project the impact of climate change on species' distributions. Without incorporating mechanisms that limit the focal species' realized distribution, these models are likely over-estimating the range of conditions within which the species will survive and reproduce under different climate change scenarios [58,59]. At the same time, we may be ignoring important local ecological processes by examining patterns of distribution at range-wide spatial scales [60]. In this study, GKR potential distribution was limited to areas with little to no rain in the driest months of the year; therefore, future increases in precipitation might be expected to reduce GKR distribution. However, we found that within the area of potential distribution, GKR were positively correlated with primary productivity. It is possible, then, that an increase in mean annual precipitation would decrease the potential distribution range-wide (where agriculture has already rendered suitable habitat uninhabitable), but increase the realized distribution within a core conservation area.

Within the potential distribution, we found that the best model of factors limiting the realized distribution of GKR showed a clear, positive correlation between primary productivity (measured as peak NDVI) and the presence of GKR in the previous year. This result conforms to recent findings from studies of other species that show rapid changes in distribution in response to temporal and

Table 2. Variable importance for range-wide model of giant kangaroo rat distribution reported by Maxent.

Variable	Percent Contribution	Permutation Importance
Precipitation of Driest Quarter	33.5	29.8
Annual Mean Temperature	21.1	0.4
Precipitation of Driest Month	16.2	33.2
Minimum Temperature of Coldest Month	15.8	22.9
Slope	5.8	5.8
Annual Precipitation	3.9	6.8
Soil Particle Size	3.7	1

spatial variability of NDVI (*e.g.*, Mongolian gazelles (*Procapra gutturosa*; [19,20]); and African buffalo (*Syncerus caffer*; [61,62]).

A key step in our approach required a selection of threshold to convert a continuous model of habitat suitability at the range-wide scale to a binary presence-absence map of potential distribution (or "available habitat" *sensu* Johnson [1980]). An alternative approach that might prove useful to explore would utilize the continuous distribution of habitat suitability as an informative prior in modeling habitat selection at finer scales. However, it is unclear whether animals select habitat in this manner. We suggest that, for GKRs, climate, soil, and topography serve as a simple filter to defining the potential distribution. That is, for example, either a GKR can construct a burrow in a particular soil type or it can't – we do not expect GKRs to have a continuous response to resource abundance in relation to soil particle size. Nevertheless, additional research on the relationship across habitat selection at multiple scales is warranted.

This study of GKR distribution in Carrizo, while conducted at a relatively small spatial scale, focused on the temporal dynamics of species' distributions. Niche and distribution theory tend to assume a species is at equilibrium, but this study and others (e.g., [21,63]) show that for many applications, considering the temporal dynamics of a species' distribution is essential. Although the importance of non-static suitability models in grassland systems has been recognized [64], the difficulties in addressing such variability have thus far limited research in this area [65].

This study focused specifically on resource abundance as a limiting factor for GKR.

While the approach presented here combining distribution models at different scales allows new insights, it is not without its shortcomings. One particular problem is our inability to identify the "true" potential distribution. By its very nature, it may be impossible to know a species' potential distribution; in fact the potential distribution may only be a theoretical construct. We can only measure the realized distribution and estimate the potential distribution from those measurements. This issue is highlighted regularly in the invasive species modeling literature. Species that appear to have a limited distribution in their native range often show a spectacular ability to live in "unsuitable" conditions when introduced to new areas (e.g., [66,67]). In these cases, the species' realized distribution in its native range is so limited by competition, dispersal, and other ecological factors that any estimate of its potential distribution will be woefully inadequate for predicting the spread of a species. Oftentimes, ecological limits to the realized distribution may be correlated with environmental conditions, thereby preventing true knowledge of the species' limits of its potential distribution. In this case, additional steps (*e.g.* physiological tests) may be required to estimate its potential distribution.

As for GKR's competitive dominance, the relationship between precipitation limitation and competition may be impossible to untangle. The *Heteromyidae* in general appear to have evolved to claim a desert-grassland niche unfilled by other small mammals. The observed relationship between dry summer months and GKR presence may be as much related to the *lower* limit for larger rodents (*e.g.*, the California ground squirrel, *Otospermophilus beecheyi*) than an upper limit for GKR. Again, this illustrates the conceptual difficulty surrounding niche theory, but the temporal mechanisms outlined in this study ought to remain relevant. GKR display differing responses to precipitation at range-wide and local

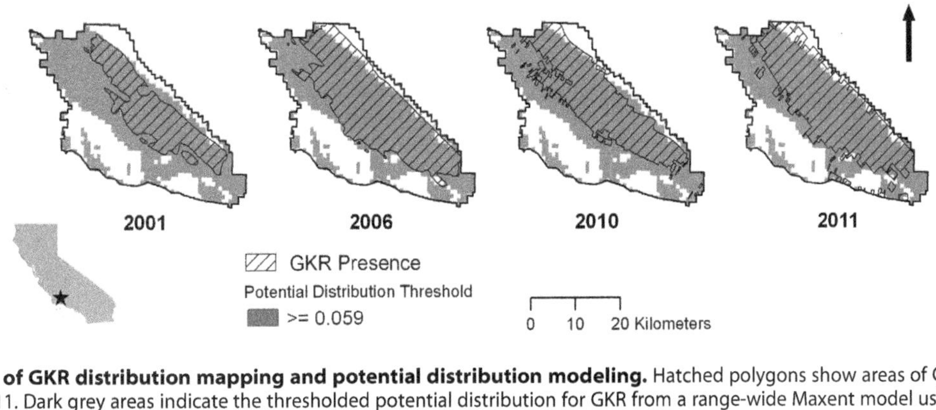

Figure 3. Results of GKR distribution mapping and potential distribution modeling. Hatched polygons show areas of GKR activity in 2001, 2006, 2010 and 2011. Dark grey areas indicate the thresholded potential distribution for GKR from a range-wide Maxent model using presence points from confirmed GKR trapping locations and museum records. A fixed model is unsuitable for predicting annual changes in population extent.

Table 3. Logistic regression of GKR presence in relation to NDVI.

Model	AIC	ΔAIC	w_i
−1.26+6.11*MaxNDVI$_{T1}$+4.15*MaxNDVI$_{T0}$ − 23.79*MinNDVI$_{T0}$ − 3.19*SlopeNDVI$_{T0}$ −0.25*year2006 − 0.55*year2010+0.44*year2011	**2,425.8**	**0**	**1**
MaxNDVI$_{T1}$+MaxNDVI$_{T0}$+MinNDVI$_{T0}$+year	2,439.3	13.50	0
MaxNDVI$_{T1}$+MinNDVI$_{T0}$+SlopeNDVI$_{T0}$+year	2,452.7	26.91	0
MaxNDVI$_{T1}$+MinNDVI$_{T0}$+year	2,461.5	35.72	0
MaxNDVI$_{T0}$+MinNDVI$_{T0}$+SlopeNDVI$_{T0}$+year	2,502.9	77.05	0
MaxNDVI$_{T0}$+MinNDVI$_{T0}$+year	2,514.7	88.94	0
MaxNDVI$_{T1}$+year	2,623.9	198.15	0
MaxNDVI$_{T1}$+SlopeNDVI$_{T0}$+year	2,624.8	199.03	0
MaxNDVI$_{T1}$+MaxNDVI$_{T0}$+year	2,625.2	199.35	0
MaxNDVI$_{T1}$+MaxNDVI$_{T0}$+SlopeNDVI$_{T0}$+year	2,626.1	200.34	0
MaxNDVI$_{T0}$+year	2,702.4	276.56	0
MaxNDVI$_{T0}$+SlopeNDVI$_{T0}$+year	2,703.6	277.82	0

The full model performed the strongest. In this model, GKR presence is positively correlated with peak primary productivity in the current year and previous year, and negatively correlated with minimum NDVI and NDVI slope in the previous year. This suggests that the best predictor of GKR presence in a given year is a positive correlation with resource abundance over two years, and presence in the area the previous year.

scales. This fact is a crucial finding for those interested in modeling ecologically relevant species' distributions.

Incorporating detailed mechanisms into species distribution models, at ecologically relevant scales and informed by ecological theory is an important next step in the field of spatial ecology. We have presented an approach to estimating a species' potential distribution and address questions about the ecological limits to its realized distribution. We presented further evidence that non-equilibrium populations are often limited not just by fixed, environmental conditions, but also other ecological conditions that vary spatially and temporally. Such research will be important as distribution modeling moves from the "how" to the "why."

Acknowledgments

We thank J. Hurl, K. Sharum and L. Saslaw for logistical field support; S. Beissinger, C. Burton, K. Fiorella, C. Golden, C. Gurney, M. Kelly, W. Lidicker, L. Prugh, E. Rubidge, and S. Sawyer for comments on multiple drafts. We are indebted to the thoughtful comments of M. Basille and three anonymous reviewers on earlier versions of this manuscript.

Author Contributions

Conceived and designed the experiments: WB JB RS HSB. Performed the experiments: WB RS. Analyzed the data: WB. Wrote the paper: WB JB RS HSB.

References

1. Anderson BJ, Arroyo BE, Collingham YC (2009) Using distribution models to test alternative hypotheses about a species' environmental limits and recovery prospects. Biological Conservation 142: 488–499. doi:10.1016/j.biocon.2008.10.036.
2. Kremen C, Cameron A, Moilanen A, Phillips SJ (2008) Aligning conservation priorities across taxa in Madagascar with high-resolution planning tools. Science. doi:10.1126/science.1155193.
3. Franklin J (2013) Species distribution models in conservation biogeography: developments and challenges. Diversity and Distributions 19: 1217–1223. doi:10.1111/ddi.12125.
4. Guisan A, Zimmermann NE (2000) Predictive habitat distribution models in ecology. Ecological Modelling 135: 147–186.
5. Peterson AT, Soberon J, Pearson RG, Anderson RP, Martinez-Meyer E, et al. (2011) Ecological Niches and Geographic Distributions (MPB-49). Princeton University Press. 1 pp. doi:10.2307/j.ctt7stnh.
6. Hanski I (1999) Metapopulation ecology. Oxford; New York: Oxford University Press
7. Guisan A, Thuiller W (2005) Predicting species distribution: offering more than simple habitat models. Ecology Letters 8: 993–1009. doi:10.1111/j.1461-0248.2005.00792.x.
8. Soberon J, Peterson T (2005) Interpretation of Models of Fundamental Ecological Niches and Species' Distributional Areas. Biodiversity Informatics. Available: https://journals.ku.edu/index.php/jbi/article/view/4/2.
9. Hutchinson GE (1957) Concluding remarks.: Cold Spring Harbor Symposia on Quantitative Biology. 13 pp.
10. Johnson DH (1980) The comparison of usage and availability measurements for evaluating resource preference. Ecology 61: 65–71.
11. Wiens JA, Rotenberry JT, Van Horne B (1987) Habitat occupancy patterns of North American shrubsteppe birds: the effects of spatial scale. Oikos 48: 132–147.
12. Thomas DL, Taylor EJ (2006) Study designs and tests for comparing resource use and availability II. Journal of Wildlife Management. doi:10.2193/0022-541X(2006)70[324:SDATFC]2.0.CO;2.
13. Boyce M, McDonald L (1999) Relating populations to habitats using resource selection functions. Trends in Ecology & Evolution 14: 268–272.
14. Garshelis DL (2000) Delusions in habitat evaluation: measuring use, selection, and importance. In: Fuller TK, Boitani L, editors. Research Techniques in Animal Ecology. New York, NY. pp. 111–157.
15. Pulliam HR (2000) On the relationship between niche and distribution. Ecology Letters 3: 349–361. doi:10.1046/j.1461-0248.2000.00143.x.
16. Pearson RG, Dawson TP, Liu C (2004) Modelling species distributions in Britain: a hierarchical integration of climate and land-cover data. Ecography 27: 285–298. doi:10.1111/j.0906-7590.2004.03740.x.
17. Grinnell J (1917) Field Tests of Theories Concerning Distributional Control. The American Naturalist 51: 115–128. doi:10.2307/2456106.
18. Pettorelli N, Vik JO, Mysterud A, Gaillard J-M, Tucker CJ, et al. (2005) Using the satellite-derived NDVI to assess ecological responses to environmental change. Trends in Ecology & Evolution 20: 503 510. doi:10.1016/j.tree.2005.05.011.
19. Ito TY, Miura N, Lhagvasuren B, Enkhbileg D, Takatsuki S, et al. (2006) Satellite tracking of Mongolian gazelles (Procapra gutturosa) and habitat shifts in their seasonal ranges. Journal of Zoology 269: 291–298. doi:10.1111/j.1469-7998.2006.00077.x.
20. Mueller T, Olson KA, Fuller TK (2008) In search of forage: predicting dynamic habitats of Mongolian gazelles using satellite-based estimates of vegetation productivity. Journal of Applied Ecology 45: 649–658.
21. Bissonette JA, Storch I (2007) Temporal dimensions of landscape ecology: wildlife responses to variable resources. New York, NY: Springer.
22. Basille M, Fortin D, Dussault C, Ouellet JP (2013) Ecologically based definition of seasons clarifies predator–prey interactions. Ecography 36: 220–229.

23. Falcucci A, Ciucci P, Maiorano L (2009) Assessing habitat quality for conservation using an integrated occurrence-mortality model. Journal of Applied Ecology 46: 600–609.
24. Pettorelli N (2013) The Normalized Difference Vegetation Index. Oxford University Press. 1 pp.
25. Kawamura K, Akiyama T, Yokota H, Tsutsumi M, Yasuda T, et al. (2005) Monitoring of forage conditions with MODIS imagery in the Xilingol steppe, Inner Mongolia. International Journal of Remote Sensing 26: 1423–1436. doi:10.1080/01431160512331326783.
26. Andreo V, Lima MP, Provensal C (2009) … Polop, J.(2009) Population dynamics of two rodent species in agroecosystems of central Argentina: intra-specific competition, land-use, and climate effects. 10 pp.
27. Hamel S, Garel M, Bianchet MF (2009) Spring Normalized Difference Vegetation Index (NDVI) predicts annual variation in timing of peak faecal crude protein in mountain ungulates. Journal of Applied Ecology 46: 582–589.
28. Cao L, Cova TJ, Dennison PE (2011) Using MODIS satellite imagery to predict hantavirus risk. Global Ecology and Biogeography 20: 620–629.
29. Pettorelli N, Ryan S, Mueller T, Bunnefeld N, Jędrzejewska B, et al. (2011) The Normalized Difference Vegetation Index (NDVI): unforeseen successes in animal ecology. Climate Research 46: 15–27. doi:10.3354/cr00936.
30. Williams DF (1992) Geographic distribution and population status of the giant kangaroo rat, Dipodomys ingens (Rodentia, Heteromyidae). In: Williams DF, Byrne S, Rado TA, editors. Endangered and sensitive species of the San Joaquin Valley, California their biology, management, and conservation. Sacramento: Endangered and sensitive species of the San Joaquin Valley. pp. 301–327. doi:10.1111/j.1365-2656.2011.01930.x/full.
31. Grinnell J (1932) Habitat Relations of the Giant Kangaroo Rat. Journal of Mammalogy 13: 305–320. doi:10.2307/1374134.
32. Williams DF, Kilburn KS (1991) Dipodomys ingens. Mammalian Species: 1–7.
33. Phillips SJ, Anderson RP, Schapire RE (2006) Maximum entropy modeling of species geographic distributions. Ecological Modelling 190: 231–259.
34. Hawbecker AC (1944) The giant kangaroo rat and sheep forage. The Journal of Wildlife Management 8: 161–165.
35. Goldingay RL, Kelly PA, Williams DF (1997) The Kangaroo Rats of California: Endemism and Conservation of Keystone Species. 3: 47.
36. Prugh LR, Brashares JS (2012) Partitioning the effects of an ecosystem engineer: kangaroo rats control community structure via multiple pathways. Journal of Animal Ecology 81: 667–678. doi:10.2307/41496035.
37. Shaw WT (1934) The ability of the giant kangaroo rat as a harvester and storer of seeds. Journal of Mammalogy 15: 275–286. doi:10.2307/1374514.
38. Bean WT, Stafford R, Prugh LR, Scott Butterfield H, Brashares JS (2012) An evaluation of monitoring methods for the endangered giant kangaroo rat. Wildlife Society Bulletin 36: 587–593. doi:10.1002/wsb.171.
39. Germano DJ, Rathbun GB (2012) Effects of grazing and invasive grasses on desert vertebrates in California. The Journal of Wildlife Management 76: 670–682. doi:10.2307/41519406.
40. Hijmans RJ, Phillips SJ, Leathwick JR, Elith J (n.d.) Package "dismo." Available: http://cran.r-project.org/web/packages/dismo/index.html.
41. R Core Team (2012) R A language and environment for statistical computing. Available: http://www.R-project.org/.
42. Bean WT, Prugh LR, Stafford R, Butterfield HS, Westphal M, et al. (2014) Species distribution models of an endangered rodent offer conflicting measures of habitat quality at multiple scales. Journal of Applied Ecology 51: 1116–1125. doi:10.1111/1365-2664.12281.
43. Hijmans RJ, Cameron SE, Parra JL, Jones PG, Jarvis A (2005) Very high resolution interpolated climate surfaces for global land areas. International Journal of Climatology 25: 1965–1978.
44. Hijmans RJ, Graham CH (2006) The ability of climate envelope models to predict the effect of climate change on species distributions. Global Change Biology 12: 2272–2281.
45. United States Geological Survey (n.d.) Shuttle Radar Topography Mission.
46. Soil Survey Staff, Natural Resources Conservation Service, United States Department of Agriculture (2012) Soil Survey Geographic (SSURGO) Database. Available: http://sdmdataaccess.nrcs.usda.gov/.
47. Bean WT, Stafford R, Brashares JS (2012) The effects of small sample size and sample bias on threshold selection and accuracy assessment of species distribution models. Ecography. doi:10.1111/j.1600-0587.2011.06545.x.
48. Liu C, Berry PM, Dawson TP, Pearson RG (2005) Selecting thresholds of occurrence in the prediction of species distributions. Ecography 28: 385–393.
49. Liu C, White M, Newell G (2013) Selecting thresholds for the prediction of species occurrence with presence-only data. Journal of Biogeography 40: 778–789.
50. Carroll ML, M DC, Sohlberg RA, Townshend JR (n.d.) 250m MODIS Normalized Difference Vegetation Index.
51. Huete A, Didan K, Miura T, Rodriguez EP, Gao X (2002) Overview of the radiometric and biophysical performance of the MODIS vegetation indices. Remote Sensing of Environment 83: 195–213.
52. Butterfield HS, Malmström CM (2009) The effects of phenology on indirect measures of aboveground biomass in annual grasses. International Journal of Remote Sensing 30: 3133–3146.
53. Burnham KP, Anderson DR (2002) Model selection and multimodel inference: a practical information-theoretic approach. New York: Springer-Verlag.
54. Freeman EA, Moisen G (2008) PresenceAbsence: An R package for presence absence analysis. Journal of Statistical Software 23.
55. Allouche O, Tsoar A, Kadmon R (2006) Assessing the accuracy of species distribution models: prevalence, kappa and the true skill statistic (TSS) - ALLOUCHE - 2006 - Journal of Applied Ecology - Wiley Online Library. Journal of Applied Ecology 43: 1223–1232.
56. Rushton SP, Ormerod SJ, Kerby G (2004) New paradigms for modelling species distributions? - Rushton - 2004 - Journal of Applied Ecology - Wiley Online Library. Journal of Applied Ecology 41: 193–200.
57. Austin M (2007) Species distribution models and ecological theory: a critical assessment and some possible new approaches. Ecological Modelling 200: 1–19.
58. Pearson RG, Dawson TP (2003) Predicting the impacts of climate change on the distribution of species: are bioclimate envelope models useful? Global Ecology and Biogeography 12: 361–371.
59. Franklin J (2010) Moving beyond static species distribution models in support of conservation biogeography. Diversity and Distributions 16: 321–330. doi:10.2307/40604232.
60. Morin X, Thuiller W (2009) Comparing niche- and process-based models to reduce prediction uncertainty in species range shifts under climate change. Ecology 90: 1301–1313.
61. Ryan SJ, Knechtel CU, Getz WM (2006) Range and habitat selection of African buffalo in South Africa. Journal of Wildlife Management 70: 764–776. doi:10.2193/0022-541X(2006)70%5B764:RAHSOA%5D2.0.CO;2.
62. Viña A, Bearer S, Zhang H, Ouyang Z, Liu J (2008) Evaluating MODIS data for mapping wildlife habitat distribution. Remote Sensing of Environment 112: 2160–2169.
63. Suárez-Seoane S, García de la Morena EL, Morales Prieto MB, Osborne PE, de Juana E (2008) Maximum entropy niche-based modelling of seasonal changes in little bustard (Tetrax tetrax) distribution. Ecological Modelling 219: 17–29. doi:10.1016/j.ecolmodel.2008.07.035.
64. Fryxell JM, Wilmshurst JF, Sinclair ARE (2004) Predictive Models of Movement by Serengeti Grazers. Ecology 85: 2429–2435. doi:10.1890/04-0147.
65. Fernandez Gimenez ME, Allen Diaz B (1999) Testing a non-equilibrium model of rangeland vegetation dynamics in Mongolia. Journal of Applied Ecology 36: 871–885. doi:10.1046/j.1365-2664.1999.00447.x.
66. Beaumont LJ, Gallagher RV, Thuiller W, Downey PO, Leishman MR, et al. (2009) Different Climatic Envelopes among Invasive Populations May Lead to Underestimations of Current and Future Biological Invasions. Diversity and Distributions 15: 409–420. doi:10.1111/j.1472-4642.2008.00547.x.
67. Steiner FM, Schlick-Steiner BC, VanDerWal J, Reuther KD, Christian E, et al. (2008) Combined modelling of distribution and niche in invasion biology: a case study of two invasive Tetramorium ant species. Diversity and Distributions 14: 538–545. doi:10.1111/j.1472-4642.2008.00472.x.

18

Hydrologic Landscape Regionalisation Using Deductive Classification and Random Forests

Stuart C. Brown[1]*, Rebecca E. Lester[1], Vincent L. Versace[2,3], Jonathon Fawcett[4], Laurie Laurenson[1]

1 School of Life and Environmental Sciences, Deakin University, Warrnambool, Victoria, Australia, 2 Greater Green Triangle University, Department of Rural Health, Flinders University, Warrnambool, Victoria, Australia, 3 Centre for Environmental Management, Faculty of Science and Technology, Federation University Australia, Ballarat, Victoria, Australia, 4 Sinclair-Knight-Merz, Bendigo, Victoria, Australia

Abstract

Landscape classification and hydrological regionalisation studies are being increasingly used in ecohydrology to aid in the management and research of aquatic resources. We present a methodology for classifying hydrologic landscapes based on spatial environmental variables by employing non-parametric statistics and hybrid image classification. Our approach differed from previous classifications which have required the use of an *a priori* spatial unit (e.g. a catchment) which necessarily results in the loss of variability that is known to exist within those units. The use of a simple statistical approach to identify an appropriate number of classes eliminated the need for large amounts of *post-hoc* testing with different number of groups, or the selection and justification of an arbitrary number. Using statistical clustering, we identified 23 distinct groups within our training dataset. The use of a hybrid classification employing random forests extended this statistical clustering to an area of approximately 228,000 km^2 of south-eastern Australia without the need to rely on catchments, landscape units or stream sections. This extension resulted in a highly accurate regionalisation at both 30-m and 2.5-km resolution, and a less-accurate 10-km classification that would be more appropriate for use at a continental scale. A smaller case study, of an area covering 27,000 km^2, demonstrated that the method preserved the intra- and inter-catchment variability that is known to exist in local hydrology, based on previous research. Preliminary analysis linking the regionalisation to streamflow indices is promising suggesting that the method could be used to predict streamflow behaviour in ungauged catchments. Our work therefore simplifies current classification frameworks that are becoming more popular in ecohydrology, while better retaining small-scale variability in hydrology, thus enabling future attempts to explain and visualise broad-scale hydrologic trends at the scale of catchments and continents.

Editor: Rachata Muneepeerakul, Arizona State University, United States of America

Funding: The authors received no specific funding for this work. Jonathon Fawcett is employed by Sinclair-Knight-Merz. Sinclair-Knight-Merz provided support in the form of salary for author JF, but did not have any additional role in the study design, data collection and analysis, decision to publish, or preparation of the manuscript. The specific role of this author is articulated in the 'author contributions' section.

Competing Interests: Jonathon Fawcett is employed by Sinclair-Knight-Merz. There are no patents, products in development or marketed products to declare. This does not alter the authors' adherence to all the PLOS ONE policies on sharing data and materials.

* Email: stuart.brown@research.deakin.edu.au

Introduction

Flow variability and ecological controls

Long-term trends in flow variability in streams have the ability to create and maintain ecosystem dynamics for a range of ecologically-important conditions [1,2] and can therefore influence biotic communities and abiotic conditions [1] at local to regional scales, both temporally and spatially [1,3]. These long-term trends are controlled by the same factors influencing the hydrologic cycle in a landscape and ultimately influence physical habitat and refuge availability, food distribution and abundance, and opportunities for migration, reproduction and recruitment [4]. Given this ability for hydrologic variability to control the ecological and biophysical attributes of in-stream and riparian systems, landscapes that have similar hydrologic properties should have similar biological and ecological assemblages [5]. Furthermore, if the same or similar hydrologic landscapes can exist in multiple spatial locations within bioregions, it stands to reason that the ecology of these systems should also be similar, regardless of spatial location. The ability to identify, classify, and validate spatial patterns in hydrologic landscapes is an important step in creating a solid foundation to assess the impact of natural flow variability, associated ecological conditions and management of water resources across a range of spatial scales. As such, hydrologic classification has been identified as a critical step in providing a spatially-explicit understanding of the magnitude and timing of flow regime variation within and between rivers and regions [2,6].

Landscape and hydrologic units

Landscape characteristics affecting the quality, quantity, and movement of water are extremely complex [7]. The earth is made up of a number of different landforms, geological settings and climatic conditions, and the idea of a simple, unifying conceptual hydrologic framework may seem impossible to achieve [7]. However, landscapes that appear unique and diverse often actually have a common set of attributes (e.g. governing the movement of water). Winter [7] introduced the concept of hydrologic landscape units, which suggests that the complete hydrologic system (i.e. incorporating surface runoff, groundwater flow and atmospheric water) interacts with simple physiographic features, and that these features then become the building blocks of all hydrologic landscapes. Therefore, by this rationale, the movement, storage and release of surface and subsurface water are controlled by a common set of physical principles regardless of the geographic location of the landscape [8]. Winter [7] termed these 'fundamental hydrologic landscape units' (FHLU), and defined the conceptual unit as a land surface form which includes an upland, an adjacent lowland and the valley side that separates them. The hydrologic system of an FHLU consists of: 1) the movement of surface water (controlled by the slopes and permeability of the landscape); 2) the movement, storage and release of groundwater (a function of the geologic setting); and 3) atmospheric water exchange (controlled by climate) [7]. Much peer-reviewed research supports the idea that all hydrologic landscapes can be considered to be variations and multiples of FHLUs, and that these can then be used to describe major, spatially-contiguous and discrete landscape types that should have similar hydrologic conditions (e.g. [6,8,9]). Since the concept was first introduced, further research has been conducted to delineate hydrologic landscape regions based on a number of different approaches and across a variety of scales (see Olden et al. [9] and Kennard et al. [6] for an extensive list of examples).

Deductive and inductive landscape classification

Classification is the process of systematically placing objects into classes that are similar with respect to a set of variables or characteristics. Hydrologic classification is therefore the process of systematically arranging streams, rivers or catchments into classes that are similar with respect to their flow regime [6,9]. While hydrologic classification can refer to a broad assortment of methods, a review by Olden et al. [9] recognises two broad approaches to hydrologic classification; deductive and inductive approaches (not to be confused with top-down and bottom-up logic; see below). The inductive approach uses the emergent properties of discharge time series data to generate classes. In contrast, the deductive approach to classification is used when attempting to describe broad spatial patterns in flow regime variability where there is a lack of gauged or modelled streamflow data available. Deductive methods of environmental classification are commonly used when the objective is to quantify and describe spatial variation in flow regime attributes. This approach to classification identifies groups on the basis of physical and climatic attributes that, over broad scales, produce similar hydrologic responses in stream systems [9]. The increased availability of high-quality, hydrologically-relevant spatial datasets (e.g. climate, topography, land use) makes deductive reasoning an appealing method when attempting to define spatial similarities or dissimilarities in hydrological characteristics [9]. It has been demonstrated that the deductive approach to hydrological classification can help in the prediction of streamflow metrics [10,11], and that it improves predictive streamflow models when those models are stratified by hydrologic regions [12]. However, some facets of flow regimes (e.g. low flow magnitude and duration) are difficult to accurately characterise and quantify with this approach due to limitations in data quality and conceptual knowledge of the systems, and spatial variability of hydrological processes in many regions [6,9,13].

Wolock et al. [8] used the concept of hydrologic landscapes introduced by Winter [7] to classify nearly 44,000 catchments (~200 km^2 in area each) using a combination of multivariate ordination and cluster analyses. Kennard et al. [6] presented a method combining non-hierarchical clustering of climate, topography, soils and geology, vegetation and flow data to group Australian streams at a continental scale with mixed success. Sawicz et al. [14] employed the use of precipitation-temperature-streamflow signatures and Bayesian clustering to characterise 280 non-contiguous catchments located in eastern USA so as to understand similarities in climatic and landscape attributes across the region. Their work found that signatures which vary along climatic gradients exerted a stronger influence on cluster separation than those signatures which may vary as a result of geology or land cover. It has also been shown by McManamay et al. [13] that hydrological regionalisations [15] can be severely lacking in their ability to explain variation in a number of streamflow metrics.

The approaches by Wolock et al. [8], Kennard et al. [6], Sawicz et al. [14] and others all require the use of catchments or some choice of arbitrary spatial unit (e.g. eco/bio-region) to delineate and display the results of the clustering. However, there is evidence of significant flow variability within river catchments [6,16] and significant spatial variability in climate and land use within sub-catchments that affect wetland extent [17]. The approach of delineating spatial units *a priori* leads to a loss of spatial variability, particularly as the catchment or spatial units become larger. Olden et al. [9] state that while deductive classification is common in the literature, hydrologic landscape regions and other similar concepts that are founded on physical principles have rarely been tested with this approach. The *a priori* (or 'top-down') specification of boundaries between classes has been criticised, while alternative 'bottom-up' approaches, where groups are developed as an emergent property of the data [18] (not to be confused with inductive reasoning which relies on time series hydrologic data) have been considered to be in keeping with physical ecohydrological principles [9]. Using a bottom-up approach, spatial and group clustering patterns are generated based on the analysis of a large number of units, such as pixels or micro-catchments. These units are then allocated into clusters based on their multivariate similarity [18]. However, a number of subjective choices as to which datasets to include, classification strategies and the number of groups in the classification process still need to be made. Such decisions could affect the quality and repeatability of the classification process when applied to different regions and datasets [9,14,18,19]. Despite the potential limitations, the routine availability of these datasets and the application of statistical clustering and analyses have allowed scientists to begin to link spatial patterns to ecohydrological processes.

Statistical clustering and multivariate analyses

Statistical clustering and multivariate analyses are important and powerful tools in the identification of spatial and temporal gradients. There is a multitude of variations on the theme of statistical clustering [20–22], but the most commonly used are hierarchical agglomerative methods [23] which fuse individual samples into like groups, gradually increasing the similarity within groups while lowering the similarity level between groups; i.e. each sample starts as its own group and pairs of groups are merged moving up a hierarchy. The process is considered complete when all samples are contained within a single group or cluster. Unlike hierarchical clustering, non-hierarchical clustering places samples into groups that are not related hierarchically, but differ from each other significantly in multivariate space. Described simply, non-hierarchical clustering tends to work by assigning each sample (n) into a pre-defined number of clusters (k) and then cluster membership of the samples is iteratively reassessed, usually with the criterion of maximising between-cluster variance while simultaneously minimising within-cluster variance. The most common example of non-hierarchical clustering is the *k*-means algorithm [24]. In some instances, the groups extracted by hierarchical and non-hierarchical algorithms do not differ significantly [25], but non-hierarchical methods can be much more efficient at extracting groups from large datasets [26].

Once statistical clustering has occurred, an analysis of the performance of the clustering can be conducted through the use of multivariate statistics and, in particular, ordination plots. There are a number of ordination techniques (e.g. principal components analysis (PCA), and multidimensional scaling (MDS)) that, broadly speaking, reduce multidimensional space so that objects can be compared graphically in two or three dimensions, without a significant loss of explanatory information. It is also possible to use the clustering information to train predictive models to classify samples not included in the original classification. This is where supervised classification algorithms, such as random forests (RF) [27,28], coupled with geographic information systems (GIS) and image processing software can be applied to extend the applicability of deductive landscape classification approaches to regionalisation studies.

Supervised classification of landscapes

One of the most common applications of remotely-sensed images and data is the creation of maps of vegetation type, soil properties or other discrete classes. In supervised classification, the location of known classes on those maps (i.e. training sites) is used by the software to determine the spectral signature of the pixels belonging to each of those classes. Each pixel in the image (i.e. outside the training sites) is then assigned, based on its spectral signature, to the class it most closely matches. Supervised classification can be applied at the individual pixel level or to groups of adjacent, similar pixels for the creation of contiguous regions. However, for the classification to work effectively, *a priori* knowledge of where the classes of interest (e.g. land cover types) are located is required. When supervised classification is combined with, for example, an unsupervised statistical classification, the process is referred to as hybrid (or semi-supervised) classification [29]. A major benefit of hybrid classifications for landscape regionalisation is that they permit the bottom-up approach to deductive classification as recommended by Mackey et al. [18] and Olden et al. [9]. This eliminates the need for survey approaches to develop *a priori* knowledge of the location of classes of interest which require expert opinion and substantial amounts of qualitative evidence which is not always available or suitable. The hybrid approach also eliminates the need to define a spatial unit a priori (e.g. a catchment) and allows small-scale (e.g. intra-catchment) variability to be identified and preserved where it may otherwise be lost.

Aim of study

The aim of this study was to create a hydrologic landscape regionalisation using deductive reasoning and a bottom-up approach to statistical clustering combined with a hybrid classification. The regionalisation was then assessed based on its ability to discriminate between groups (regions) based on a number of streamflow indices. In this research, we used unsupervised classification (i.e. the statistical clustering) to first determine class membership based on multivariate space and then used supervised image classification to classify the remaining pixels from a number of ecohydrologically-important layers into the classes of interest as defined by the statistical clustering. This approach will permit the regionalisation of spatially non-contiguous regions, while maintaining small-scale intra-catchment variability that would be lost using catchments as the unit of classification as has often been done in the past. The assessment of the ability of the regions to differentiate among streams based on a number of flow indices provides insight into the utility of the method in predicting streamflow characteristics in ungauged catchments.

Materials and Methods

To clarify the process used in the creation of the hydrological regionalisation and the validation and training methods for the RF models, a graphical overview of the methods is presented in Figure 1.

Site description

Victoria is the southernmost state of mainland Australia, comprising an area of 227,594 km^2, and bordered by the southern bank of the Murray River to the north, South Australia to the west and separated from Tasmania by Bass Strait to the south. Topographically, geologically, and climatically, Victoria is diverse, varying from wet temperate climates in the southeast to alpine areas rising to ~2000 m altitude in the northeast (Figure 2). To the west and northwest are extensive, flat areas of semi-arid plains, while most of the rest of the state experiences a Mediterranean climate consisting of hot, dry summer and cool, wet winters [30]. Median annual rainfall in Victoria exceeds 2,500 mm in some parts of the mountainous northeast but is less than 300 mm in a large part of the west and northwest [31]. Generally, snowfall is only observed in the mountains and hills to the east and centre of the state. Victoria has an extensive wetland system, with nearly 17,000 wetlands larger than 0.01 km^2 in surface area [32], and a large river network, with the largest being the Murray River system.

Site description of case-study area in western Victoria

The Glenelg-Hopkins region of western Victoria covers approximately 27,000 km^2 and the regional cities of Warrnambool, Ararat, Hamilton, Portland and the western fringes of Ballarat are within its boundary. The region contains the Grampians Ranges in the north but is generally a low-lying series of catchments across three major catchments – Glenelg, Hopkins and Portland. The region has been previously studied with respect to land-use and land-cover changes [33] and the associated impacts on nutrient exports, in-stream salinity and dryland salinity [34,35], while recent work has examined the spatio-temporal variability between land cover, climate and wetland extent [17], the impact of land-cover changes on groundwater levels [36] and empirically modelled streamflow response to land-use change [37].

Variable selection and Processing

The first phase of the classification involved selecting suitable variables upon which to base our classification. Steps associated with variable selection and processing, described in this section, are outlined in green in Figure 1. Based on the concept of FHLUs, 25 variables were chosen that could explain the storage, movement, and quality of surface water, groundwater, and atmospheric water. A full list and brief description of each of the variables are presented in Table S1. All raster calculations and raster analysis for the processing of variables was conducted in ArcGIS 10.1 [38].

The raster datasets employed in the study covered a wide range of resolutions (30 m–10 km). Typically, with GIS, analyses are only considered to be suitable if all rasters are resampled to the coarsest resolution. However, this can result in the loss of a

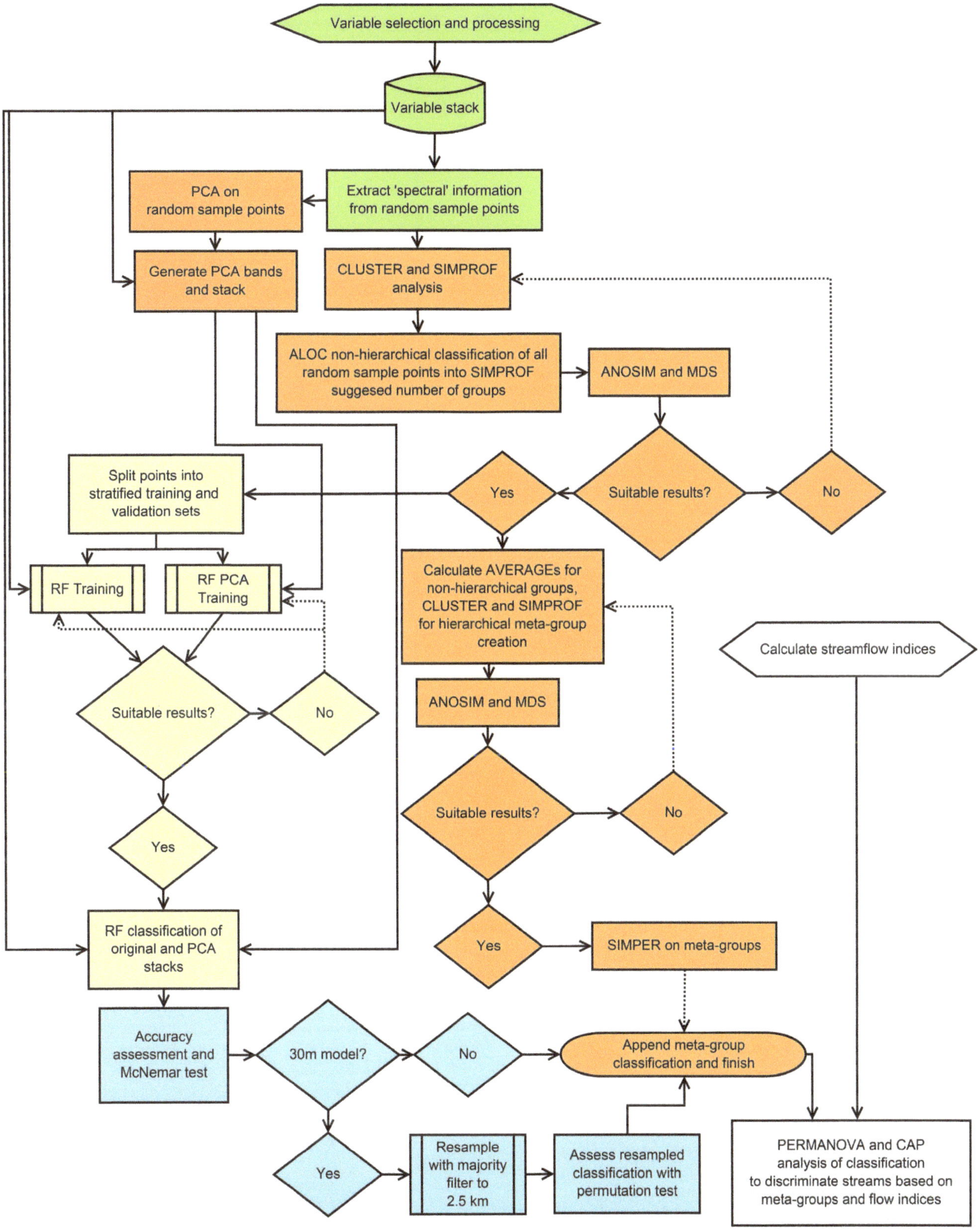

Figure 1. Graphical overview of the methods employed in the creation of the hydrological regionalisation; the training and validation of the RF models used to extend the statistical clustering to the state of Victoria; and of the validation of the classification with hydrological data. Each colour coded section of the figure corresponds to a section in the methods. Green: Variable Selection and Processing; Orange: Development of Classification Groups; Beige: Hybrid Classification with Random Forests; Light Blue: Accuracy Assessment; Grey: Relationship between the Regionalisation and Hydrologic Indices. The process for the ALOC 20 100% models did not involve splitting the ALOC classified random sample points into training and validation subsets and model accuracy was only assessed with OOB accuracy from EnMap-Box.

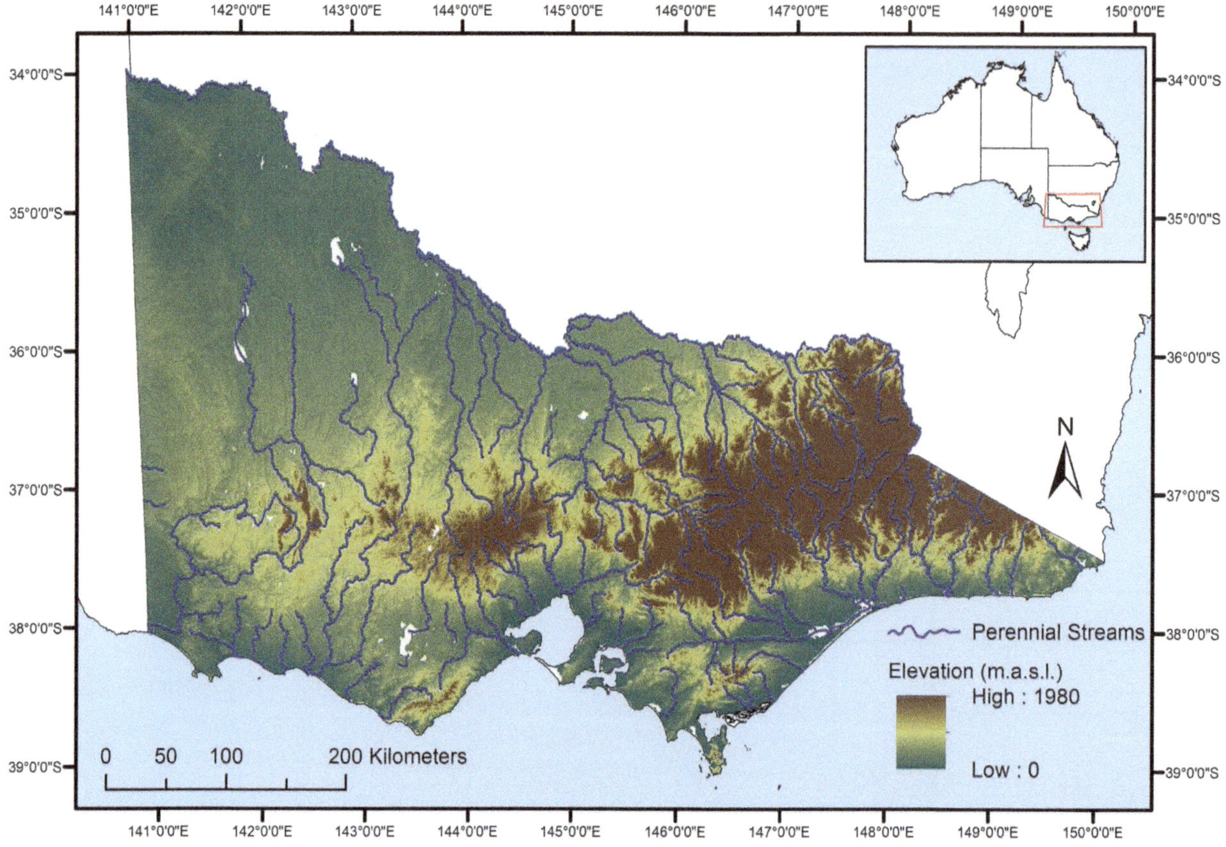

Figure 2. Location of the study area in south-eastern Australia. Dark blue lines represent perennial rivers, while the colour gradient represents elevations, with darker browns indicating higher elevations.

substantial amount of detail and information and can affect the ability of supervised classification methods to successfully classify pixels (See Figure 3 for a comparison between the 30-m and 10-km Landscape Development Index (LDI)). Therefore, in this study, two approaches were used to standardise the scale of our raster data. The first approach was to resample all datasets to the finest resolution (30 m); and the second involved re-sampling all the raster datasets to the coarsest resolution found in our datasets (10 km). All rasters were continuous in their spatial coverage with the exception of the soil hydrological properties (KSAT, PAWC and soil horizon thickness) which had significant gaps where large lakes and wetlands were found. There was also a significant gap in coverage on the eastern headland of Port Phillip Bay. To ensure that all datasets aligned correctly and had the same degree of spatial continuity, the digital elevation model (DEM) was used as a snap raster for the resampling. Once the resampling had been completed using a nearest neighbour algorithm, the now 30-m soil properties were used as a mask to extract all other raster values. The result of this was that all of the datasets used in the analysis had a 30-m spatial resolution and all had corresponding areas of missing data that would be excluded from any analysis.

For the second approach, all of the original datasets were resampled to 10 km using a nearest neighbour algorithm and the mean annual evapotranspiration raster as the snap raster. Two different datasets were used as snap rasters so that the pixels of the resampled rasters (at either 30 m or 10 km) would be aligned correctly at the respective resolution. If this study had been conducted at a continental scale, then a coarser resolution would be more suitable, however as it was conducted on a relatively small scale, we considered that resampling to a finer resolution was both suitable and justifiable. Furthermore, limiting our analysis to a coarser resolution (on the basis of a single coarse-scaled dataset; mean annual evapotranspiration), would have significantly affected the applicability and usefulness of the method presented here, specifically, the ability of the RF model to accurately recover and reproduce allocated class information. Previous research has shown that the accuracy of supervised classifications, at both an overall and per class level, can be affected by the spatial resolution of the input images [39,40]. As such, we assessed this issue through the accuracy of the RF model and the ability of the hybrid classifications to accurately recover the class information using the coarser dataset (see section on accuracy assessment). A layer stack of both sets of variables (30-m and 10-km resolution) was produced in ENVI 4.8 [41] for later use with the RF model.

Two random distributions of sampling points ($n = 10{,}000$ and $n = 410$; with minimum distances between points of 30 m and 10 km respectively) were then created. Ripley's K function [42], which determines whether features are significantly clustered or dispersed over a range of distances, was then used to assess whether both sets of points were distributed across our sampling area. Using the random-sampling points, raster values were extracted from each of the raster layers for later use in clustering, and then the training and validation of the RF model.

Development of classification groups

The selected variables were then statistically analysed to develop the groups (known as 'regions') to be used as the basis for

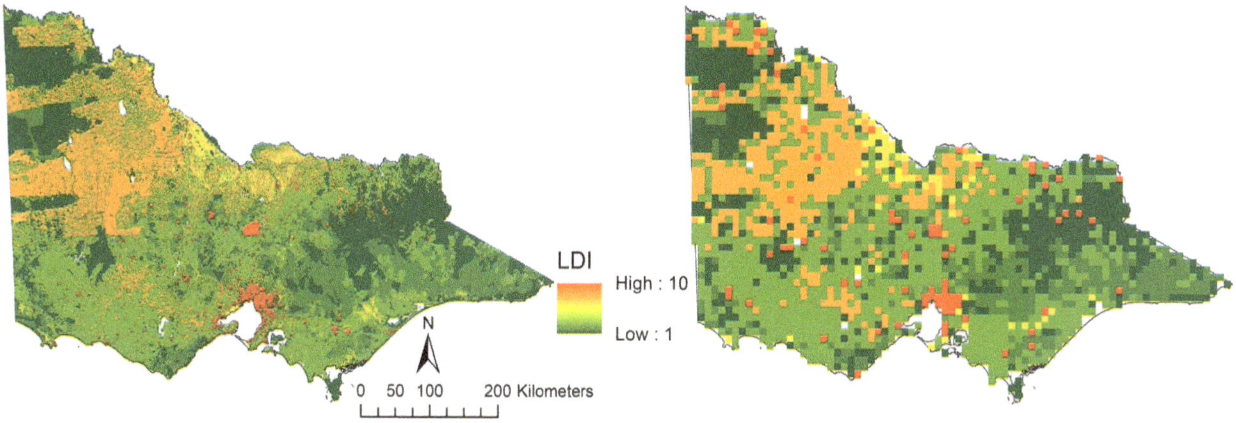

Figure 3. An example of the differences in resolution identified when working in relatively small study areas. The left hand image is the Landscape Development Intensity index (LDI) at a 30-m resolution while the image on the right is the LDI at a 10-km resolution. The accuracy of supervised classifications can be affected by the spatial resolution of the input images and as such we developed models at both resolutions.

classification. The steps involved are outlined in orange in Figure 1. All sampling points that were found to contain missing data were removed prior to analysis, resulting in $n = 9{,}958$ (30-m data) and $n = 406$ (10-km data) for the two sets of data points. For the 30-m data, three 1000-point subsamples were taken for initial statistical analysis, while all data points were used for the 10-km dataset.

In order to develop a bottom-up approach to classification, where groups are an emergent factor of the data rather than defined *a priori*, we cluster-analysed each of the subsamples individually. Due to the large ranges and different scales used across our variable set, Gower similarity matrices were constructed for each of the initial analysis datasets using PRIMER 6 [43]. The CLUSTER [44] function was then used with a SIMPROF [45] test, to identify the number of statistically-significant ($\alpha = 0.05$) groups within the datasets. Essentially, SIMPROF determines the number of significant groups with the assumption of no *a priori* groups by calculating similarities between every pair of samples using the chosen resemblance matrix and a hierarchical cluster dendrogram. Beginning at the top of an already-defined hierarchy (i.e. by the CLUSTER function), progression down the divisions or branches of the dendrogram is only permitted if the current set of samples is deemed to still have statistically significant dissimilarity. Upon encountering a non-significant result (i.e. the samples are similar), no further tests are performed down that branch of the dendrogram and all samples below are considered part of the same group [45]. A limitation of the SIMPROF test is that groups identified by the test may be at too fine a level of detail for practical purposes. However, if the resulting clusters are super-sets of the SIMPROF-defined groups, it is appropriate to define coarser groupings based on an arbitrary slice at some chosen level of similarity [45]. As the 30-m SIMPROF tests were conducted on three 1000-point subsamples, the subsample that produced the largest number of groups was used to determine the number of groups for the 30-m data. Even though SIMPROF uses a hierarchical relationship between sampling points to determine the number of clusters present in the data, we believe this approach is suitable for estimating an appropriate number of non-hierarchical groups as opposed to choosing an arbitrary *k* number of groups.

We then classified the full datasets ($n = 9{,}958$ and $n = 406$) into the number of classes suggested by the SIMPROF tests using the non-hierarchical clustering algorithm ALOC [26] and the Gower metric in PATN v3.1.2 [46]. Group allocations were then exported from PATN and joined to the original datasets as factors for further analysis. Once group membership information was in PRIMER, the ANalysis Of SIMilarities (ANOSIM; [44]) routine was used to test for statistically-significant differences among sample groups. Based on the *R* statistic, which is scaled to be between -1 and $+1$, global *R* values >0 indicate greater dissimilarity between groups than within groups [23]. Group averages were calculated using the AVERAGE [43] tool in PRIMER to visually analyse group separation using MDS. MDS is useful in providing a visual representation of the pattern of similarities between objects or groups while reducing the multidimensional space to be more readily interpretable (i.e. reducing data to two or three dimensions). The ability of MDS to reduce the degree of multidimensional space is measured with a stress value. Essentially, stress is the mismatch between distances between all samples in the plot in multidimensional space and the calculated estimate of their respective locations in two or three dimensions, with lower values indicating better representation [23]. The CLUSTER and SIMPROF routines were then used to hierarchically cluster the ALOC generated group averages into 'meta-groups'. By definition, non-hierarchical groups are not linked based on their hierarchical multivariate relationship to each other, but rather are defined by their multivariate dissimilarity. As such, group *x* may not be closely related to group *y* but could be more closely related to group *z*. By hierarchically clustering our ALOC generated groups, were we able to determine which ALOC groups were more closely related to each other based on their multivariate means. A standardised Euclidean distance similarity matrix was then created and the SIMilarity PERcentage (SIMPER; [44]) routine was then used to analyse variable contribution to each of the meta-groups and to examine between meta-group similarity, while the Kruskal-Wallis [47] statistic was used to assess the ability of each of the variables to differentiate between clusters.

Previous studies (e.g. Wolock et al. [8]) have employed PCA [44] to reduce dimensionality and reduce multi-collinearity among variables. Our method relied firstly on using 'raw' data (i.e. the data was not transformed in any way) to extract spectral information for the statistical clustering and then classification. The results of this method were then compared against a classification based on PCA-transformed data. PCA is a procedure where possibly correlated variables are orthogonally transformed into a new set of linear, uncorrelated variables known as principal components [23]. The transformation results in the first component explaining most of the data variance (i.e. the first component explains as much of the multivariate data as possible) while each additional component in turn is then created to explain the remaining variance. The number of components is less than or equal to the number of original variables and each additional component is created under the condition that it is uncorrelated with all of the preceding components [23]. Here, a PCA was conducted in PRIMER on the same standardised Euclidean distance matrix used for the SIMPER analysis. The eigenvectors, for the first five principal components (eigenvalues ≥ 1) were used in ArcMap to generate PCA bands and new PCA raster stacks were created in ENVI.

Hybrid classification with Random Forests

The regions that were developed in the previous step were then used to classify the raster stack of the variables for the entire study area. Steps in this section are outlined in beige in Figure 1. RF is an ensemble machine-learning method used in classification and regression [27]. The RF method is relatively unknown in land remote sensing and has not been thoroughly evaluated by the remote sensing community, although it has been shown to be more accurate than single decision-tree classifiers [48]. RF requires two parameters for generating a predictive model: the number of trees (k) and the number of variables used for growing the trees (m). Therefore, a dataset can be classified by defining a constant number of m variables, while each of the training samples is classified by k trees. Classification is determined by using the mode of the classes output by individual trees for each training site (x), $\hat{C}^{B}_{rf} = majority\ vote \left\{ \hat{C}_b(x) \right\}_1^B$, where $\hat{C}_b(x)$ is the class prediction of the bth RF tree from a possible B classes [48]. RF increases the diversity of the constituent trees by making them grow from different training data through bootstrap aggregation which involves random re-sampling (without deletion) of the original training dataset [27]. Therefore some data may be used more than once in the training of the model, while some may not be used at all [48]. Being an ensemble method, multiple models (trees) are used allowing the algorithm to obtain better predictive performance than that which would be obtained by using any of the constituent models individually. RF is becoming increasingly popular in data mining, remote sensing and landscape ecology as it is non-parametric, can generate internal, unbiased error estimates and variable importance, is robust to training data reduction and noise, and is highly accurate [27,48].

In order to develop the RF models, the sampling points were randomly split into independent training (80%) and validation (20%) datasets and then stratified using the cluster-membership allocations from PATN. Due to the small sample size of the 10-km dataset, an additional RF model was created using 100% of the sample for RF training. Using ENVI, layer stacks of the raster data (30 m, 10 km, and PCA at both resolutions) were constructed and masked. Layer stacks, masks and training regions of interest were imported in to EN-Map Box 1.4 [28] to permit the building of RF models and the classification of the image stacks. EN-Map Box was set to use 200 trees (k) per sample, and the square root of the number of input variables on the non-PCA transformed data ($m = \sqrt{25} = 5$), or m-1 variables ($m = 4$) for the PCA models. The Gini coefficient [49] was used to calculate impurity, which is one method used to evaluate the best split decision for each tree.

Accuracy assessment

The ability of the classification to accurately represent the information across a range of resolutions was then tested, with the relevant steps outlined in light blue in Figure 1. The accuracy of the RF models in recovering and classifying the image stacks into the ALOC classes was assessed with out-of-bag (OOB) error rates generated in EN-MAP box, and using the independent 20% validation dataset to calculate percent agreement between classified and validation data, user and producer accuracies, and Kappa (κ) coefficients [50] in ENVI and R 3.0.1 [47] with the *psych* [51] and *irr* packages [52]. User's accuracy refers to the probability that a pixel classified into a certain class really belongs to that class, while producer's accuracy refers to the probability that a certain class is classified correctly. The locations of the validation pixels were used to extract class information from the original and PCA classifications and percent agreement and κ coefficients between model types were examined. High levels of agreement between the original and PCA classifications would indicate an insignificant amount of multivariate information loss by PCA and further support the use and application of methods that reduce data dimensionality and multi-collinearity between variables in regionalisation studies. The 10-km sample that used 100% of the data for RF training could not be assessed for accuracy independently and was therefore only assessed with OOB error and class distributions.

We also decided that, due to the resampling of the original data (to 30-m from a range of resolutions), it was worth investigating the effect of resampling the 30-m classifications to a coarser resolution (using the majority filter) to help remove some of the finer-scale variability in the data. A resolution of 2.5 km was chosen as a suitable pixel resolution, to ensure that our resampled assessment points were further apart than the mean distance of the original samples (see below), this meant our resampled classifications were equal to or larger than the resolution of the majority of the datasets while still being finer than the soil erosivity index. To assess agreement between the 30-m and 2.5-km resampled classifications (i.e. original and PCA both at 30-m and resampled 2.5-km), 300 random validation points were selected from the 20% validation datasets, with minimum distances between points of 2.5 km, and used to extract class information from the 30-m and 2.5-km resampled classifications. Agreement between the validation data and the 30-m and 2.5-km resampled classifications was assessed with percent agreement and κ coefficients between model resolutions. Visual inspection of the class distributions was also conducted; for this the 300 random validation points were used for the 30-m and 2.5-km classifications, while all 81 validation points for the 10-km classifications were used. McNemar's chi-square test [53,54] was used to formally test for statistically-significant differences between model types (e.g. 30-m original and 30-m PCA) and resolutions (e.g. 30-m original and resampled 2.5-km original) and the validation dataset using the same 300 random validation points; however we recognised *a priori* that the relatively large number of random validation points would likely make any difference statistically significant and that the resolution tests would not be independent of one another. As such, a permutation- based method ($n = 9999$) was also used to assess for statistically-significant differences in the κ values of the classifications [55,56]. The algorithm worked as follows:

1. If V = validation data; A = classification 1; B = classification 2
2. Let test(x, y), be a function that calculates the test statistic (κ) for the classifications,
3. H_0 = if A and B are approximately equal in their classification accuracy, observations in A and B can be exchanged without affecting κ, given by test(V, A) and test(V, B),

4. Randomly exchange data between A and B, n times, and observe how these changes affect the κ of A and B, i.e. test(V, A) – test(V, B),
5. n permutations would result in N data points. Rank the N data points and observe where the κ from the original test (i.e. not the permutated data) is located among the κ values from the permutated data points. If the original κ is outside the 0.975 percentile (or below the 0.025 percentile) then we can claim that the two classifications are different at $\alpha = 0.05$.

If a high level of agreement was found, as indicated by high percent agreement and high κ, then resampling to 2.5-km *a posteriori* could be considered a suitable and justifiable method for smoothing the finer-scale variability.

Relationships between the regionalisation and hydrologic indices

Finally, we tested the results of our classification against a traditional classification based on hydrologic indices. Steps in this section are outlined in grey in Figure 1. As a preliminary assessment of the ability of the regionalisation to differentiate among streams with differing hydrology, we calculated a range of streamflow indices based on the recommendations of Olden and Poff [57] and then explored the relationships between the regions and streams with a permutation-based ANOVA (PERMANOVA) [58] and a constrained discriminant ordination (Canonical Analysis of Principal Coordinates (CAP); [59,60]).

Streamflow gauge locations were downloaded from the Water Measurement Information System (data.water.vic.gov.au/monitoring.htm). These locations were then overlayed on the regionalisation (30-m ALOC 23 meta-group classification, see results) and had group (region) information appended to them. In regions where there were more than 50 stream-gauges present, 50 were chosen at random to be included in the analysis. Daily streamflow data between 1980 and 2010 was then downloaded for 564 gauges throughout Victoria. A minimum record length of 15 years within the 30-year temporal window was required for a gauge to be included in the analysis [61]. Stream gauges that were potentially subject to modification by weirs, dams or water extractions were not specifically excluded from the analysis. Where there were missing periods of flow information (to a maximum of 20 days in any single event) the record was in-filled using linear interpolation [62] with the Time Series Manager module of the River Analysis Package (RAP) [63]. Gauges that had a single period of missing data greater than 20 days were excluded from the analysis. Thirty-two indices characterising different aspects of the flow regime (Table III, All Streams; [57]) for each stream were calculated using the Time Series Analysis module of RAP. Indices that were related to discharge (i.e. those divided by catchment area) were not included in the analysis.

To test for differences among our groups, the PERMANOVA+ add-on [64] for PRIMER was used. One-way PERMANOVAs, (999 permutations) using Group as a fixed factor, were run for the dataset of flow indices based on the original gauges ($n = 201$) and for an additional dataset consisting of a bootstrapped sample of those flow indices ($n = 383$), based on normalised Euclidean distance matrices. Analyses tested both for main effects and pairwise differences among groups. Traditional Analysis of Variance (ANOVA) is powerful for univariate data however the traditional multivariate analogues (e.g. MANOVA), are too stringent in their assumptions (in particular, that of multivariate normality which is frequently untrue in ecological data [25,58]), for use in ecology. As such, permutation-based non-parametric methods are preferred [58]. PERMANOVA uses permutation methods to test the simultaneous response of one or more variables to one or more factors in an analysis of variance. The use of permutations in PERMANOVA removes the assumption of normal distributions which are required for traditional ANOVA/MANOVA testing and, as such, the only assumption of the test other than independence is that the observations can be exchanged under a true null hypothesis [58]. Another benefit of using a permutation approach is that the permutated P-values provide an exact test of each *individual* null hypothesis of interest, and as such ad-hoc pairwise corrections (e.g. Bonferroni) are not strictly necessary [64].

When data are classified into *a priori* groups, unconstrained ordinations (PCA, MDS) are extremely useful for visualising patterns from a multivariate space. However, the overall dispersion of points (when reduced to two or three dimensions) can often hide the true multivariate differences among those groups and it may be very possible to discriminate among groups through another direction or dimension of the multivariate data [64]. Unlike unconstrained ordinations, constrained ordinations have an *a priori* hypothesis which controls the way the multivariate data can be interpreted in an attempt to relate predictor variables (streamflows indices) to response variables (groups) [60]. In a discriminant analysis, the ordination axes are interpreted in such a way as to maximise the differences between *a priori* groups, while in a canonical correlation, the axes are interpreted to maximise correlations among variables. CAP first calculates the principal coordinate axes (PCO) among N samples, and then chooses an appropriate number of PCO axes (m) for interpretation based on a number of criteria (see Anderson et al. [64] for details), including a leave-one-out cross validation procedure which attempts to maximise classification success. The benefits of CAP over other constrained ordination methods are its ability to use any distance or dissimilarity measure, conduct permutation tests for significance of relationships among variables and predict group membership of new samples [60]. To assess the ability of our regionalisation to discriminate groups based on streamflow indices, three CAPs were conducted using the PERMANOVA+ add-on for PRIMER. Two of the analyses were performed against the group information extracted for each stream gauge using the same normalised Euclidean distance matrices that were used in the PERMANOVA tests (i.e. the original dataset and the bootstrapped dataset of hydrologic indices). The number of axes (m) was not specified and a permutation test ($n = 999$) was conducted to test the strength of the relationship. In addition, a third CAP was conducted using the bootstrap dataset where a stratified random sampling approach was used to remove 20% of the group information as a validation sample. The CAP was conducted as before, with the exception being that the validation cases were allocated groups based on the results of the CAP. Allocation accuracy by CAP was assessed by calculating percent agreement between the CAP allocated group and the original group information, and κ coefficients with the *irr* and *psych* packages in R. Pearson's r [47] was calculated between the between the number of gauges in each group and the number of gauges classified correctly from each group to assess for thresholds at which a specified level of accuracy could be achieved.

Results

Spatial distribution of sample points

Analysis of spatial distribution of the sampling points concluded that the mean, minimum and maximum distances to the nearest neighbour among the 30-m points were 2.4 km, 202 m and 8.8 km, respectively. The corresponding values for the 10-km points were 15 km, 10 km, and 47.4 km, respectively. Ripley's K, based on 999 permutations, indicated significant over-dispersion of the 30-m sample points to a distance of ~250 m and a significant clustering at distances greater than ~650 m, while the 10-km sample points were significantly over-dispersed at distances less than ~12 km, but were significantly clustered at distances greater than ~18 km. Based on these values, the spatial distribution of our sampling points was considered suitable for the analyses as no significant clustering was displayed by the sample points at the resolution of the datasets they would be sampling (e.g. there was no significant over-dispersion or clustering at the mean distance of 15 km for the 10-km points).

Clustering and ordination

Twenty-three groups were identified within the 30-m data points and 20 groups for the 10-km data. As a result, the data were allocated into 23 (ALOC 23: 30-m data) and 20 (ALOC 20: 10-km data) non-hierarchical groups. There were wide variations in the ability of the model variables to differentiate among clusters (Table S2) and the group populations had differing multivariate distributions. The allocated groups were well separated (Figure S1) with global-R values of 0.852 (p = 0.001) for the ALOC 23 clustering and 0.908 (p = 0.001) for the ALOC 20 clustering, indicating that cluster membership was highly unlikely to be a result of chance alone. This was supported by the fact that neither ANOMSIM resulted in any permutations that had R-statistics greater than or equal to the global-R value.

The allocated groups were then further clustered into hierarchical meta-groups. Using the group averages for ALOC 23 and ALOC 20, 11 and ten meta-groups were generated (Figure S1 and Figure S2). ANOSIM analysis again indicated that cluster membership was highly unlikely to be a result of chance alone and suggested that the meta-groups for both the ALOC 23 and ALOC 20 models were well separated with global-R values of 0.668 and 0.762 (p = 0.001). Again no permutations had R-statistics that were greater than or equal to the global-R value. Some variables exhibited no relationship between the observed values and the meta-groups, while other variables show very clear relationships to the meta-groups (Figures S3–S7). For example, the values for the aridity index (low aridity index values represent drier regions) and rainfall decreased from groups A to K. The opposite relationship was observed for maximum and minimum temperatures, again suggesting as we move through regions A to K, the environment becomes drier and warmer. The BioClim variables 8, 9, 15 and 16 also supported this relationship, with increases in BIO08 and BIO09 (mean temperature of the wettest and driest quarter, respectively), and decreases in BIO16 and BIO17 (precipitation of the wettest and driest quarters, respectively). Variable percentage contributions to the meta-groups for both the ALOC 23 and ALOC 20 classifications differed markedly (Table S2, Figure S8) with, for example, saturated hydraulic conductivity of the B soil horizon (B_KSAT) contributing 0% to a number of ALOC 23 meta-groups, while contributing 46% to ALOC 23 meta-group F. The within-group variation was highly variable among groups, with average squared distances of groups ranging from 4.18 to 11.77 for the ALOC 23 meta-groups, and 4.44 to 13.45 for the ALOC 20 meta-groups (Table S2).

Ordination by PCA resulted in the creation of five principal components (PC) for both ALOC 23 and ALOC 20, using a minimum eigenvalue of one. The first PC for each of the ALOC 23 and ALOC 20 clustering explained 45% and 53% of the data variance, with eigenvalues of 11.2 and 13.2, respectively (Table 1). The five PCs combined explained 78% and 87% of the variance, with the final PC having eigenvalues of 1.39 and 1.25, respectively.

PC 1 can be interpreted to represent those areas that are wet, cool, heavily vegetated, high elevation environments with steep slopes and low erodibility soils (Table 1, $0.2<r<-0.2$). With the exception of PC 1, none of the PCs for either of the datasets can be interpreted to represent similar environments across the two classifications. For example, PC 2 of the ALOC 23 dataset can be interpreted to represent areas with thick, weathered, A-horizon soils with high levels of plant available water, with relatively small variations in temperature seasonality and poorly developed B-horizon soils. For the ALOC 20 dataset, on the other hand, PC 2 suggests areas of poorly developed A-horizon soils, but with thick B horizon soils, and cooler, wetter summers.

Random Forests

Training and validation. Two hundred trees were sufficient for the RF models to achieve acceptable accuracies. Exploratory analysis, using models with 5000 trees (not presented) showed little improvement in OOB error (<1%). OOB error rates were low for the ALOC 23 and ALOC 23 PCA RF models, with estimated maximum accuracies of 95% and 92%, respectively. The OOB error rate of the ALOC 20, ALOC 20 PCA and the 100% ALOC 20 and 100% ALOC 20 PCA models (the latter two were created due to the small sample size for the 10-km dataset as described above) was significantly worse, with estimated accuracies of 59%, 56%, 56% and 54%, respectively (Figure 4).

Classification accuracy was 95% ($\kappa = 0.94$) for the ALOC 23 classification and 92% ($\kappa = 0.92$) for the ALOC 23 PCA classification (Table S3) when tested against the validation dataset. The accuracy of the ALOC 20 and ALOC 20 PCA classifications decreased relative to those estimated by the RF OOB error, with accuracies of 46% ($\kappa = 0.42$) and 47% ($\kappa = 0.44$) (Table S3). The producer accuracies differed significantly for each of the classifications (Table S3), with observed minimum producer classification accuracies of 81% for the ALOC 23 classification and 59% for the ALOC 23 PCA classification. Likewise, the ALOC 20 and ALOC 20 PCA classification also exhibited low producer accuracies with minima of 0% observed for a number of classes in each classification. Visual inspection of the resulting classifications showed few obvious differences among the various ALOC 23 classifications (Figure 5), but more differences were apparent among the ALOC 20 classifications (Figure 6).

Comparisons between original, PCA and resampled classifications. The class distributions of the RF and resampled classifications illustrate that the RF classifications at 30-m and resampled to 2.5-km (bottom row, Figure 5) were quite successful in maintaining the distribution of the validation dataset (Figure 7). All 10-km models were missing classes 7, 8, 12, 18, 19 and 20 from the validation dataset (Figure 7), meaning that accuracy assessment of these classes was not possible, although classes 7, 8, 12, and 18 were present in the final classification (Figure 6). Class 1 tended to be over-classified by the ALOC 20 RF models, as evidenced by the large number of validation points classified as such (Figure 7). The ALOC 20 100% classification was also the only classification to have classes 6 and 15 represented at the validation locations.

Agreement between the ALOC 23 and ALOC 23 PCA classifications was high at 93% ($\kappa = 0.93$). There was no statistically-significant difference between the two classifications ($\chi^2 = 2.72$, $p = 0.1$). Agreements for the ALOC 20 and ALOC 20 PCA classifications was lower than that observed for the ALOC 23 classifications, at 72% ($\kappa = 0.67$), which was also non-significant ($\chi^2 = 0.24$, $p = 0.63$). This indicates that the classification based on the five principal components captured the vast majority of the

Table 1. Results of Principal Components Analysis.

ALOC 23						ALOC 20				
	PC1	PC2	PC3	PC4	PC5	PC1	PC2	PC3	PC4	PC5
Eigenvalue	**11.2 (45)**	**3.26 (58)**	**2.07 (66.3)**	**1.58 (72.6)**	**1.39 (78.2)**	**13.2 (52.7)**	**4.04 (68.9)**	**1.68 (75.6)**	**1.58 (81.9)**	**1.25 (86.9)**
Variable										
A_KSAT	0.14	0.14	**−0.42**	**−0.3**	0.18	0.13	**−0.22**	0.15	**0.53**	0.03
A_PAWC	0.01	**0.48**	−0.18	−0.02	−0.12	0.05	**−0.45**	0.12	0.1	−0.07
A_THICK	0.1	**0.45**	**−0.21**	−0.12	−0.02	−0.04	**−0.45**	0.1	−0.04	−0.12
ARIDITY_INDEX	**0.29**	0.01	0.04	0.02	0.03	**0.26**	−0.09	0	−0.03	−0.02
B_KSAT	0.15	0.05	**−0.4**	**−0.31**	0.2	0.16	**−0.22**	**0.26**	**0.29**	−0.08
B_PAWC	0.12	**−0.34**	−0.09	**−0.41**	**0.26**	0.17	0.18	0.01	**0.47**	**0.29**
B_THICK	0.06	**−0.38**	0.14	**−0.38**	0.16	0.1	0.3	−0.1	**0.35**	**0.42**
BIO04	−0.08	**−0.34**	**−0.29**	0.08	**−0.39**	**−0.21**	−0.01	**−0.26**	**0.26**	**−0.36**
BIO08	**−0.22**	0.05	−0.1	0.2	**0.44**	**−0.25**	−0.01	0.19	0.09	0.07
BIO09	**−0.24**	−0.1	−0.17	−0.04	−0.19	0.02	**−0.26**	**0.35**	**−0.26**	**0.52**
BIO15	0.08	0.14	0.16	**−0.43**	**−0.53**	**0.26**	−0.09	0	−0.11	0.1
BIO16	**0.28**	0.03	0.07	−0.05	−0.05	**0.26**	0.01	−0.19	0.03	−0.13
BIO17	**0.27**	−0.02	0.03	0.18	**0.23**	−0.04	**0.38**	**0.38**	−0.08	−0.01
ELEVATION	**0.25**	−0.14	−0.08	0.04	−0.16	**0.24**	0.18	0.11	−0.11	−0.01
ET_ANNUAL	**0.24**	0.14	**0.28**	0.06	0.1	**0.21**	−0.1	**−0.41**	−0.09	0.1
GW_SWL	0.17	−0.08	**−0.29**	0.09	−0.14	**0.23**	0.11	0.2	0.05	**−0.24**
GW_TDS	−0.17	−0.1	**−0.32**	0.08	0.02	−0.16	0.11	**0.41**	0.13	**−0.25**
LDI	−0.11	−0.11	0.16	−0.07	0.05	−0.15	−0.08	**−0.26**	0.18	0.08
MAX_TEMP	**−0.27**	−0.05	−0.18	0	−0.07	**0.26**	−0.09	0.01	−0.03	0
MIN_TEMP	**−0.25**	0.08	−0.05	0	0.15	**0.25**	0.02	0.07	0.08	**−0.21**
RAIN_ANNUAL	**0.29**	0	0.04	0.03	0.02	**0.27**	−0.02	0.04	0.01	−0.07
SLOPE_RAD	**0.23**	−0.1	−0.2	**0.22**	−0.03	**−0.26**	0.09	0.05	0.1	−0.04
SOIL_EROS	**0.26**	−0.07	−0.05	0.09	0.03	**−0.24**	−0.16	−0.07	0.13	0.01
TWI	−0.18	−0.01	0.16	**−0.27**	0.03	**−0.23**	0	0.03	−0.06	0.1
WEATH_IND	−0.16	**0.23**	0.1	**−0.26**	0.14	−0.19	−0.17	−0.03	0.11	**0.27**

Eigenvalues are presented in the top row while the numbers in brackets represents the cumulative % variance explained by the PCA. Bold numbers indicate variables with a correlation ≥0.2 or ≤−0.2 which was used for PC interpretation.

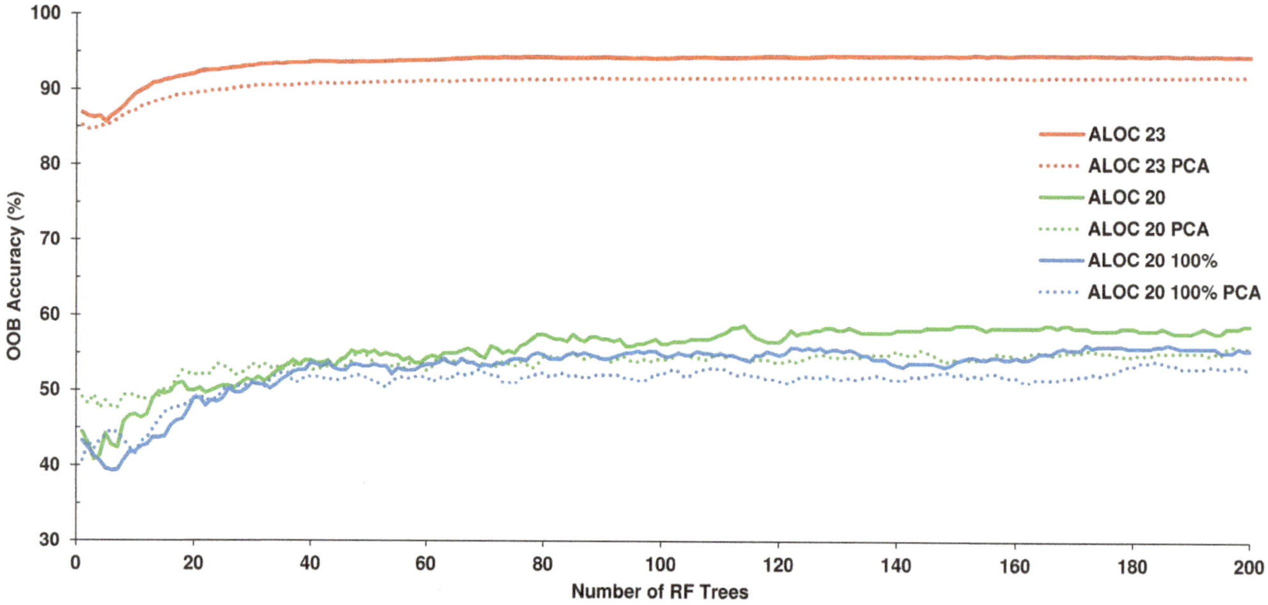

Figure 4. Out-Of-Bag (OOB) percent accuracies for the ALOC clusterings as classified by Random Forests. The 30-m ALOC 23 and ALOC 23 PCA models were significantly more accurate than the 10-km ALOC 20 classifications.

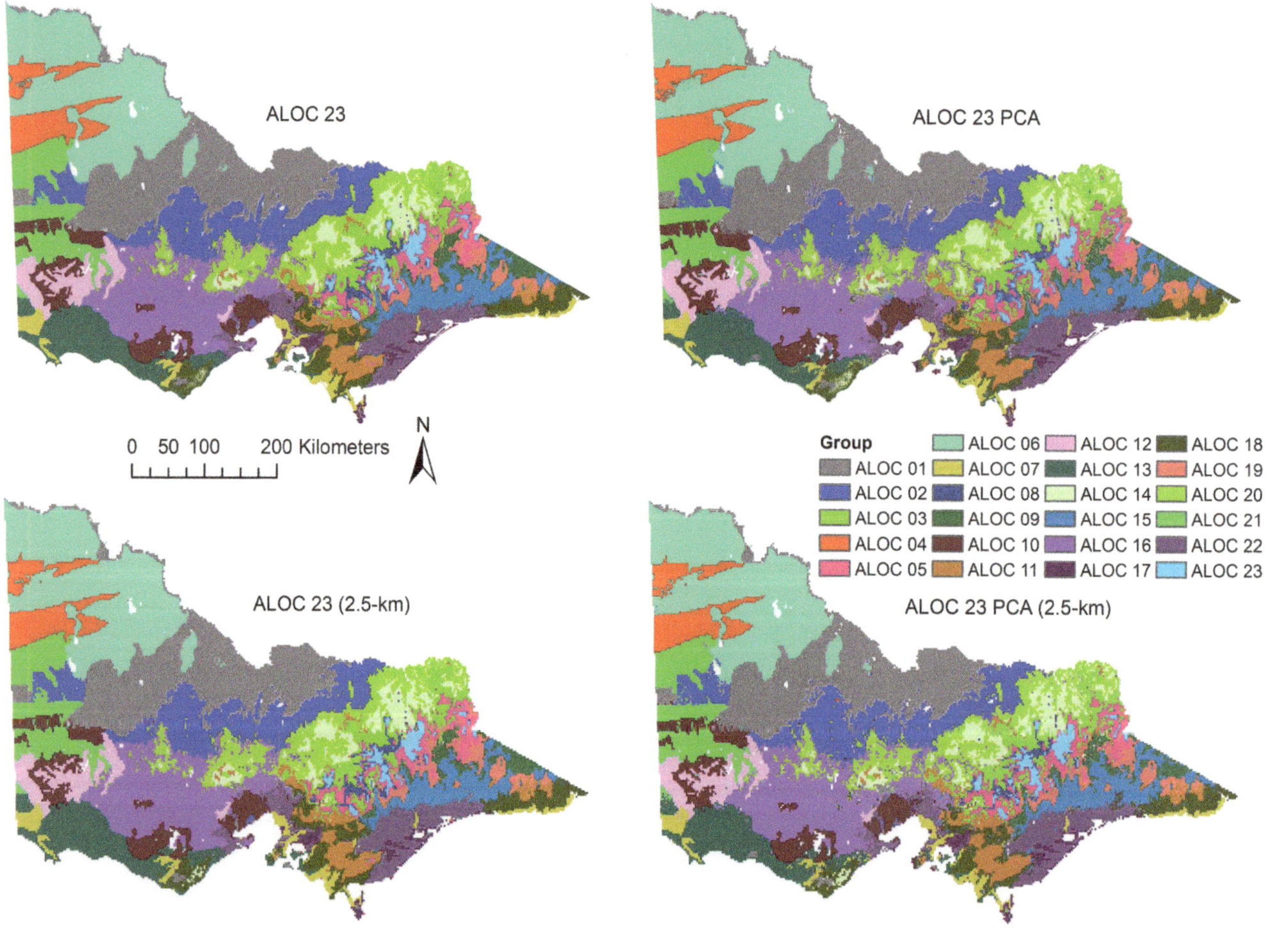

Figure 5. Results of the ALOC 23 (30 m) classifications. Top row - ALOC 23 and ALOC 23 PCA; Bottom row - ALOC 23 and ALOC 23 PCA resampled to 2.5 km. Colours represent each of the ALOC non-hierarchical clusters. Similar colours and cluster numbers do not necessarily represent related groups.

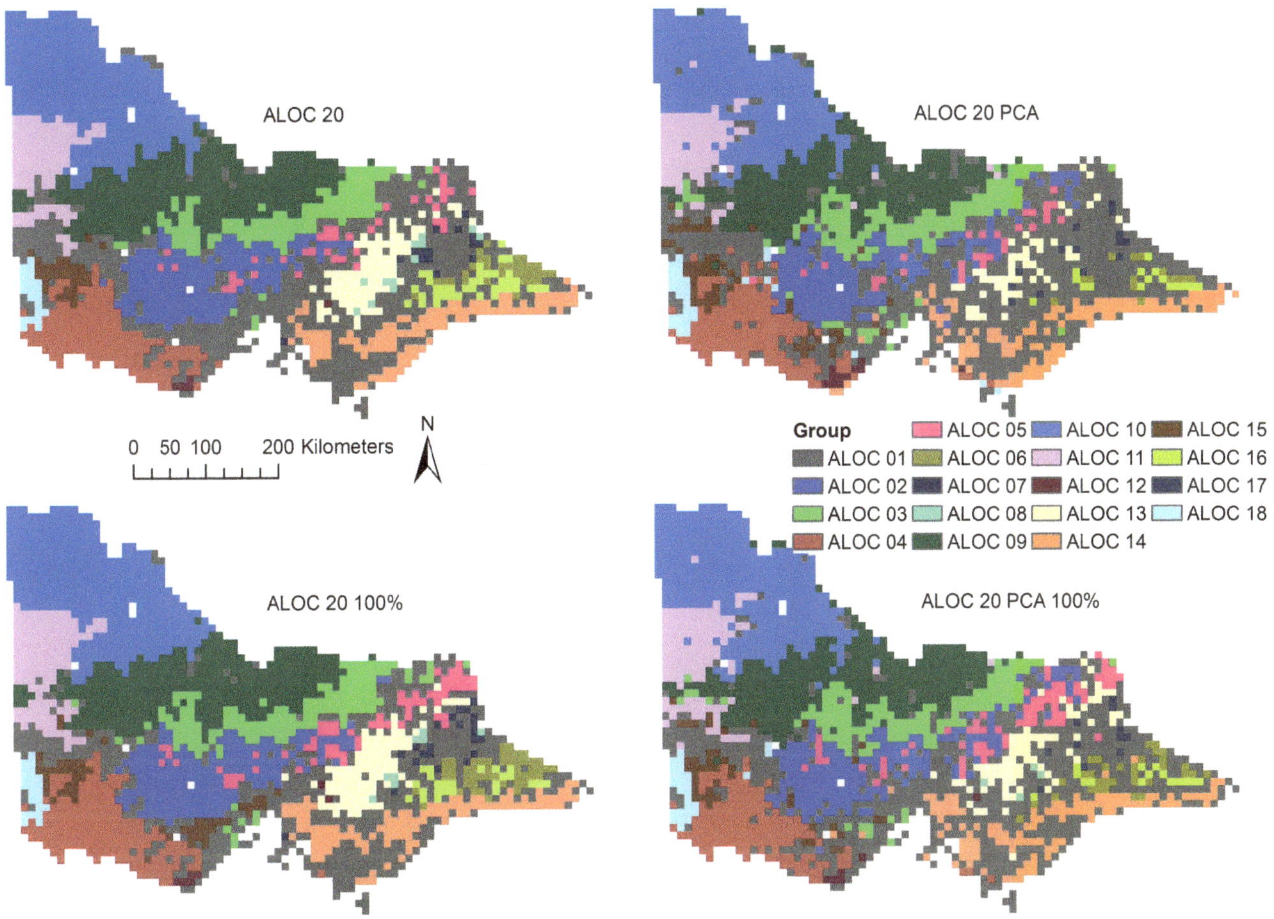

Figure 6. Results of the ALOC 20 (10 km) classifications. Top row - ALOC 20 and ALOC 20 PCA; Bottom row - ALOC 20 (100%) and ALOC 20 PCA (100%). Note that not all ALOC clusters are present in the final classifications. Colours represent each of the ALOC non-hierarchical clusters. Similar colours and cluster numbers do not necessarily represent related groups.

variability among points at the 30-m scale, and most of the variability at the 10-km scale.

Visual inspection of the resampled classifications (bottom row, Figure 5) showed very similar results to the original 30-m classifications. Agreement between the resampled ALOC 23 classification and the original 30-m classification was high (92%, $\kappa = 0.92$) and an accuracy of 91% ($\kappa = 0.9$) was observed with the 300 randomly chosen points from the validation dataset. This was a 4% difference in accuracy to the original 30-m ALOC 23 classification. However, McNemar's test suggested the difference in classification accuracy was statistically significant ($\chi^2 = 12.56$, $p<0.001$). As the two classifications are not independent, the permutation test was also used to compare the two. This supported the results of the McNemar's test and a permutated κ difference of 0.062 was deemed significant ($p<0.001$). The ALOC 23 PCA resampling results were similar, with a relatively high agreement with the original 30-m PCA classification (88%, $\kappa = 0.87$), and 87% ($\kappa = 0.86$) agreement with the validation dataset, a 5% difference compared to the original ALOC 23 PCA classification. These differences were also statistically significant (McNemar's test, $\chi^2 = 11.76$, $p<0.001$; permutation test, κ difference $= 0.073$, $p<0.001$). Agreement was quite high between the 2.5-km resampled ALOC 23 and 2.5-km resampled ALOC 23 PCA classifications (93%, $\kappa = 0.92$), however unlike the 30-m classification results, when comparing the agreement between the two resampled classifications McNemar's test suggested the results were significantly different ($\chi^2 = 5.88$, $p = 0.015$). As these two classifications were independent of one another the permutation test was not required.

Once the classification assessment was finalised, the meta-group allocations were appended to the classifications. The ALOC 23 meta-group allocations were examined visually (Figure 8) and were deemed suitable given that regions having similar groups (e.g. A and B, represented by distinct colours in Figure 8) exist in similar areas among the two classifications and show quite obvious spatial relationships. While the performance of the ALOC 20 models was weaker, the meta-group assignment results were similar to those observed for the ALOC 23 classification in that similar meta-groups existed closer spatially (not presented).

Spatial variability in western Victoria

The variability of model results was not uniform across the entire Glenelg-Hopkins case-study region (Figure 9). While the eastern and central parts of the region appeared to be relatively spatially uniform (i.e. they do not show a significant amount of variation in ALOC groupings); the western and north-western parts of the region are comprised of a number of ALOC classes. This suggests that the hydrological system varies spatially across

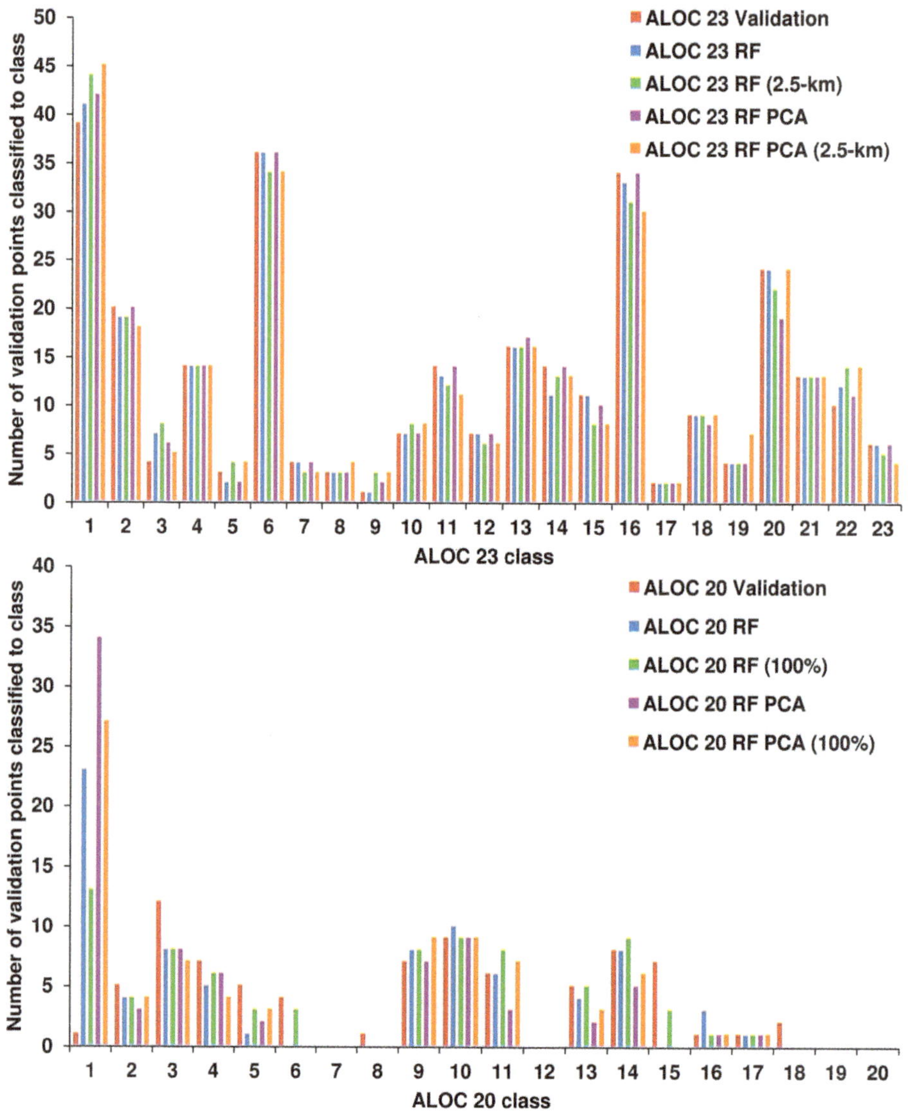

Figure 7. Class distributions from the final classified images. The 20% validation points were used to extract class information from the classified images. The resulting figure shows that the ALOC 23 classifications were much better at maintaining the class distribution of the validation dataset (shown above in red) than the ALOC 20 classifications.

the region, with the most variability likely occurring in the western half of the catchment.

The meta-group classifications support the idea that there was a difference in the hydrological systems of the three major catchments of the Glenelg Hopkins region (Figure 10). Interestingly, the amount of spatial variation did not decrease when the meta-group assignments of the original ALOC classes were examined, suggesting that hydrologic responses could be very different in areas that are quite close together. In the Glenelg River catchment (Figure 10, shown in red), there is obvious spatial variation in the assigned hydrological classes. Of particular interest are the two catchments in the north-east of the catchment that contain Rocklands Reservoir and the majority of the Grampians ranges, as they each consist of five different hydrological meta-groups (E, F, H, I & J). Even though the meta-groups represented in those particular catchments occupy a small area, they are still all present in the resampled classification (Figure 10, bottom row). While the differences were less pronounced in the Hopkins River catchment (Figure 10, outlined in blue), there was some spatial variability in classes in the north (dominated by meta-group I, with some small patches comprised of meta-group E), while most of the catchment belongs to group I and the two southernmost catchments belong to meta-group G. There was even less variation in the classification of the Portland catchment (Figure 10, outlined in green) with meta group I dominating that catchment. Nonetheless, there were small areas of groups F and J in the south-west catchments of the catchment. Visual examination of the PCA and the resampled classifications showed little difference to that observed in the original 30-m classifications. The most obvious change was the small area in the southern catchment of the Hopkins River catchment (outlined in blue), and the easternmost parts of the Portland catchment (outlined in green), that was classified as meta-group J in the PCA classifications.

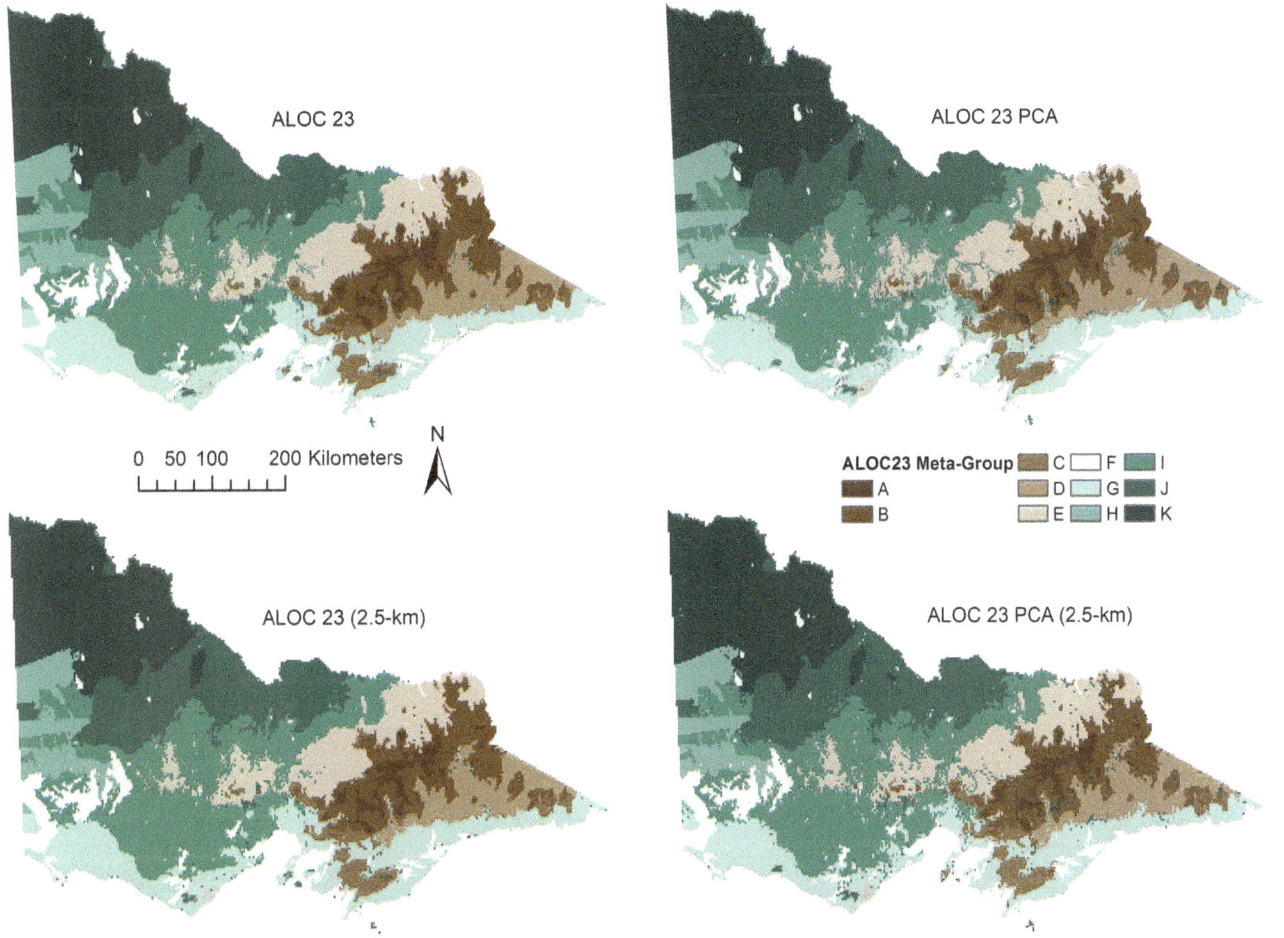

Figure 8. Results of the ALOC 23 (30 m) meta-group allocations. Top row - ALOC 23 and ALOC 23 PCA; Bottom row - ALOC 23 and ALOC 23 PCA resampled to 2.5 km. Colours represent each of the hierarchical meta-groups as defined by SIMPROF. Similar colours and group letters indicate a closer relationship than those further apart.

Relationships between the regionalisation and hydrologic indices

A total of 201gauges were deemed suitable for inclusion in the analysis based on the criteria specified above. With the exception of groups A, K and H, all regions had multiple suitable gauges.

There were significant differences among groups based on their hydrologic indices for both the original (pseudo-$F_{1,7} = 3.655$, $p = 0.001$) and the bootstrapped datasets (pseudo-$F_{1,7} = 9.304$, $p = 0.001$). *A posteriori* pairwise comparisons of groups using the original dataset indicated that the differences were significant ($p \leq 0.05$) between all groups with the exception of pairs B:E, C:D, C:E and F:I. The pairwise comparisons on the bootstrapped dataset were all significant ($p < 0.05$).

The ability of CAP to correctly classify cases within the original dataset based on their hydrologic indices was relatively poor, with only 92 samples correctly classified, but the model was statistically significant (48%, $m = 30$, $p = 0.001$). Stream gauges from meta-group C had the lowest classification accuracy with only 10% of gauges being successfully allocated. The highest classification accuracy was observed for both meta-groups B and G, with 67% of gauges correctly allocated to each. The bootstrap dataset performed better with 253 samples being correctly classified (66%, $m = 28$, $p = 0.001$). The lowest classification accuracy was observed for meta-group E with only 47% of gauges being correctly allocated. The highest classification accuracy was meta-group I with 74% of gauges correctly allocated.

When 20% of cases were used as a validation sample within the bootstrapped dataset, CAP performed reasonably, with 208 of the gauges correctly classified (67%, $m = 32$, $p = 0.001$). The lowest classification accuracy was observed for meta-group J with 58% of gauges being correctly allocated. The highest classification accuracy was meta-group F with 77% of gauges correctly allocated. Agreement between the samples that were allocated to new groups and their original group was relatively high at 60% ($\kappa = 0.53$). The lowest successful group allocation was observed for meta-group E where only 20% of gauges were correctly classified. The highest allocation accuracy was observed for meta-group D where 100% of gauges were correctly classified.

Pearson's correlation indicated there was a statistically-significant positive relationship between the number of gauges in each group and the number of gauges classified correctly from each group (Figure S9) for the original ($r = 0.91$, $p = 0.002$, $n = 8$), the bootstrapped dataset ($r = 0.93$, $p = 0.001$, $n = 8$), and the bootstrapped dataset with the validation samples ($r = 0.96$, $p < 0.001$, $n = 8$). This indicated that the model was most likely to correctly classify groups that were common in the dataset, with uncommon groups being correctly classified less often. All classifications performed highly favourably compared to a random allocation of

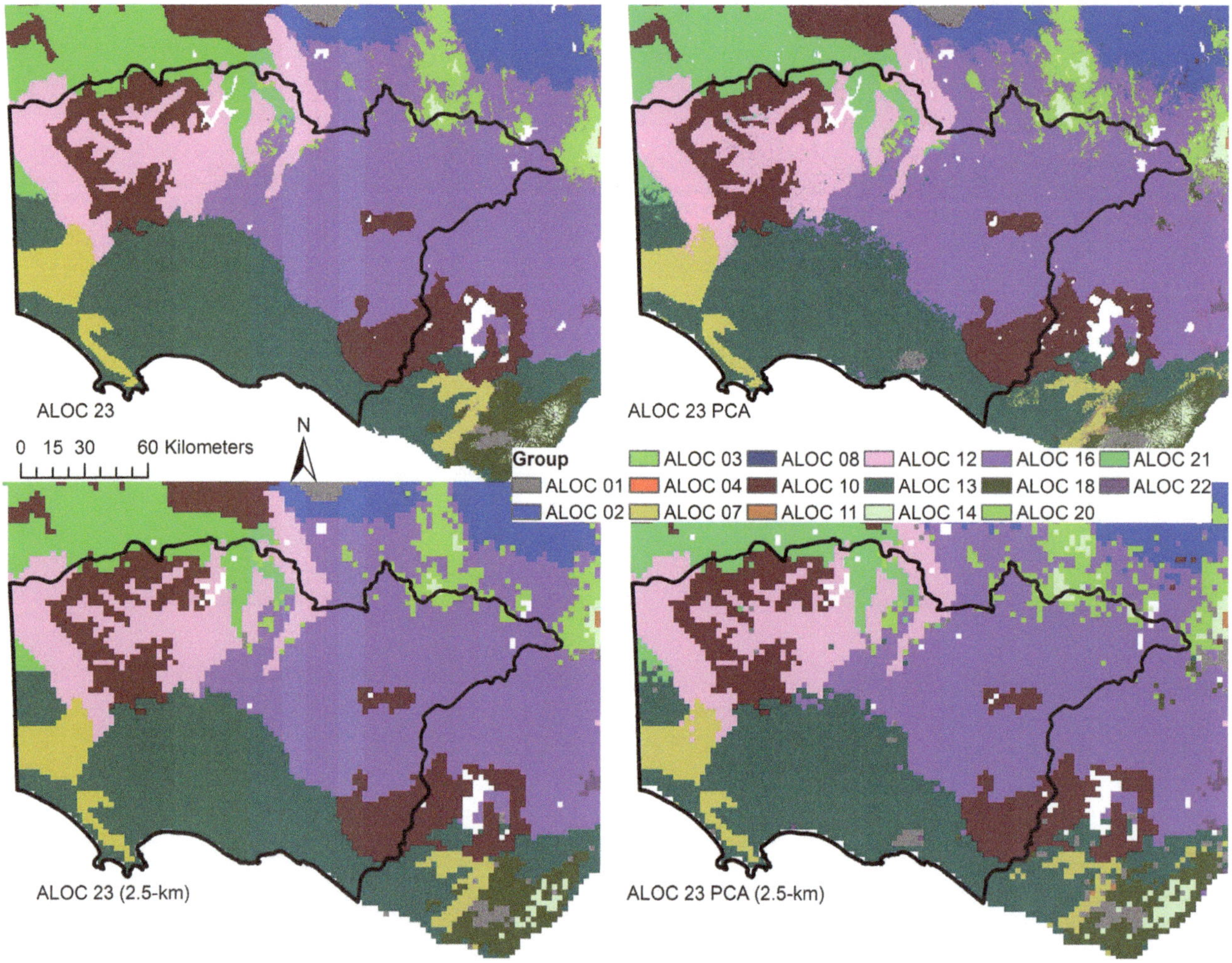

Figure 9. Results of the ALOC 23 (30 m) classifications for the Glenelg Hopkins region. Top row - ALOC 23 and ALOC 23 PCA; Bottom row - ALOC 23 and ALOC 23 PCA resampled to 2.5 km. Colours represent each of the ALOC non-hierarchical clusters. Similar colours and cluster numbers do not necessarily represent related groups. More spatial variability in the ALOC clusters is obvious in the Glenelg catchment (western and NW side), than that in the Hopkins (eastern) and Portland (south central) catchments.

cases among 11 groups which, assuming equal sample sizes, results in 9% of cases being correctly classified.

Discussion

By incorporating a number of environmental variables likely to influence regional hydrology and a range of non-parametric statistical and classification methods, this study aimed to generate a hydrologic landscape classification that did not require the use of an *a priori* selection of a spatial unit such as a catchment. The main objective of the approach was to see whether it was possible to create a classification that could preserve the environmental and hydrological variability that are known to influence streamflows within and between catchments that has typically been lost in previous regionalisation studies. An analysis of the ability of the classification to differentiate between streams from each group based on a number of streamflow indices was also undertaken.

Difference to previous regionalisation studies

Inductive methods of hydrologic regionalisation have been popular in the past (see Table 2 in Olden et al. [9]) and, while there have been a number of studies that have focused on deductive methods (see Table 1 in Olden et al. [9]), the choice of variables, their resolution (temporal and spatial), the classification method, the spatial scale of the classification and the number of groups are all known to influence deductive classifications [9,14]. While the final product of deductive methods is a spatial mosaic of independent hydrologic types, the final classifications do not always only identify hydrologic variation [10,12]. Inter-catchment variability can limit the applicability of hydrologic regionalisations to generalise and predict catchment behaviour as a function of climatic and environmental gradients [14]. As previous studies have relied on catchments, landscape units or stream sections [8,9,14,19], an issue that is more apparent in deductive regionalisation studies is the loss of small-scale spatial hydrologic variability [6,16] as the unit of analysis gets larger. Our method relied on using an accurate supervised image classification method to extend our statistical clustering to an area covering ~228,000 km^2 without the need to rely on catchments, landscape units or stream sections. We opted for this approach as it is well known in the literature that there is significant flow variability within and among catchments and that the variables governing flow

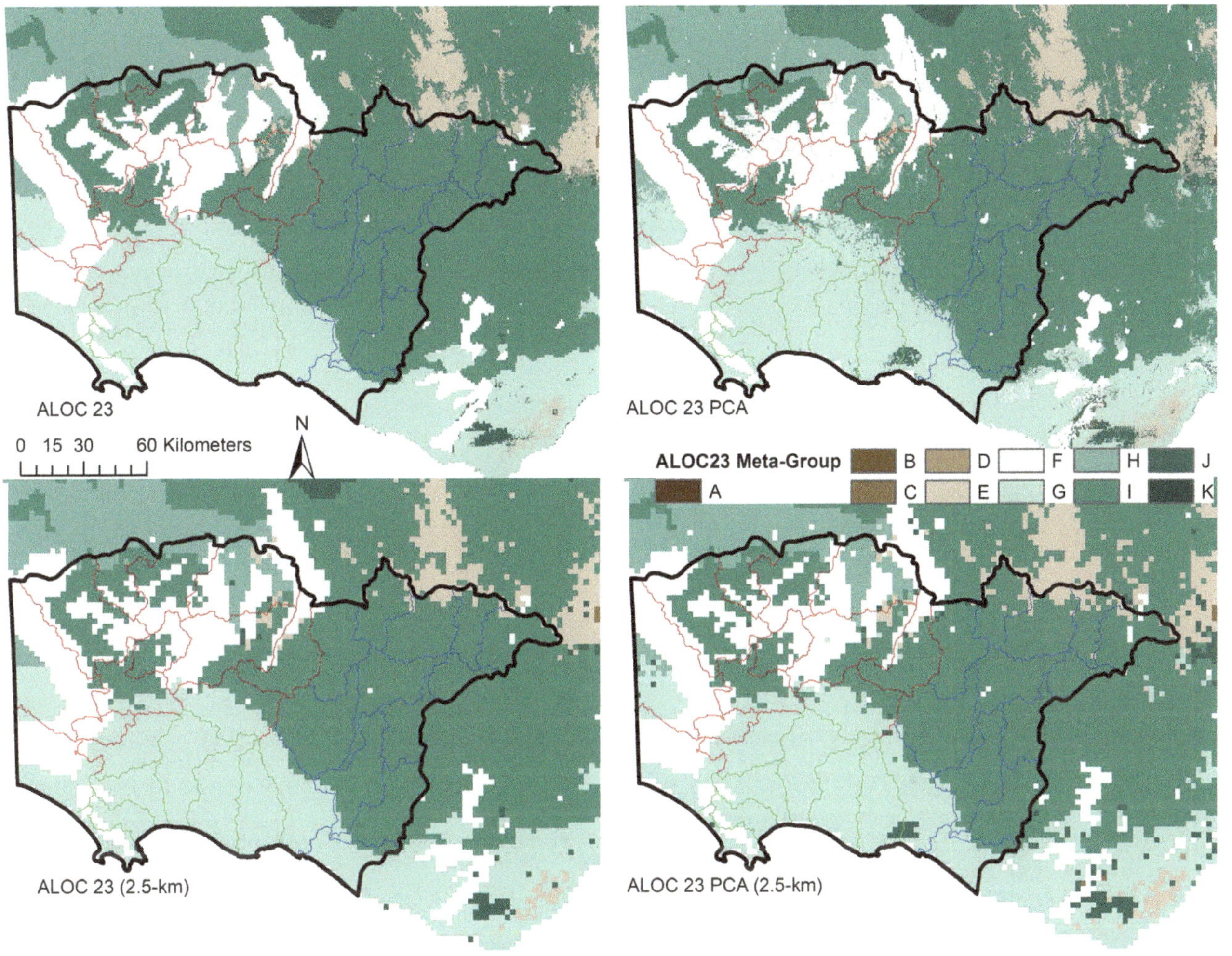

Figure 10. Results of the ALOC 23 (30 m) meta-group allocations for the Glenelg Hopkins region. Top row - ALOC 23 and ALOC 23 PCA; Bottom row - ALOC 23 and ALOC 23 PCA resampled to 2.5 km. Colours represent each of the hierarchical meta-groups as defined by SIMPROF. Similar colours and group letters indicate a closer relationship than those further apart. More spatial variability in the meta-groups is obvious in the Glenelg catchment (shown in red), than that in the Hopkins (blue) and Portland (green) catchments.

variability are dependent on scale [13]. While other regionalisation studies have identified that the primary drivers of catchment function are largely related to climatic gradients [14], our results suggest that a mixture of climatic, geological and environmental functions are driving catchment, and thus hydrologic, variability (Table S2, Figure S8) in our regionalisation. It would be expected, in a traditional approach to hydrologic regionalisations, that some of this variability would be lost – which could explain why other studies have largely identified climatic gradients that vary slowly with space to be the primary drivers of catchment function. Our approach allows for different classes to be represented within a single catchment, thus preserving intra-catchment variability.

Statistical evaluation, clustering, and PCA

Traditional parametric statistical analyses and clustering algorithms such as *k*-means tend to have restrictive assumptions regarding independence of samples, multivariate normality and collinearity. The assumption of samples being distributed normally through multivariate space, for example, is unlikely to be true in most ecological and environmental datasets [25]. The approach that we employed relied on the use of non-parametric and permutation-based statistical methods in conjunction with the RF classification algorithm. This approach had far fewer assumptions relating to data normality and collinearity [27,44,45]. The approach also allowed the decision regarding the number of groups used in the analysis to be statistically-justified, when this decision is typically arbitrary. Our method was supported by the application of both ANOSIM and MDS to assess group separation, with each suggesting that the groups were distinct and that we had chosen an appropriate number of groups for our dataset. While not perfect (SIMPROF, by design, tests for hierarchically-related groups and we were after non-hierarchical groups), we believe this approach to be simpler, more statistically sound and more efficient than methods employed in the past which require large amount of *a posteriori* or *post-hoc* statistical testing [10,19].

The use of PCA-transformed data was shown not to significantly affect the classification accuracy of the model, even though the PCA was only able to explain <80% of the data variability (Table 1, ALOC 23 PCA). This suggests that future classifications could be conducted on PCA-transformed datasets while still producing accurate classification schemes. Our method has essentially shown that it is possible to extract the same number of groups from PCA-transformed data as it was from the non-PCA transformed data. However, using PCA from the beginning has the potential to influence the number of groups identifiable by SIMPROF (as there are fewer data and less variance in the remaining data) and therefore influence the overall classification process. Using PCA could, however, make the process more efficient in that having a reduced number of groups to begin with could remove the need to first use non-hierarchical classification before hierarchically classifying the groups. The major benefit of using PCA-transformed data in this study was that the time to parameterise and classify the raster data with RF was substantially reduced.

Classification by Random Forests

The non-parametric and highly accurate [27,48] RF classifier was very successful in recovering and classifying the ALOC class information of the remaining pixels in the raster datasets. While the ALOC 23 RF models had very high classification accuracies (95%, $\kappa = 0.94$ and 92%, $\kappa = 0.92$), our hypothesis that the classification of the coarser datasets would be inferior was confirmed by the low accuracies of the ALOC 20 classifications (46%, $\kappa = 0.42$ and 47%, $\kappa = 0.44$). As our sampling density was severely limited by pixel size in the 10-km models, this further supports previous research showing that the overall and per-class accuracy of supervised classifications can be limited by pixel size [39,40]. While the RF classifier has been shown to be robust against statistical noise and training data reduction [48], it is possible that, in this case, there were simply not enough training sites to allow for the creation of an appropriate model at the 10-km scale. This analysis used only 325 training sites (80% of $n = 406$ 10-km sample points) to try to produce a classifier for 20 classes and another 33% of the training site data was excluded for OOB accuracy assessment [48]. This left the RF algorithm with only 215 points to generate the required classification trees. The two ALOC 20 100% models used 268 points (66% of $n = 406$ 10-km sample points) to create an RF model but still performed poorly with OOB error rates limiting accuracies to 56% and 54%. As all classes were included in training data for the ALOC 20 100% models, their omission from the final classification (Figure 6 & 7) suggests that they, by chance, happened to be excluded from the training data selected by the bootstrapping step used to calculate OOB error. This could partly explain the high OOB error rates observed for those models. The missing classes from the ALOC 20 classifications (Figure 7, Table S3) could be similarly explained, although it is also possible that they were excluded randomly from the 80% training data at the previous step. As expected, the exclusion of data, either manually for validation purposes or automatically by RF to enable an OOB estimate, appeared to severely limit the classification accuracy when using small amounts of training data [48]. This further supports our hypothesis that limiting the study to a 10-km resolution based on a single, coarse dataset would influence the results presented here, particularly as this study was conducted over a relatively small area. If the study had been conducted at a continental scale, for example, it would be possible to generate more than 15,500 random points at a minimum distance of 10 km and, therefore, resampling the same datasets to a finer resolution (to avoid sub-sampling of pixels) would not be necessary.

The resampling of the ALOC 23 models from a 30-m to 2.5-km resolution appeared to remove some of the finer-scale spatial variability in the classifications (Figures 5 & 8). While not appearing to constitute a significant change between the original and resampled classifications (percent agreements between the resampled and original classifications were 92% ($\kappa = 0.92$) for the ALOC 23 model and 88% ($\kappa = 0.87$) for the ALOC 23 PCA model), the results of both McNemar's test and the permutation test indicated that the resampled classifications were significantly different from their equivalent 30-m classifications. This may seem like a serious drawback to the method, however, when compared to the validation dataset the accuracy of the 2.5-km ALOC 23 classification was only 4% less than that observed for the 30-m model, while the accuracy of the 2.5-km ALOC 23 PCA classification was only 5% less than that of the 30-m PCA model. While statistically the difference may be deemed significant, we contend that in reality a difference of $\leq 5\%$ would likely not be ecologically or environmentally important and thus, would not affect the ability of the method to create a hydrological landscape classification that could be used to explain spatial differences in streamflow metrics. Additionally, the resampling step was performed *a posteriori* and therefore may not actually be necessary in all cases.

Case study on spatial variability in western Victoria

The spatial variability in the classification of the Glenelg Hopkins region was most evident in the Glenelg catchment, and less so in the Hopkins and Portland catchments (Figure 9 & 10). Spatial hydrologic variability has been observed in a number of studies in the past and the strongest and most variable relationships between environmental factors and water quality and quantity have consistently been found in the Glenelg catchment. Brown et al. [17] found that the Glenelg catchment exhibited the most variability in the relationships between climate, land use and wetland extent, which may help to explain some of the variability observed here. Relationships explored in the past relating the proportion of native vegetation to in-stream salinity [35] showed strong relationships in the Glenelg and Portland catchments, but less evidence for the same relationships in the Hopkins catchment and it was suggested that this may have been due to the degraded nature of the catchment. The degraded nature of the Hopkins catchment could also explain the lack of variability observed in that catchment in this study. A spatially-varying relationship between nutrient exports and land use has also been observed in the Glenelg Hopkins region [34] although, due to the lack of suitable stream-gauge data, it is not clear whether this relationship holds for the streams and rivers of the region. However, the relationships in the region that have been described in the past [17,34,35] are likely to be complicated due to variations in geomorphology, groundwater levels and salt concentrations [36]; conditions that we have attempted to account for in this study. Water resource managers in the region need to take into account possible differences in intra- and inter-catchment hydrology that could drastically affect river management and restoration plans and regionalisation studies such as the one presented here could assist in identifying that variability.

Relationships between the regionalisation and hydrologic indices

Understanding, and being able to accurately predict, streamflow characteristics in ungauged locations is crucial for ecohydrological and other studies [2,6]. Our method set out to test a new approach to hydrological regionalisation that removed the need for catchments as a spatial unit of analysis for our statistical clustering [6,8,14]. However, the ability to link the results of the regionalisation to streamflow indices could have presented an issue given that we did not rely on catchments as a spatial unit.

The results supported our hypothesis that our method would be able to identify and preserve inter- and intra-catchment variability. Pairwise comparisons suggested that, even in the original dataset ($n = 201$), there was enough variability in our 32 streamflow indices to separate all but four pairs (of a total of 28 pairs). A simple bootstrap (with replacement) to $n = 383$ gauges was sufficient for all groups to be easily identified as distinct suggesting there is a minimum number of gauges for the approach to be reliable (likely to relate to the number of each individual class in the dataset). In addition, CAP demonstrated that the regionalisation was able to discriminate among streams from different groups. While the classification was not perfect, the analysis conducted on the bootstrap datasets indicated that the stream gauges could be classified correctly significantly better than chance alone, and that gauges from one class could be correctly classified 100 percent of the time. In general, common classes were correctly classified more frequently than uncommon classes. While the correlation between the number of gauges in each class and the number of correctly-allocated samples was significant, there was no clear threshold (i.e. the relationship was linear), so it is difficult to identify a single minimum number of gauges that could be implemented to ensure that results met pre-defined criteria for reliability. Therefore, future users should interpret results for rare groups with caution. However, we believe that, based on this preliminary assessment of our method, the results illustrate that there is promise in the method for categorising regions, particularly in the absence of comprehensive streamflow data as is the case in many regions in Australia and elsewhere. Further investigation of the validation (i.e. linking the regionalisation to streamflow indices) using more in-depth data mining approaches (e.g. decision trees [49]) is likely to produce even greater classification success.

Conclusions

Hydrologic classifications are increasingly being employed in the management and research of aquatic resources. Our approach differed from inductive hydrological regionalisation where membership is defined quantitatively based on metrics of stream flow, and traditional deductive regionalisation which require the use of catchments or other appropriate spatial units. Instead, membership of pixels was defined qualitatively with the random forest classifier based on a statistical classification of a number of environmental variables that we believe could have a direct influence on the hydrologic cycle. In essence, we present a method that allowed the creation of spatially-independent hydrological regions; these regions represent a series of fundamental hydrologic landscape units that exist in multiple locations depending on environmental similarity rather than a combination of environment and streamflow-metric response similarity. To our knowledge, the application of deductive reasoning and hybrid classification is a novel approach in hydrological regionalisation. This method has removed the need to rely on a spatial unit specified *a priori*, such as a catchment or ecoregion, and has allowed the preservation of intra-catchment variability. Thus, it should be useful in the spatial explanation and prediction of streamflow responses that are known, or suspected, to vary within catchments. The ability of our regionalisation to discriminate among streams from different groups based on their range of flow indices highlights the value of this approach, particularly in regions where streamflow data are lacking.

Supporting Information

Figure S1 MDS Analysis plots for the ALOC 23 and ALOC 20 models. Top row shows the ALOC 23 and ALOC 20 groups, while the bottom row shows the ALOC 23 and ALOC 20 meta-group plots. Legends in the top row represent the meta-groups.

Figure S2 ALOC 23 and ALOC 20 dendrogram demonstrating the hierarchical relationships between the non-hierarchical groups as defined from the ALOC group averages using SIMPROF. Letters in green represent the meta-groups each combination of non-hierarchical groups belongs to.

Figure S3 BioClim variable distributions across each of the ALOC 23 meta-groups.

Figure S4 Groundwater variable distributions across each of the ALOC 23 meta-groups. Note that observations >>30,000 are missing from GW_TDS.

Figure S5 Landscape variable distributions across each of the ALOC 23 meta-groups.

Figure S6 Soil variable distributions across each of the ALOC 23 meta-groups.

Figure S7 Climate variable distributions across each of the ALOC 23 meta-groups.

Figure S8 Variable contribution to each of the hierarchical meta-groups calculated using SIMPER on a standardised Euclidean distance matrix. Any variables contributing <5% to the variance were pooled together and are represented by ALL_OTHER_VARS. Missing groups contained only one ALOC cluster and therefore % variable contribution could not be calculated with SIMPER.

Figure S9 The correlation between the number of samples from each class and the number of samples correctly classified by CAP was highly significant, with the linear relationship between the variables for each sample illustrated in the figure. However, as the relationship was linear there was no clear threshold suggesting a minimum number of gauges needed to guarantee an acceptable level of accuracy. The CAP on the original dataset ($n = 201$) was quite poor (classification accuracy = 48%, $m = 30$, $p = 0.001$), while the bootstrapped dataset ($n = 383$) was a significant improvement (classification accuracy = 66%, $m = 28$, $p = 0.001$). CAP on the bootstrap dataset with a 20% validation sample also performed reasonably (classification accuracy = 67%, $m = 32$, $p = 0.001$).

Table S1 The variables used in the creation of the hydrological regionalisation. A number of variables describing the storage, transport and release of surface water, groundwater and atmospheric water were included in the analysis. * The DTM data was resampled to 30 m to enable geo-TIFF compatibility with ENVI 4.8.

Table S2 SIMPER results from PRIMER 6. Numbers represent % contribution of each of the variables to the ALOC 23 and ALOC 20 meta-groups on a standardised Euclidean distance matrix. Blank columns are meta-groups that only contained 1 ALOC cluster and therefore % contribution could not be calculated using SIMPER. KW = Kruskal-Wallis statistic, with higher values indicating a better ability of that variable to discriminate between clusters. All KW values were significant at $p<<0.001$. See Figure S9 for a graphical representation of this table.

Table S3 Total accuracies and kappa statistics for the 4 RF classified models, and producer and user accuracies for each of the classes defined by the ALOC algorithm as classified by RF. N/A indicates groups that were missing from the classified dataset as a result of exclusion from the samples used to train the RF model. In some cases, groups were absent from the 80% training data, while others were excluded by the bootstrap aggregation step used to train the RF models.

Code S1 S_perm_test.R: R code for running the permutation test for testing for differences between the resampled and original classifications.

Data S1 S_Kappa_ALOC23.csv: CSV file with 4 columns. The first column is the point ID from the random subsample used for comparisons between the original and resampled classifications, the remaining columns are the "true" class from the ALOC clustering, the class extracted from the location of the 30-m RF classification, and the class extracted from the 2.5-km resampled RF classification.

Author Contributions

Conceived and designed the experiments: SCB REL JF. Performed the experiments: SCB. Analyzed the data: SCB REL JF. Wrote the paper: SCB REL VLV JF LL.

References

1. Bunn SE, Arthington AH (2002) Basic principles and ecological consequences of altered flow regimes for aquatic biodiversity. Environmental Management 30: 492–507.
2. Poff NL, Richter BD, Arthington AH, Bunn SE, Naiman RJ, et al. (2010) The ecological limits of hydrologic alteration (ELOHA): A new framework for developing regional environmental flow standards. Freshwater Biology 55: 147–170.
3. Poff NL, Allan JD, Bain MB, Karr JR, Prestegaard KL, et al. (1997) The natural flow regime: A paradigm for river conservation and restoration. BioScience 47: 769–784.
4. Naiman RJ, Latterell JJ, Pettit NE, Olden JD (2008) Flow variability and the biophysical vitality of river systems. Comptes Rendus - Geoscience 340: 629–643.
5. LeRoy Poff N, Allan JD (1995) Functional organization of stream fish assemblages in relation to hydrological variability. Ecology 76: 606–627.
6. Kennard MJ, Pusey BJ, Olden JD, MacKay SJ, Stein JL, et al. (2010) Classification of natural flow regimes in Australia to support environmental flow management. Freshwater Biology 55: 171–193.
7. Winter TC (2001) The concept of hydrologic landscapes. Journal of the American Water Resources Association 37: 335–349.
8. Wolock DM, Winter TC, McMahon G (2004) Delineation and evaluation of hydrologic-landscape regions in the United States using geographic information system tools and multivariate statistical analyses. Environmental Management 34: S71–88.
9. Olden JD, Kennard MJ, Pusey BJ (2012) A framework for hydrologic classification with a review of methodologies and applications in ecohydrology. Ecohydrology 5: 503–518.
10. Snelder TH, Biggs BJF, Woods RA (2005) Improved eco-hydrological classification of rivers. River Research and Applications 21: 609–628.
11. Santhi C, Allen PM, Muttiah RS, Arnold JG, Tuppad P (2008) Regional estimation of base flow for the conterminous United States by hydrologic landscape regions. Journal of Hydrology 351: 139–153.
12. Carlisle DM, Falcone J, Wolock DM, Meador MR, Norris RH (2010) Predicting the natural flow regime: Models for assessing hydrological alteration in streams. River Research and Applications 26: 118–136.
13. McManamay RA, Orth DJ, Dolloff CA, Frimpong EA (2012) Regional Frameworks Applied To Hydrology: Can Landscape-Based Frameworks Capture The Hydrologic Variability? River Research and Applications 28: 1325–1339.
14. Sawicz K, Wagener T, Sivapalan M, Troch PA, Carrillo G (2011) Catchment classification: Empirical analysis of hydrologic similarity based on catchment function in the eastern USA. Hydrology and Earth System Sciences 15: 2895–2911.
15. Poff NL (1996) A hydrogeography of unregulated streams in the United States and an examination of scale-dependence in some hydrological descriptors. Freshwater Biology 36: 71–91.
16. Poff NL, Olden JD, Pepin DM, Bledsoe BP (2006) Placing global stream flow variability in geographic and geomorphic contexts. River Research and Applications 22: 149–166.
17. Brown S, Versace V, Laurenson L, Ierodiaconou D, Fawcett J, et al. (2012) Assessment of Spatiotemporal Varying Relationships Between Rainfall, Land Cover and Surface Water Area Using Geographically Weighted Regression. Environmental Modeling and Assessment 17: 241–254.
18. Mackey BG, Berry SL, Brown T (2008) Reconciling approaches to biogeographical regionalization: A systematic and generic framework examined with a case study of the Australian continent. Journal of Biogeography 35: 213–229.
19. Stein JL, Hutchinson M, Pusey BJ, Kennard MJ (2009) Ecohydrological classification based on landscape and climate data. Appendix 8. In: Pusey BJ, Kennard MJ, Stein JL, Olden JD, editors. Ecohydrological regionalisation of Australia: A Tool for Management and Science. Canberra: Land and Water Australia.
20. Clifford DHT, Stephenson W (1975) An introduction to numerical classification. New York: Academic Press.
21. Cormack RM (1971) A Review of Classification. Journal of the Royal Statistical Society Series A (General) 134: 321–367.
22. Everitt B (1980) Cluster analysis. 2nd Edition. London: Heinemann.
23. Clarke KR, Warwick RM (2001) Change in Marine Communities: An Approach to Statistical Anlysis and Interpretation, 2nd Edition. Plymouth: PRIMER-E.
24. Legendre P, Legendre L (1998) Numerical Ecology, 2nd Edition. Amsterdam: Elsevier Science.
25. Belbin L, McDonald C (1993) Comparing three classification strategies for use in ecology. Journal of Vegetation Science 4: 341–348.
26. Belbin L (1987) The use of non-hierarchical allocation methods for clustering large sets of data. Australian Computer Journal 19: 32–41.
27. Breiman L (2001) Random forests. Machine Learning 45: 5–32.
28. Held M, Jakimow B, Rabe A, van der Linden S, Wirth F, et al. (2012) EnMAP-Box Manual, Version 1.4, Humboldt Universität zu Berlin, Germany. Available: www.enmap.org. Accessed 2013 Aug.
29. Lillesand TM, Kiefer RW, Chipman JW (2008) Remote Sensing and Image Interpretation: 6th edition. New Jersey: Wiley.
30. Department of Environment and Primary Industries (2013) Victorian Resources Online - Temperature, Available: http://vro.dpi.vic.gov.au/dpi/vro/vrosite.nsf/pages/climate-temperature. Accessed 2013 Aug.
31. Department of Environment and Primary Industries (2013) Victorian Resources Online - Rainfall, Available: http://vro.dpi.vic.gov.au/dpi/vro/vrosite.nsf/pages/rainfall. Accessed 2013 Aug.
32. Corrick A (1992) An assessment of Victoria's wetlands. Melbourne: Department of Conservation and Environment.
33. Versace VL, Ierodiaconou D, Stagnitti F, Hamilton AJ (2008) Appraisal of random and systematic land cover transitions for regional water balance and revegetation strategies. Agriculture, Ecosystems & Environment 123: 328–336.
34. Ierodiaconou D, Laurenson L, Leblanc M, Stagnitti F, Duff G, et al. (2005) The consequences of land use change on nutrient exports: a regional scale assessment in south-west Victoria, Australia. Journal of Environmental Management 74: 305–316.
35. Versace VL, Ierodiaconou D, Stagnitti F, Hamilton AJ, Walter MT, et al. (2008) Regional-scale models for relating land cover to basin surface-water quality using remotely sensed data in a GIS. Environmental Monitoring and Assessment 142: 171–184.
36. Yihdego Y, Webb JA (2011) Modeling of bore hydrographs to determine the impact of climate and land-use change in a temperate subhumid region of southeastern Australia. Hydrogeology Journal 19: 877–887.

37. Yihdego Y, Webb J (2013) An Empirical Water Budget Model As a Tool to Identify the Impact of Land-use Change in Stream Flow in Southeastern Australia. Water Resources Management 27: 4941–4958.
38. ESRI (2012) ArcGIS Desktop 10.1, Environmental Systems Research Institute, Redlands. Available: http://www.esri.com
39. Sprintsin M, Karnieli A, Berliner P, Rotenberg E, Yakir D, et al. (2007) The effect of spatial resolution on the accuracy of leaf area index estimation for a forest planted in the desert transition zone. Remote Sensing of Environment 109: 416–428.
40. Mumby PJ, Edwards AJ (2002) Mapping marine environments with IKONOS imagery: Enhanced spatial resolution can deliver greater thematic accuracy. Remote Sensing of Environment 82: 248–257.
41. Exelis Visual Information Solutions (2010) ENVI version 4.8, Exelis Visual Information Solutions, Boulder. Available: http://www.exelisvis.com/ProductsServices/ENVI/ENVI.aspx
42. Dixon PM (2006) Ripley's K function. In: El-Shaarawi A, Piegorsch WW, editors. Encyclopedia of Environmetrics. New York: Wiley.
43. Clarke KR, Gorley RN (2006) PRIMER v6: User Manual/Tutorial. Plymouth: PRIMER-E.
44. Clarke KR (1993) Non-parametric multivariate analyses of changes in community structure. Australian Journal of Ecology 18: 117–143.
45. Clarke KR, Somerfield PJ, Gorley RN (2008) Testing of null hypotheses in exploratory community analyses: similarity profiles and biota-environment linkage. Journal of Experimental Marine Biology and Ecology 366: 56–69.
46. PATN Website. Blatant Fabrications Pty. Ltd. (2009) PATN Version 3.1.2, Available: http://www.patn.com.au/default.htm. Accessed 2012 Apr.
47. R Core Team (2013) R: A language and environment for statistical computing, R Foundation for Statistical Computing, Vienna. Available: http://www.R-project.org/
48. Rodriguez-Galiano VF, Ghimire B, Rogan J, Chica-Olmo M, Rigol-Sanchez JP (2012) An assessment of the effectiveness of a random forest classifier for land-cover classification. ISPRS Journal of Photogrammetry and Remote Sensing 67: 93–104.
49. Breiman L (1993) Classification and regression trees: Chapman and Hall/CRC press.
50. Cohen J (1960) A coefficient of agreement for nominal scales. Educational and Psychological Measurement 20: 37–46.
51. Revelle W (2013) psych: Procedures for Psychological, Psychometric, and Personality Research, Available: http://cran.r-project.org/web/packages/psych/index.html. Accessed 2013 Aug.
52. Gamer M (2012) irr: Various Coefficients of Interrater Reliability and Agreement, Available: http://cran.r-project.org/web/packages/irr/index.html. Accessed 2013 Aug.
53. Fay MP (2010) Two-sided Exact Tests and Matching Confidence Intervals for Discrete Data. R journal 2: 53–58.
54. McNemar Q (1947) Note on the sampling error of the difference between correlated proportions or percentages. Psychometrika 12: 153–157.
55. McKenzie DP, Mackinnon AJ, Péladeau N, Onghena P, Bruce PC, et al. (1996) Comparing correlated kappas by resampling: Is one level of agreement significantly different from another? Journal of Psychiatric Research 30: 483–492.
56. Foody GM (2004) Thematic Map Comparison: Evaluating the Statistical Significane of Differences in Classification Accuracy. Photogrammetric Engineering & Remote Sensing 70: 627–633.
57. Olden JD, Poff NL (2003) Redundancy and the choice of hydrologic indices for characterizing streamflow regimes. River Research and Applications 19: 101–121.
58. Anderson MJ (2001) A new method for non-parametric multivariate analysis of variance. Austral Ecology 26: 32–46.
59. Anderson MJ, Robinson J (2003) Generalized discriminant analysis based on distances. Australian and New Zealand Journal of Statistics 45: 301–318.
60. Anderson MJ, Willis TJ (2003) Canonical analysis of principal coordinates: A useful method of constrained ordination for ecology. Ecology 84: 511–525.
61. Kennard MJ, Mackay SJ, Pusey BJ, Olden JD, Marsh N (2010) Quantifying uncertainty in estimation of hydrologic metrics for ecohydrological studies. River Research and Applications 26: 137–156.
62. Kennard MJ, Pusey BJ, Olden JD, Mackay S, Stein JL, et al. (2009) Ecohydrological classification of Australia's flow regimes. Appendix 5. In: Pusey BJ, Kennard MJ, Stein JL, Olden JD, editors. Ecohydrological regionalisation of Australia: A Tool for Management and Science. Canberra: Land and Water Australia.
63. Marsh NA, Stewardson MJ, Kennard MJ (2003) River Analysis Package. Monash University Melbourne: Cooperative Research Centre for Catchment Hydrology.
64. Anderson MJ, Gorley RN, Clarke KR (2008) PERMANOVA+ for PRIMER: Guide to Software and Statistical Methods. Plymouth, UK: PRIMER-E.

Spatial Scale, Means and Gradients of Hydrographic Variables Define Pelagic Seascapes of Bluefin and Bullet Tuna Spawning Distribution

Diego Alvarez-Berastegui[1]*, Lorenzo Ciannelli[2], Alberto Aparicio-Gonzalez[3], Patricia Reglero[3], Manuel Hidalgo[3], Jose Luis López-Jurado[3], Joaquín Tintoré[1,4], Francisco Alemany[3]

1 Balearic Islands Coastal Observing and Forecasting System (SOCIB), Palma de Mallorca, Balearic Islands, Spain, **2** College of Earth, Ocean, and Atmospheric Sciences (CEOAS), Oregon State University, Corvallis, Oregon, United States of America, **3** Instituto Español de Oceanografía - Centre Oceanogràfic de les Balears (COB-IEO), Palma de Mallorca, Balearic Islands, Spain, **4** Institute for Advanced Studies (IMEDEA), Consejo Superior de Investigaciones Científicas (CSIC) and the University of the Balearic Islands (UIB). Esporles, Spain

Abstract

Seascape ecology is an emerging discipline focused on understanding how features of the marine habitat influence the spatial distribution of marine species. However, there is still a gap in the development of concepts and techniques for its application in the marine pelagic realm, where there are no clear boundaries delimitating habitats. Here we demonstrate that pelagic seascape metrics defined as a combination of hydrographic variables and their spatial gradients calculated at an appropriate spatial scale, improve our ability to model pelagic fish distribution. We apply the analysis to study the spawning locations of two tuna species: Atlantic bluefin and bullet tuna. These two species represent a gradient in life history strategies. Bluefin tuna has a large body size and is a long-distant migrant, while bullet tuna has a small body size and lives year-round in coastal waters within the Mediterranean Sea. The results show that the models performance incorporating the proposed seascape metrics increases significantly when compared with models that do not consider these metrics. This improvement is more important for Atlantic bluefin, whose spawning ecology is dependent on the local oceanographic scenario, than it is for bullet tuna, which is less influenced by the hydrographic conditions. Our study advances our understanding of how species perceive their habitat and confirms that the spatial scale at which the seascape metrics provide information is related to the spawning ecology and life history strategy of each species.

Editor: Athanassios C. Tsikliras, Aristotle University of Thessaloniki, Greece

Funding: The development of this study was supported by the Project "PERSEUS" financed by the Seventh Framework Program for Research (FP7) under theme "Oceans Tomorrow", grant Agreement No. 287600 (http://www.perseus-fp7.eu/), the Spanish Foundation for Science and Technology, under the project "BALEARES" Grant number: CTM2009-07944/MAR (http://www.fecyt.es/fecyt/home.do) and the Science, Engineering and Education for Sustainability NSF-Wide Investment (SEES) Research Coordination Network, Grant 1140207 "Sustainability of Marine Renewable Resources in Subarctic Systems Under Incumbent Environmental Variability and Human Exploitation" (http://grantome.com/grant/NSF/OCE-1140207). Hydrographic data used in the study were provided by IBAMAR (http://www.ba.ieo.es/ibamar) a public data base founded by the Council for Economy and Competitiveness of the Balearic Islands Government (http://www.caib.es/govern/organigrama/area.do?coduo=6&lang=es), grant number AAEE0022/08. The research was developed under the framework of the BLUEFIN TUNA Project driven by the Balearic Island Observing and Forecasting System (www.socib.es) and the Spanish Institute of Oceanography (www.ieo.es). The funders had no role in study design, data collection and analysis, decision to publish, or preparation of the manuscript.

Competing Interests: The authors have declared that no competing interests exist.

* Email: dalvarez@socib.es

Introduction

Seascape ecology represents an emerging field in the study of how the habitat structure shapes the spatial distribution of marine species and influences key ecological processes [1],[2]. This discipline initiated applying techniques and metrics from the traditional landscape ecology to characterize and quantify spatial structure of benthic habitats, observed as a mosaic of patches of different habitat classes [1],[3],[4], [5],[6]. Nevertheless, there is still a gap in the development of concepts and techniques providing metrics to characterize the spatial structure of the seascape in the pelagic environments, where there are no clear boundaries delimitating different habitats [1],[2],[6]. In the framework of landscape ecology spatial gradients have been recently proposed as more appropriated metric than traditional categorical patch mosaic based metrics to characterize continuous habitats [7]. Accordingly, a location in a pelagic seascape would be better characterized by the combination of the value of a particular hydrographic variable and its spatial gradient.

Several studies have applied gradients of hydrographic parameters to characterize the spatial distribution of marine species during various life history stages, as nursery, foraging or spawning [8],[9],[10],[11],[12]. It is likely that the scale at which an individual perceive a change in the environment (i.e., a gradient) varies according to life history, ontogeny and to the hydrographic variable in exam. For instance, while large-scale gradients associated with the North Pacific transition zone drive the location of many pelagic predators including albacore tuna (*Thunnus alalunga*) during their feeding migratory stages [13], once off the west coast of the US, albacore tuna distribution is associated with smaller scale features linked to upwelling fronts [14]. In the Mediterranean Sea during spawning, bluefin tuna distribution is regulated by oceanographic variables that can change at relatively small scales [15],[16]. In spite of the expected importance of gradient scales, to our knowledge there are no studies that have evaluated the effect of changing the spatial scales at which environmental gradients are calculated to model the spatial distribution of fish. The goal of our study is to examine the distribution of large pelagic predators during spawning by explicitly considering mean, gradients and scale of gradients of hydrographic variables.

Atlantic bluefin tuna (*Thunnus thynnus*) and bullet tuna (*Auxis rochei rochei*) are two species of pelagic predators showing different spawning strategies. We target these species in the Balearic Sea (Figure 1), known as a recurrent spawning area for large pelagic species in the Western Mediterranean [17]. Bluefin tuna is a highly migratory species; the Eastern population enters in the Mediterranean Sea from the North Atlantic at the end of spring and early summer [18],[19]. Their spawning activity at the Balearic Sea is linked to the regional oceanography with spawning grounds located in the vicinity of frontal structures formed when the recent Atlantic water mass encounters the more saline resident surface Atlantic waters [15]. The area is characterized by highly dynamic processes that promote a seascape shaped by filaments, fronts and eddies whose location varies between years [20],[21]. Bullet tuna, by contrast, is smaller and more frequent in near coastal areas [22]. The spawning of bullet tuna is associated with the geography. Young larvae are found mainly in coastal areas and with little influence of the local oceanography in comparison with bluefin [16].

We expect that, when selecting spawning locations, a large-bodied and long-distance migratory pelagic fish, such as the bluefin tuna explores its environment at larger spatial scales than bullet tuna, a small-bodied and non-migratory pelagic fish. Therefore, we expect that pelagic seascape metrics based on the combination of hydrographic parameter values and their gradients calculated at appropriate spatial scales provide relevant information for bluefin tuna but not for bullet tuna, where we expect a greater reliance on geographic and hydrographic parameters, calculated at comparatively small-scale.

In this work we analyze the influence of the pelagic environment by depicting the spatial scales at which gradients of hydrographic variables are linked to the spawning ecology of these tuna species. We investigate the two most relevant hydrographic variables describing their spawning spatial distribution: salinity and geostrophic currents velocity [16], already determined in previous studies. Our analytical approach has two steps. Firstly, we identify the scale at which each hydrographic variable (Figure 2) maximizes the performance of a model fitted to larval distribution. Second, we evaluate whether components of the seascape (i.e., mean and gradients of oceanographic variables) are interactively affecting the spatial distribution of tuna larvae. By performing this analysis on two species we evaluate how fish with contrasting life history strategies perceive their environments when deciding for spawning locations.

Materials and Methods

Data acquisition

Bluefin and bullet tuna larvae were collected during ichthyoplankton surveys using Bongo nets from 2001 to 2005. The surveys were conducted by the Instituto Español de Oceanografía (www.ieo.es), a Spanish Government marine research institution. The sampling scheme was communicated and approved by the Spanish Directorate of Fisheries before the sampling was conducted. No specific ethical approval was required and the survey of biological data was conducted using Bongo nets,which are accepted standard

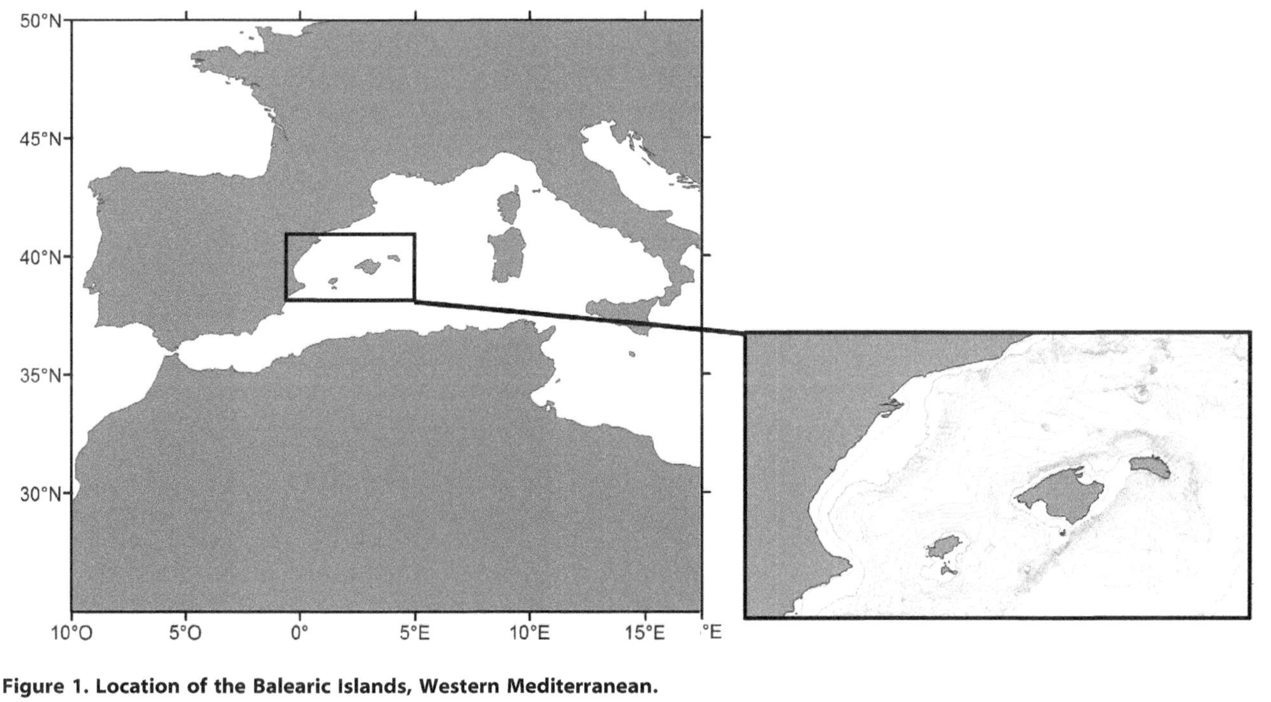

Figure 1. Location of the Balearic Islands, Western Mediterranean.

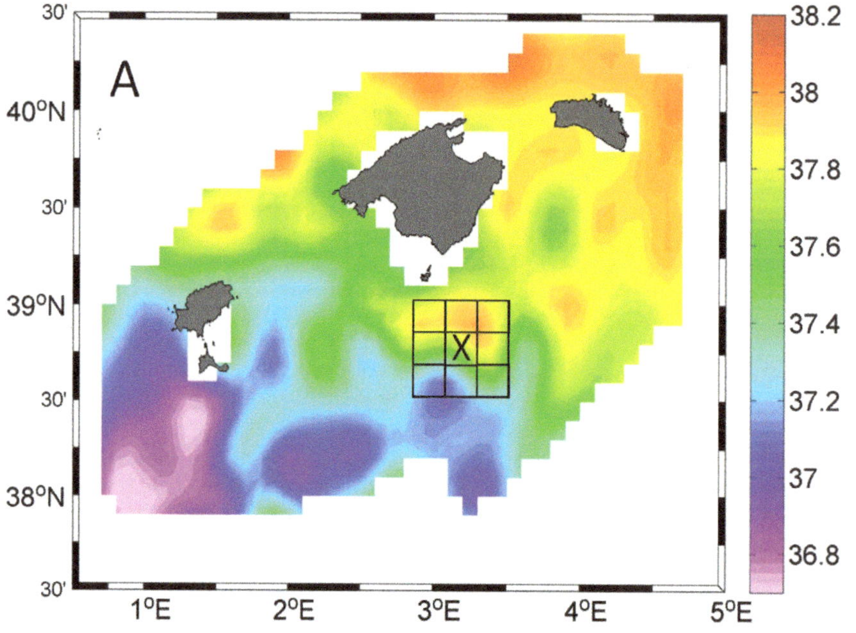

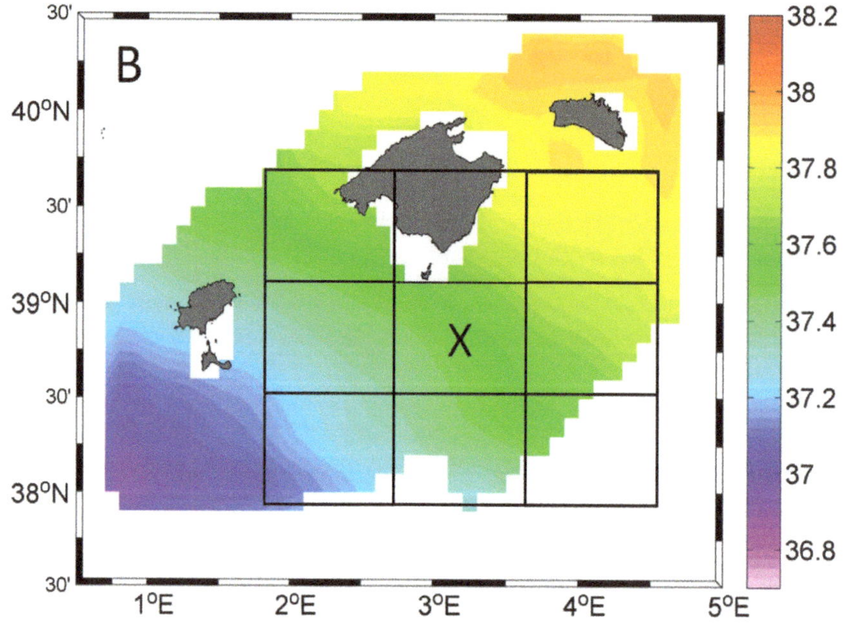

Figure 2. Sea surface salinity field in 2003. Spatial means of sea surface salinity processed at two different scales A) 0.15 degrees and B) 0.75 degrees. Spatial means were interpolated following an objective analysis onto a regular grid by using minimum error variance methods (Bretherton, 1976). The squares in each figure show the polygons used for the calculation of the spatial gradients at the two scales at station X.

techniques for this type of surveys, used worldwide for the collection of plankton samples, including billfish and tuna larvae [15],[17], [23], [24]. The nets were towed at low speeds, around 2 knots, during 8–10 minutes, and plankton samples were immediately preserved with 4% formalin buffered with borax onboard.

Around 200 stations were sampled every year, in a regular sampling grid of 10×10 nm located between 37.85°–40.35° N and 0.77°–4.91°E, covering a total area of 101360 km2 (=280×362 km) around the Balearic archipelago. The sampling was conducted during June–July coinciding with the spawning period of bullet and bluefin tuna (see [15] for details of the sampling procedures). Tuna larvae were identified to the species level and measured in standard length. All larvae identified as yolk sac and preflexion stages (<4.5 mm) were classified as "stage 1".

Mean and maximum age of larvae under this size are 6 days and 11 days old respectively, accounting for growth [25] and hatching time [26]. Displacement from spawning areas during these intervals in the study area are below 25 kilometers for the mean ages, which are in the range of 1.4 sampling stations, and 46 kilometers for the maximum ages, what is in the range of 2.6 sampling stations. These values have been calculated following methods in [27]. Considering these values, abundances of stage 1 larvae have been defined to get a proxy of spawning locations as in previous research in the study area [28].

Vertical profiles of conductivity, temperature and pressure data were recorded at all stations, by means of Sbe911 CTD. Sea surface salinity at each station (SAL) was calculated as the mean salinity over the mixed layer depth. Geostrophic velocities (GVEL) were calculated at sea surface from the first-derivative of the sea surface height between adjacent points, which was obtained by vertical integration of the specific volume, using 600 m as the level of no motion [21].

These two variables (SAL and GVEL) were selected, since they have been demonstrated to be the two most relevant environmental variables describing the spawning spatial distribution of tuna [16]. Sea surface temperature was also included in the models since in this area it is a secondary but relevant variable mainly related to the phenology of the spawning process [29]. However, the spatial gradient was not explored because sea surface temperature during the summer changes relatively fast due to solar irradiance [21].

Processing of spatial gradients along continuous spatial scales

Spatial gradients from the sea surface salinity field (gSAL) and geostrophic velocity field (gGVEL) within the sampled region were calculated at six spatial scales, from 0.15° to 0.90° with a spatial increment of 0.15°. The minimum (0.15°) and the maximum scale (0.90°) were in the range of the smallest (from 0.13° to 0.27°) and largest (up to 0.92°) mesoscale oceanographic structures in the area [21]. For the computation of the gradients, nine square polygons at every scale were delimitated around each sampling position (see examples for scale 0.15° and 0.75° in figure 2A, 2B respectively). The gradient was then computed as the maximum absolute difference between the mean hydrographic variable at the center polygon and each of the eight surrounding polygons standardized to distance [9]. The software for the spatial processing was developed in R language [30].

Identification of spatial scales

Comparison of how models perform along scales allowed identifying the spatial scales at which information provided by gradients is maximized. The effect of gGVEL and gSAL at each scale on the abundance of bullet and bluefin tuna larvae was assessed using nonparametric regression statistical models (generalized additive models, GAMs, [31]). A base model was formulated to describe inter-annual variability (variable YEAR), sampling location (latitude and longitude variables), and the hour of the day on the catch of tuna larvae. Over-dispersed Poisson distribution family and a natural-log link were selected to model larval data. The volume of water filtered was included as an offset (after natural log transformation), to account for the effort expanded in catching the sample (Equation 1). The effects of these variables on the base model have been already analyzed in previous studies [16]. Here, the base model represented the null hypothesis of no gradient effect on tuna larvae distribution, against which all other more complex formulations will be compared.

Equation 1: Base model

$$Larvae\ abundance = offset(log(m3)) + factor(year) + sm_1(long, lat) + sm_2(hour)$$

m3 = volume filtered by the bongo nets (m3); long = longitude; lat = latitude; hour = hour of the day expressed from 0 to 1, sm_1 and sm_2 the smoothing functions.

At each spatial scale a GAM model was processed including the gradient of one hydrographic variable (gSAL, gGVEL) as a new additive term (s3) in the standardization model. The number of knots for the new smoother was always set to a maximum of three (i.e. two degrees of freedom) in order to avoid over fitting in the responses.

The identification of characteristic spatial scales (cgSAL, cgGVEL) was assessed with scalogram where the scale of the covariate is plotted against a measure of the model goodness of fit, which in our case were represented by the adjusted R-squared (Rsq, the higher the better), and the Generalized Cross Validation (GCV, the lower the better) [31]. We selected the scale that maximize Rsq and minimize the GCV. Results of the base model (when a seascape covariate was not included) were presented in the same graphics. Note that due to the greater complexity of the gradient model higher Rsq values do not necessarily imply an improvement in relation to the base model, while they do represent a better performance when compared to other gradient models.

Significant differences of Rsq values between models, or GCVs, were obtained from t-test of these parameters obtained from 500 iterations where 10% of the data was excluded. For all cases, alternative hypothesis (difference in means is not equal to 0) was accepted only if the P value was lower than 0.001, with a confidence level of 0.99. When one variable presented similar Rsq and GCV values at various scales, selection was assessed by inspection of the plot showing the response of the abundance in relation to the gradient processed at those scales.

Once the characteristic scale of the gradients was identified, we tackled the questions of whether the information provided by the gradients is different and complementary to the information provided by the hydrographic variables from which they were calculated, and in that case, how the information from these two variables (spatial mean and gradient) should be combined to maximize the goodness of fit of the models and the ecological information they provide. To assess these questions we analyzed the performance of models with different complexity:

i) The base model from equation 1.
ii) Hydrographic models combining the sea surface salinity, geostrophic velocity and sea surface temperature at the sampling station (stSAL, stGVEL, stSST).
iii) Seascape models combining the gradients at characteristic scales (cgSAL, cgGVEL) and the hydrographic variable at the sampling location (stSAL, stGVEL, stSST). Different seascape models were constructed including the two components of the seascape (values at stations and gradients) as additive and interactive terms. An interaction may be ecologically meaningful when a species is selecting its spawning habitat on a specific side of a frontal region, for example. In such case, it is the combination of both the gradient and the mean that provide the suitable conditions for spawning. The performance of different model configurations for each species was assessed by the delta AIC (ΔAIC), calculated as the difference between model AICs and the base model AIC. The AIC in this case is best suited for model comparisons because each model had different number of variables [32]. Rsq, GCV and explained deviances were used to compare how models perform between the two species, as AIC values among models with different dependent variables are not comparable.

Results

Identification of characteristic spatial scales

In all years considered, the recent Atlantic water masses encountered the more saline resident water masses forming an oceanic frontal zone inside the study area. The size of such frontal zone was bigger than other oceanographic phenomena as small eddies and meanders derived from the instabilities along the haline front and the effect of strong bathymetric changes (Figures S3, S4).

The scalogram of gGVEL for bluefin showed that Rsq values gradually improve as the spatial scales increased to a maximum at 0.6° (Rsq = 0.44, Figure 3A), which was chosen as the geostrophic velocity gradient characteristic scale for bluefin tuna. Values of GCV showed a similar pattern of model improvement, being significantly better than the base model at 0.6 degrees (Figure 3B). At this characteristic scale the response of the larvae abundance is positively related to the gradient of geostrophic velocity (Figure 3C).

The scalograms of gSAL for bluefin tuna showed also an increment of Rsq with higher values at 0.6° and 0.75° that also coincide with lower values of GCV (Figure 3D, 3E). Differences of R-sq between these two scales (0.6° and 0.75°) were not significant. The characteristic scale for gSAL was chosen at 0.6° as the model response at this scale presented a less ambiguous effect on larval abundance (Figure 3F). The gSAL at 0.75° spatial gradients displays a dome-shaped response with a less clear ecological interpretation (Figure S1). At this scale GCV was lower than the base model.

In contrast to the results obtained for salinity, the gradients of geostrophic velocities (gGVEL) did not show any single scale that maximizes R-square and minimizes GCV for bullet tuna (Figure 4A, 4B). The Rsq scalogram showed a flat trend with the highest value at 0.15 degrees. The Rsq value at this scale (= 0.18) showed similar values than other scales (values between 0.170 and 0.173) or when compared to the base model (= 0.166). On the contrary the GCV scalogram showed significantly lower values than the base model at 0.45 and 0.6 degrees, scales at which Rsq were not even significantly higher than the base model. Therefore, the contradictory response of the model performance indicators, the flat trend of Rsq scalogram and their very low values (despite the higher complexity of the gradient models in relation to the base model), may indicate that the spatial gradient of geostrophic velocity is not a valid seascape metric for the spawning locations of bullet tuna. Consequently, gGVEL was excluded from further analysis in relation to this species.

The gSAL scalogram of bullet tuna showed a moderate effect of the scale at which gradients were calculated. The shape of the scalogram did not show a peak scale at which model performs better (Figure 4C). Scales from 0.45°, 0.6° and 0.75° for gSAL seemed to maximize Rsq, but not being different from each other. In this case GCV-scalograms showed the lowest at 0.75° (Figure 4D), being significantly lower than the base model, which

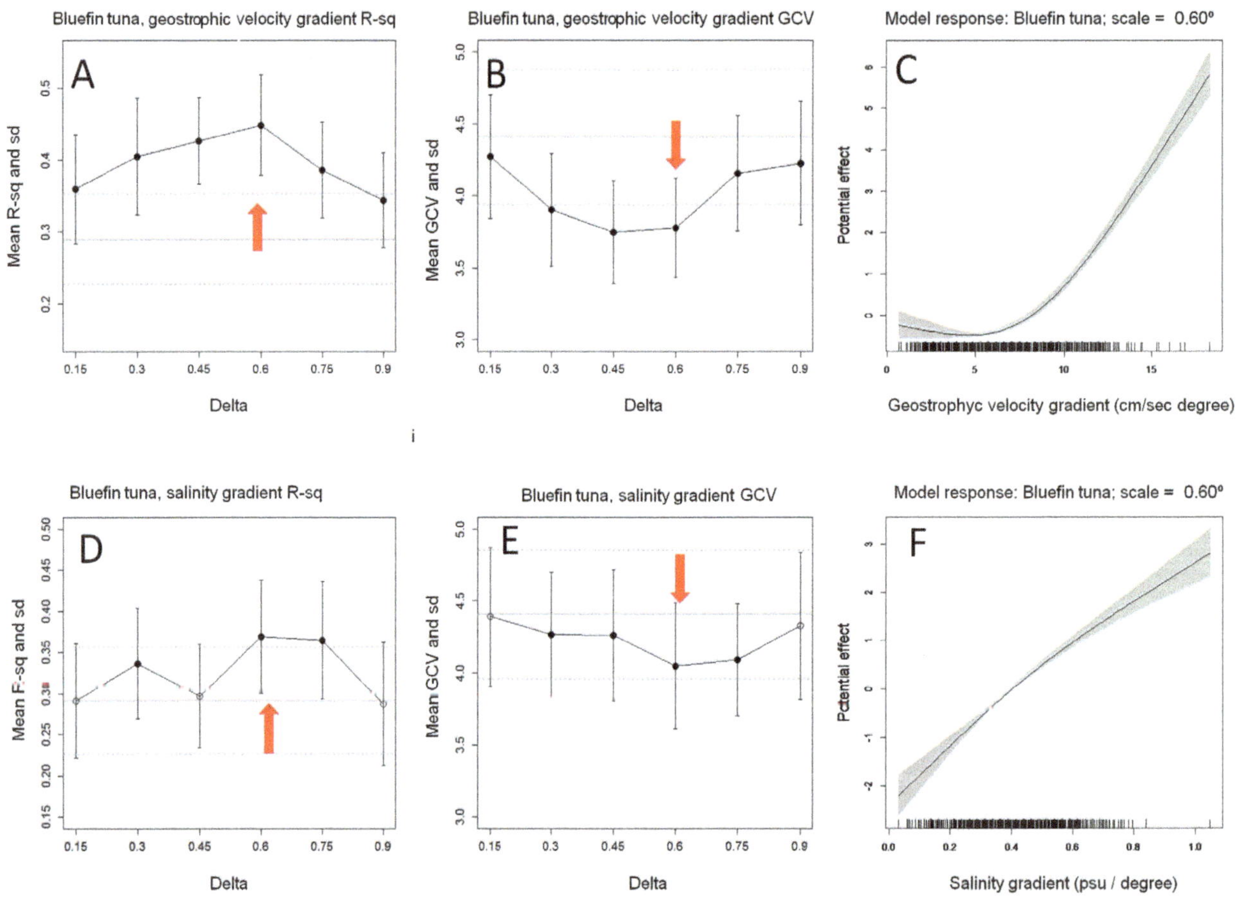

Figure 3. Rsq and GCV Scalograms of bluefin tuna larva abundance models along spatial scales, standard deviations. Horizontal grey lines indicate statistics from the base model (Straight line = mean, dashed line = Sd). Black dots show scales at which values are significantly different from the standardization model. Red arrows indicate the selected characteristic scale.

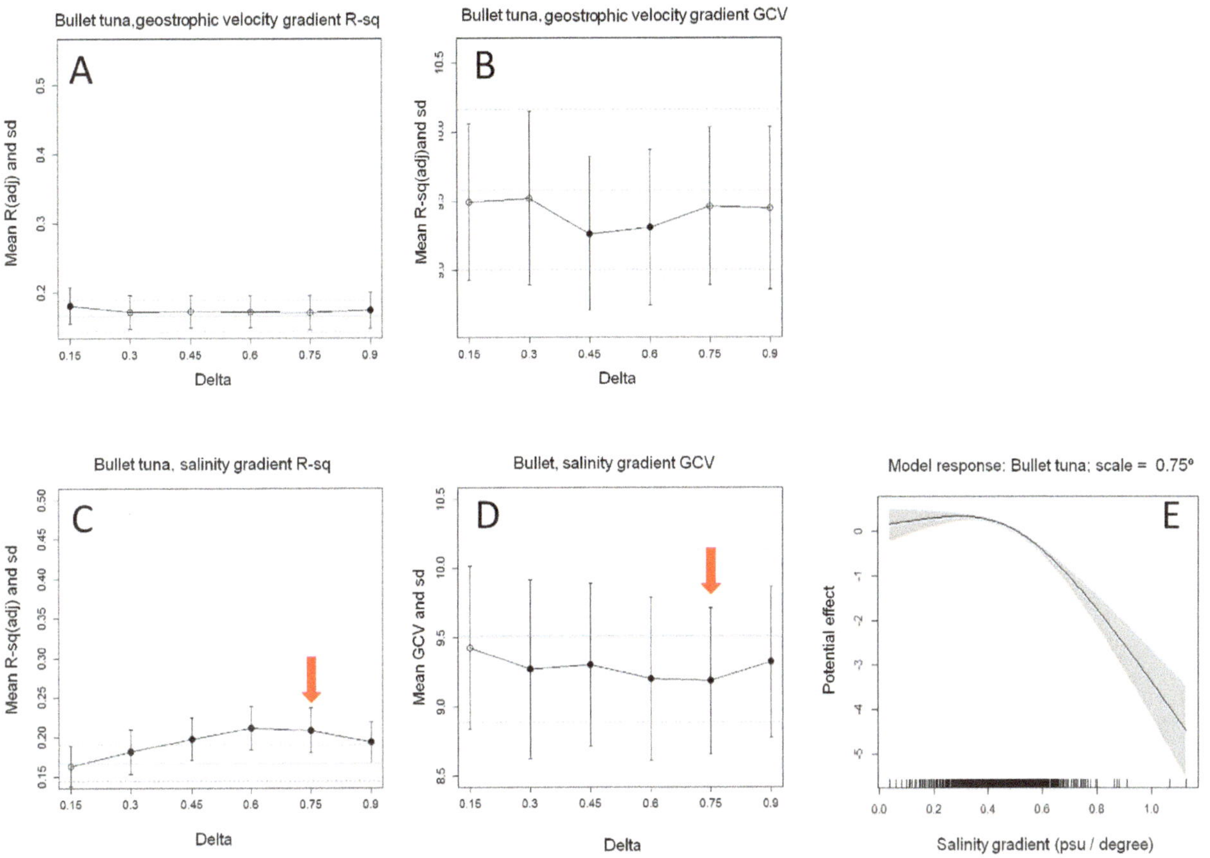

Figure 4. Rsq and GCV scalograms of bullet tuna larva abundance models along spatial scales, standard deviations. Horizontal grey lines indicate statistics from the standardization model (Strait line = mean, dashed line = Sd). Black dots show scales at which values are significantly different from the standardization model. Red arrows indicate selected characteristic scale.

was selected as characteristic scale. The Rsq at this scale was higher (= 0.21) than that of the base model (Rsq = 0.16). At this scale, gradients displayed a negative effect on the bullet tuna larvae abundance (Figure 4E) showing that bullet tuna spawning locations are found with more probability in areas where salinity is spatially homogeneous– a result that contrasted to that obtained for bluefin tuna.

Species-specific seascape characterization

The best model for each species included a gradient and a mean term (Tables 1 and 2). Note however that the hydrographic model already represents an improvement respect to the standardization model (Tables 1 and 2). For bluefin tuna the best seascape model had an improvement of 186% of the Rsq when compared to the base model (Rsq base mode = 0.23; Rsq best seascape model = 0.66, Table 1). The improvement for bullet was 61%, considerably lower compared to bluefin (Rsq base mode = 0.16; Rsq best seascape model = 0.25, Table 2).

Pearson correlation coefficients between hydrographical variables and their gradients at characteristic scales were in all cases below 0.50 and pair plots showed no clear tendencies on the correlations (Figure S2), indicating that selected gradients provided complementary information to that of the hydrographical variable. Models showed a better performance (i.e. lower GCV and higher ΔAIC; Tables 1 and 2) in all the cases when the gradient and the hydrographic value were considered as an interaction term, suggesting dependence in their effect on larvae abundance rather than an additive response. However, larvae abundance of each species responded differently to the interaction of seascape components (Figures 5A, 5B, 5C).

For Bluefin tuna, higher probability of spawning was associated to higher values of geostrophic velocity gradients, but where velocities at station may present either high or low values (Figure 5A), (Figures S7, S8). Considering that a gradient is characterized by an area with high current speed near an area of low current speed, this result indicates that spawning locations were not associated to a particular side of the gradient, but in an area around the location where maximum velocities occurs. The extension of this area would be around a circle of 0.6 degrees of radius (aprox. 65 km in the study area), the characteristic scale at which the gradients were more relevant. In contrast, the interaction of the salinity seascape variables showed high larvae abundances in areas with high salinity gradients and intermediate-high salinity levels (Figure 5B), indicating an effect of the location of the main haline front and a preference for spawning at the high salinity values of that front (Figures S5, S6). The characteristic spatial correlation scale of local oceanographic structures in the area is around 18 nmi (aprox. 0.15 meridian degrees) [21] and therefore surface oceanographic structures at smaller spatial scales are ephemeral. The oceanographic structures at larger spatial scales, relevant for bluefin tuna, are linked to the Med-Atlantic salinity front and last longer.

The functional form of the interaction terms was different for bullet tuna. The relation between the bullet tuna larvae

Table 1. Summary of GAM models of larvae abundance for Atlantic bluefin tuna (*Thunnus thynnus*).

Model group	Model variables	R2	Dev(%)	GCV	AIC	delta AIC
Base model	(latitude, longitude) + filtered volume + hour	0.232	40.8	4,596	3985,54	0
One additive variable models	base model + stGVEL0.15	0,271	43,4	4,412	3827,79	157,75
	base model + gGVEL0.6	0,39	49,4	3,947	3465,03	520,50
	base model + stSAL0.15	0,222	41,8	4,534	3924,70	60,84
	base model + gSAL0.6	0,301	45,1	4,275	3723,53	262,01
Hydrographic model	base model + stGVEL + stSAL + stTEMP	0,472	51,6	3,814	3338,62	646,92
GVEL seascape models	Hydrographic model + stGVEL + gGVEL0.6	0,53	55,8	3,500	3087,14	898,40
	Hydrographic model + (stGVEL,gGVEL0.6)	0,666	59,3	3,251	2881,29	1104,25
SAL seascape models	Hydrographic model + stSAL + gSAL0.6	0,506	54,8	3,571	3145,80	839,74
	Hydrographic model + (stSAL,gSAL0.6)	0,533	57,1	3,417	3009,58	975,96

Interaction terms included in parenthesis.

abundance and the interaction of spatial distribution of sea surface salinity and its gradients is presented in Figure 5C. Spawning locations were associated to areas where the salinity at the station were lower and gradients presented intermediate values, but the interaction plot revealed that spawning also appears in areas of higher salinities associated to very low gradients. Areas defined by this twofold combination were located at both sides of the front avoiding more mixed waters. This spatial distribution was more evident in years 2001, 2003 and 2005 (Figures S9, S10)

Discussion

We have found that the combination of sea surface current velocities, salinities and their gradients calculated at characteristic spatial scales are relevant for the parameterization of the pelagic seascape affecting a key ecological process of bluefin tuna. For bullet tuna only salinity and their gradient provided a valid seascape metric not being relevant the gradients of sea surface current velocities. In agreement with our expectations, the importance of these metrics was much higher for bluefin, a large-bodied, long distance migratory and more dependent on local oceanography than were for bullet tuna a smaller coastal species with shorter migration distance.

Previous studies have documented the links between the frontal activity and the spawning of bluefin tuna [15],[16],[33]. In this study we add to these results by examining the effect of gradients and their interactions with hydrographic mean. These metrics improved our understanding of the conditions for bluefin and bullet tuna spawning when compared to models using just the hydrographical values but not the gradients. Furthermore the identification of characteristic scales of gradients provided a new source of information for the interpretation of how local oceanography determines the selection of the site to spawn.

For bluefin tuna larvae, spatial salinity and geostrophic velocity gradients maximize spatial model performance when calculated at 0.6 degrees. The higher abundance of bluefin tuna larvae in areas with intermediate to high salinities and with high gradients of velocity is consistent among years. Higher abundance occurs around the location of the main frontal area, at the side of higher salinity of the front and where current speed presents high values. Higher salinity water likely has higher resident time near the islands than the less saline water, which may run along the front towards east getting farther from the archipelago. Spawning at the higher salinity side may favor spatial overlap with other larval species that are also located in this water mass [17],[34].

Results for bullet tuna showed that pelagic seascape metrics are not as relevant to explain the spawning distribution as they are for bluefin. In the western Mediterranean, bullet tuna spawning has been associated to near coastal areas [22], being less influenced by the local oceanography than bluefin tuna [16], which is consistent with our results.

Table 2. Summary of GAM models of larvae abundance for bullet tuna (*Auxis rochei rochei*). Interaction terms included in parenthesis.

Model group	Model variables	R2	Dev(%)	GCV	AIC	delta AIC
Base model	(latitude, longitude) + filtered volume + hour	0,158	32,8	9,615	8857,17	0
One additive variable models	base model + GVEL st0.15	0,177	35,3	9,281	8572,91	284,26
	base model + stSAL0.15	0,16	36,5	9,130	8440,06	417,11
	base model + gSAL0.75	0,207	39,6	8,741	8098,28	758,89
Hydrographic model	base model + stGVEL + stSAL + stTEMP	0,192	38,8	8,847	8183,39	673,78
SAL seascape models	Hydrographic model + stSAL + gSAL0.75	0,215	40,6	8,617	7984,18	872,99
	Hydrographic model + (stSAL,gSAL0.75)	0,255	43,1	8,327	7708,82	1148,35

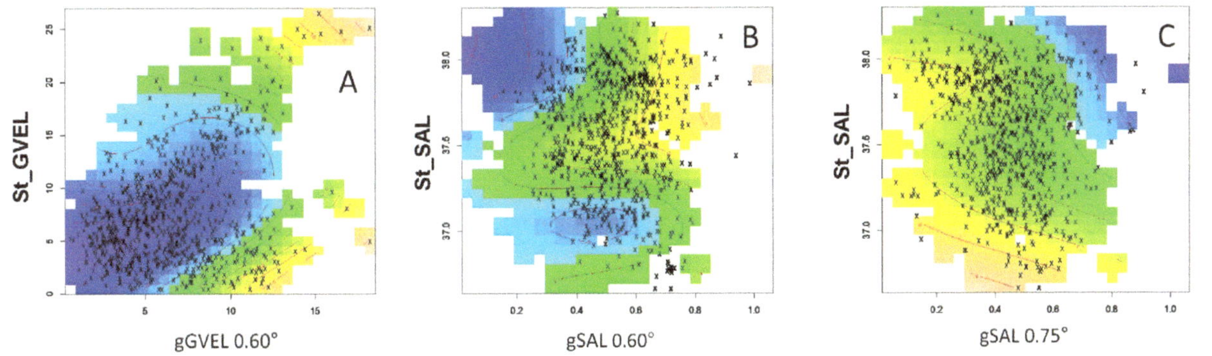

Figure 5. The effect of the interactions of the seascape components on the larval abundance as estimated from the seascape generalized additive model. The effects are shown for bluefin tuna (A–B) and bullet tuna (C). For bluefin tuna: A) the effect of the gGVEL and st_GVEL interaction. B) the effect f the gSAL and st_SAL interaction. For bullet tuna C) the effect of the gSAL and st_SAL interaction. Isolines indicate larval abundances predicted by the model. Peak of abundances are indicated in pink-yellow. Low and very low abundances are indicated in green and blue, respectively.

Despite the lower importance seascape metrics in bullet tuna, the inclusion of salinity gradients provided additional information for the identification of spawning sites. The analysis indicated that bullet tuna spawning areas are mostly found in areas where salinity gradients are low. Bullet tuna was found at both sides of the front but avoiding more mixed waters, located closer to the front. This was verified when observing the spatial distribution of larvae in relation to the salinity seascapes among the different years. For instance, in 2001, 2003 and 2005 high larvae abundances were observed North of the archipelago (high salinity waters with very low gradients), but intermediate abundances, indicating spawning, also occurs in Southern areas (low salinities and intermediate gradients). In 2002 and 2004 higher abundances were linked to low salinity and intermediate gradients shown in the south of the archipelago (Fig. S9 and S10). These results reinforce the theory of bullet tuna spawning occurs in widespread geographic areas, and not only close to the coast and suggest that the location of the main haline front negatively affects the spawning of this species.

Overall results related to bullet tuna point to the fact that, besides the avoidance of areas near strong surface haline gradients, other factors not considered in this study may also be relevant for spawning site selection in this species. It is also relevant that the spatial pattern in relation to the salinity is the opposite to that shown by bluefin tuna, located in areas near the front, suggesting possible avoidance of predators by bullet spawners [35].

The application of seascape metrics derived from salinity and geostrophic currents to characterize the spawning habitat provides new descriptors for environmental variables that improve model quality and predictions. This improvement allows a more precise identification of the relationships between the spatial location of the spawning grounds and the local oceanographic processes. Moreover, our study demonstrates that seascapes must be characterized at specific spatial scales to provide useful information as proposed in previous studies [36] and supporting results on terrestrial landscapes [37],[38] and bottom seascapes [39]. Therefore, the relations between the location of spawning sites and the mesoscale oceanographic processes may prove to be non significant if seascape metrics are not processed at the right spatial scales.

Seascape ecology is an emerging field generally being applied for the analysis of how benthic habitats pattern in coastal areas drives different aspects of marine species ecology [39]. Techniques are applied following categorical approach where the seascape is composed by a number of patches of different type of habitats [40],[7]. However, very little attention has been paid to the techniques and concepts to investigate pelagic seascape ecology due to the complex spatiotemporal dynamics of this system [2]. Thus, the work presented here sheds new light to modeling spatial distribution and investigating key ecological processes of species highly dependent on the variability of the pelagic environment, like spawning ecology of many of the big tuna species are [41]. In areas as the Balearic Sea, for which new operational oceanography platforms provide near real time data of hydrography [42] and also in combination with remote sensing data (e.g. altimetry [43]) and modeling [44],[45] these metrics will improve the species spatial distribution forecast that has proved effective for management [46].

In contrast to seascapes, landscape metrics have a long history in terrestrial ecology, and have improved over time. For instance, the effect caused on the habitat analysis derived from the spatial definition of the input habitat maps or the extent of the study area are common studied topics, [47],[48]. Likewise, calculation of seascape metrics and the final results from their application in ecological studies may be affected by different issues, like the different ways of computing the hydrographic variables and their gradients, or the origin of the input data source like from in situ measurements, remote sensing or hydrodynamic models, each with different sources of uncertainty. A relevant question is how seascapes can provide information for other type of species and ecological processes. Addressing all these challenges and developing comparative studies between different data sources, processing methods, species and ecological processes will allow advancing towards the understanding of how seascape metrics can provide information about how ecological processes and oceanography are linked together.

In summary, pelagic seascapes based on gradients and characteristic scales allow improving spatial distribution models and the identification of essential fish habitat of pelagic species. They also provide a tool for analyzing the links between particular ecological processes and local oceanography going far beyond than stochastic models based on just hydrographic parameters as salinity, temperature or geostrophic velocities. As a consequence these metrics will provide an improvement in all the management approaches and tools pending on the capability of models to identify essential habitats as near real-time spatial management based on habitat predictions [46],[10], pelagic species distribution from deterministic models [49] or the standardization of larvae indices to assess adult stock, [50],[51],[52].

Supporting Information

Figure S1 Model response of bluefin tuna in relation to salinity gradient processed at 0.75 degrees. Fitted line (solid line) and 95% confidence intervals (grey shaded areas) are shown. Whiskers on the x-axis show the locations of measurements.

Figure S2 Correlation between the gradients at the characteristic scales and the hydrographical variables at the sampled station. A) Current velocity and B) salinity for Atlantic Bluefin tuna. C) Salinity for Bullet tuna.

Figure S3 Sea surface salinities in the area during the five years analyzed (2001 to 2005).

Figure S4 Sea surface geostrophic currents in the area during the five years analyzed (2001–2005).

Figure S5 Spatial distribution of bluefin tuna (*Thunnus thynnus*) larvae in relation to the salinity mean calculated at its characteristic scale (0.6 degrees). Relative stage-1 larval abundances are shown in the maps such as dots.

Figure S6 Spatial distribution of bluefin tuna (*Thunnus thynnus*) larvae in relation to the salinity gradient calculated at 0.6 degrees. Relative stage-1 larval abundances are shown in the maps such as dots.

Figure S7 Spatial distribution of bluefin tuna (*Thunnus thynnus*) larvae in relation to the geostrophic velocity mean calculated at its characteristic scale (0.6 degrees). Relative stage-1 larval abundances are shown in the maps such as dots.

Figure S8 Spatial distribution of bluefin tuna (*Thunnus thynnus*) larvae in relation to the geostrophic velocity gradient calculated at the characteristic scale (0.6 degrees). Relative stage-1 larval abundances are shown in the maps such as dots.

Figure S9 Spatial distribution of bullet tuna (*Auxis rochei rochei*) in relation to the salinity mean calculated at 0.75 degrees. Relative stage-1 larval abundances are shown in the maps such as dots.

Figure S10 Spatial distribution of bullet tuna (*Auxis rochei rochei*) in relation to the salinity gradient calculated at 0.75 degrees. Relative stage-1 larval abundances are shown in the maps such as dots.

Acknowledgments

We thank the scientists and crew participating in the TUNIBAL cruises. We also thank P. Tugores and L. Rueda for their help in software development, and R. Balbín for their comments about methodological approaches.

Author Contributions

Conceived and designed the experiments: DAB LC AAG PR FA MH JLLJ JT. Performed the experiments: DAB LC FA. Analyzed the data: DAB LC AAG PR FA MH. Contributed reagents/materials/analysis tools: DAB LC AAG PR FA MH JLLJ JT. Wrote the paper: DAB LC AAG PR FA MH JLLJ JT. Sample collection: FA JLLJ PR. Laboratory sample processing: FA JLLJ AAG. Software development: DAB LC AAG. Numerical modelling design: DAB LC PR. Environmental variables calculation: DAB LC AAG PR FA MH JLLJ JT. Relevant literature review in different fields: DAB LC AAG PR FA MH JLLJ JT. Final manuscript edition: DAB LC AAG PR FA MH JLLJ JT.

References

1. Hinchey EK, Nicholson MC, Zajac RN, Irlandi EA (2008) Preface: marine and coastal applications in landscape ecology. Landscape Ecology 23: 1–5.
2. Pittman SJ, Kneib RT, Simenstad CA (2011) Practicing coastal seascape ecology. Marine Ecology-Progress Series 427: 187–190.
3. Turner MG (1989) Landscape Ecology - the Effect of Pattern on Process. Annual Review of Ecology and Systematics 20: 171–197.
4. Pittman SJ, McAlpine CA, Pittman KM (2004) Linking fish and prawns to their environment: a hierarchical landscape approach. Marine Ecology-Progress Series 283: 233–254.
5. Boström C, Pittman SJ, Simenstad C, Kneib RT (2011) Seascape ecology of coastal biogenic habitats: advances, gaps, and challenges. Marine Ecology Progress Series 427: 191–217.
6. Wedding LM, Lepczyk CA, Pittman SJ, Friedlander AM, Jorgensen S (2011) Quantifying seascape structure: extending terrestrial spatial pattern metrics to the marine realm. Marine Ecology-Progress Series 427: 219–232.
7. Cushman SA, Gutzweiler K, Evans JS, McGarigal K (2010) The gradient paradigm: a conceptual and analytical framework for landscape ecology. In: Huettmann F, Cushman S, editors. Spatial complexity, informatics and wildlife conservation. Tokyo: Springer. pp. 83–110.
8. Mannocci L, Laran S, Monestiez P, Dorémus G, Van Canneyt O, et al. (2013) Predicting top predator habitats in the Southwest Indian Ocean. Ecography 37: 261–278.
9. Worm B, Sandow M, Oschlies A, Lotze HK, Myers RA (2005) Global patterns of predator diversity in the open oceans. Science 309: 1365–1369.
10. Druon JN, Fromentin JM, Aulanier F, Heikkonen J (2011) Potential feeding and spawning habitats of Atlantic bluefin tuna in the Mediterranean Sea. Marine Ecology-Progress Series 439: 223–240.
11. Louzao M, Pinaud D, Peron C, Delord K, Wiegand T, et al. (2011) Conserving pelagic habitats: seascape modelling of an oceanic top predator. Journal of Applied Ecology 48: 121–132.
12. Hidalgo M, Gusdal Y, Dingsor GE, Hjermann D, Ottersen G, Stige LC, et al. (2012) A combination of hydrodynamical and statistical modelling reveals non-stationary climate effects on fish larvae distributions. Proceedings of the Royal Society B: Biological Sciences 279: 275–283.
13. Block BA, Jonsen ID, Jorgensen SJ, Winship AJ, Shaffer SA, et al. (2011) Tracking apex marine predator movements in a dynamic ocean. Nature 475: 86–90.
14. Phillips AJ, Ciannelli L, Brodeur RD, Pearcy WG, Childers J (2014) Spatio-temporal associations of albacore CPUEs in the Northeastern Pacific with regional SST and climate environmental variables. ICES, Journal of Marine Science. In press.
15. Alemany F, Quintanilla L, Velez-Belchi P, Garcia A, Cortés D, et al. (2010) Characterization of the spawning habitat of Atlantic bluefin tuna and related species in the Balearic Sea (western Mediterranean). Progress in Oceanography 86: 21–38.
16. Reglero P, Ciannelli L, Álvarez-Berastegui D, Balbín R, López-Jurado JL, et al. (2012) Geographically and environmentally driven spawning distributions of tuna species in the western Mediterranean Sea. Marine Ecology Progress Series 463: 273–284.
17. Torres AP, Reglero P, Balbin R, Urtizberea A, Alemany F (2011) Coexistence of larvae of tuna species and other fish in the surface mixed layer in the NW Mediterranean. Journal of plankton research 33: 1793–1812.
18. Block BA, Teo SL, Walli A, Boustany A, Stokesbury MJ, et al. (2005) Electronic tagging and population structure of Atlantic bluefin tuna. Nature 434: 1121–1127.
19. Rooker JR, Secor DH, De Metrio G, Schloesser R, Block BA, et al. (2008) Natal homing and connectivity in Atlantic bluefin tuna populations. Science 322: 742–744.
20. La Violette PE, Tintoré J, Font J (1990) The surface circulation of the Balearic Sea. Journal of Geophysical Research 95: 1559–1568.
21. Balbín R, López-Jurado JL, Flexas MM, Reglero P, Vélez-Velchí P, et al. (2013) Interannual variability of the early summer circulation around the Balearic Islands: driving factors and potential effects on the marine ecosystem. Journal of Marine Systems In Press, DOI: 10.1016/j.jmarsys.2013.12.007.

22. Sabatés A, Recasens L (2001) Seasonal distribution and spawning of small tunas (*Auxis rochei* and *Sarda sarda*) in the northwestern Mediterranean. Scientia Marina 65: 95–100.
23. Muhling BA, Lamkin JT, Roffer MA (2010) Predicting the occurrence of Atlantic bluefin tuna (*Thunnus thynnus*) larvae in the northern Gulf of Mexico: building a classification model from archival data. Fisheries Oceanography 19: 526–539.
24. Rooker JR, Simms JR, Wells RD, Holt SA, Holt GJ, et al. (2012) Distribution and habitat associations of billfish and swordfish larvae across mesoscale features in the Gulf of Mexico. PloS one 7: e34180.
25. de la Gándara F, Ortega A, Blanco E, Viguri FJ, Reglero P (2013) La flexión de la notocorda en larvas de atún rojo,*Thunnus thynnus* (L, 1758) cultivadas a diferentes temperaturas. In: Sociedad Española de Acuicultura, editors. XIV Congreso Nacional de Acuicultura. Gijón: Universidad Laboral de Gijón. pp. 181–182. http://www.repositorio.ieo.es/e-ieo/bitstream/handle/10508/1559/5905A.pdf?sequence=2&isAllowed=y. Accessed 23 July 2014.
26. Gordoa A, Carreras G (2014) Determination of Temporal Spawning Patterns and Hatching Time in Response to Temperature of Atlantic Bluefin Tuna (*Thunnus thynnus*) in the Western Mediterranean. PloS one 9: e90691.
27. Reglero P, Balbín R, Ortega A, Alvarez-Berastegui D, Gordoa A, et al. (2013) First attempt to assess the viability of bluefin tuna spawning events in offshore cages located in an a priori favourable larval habitat. Scientia Marina 77: 585–594.
28. Reglero P, Urtizberea A, Torres AP, Alemany F, Fiksen Ö (2011) Cannibalism among size classes of larvae may be a substantial mortality component in tuna. Marine Ecology Progress Series 433: 205–219.
29. Blank JM, Morrissette JM, Landeira-Fernandez AM, Blackwell SB, Williams TD, et al. (2004) In situ cardiac performance of Pacific bluefin tuna hearts in response to acute temperature change. Journal of Experimental Biology 207: 881–890.
30. RDevelopment CORE (2011) TEAM. 2008. R: A language and environment for statistical computing. R Foundation for Statistical Computing, Vienna, Austria. ISBN 3-900051.
31. Wood, Simon N. (2006) Generalized additive models: an introduction with R. CRC, Boca Raton, Forida: Chapman & Hall.
32. Burnham, Kenneth P. and Anderson, David R. (2002) Model selection and multimodel inference: a practical information-theoretic approach. New York: Springer-Verlag. 488 p.
33. Muhling BA, Reglero P, Ciannelli L, Alvarez-Berastegui D, Alemany F, et al. (2013) Comparison between environmental characteristics of larval bluefin tuna *Thunnus thynnus* habitat in the Gulf of Mexico and western Mediterranean Sea. Marine Ecology Progress Series 486: 257–276.
34. Rodriguez JM, Alvarez I, Lopez-Jurado JL, Garcia A, Balbín R, et al. (2013) Environmental forcing and the larval fish community associated to the Atlantic bluefin tuna spawning habitat of the Balearic region (Western Mediterranean), in early summer 2005. Deep Sea Research Part I: Oceanographic Research Papers 77: 11–22.
35. Bakun A (2013) Ocean eddies, predator pits and bluefin tuna: implications of an inferred 'low risk-limited payoff' reproductive scheme of a (former) archetypical top predator test. Fish and Fisheries 14: 424–438.
36. Steele JH (1989) The ocean landscape. Landscape Ecology 3: 185–192.
37. Wiens JA (1989) Spatial Scaling in Ecology. Functional ecology 3: 385–397.
38. Wu JG, Li HB (2006) Perspectives and methods of scaling. In: Jianguo Wu, K.. Bruce ones, Harbin Li, Orie L.. Loucks, editors. Scaling and Uncertainty Analysis in Ecology. Netherlands: Springer. pp. 17–44.
39. Bostrom C, Pittman SJ, Simenstad C, Kneib RT (2011) Seascape ecology of coastal biogenic habitats: advances, gaps, and challenges. Marine Ecology-Progress Series 427: 191–217.
40. Forman, R. T T. (1995) Land mosaics: the ecology of landscapes and regions. Cambridge, UK: Cambridge University Press.
41. Reglero P., Tittensor D.P., Álvarez-Berastegui D., Aparicio-González A., Worm B. (2014) Worldwide distributions of tuna larvae: revisiting hypotheses on environmental requirements for spawning habitats. Marine Ecology Progress Series 501: 207–224.
42. Tintore J, Vizoso G, Casas B, Heslop E, Pascual A, et al. (2013) SOCIB: The Balearic Islands Coastal Ocean Observing and Forecasting System, Responding to Science, Technology and Society Needs. Marine Technology Society Journal 47: 101–117.
43. Pascual A, Bouffard J, Ruiz S, Buongiorno Nardelli B, et al. (2013) Recent improvements in mesoscale characterization of the western Mediterranean Sea: synergy between satellite altimetry and other observational approaches. Scientia Marina 77: 19–36.
44. Juzza M, Mourre B, Renault L, Tintoré J (2014) Assessment and intercomparison of numerical simulations in the Western Mediterranean Sea. Geophysical Research Abstracts, Vol 16, EGU2014-6893, 2014, 27 April, 2014 Vienna, Austria. http://adsabs.harvard.edu/abs/2014EGUGA..16.6893J. Accessed 23 July 2014
45. Juza M, Renault L, Ruiz S, Tintoré J (2013) Origin and pathways of Winter Intermediate Water in the Northwestern Mediterranean Sea using observations and numerical simulation. Journal of Geophysical Research: Oceans 118: 6621–6633.
46. Hobday AJ, Hartmann K (2006) Near real-time spatial management based on habitat predictions for a longline bycatch species. Fisheries Management and Ecology 13: 365–380.
47. Cushman SA, McGarigal K, Neel MC (2008) Parsimony in landscape metrics: strength, universality, and consistency. Ecological Indicators 8: 691–703.
48. Wu JJ (2013) Landscape ecology. In: Rik Leemans, editors. Ecological Systems. New York: Springer. pp. 179–200.
49. Lehodey P, Senina I, Murtugudde R (2008) A spatial ecosystem and populations dynamics model (SEAPODYM)-Modeling of tuna and tuna-like populations. Progress in Oceanography 78: 304–318.
50. Ingram Jr GW, Richards WJ, Porch CE, Restrepo V, Lamkin JT, et al. (2008) Annual indices of bluefin tuna (*Thunnus thynnus*) spawning biomass in the Gulf of Mexico developed using delta-lognormal and multivariate models. ICCAT Working Document SCRS/2008/086.
51. Ingram Jr GW, Alemany F, Alvarez-Berastegui D, García A (2013) Development of indices of larval bluefin tuna (*Thunnus thynnus*) in the Western Mediterranean Sea. ICCAT,Collect Vol Sci Pap 69: 1057–1076.
52. Muhling BA, Lee SK, Lamkin JT, Liu Y (2011) Predicting the effects of climate change on bluefin tuna (*Thunnus thynnus*) spawning habitat in the Gulf of Mexico. ICES Journal of Marine Science: Journal du Conseil 68: 1051–1062.

20

Hierarchical Levels of Seed Predation Variation by Introduced Beetles on an Endemic Mediterranean Palm

Marta Rodríguez[1], Miguel Delibes[1], José Mª. Fedriani[1,2,3]*

1 Departament of Conservation Biology. Estación Biológica de Doñana (CSIC), Sevilla, Spain, 2 Department of Ecological Modelling. Helmholtz Centre for Environmental Research GmbH-UFZ, Leipzig, Germany, 3 Centre for Applied Ecology "Prof. Baeta Neves", Institute Superior of Agronomy (University of Lisbon), Lisbon, Portugal

Abstract

Seed predators can limit plant recruitment and thus profoundly impinge the dynamics of plant populations, especially when diverse seed predators (e.g., native and introduced) attack particular plant populations. Surprisingly, however, we know little concerning the potential hierarchy of spatial scales (e.g., region, population, patch) and coupled ecological correlates governing variation in the overall impact that native and introduced seed predators have on plant populations. We investigated several spatial scales and ecological correlates of pre-dispersal seed predation by invasive borer beetles in *Chamaerops humilis* (Arecaceae), a charismatic endemic palm of the Mediteranean basin. To this end, we considered 13 palm populations (115 palms) within four geographical regions of the Iberian Peninsula. The observed interregional differences in percentages of seed predation by invasive beetles were not significant likely because of considerable variation among populations within regions. Among population variation in seed predation was largely related to level of human impact. In general, levels of seed predation were several folds higher in human-altered populations than in natural populations. Within populations, seed predation declined significantly with the increase in amount of persisting fruit pulp, which acted as a barrier against seed predators. Our results revealed that a native species (a palm) is affected by the introduction of related species because of the concurrent introduction of seed predators that feed on both the introduced and native palms. We also show how the impact of invasive seed predators on plants can vary across a hierarchy of levels ranging from variation among individuals within local populations to large scale regional divergences.

Editor: Suzannah Rutherford, Fred Hutchinson Cancer Research Center, United States of America

Funding: The Spanish Ministerio de Medio Ambiente (National Park Service, grant 070/2009) and Ministerio de Educación y Ciencia (CGL2007-63488/BOS and CGL2010-21926/BOS) supported this study. JMF was funded by a Marie Curie Intraeuropean fellowship (FP7-PEOPLE-2011-IEF-298137). The funders had no role in study design, data collection and analysis, decision to publish, or preparation of the manuscript.

Competing Interests: The authors have declared that no competing interests exist.

* Email: fedriani@ebd.csic.es

Introduction

The degradation of ecosystems has altered the strength, pattern, and outcomes of many plant-animal interactions because of modifications such as habitat fragmentation, species introduction, and defaunation [1,2,3]. As a result, for example, the dynamics of native plant populations and even communities are altered by the combined impact of native and introduced seed predators limiting plant recruitment [4,5,6]. Ecological correlates of seed predation are rather species-specific and, even within a given plant species, intensity of seed predation may vary enormously depending on a myriad of factors acting at different levels [7,8,9,10]. Surprisingly, however, we know little concerning the potential hierarchy of spatial scales (e.g. among regions, within population) and coupled ecological correlates governing variation in the combined effect of introduced and native seed predators on plant populations.

Insects are major seed predators in many tropical and temperate habitats [7,11,12,13]. In particular, pre-dispersal seed losses by boring insects are pervasive worldwide [14,15]. Because palms (Family Arecaceae) typically produce large seeds with nutritious endosperms [16,17], seed boring insects often target them, lessening palm reproductive potential [17,18,19]. Most palm seed borers are bruchids and scolytines [20] and their significance as seed predators has been well studied in New World palms [14,17,18]. However, very little is known about the ecological factors determining palm seed predation in Europe, where introduced scolitines are common, especially in human-altered habitats [21,22]. Interestingly, the only native Arecaceae in continental Europe, the Mediterranean dwarf palm (*Chamaerops humilis*), often coexists with many ornamental palm species introduced together with their respective seed predators and pests [23,24] especially at the interface between natural and human-altered habitats ([25], Authors *unpublished data*). Identifying how introduced seed predators, in conjunction with native ones, lessen plant reproductive performance is clearly necessary to forecast the outcomes of global change in human-altered areas, such as the Mediterranean Europe [2,26].

In this study, we investigated several levels of spatial variation in the intensity of seed predation by introduced and native beetles on *C. humilis* seeds across the Mediterranean Spain. Despite the ecological, cultural, and economic importance of *C. humilis* [19,25,27], we know little about the amount and ecological correlates of seed predation in natural populations. In addition, habitat degradation modifies the relative abundances of local populations and introduces new species, including ornamental palms and their seed predators (such as scolytine species; [28]). Consequently, habitat degradation can be expected to enhance *C. humilis* seed predation by non-native boring predators. On the other hand, *C. humilis* drupes are often harvested by small mammals, which feed on their fleshy pulp and leave total or partially defleshed seeds under mother plants ([19], Authors *unpublished data*). Because the flesh of many fruit species often acts as a physical and chemical barrier against invertebrate seed predators [17,19,29,30], we predict that defleshed drupes, with well exposed seeds, will be more often predated by invertebrates as compared with intact drupes. Specifically, we seek to answer the following three questions: 1) How strong and variable are the interactions between introduced and native seed predators and *C. humilis* at different spatial levels (i.e. among regions, within region, and within population) in the Iberian Peninsula? 2) Does seed predation intensity vary between human-altered and natural *C. humilis* populations? and 3) Does a high amount of persisting pulp in *C. humilis* drupes lessen seed predation?

Materials and Methods

Study system

Chamaerops humilis is a small dioecious palm endemic to the Western Mediterranean basin ([31]; Fig. 1). It is considered a thermomediterranean bioindicator and usually it is not present above an elevation of 1000 meters, being most common in coastal areas (Fig. 1). It is relatively abundant in Mediterranean scrub thickets and open pine forests. Due to its vigorous sprouting, *C. humilis* is very tolerant to disturbance (fire, herbivory, etc; [31]) and thus it is often used in restoration programs in arid areas. Besides, it has been used in ornamental landscaping as well as in traditional craft [27].

It flowers in March-May and is pollinated by wind and insects [31,32]. The fruits are tridrupes grouped in bunches 5–30 cm above ground that ripen in autumn (September-November). Drupes are single-seeded, protected by a hard endocarp and a fleshy and fibrous mesocarp [19]. Since the dispersal unit is the drupe, hereafter we will refer to each drupe as fruit. Seeds are mainly dispersed by mammals such as red foxes *Vulpes vulpes* and badgers *Meles meles* [19]. Rabbits *Oryctolagus cuniculus* and some rodent species feed on the pulp of ripe and unripe fruits, generally leaving the defleshed seeds underneath mother palms ([19], Authors *unpublished data*). Defleshed seeds are often attacked by insects, with adult and larvae of introduced seed-boring beetles were observed frequently.

Beetles that bore into *C. humilis* seeds (several of them kindly identified by Dr. Alain Roques) included at least two non-native scolytine species, *Coccotrypes dactyliperda* and *Dactylotrypes longicollis* [23,24]. *C. dactyliperda* seems to be native to the tropical areas of the Old World, but today has a circumtropical range [23]. Considered a polyphagous species, it is a regular pest of wild and ornamental palms, attacking at least 18 species [24]. When feeding on date palms *Phoenix dactylifera*, female *C. dactyliperda* bore the pulp of unripe fruits and the injured dates drop from the infrutescenses; later, female beetles reproduce mainly on the dropped fruits by boring the hard coat of the seeds to lay eggs inside [33]. Conversely, *D. longicollis* seems to be native of the Canary Islands [23]. It mainly predates Canary palm seeds [34,35,36], but also fallen *C. humilis* seeds with dry or deprived pulp [37].

Sampling sites

Predation by boring insects on *C. humilis* seeds was studied in 13 populations at four geographical areas within the Iberian Peninsula (Valencia, Sevilla, Huelva, Cádiz-Málaga; Fig. 1a). Within each region, separation between adjacent populations was ≥5.5 Km. For each population, we recorded whether it was located in a natural or human-altered site, the dominant species, and other details (Table 1). Human-altered populations were those located in urban areas or very close to human settlements, usually hosting several species of ornamental palms (mainly date palm, Canary palm *Phoenix canariensis*, *C. humilis*, and *Washingtonia* spp). In all localities climate is typically Mediterranean, with two well-defined seasons: a hot dry summer from June to September and a mild rainy winter from November to February [38]. Several of our sampled Huelva populations were located within the Doñana National Park; thus, we attained permits from the authorities responsible, i.e. the Spanish National Park Service as well as the Junta de Andalucía (ref. 4225/MDCG/mect). For the remaining populations permission was not required. Our study did not involve any endangered or protected species.

Seed predation estimates

Seed predation was estimated from the presence and number of entrance holes made by boring insects on *C. humilis* seed endocarps. Typically, an endocarp attacked by borer beetles showed several holes, with the embryos and most endosperms consumed. These holes were located at either the distal, narrow tip of the endocarp or at the middle of endocarps. Germination pores were generally located at the proximal base and clearly differed in size and contour from holes made by borer beetles [19]. All seeds having at least one hole were classified as predated. Also, the number of holes per depredated seed was counted and used as a second estimate of the infestation level by boring insects. In each population, about ten fruiting *C. humilis* were usually considered (range: 3–13) and up to 20 fallen drupes (range: 4–20) were haphazardly collected underneath each individual. In total, 1639 fruits of 115 individuals were collected and then characterized in the laboratory. For each fruit, we visually estimated the percent of persisting pulp. After assessing the amount of persisting pulp, all drupes were carefully defleshed in order to search for boring insect holes in the endocarp.

Statistical analysis

Data was analyzed using generalized linear mixed models using the SAS GLIMMIX macro [39]. This allowed modeling each response variable according to the particular distribution of their residues (e.g., binomial, Poisson). Target response variables were percentage of seeds predated and the average number of seed holes for each palm (considering only depredated seeds). The sample unit was always the individual plant and, thus, we estimated percentage of predated seeds and the mean number of holes per fruit for each sampled *C. humilis* in each population.

The explanatory variables considered were the geographic region (as defined above) and the type of population (human-altered or natural), and the percentage of persisting pulp (from 0–100%). In all analyses, population and plant (nested in population) were included as random factors. The adjusted means and standard errors were calculated using the LSMEANS option and were then back-transformed using the appropriate Taylor's series approach [39]. When the interaction between any two factors was significant, we performed tests for the effect of a given factor at the different levels of the other factor ("test of simple main effects") using the Slice option in the LSMEANS [39].

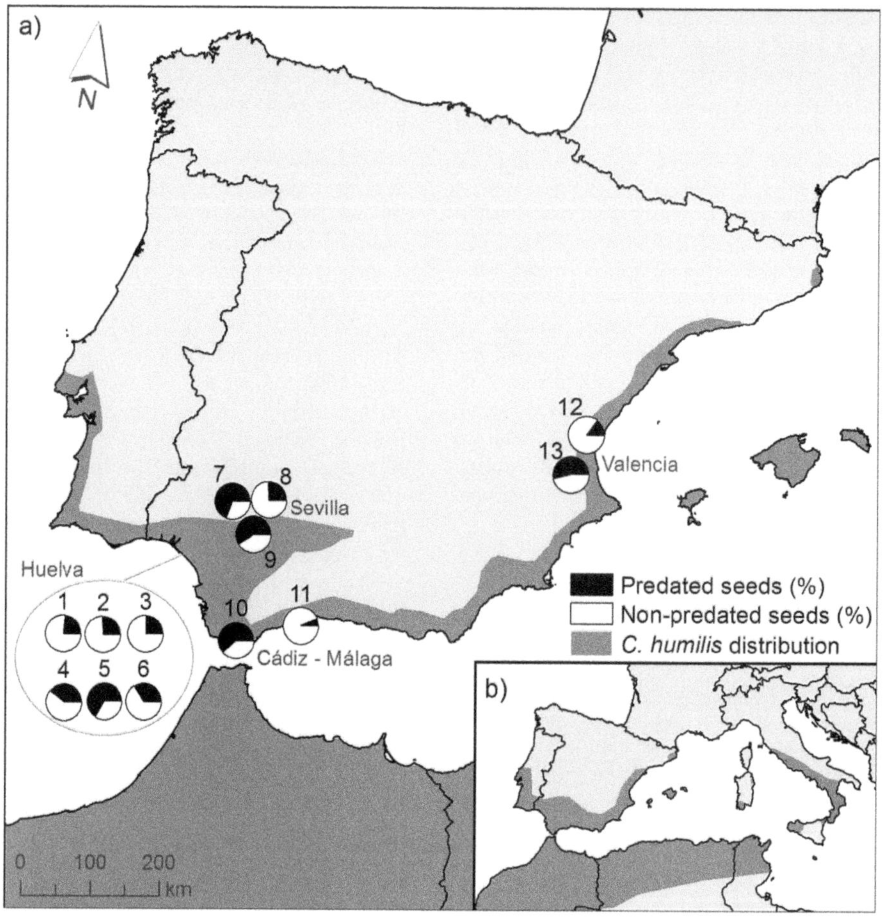

Figure 1. Distribution of *C. humilis* and location of sampled populations. a) Map showing the approximate distribution of *C. humilis* in the Iberian Peninsula as well as the percentages of seed predation in different regions using sector charts for each sampled population. b) Small inset showing *C. humilis* distribution in the Mediterranean basin (modified from Merlo et al. 1993). Populations: 1: Hinojos I; 2: Hinojos II; 3: Hato Ratón; 4: Matasgordas; 5: Chalet/Palacio; 6: Rocina; 7: Alamillo; 8: La Cantina; 9: Matallana; 10: Garganta Verde; 11: Fuengirola; 12: Saler; 13: Viveros.

Results

Frequency of seed predation

Mean percentages of *C. humilis* seed predation were highly variable among regions and also among localities in all four geographical regions (Fig. 1a). Seed predation was highest in Sevilla (56.2%±6.9; mean ±1ES), lowest in Cádiz-Málaga (25.8%±6.8), and intermediate in Valencia (37.8%±7.8) and Huelva (36.4%±4.2). Variation among populations within regions was also marked, ranging from 6.3% predation in Fuengirola to 66.1% predation in Reserva (Table 1). The mean percentage of persisting *C. humilis* fruit pulp also varied among populations, e.g. Saler (77.9%±7.2) *vs*. Alamillo (14.8%±5.1).

No significant effect was detected for the main factors in the GLM analysis (i.e., region, type of population, or percent of persisting fruit pulp; $P > 0.249$; Table 2). However, there was a strong significant interaction between population type and region ($P < 0.0001$), indicating that the effect of population type (i.e. natural *vs*. human-altered) on seed predation was not consistent across regions. Specifically, tests of "slices" revealed that population type had a significant effect in Valencia ($F_{1,93} = 7.32$, $P < 0.009$) and Huelva ($F_{1,93} = 66.17$, $P < 0.0001$). As expected, the percentages of predation in human-altered populations of Valencia and Huelva were higher (3.1 and 2.7-fold, respectively) than in natural ones (Fig. 2a). Moreover, in Sevilla, there was also a much higher (over 27-fold) percentage of seed predation in human-altered than in natural populations (Fig. 2a). However, likely due to small sample size in natural populations, differences in this region were not significant ($P = 0.286$). In contrast, in Cádiz-Málaga we found on average a 9.2-fold higher percentage of seed predation in the natural population as compared with the human-altered population (Fig. 2a). Besides, there was a significant interaction between region and amount of persisting pulp ($P < 0.001$), indicating that the effect of fruit pulp on seed predation was inconsistent across regions (Table 2). Indeed, though simple regressions showed that the relationship between amount of persisting pulp and percentage of seed predation was always negative and significant ($r < -0.383$, $P < 0.049$, $R^2 > 0.209$), the slope of the curve was higher in Cádiz/Málaga and Valencia as compared to Huelva and Sevilla (Fig. 3).

Average number of holes per predated seed

The number of holes per seed predated was highly variable among localities in all four regions. The populations with the highest average number of holes per depredated seed were Viveros (6.1±0.9) and Alamillo (4.6±0.9) and the ones with lowest number of holes were Cantina (1.0±0.0) and Hinojos II (2.0±0.3).

Table 1. Summary table showing the number of *C. humilis* individuals and number of fruits sampled in each population, the habitat type, geographical region, type of population (i.e., human-altered [H] vs. natural [N]), and the percent predation and average number of holes per predated seed.

Region	Population	Coor (N; W)	Type	Habitat	#palms	#fruits	%predation	#holes
Valencia	Saler	39.39; 0.33	N	Mediterranean scrub dominated by *Pistacia lentiscus.*	7	138	15.0±6.5	2.11±0.5
	Viveros	39.48; 0.36	H	Urban park.	10	179	53.8±6.9	6.12±0.9
	Total				**17**	**317**		
Sevilla	Alamillo	37.42; 5.99	H	Urban park.	5	78	69.4±7.1	4.6±0.9
	La Cantina	37.64; 6.07	N	Mediterranean scrub.	3	24	25.8±14.4	1±0.0
	Matallana	37.66; 5.57	N	*C. humilis* palm grove.	9	139	59.0±8.6	3.3±0.5
	Total				**17**	**241**		
Cádiz-Málaga	Fuengirola	36.52; 4.64	H	Mediterranean scrub.	10	147	6.3±3.6	3.9±1.7
	Garganta verde	36.48; 5.24	N	Mediterranean scrub dominated by *Pistacia lentiscus.*	9	136	61.6±12.2	2.5±0.2
	Total				**24**	**360**		
Huelva	Hinojos I	37.21; 6.41	N	Mediterranean scrub dominated by *Pistacia lentiscus.*	8	101	26.2±9.0	2.0±0.3
	Hinojos II	37.26; 6.39	N	Mediterranean scrub dominated by *Pistacia lentiscus.*	10	171	23.5±11.4	3.2±1.0
	Hato Ratón	37.17; 6.33	N	Mediterranean scrub dominated by *Pistacia lentiscus.*	10	96	24.4±9.3	2.4±0.4
	Matasgordas	37.13; 6.44	N	Mediterranean scrub	10	86	39.3±10.9	2.7±0.8
	Chalet/Palacio	37.07; 6.51	H	Mediterranean scrub dominated by *Halimium halimifolium*	11	129	66.1±9.6	4.4±0.6
	Rocina	37.13; 6.52	N	Mediterranean scrub dominated by *Pistacia lentiscus*	13	138	34.4±6.9	2.9±0.5
	Total				**62**	**721**		

Despite the fact the average number of holes per seed in human-altered populations (4.15±0.41) was about 2.4 times higher than in natural ones (1.71±0.39), a saturated model for such response did not revealed any significant effect likely due to its low statistical power. Indeed, after excluding the interaction between region and percentage of pulp from the saturated model, we found a significant effect ($P<0.001$) of population type on the average number of holes (Fig. 2b; Table 2). This trend was rather consistent across all regions, as indicated by the no detectable interaction between population type and region ($P = 0.183$; Table 2). Neither region ($P = 0.428$), nor percentage of persistent pulp ($P = 0.781$), nor interactions between factors had significant effects (Table 2).

Discussion

Our study provides the first evidence indicating that *C. humilis* seed predation by insects and, in particular, introduced scolytine beetles, is widespread across the Iberian Peninsula, occurring in all studied regions and populations. These results are of interest not only because of the ecological and cultural importance of *C. humilis*, but also in the context of global change and on-going species invasions in the Mediterranean basin [2,26].

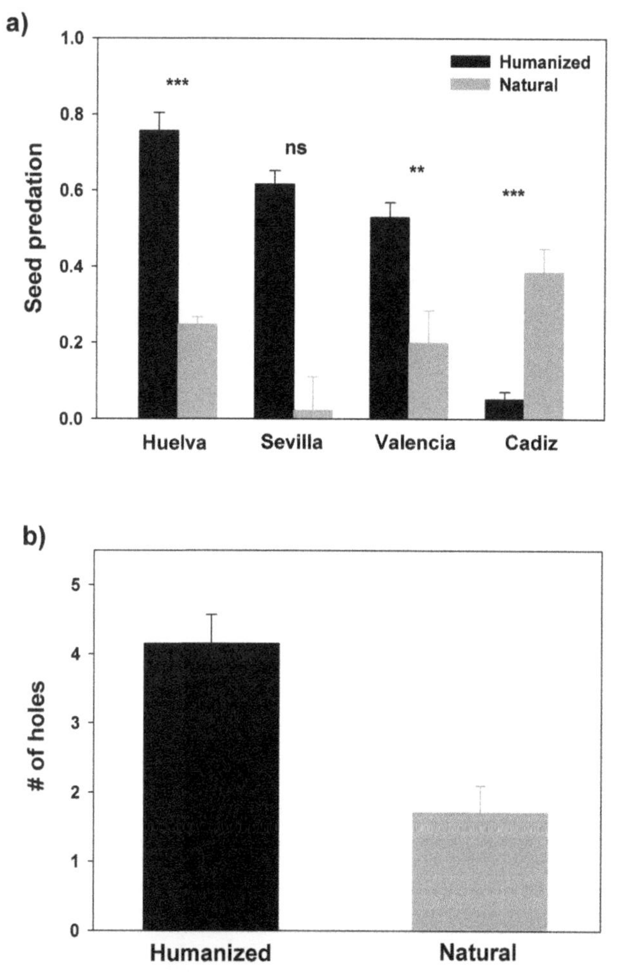

Figure 2. Factors explaining variation in *C. humilis* seed predation in the Iberian Peninsula. a) Model-adjusted mean percent seed predation in each geographical region for human-altered and natural populations. b) Model-adjusted mean number of beetle holes for human-altered and natural populations. Error bars represent standard errors.

The incidence of scolityne borers on palm crops is often high in both temperate and tropical species. Blumberg [33] quoted a 30–40% yield loss of unripe *P. dactylifera* dates attacked by *C. dactyliperda* in Israel, Zorzenon and Bergmann [40] noted about 80% reduction in germination of *Euterpe* spp. seeds attacked by *Xyleborus ferrugineous* in Brazil, and Janzen [14] found up to 99% of *Euterpe globosa* seeds damaged by *Coccotrypes carpophagus* in Puerto Rico. The only previous information about borer seed predation on wild *C. humilis* corresponds to a periurban population in Genova (Italy) and refers to a "massive infestation" by *D. longicollis* [37]. Such percentages of *C. humilis* seed predation are comparable to our estimates as approximately half of our sampled populations showed seed predation rates of 30–70%.

Spatial variation in seed predation

Our results for *C. humilis* in the Iberian Peninsula support the worldwide pervasive trend of strong spatial variation in seed predation [7,8,41]. We found high variability in estimates of seed predation in most levels considered (i.e. among regions, within region, and within population). Variation among regions was not detectable presumably because of considerable variation among populations within a region (Fig. 1). For example, in Huelva, the percentage of seed predation at Chalet/Palacio was almost triple that found at Hinojos II (Table 1). Also, even within the same population, seed predation rate varied greatly among plants. For example, up to 95% of seeds of some plants were predated at Garganta Verde, while other plants did not have any predated seed. Also, we found similar spatial variation in the average number of entrance holes per predated seed at all scales considered (geographical, interpopulation, and within population; see Table 1). The myriad of biotic and abiotic factors potentially leading to such spatial variation in this plant-animal interaction certainly deserves further research [42].

Effects of habitat degradation and amount of persisting pulp

Most identified borer beetles attacking *C. humilis* seeds belonged to two exotic species, *C. dactyliperda* and *D. longicollis*. These scolytine species probably were introduced in Spain with the date palm and the Canary palm [21,23] and now attack other palm species [21]. Thus, it should be more probable to find infested *C. humilis* seeds in human-altered areas, where these palms are used commonly as ornamentals. As predicted, predation percentages in human-altered populations were much higher than in natural ones, except at Cádiz (Fig. 2b). The higher average number of entrance holes per seed in human-altered populations also support the idea that ornamental palms tended to enhance predation of *C. humilis* seeds by boring insects.

However, the effect of habitat degradation on seed predation was inconsistent across regions. Whereas human-altered populations of Huelva, Valencia and Sevilla had more infested seeds than natural populations, the trend was inversed in Cádiz-Málaga. The higher level of seed predation in the natural than human-altered populations of Cádiz-Málaga was unexpected and may relate to the fact we only sampled one natural and one human-altered population (Garganta Verde and Fuengirola, respectively). Thus, particularities of these two populations could have led to such an unexpected result. For instance, palm seeds infested by urban scolitid beetles are subjected to frequent management [43]. Thus, in some urban habitats fallen fruits (and infested seeds) are

Table 2. Main results of our generalized mixed linear models testing the effect of region (Valencia, Sevilla, Huelva, Cádiz-Malaga,), type of population (human-altered, natural) and amount of persisting fruit pulp on *C. humilis* seed predation and the average number of invertebrate holes per preyed seed.

	Seed predation			Mean hole number		
	d.f.	F	*P*	d.f.	F	*P*
Region (R)	3, 6	0.54	0.670	3, 6	1.07	0.428
Pulp (P)	1, 93	1.34	0.249	1, 70	0.08	0.781
Type (T)	1, 93	0.91	0.342	1, 70	13.82	**0.0004**
R*T	3, 93	12.31	**<0.0001**	3, 70	1.66	0.183
P*R	3, 93	6.92	**0.0003**	3, 70	2.12	0.105
P*T	1, 93	1.24	0.268		-	
P*R*T	3, 93	2.10	0.106		-	

Significant differences ($P<0.05$) are marked in bold.

regularly removed by gardeners as yard waste, which could reduce some local beetle populations [43].

The fleshy pulp of many fruits acts as a chemical and/or physical defense against seed predators and pathogens [29,44], including seed borer insects [14,16]. Because the percentage of persisting *C. humilis* pulp was always negatively related to seed predation, our study supports the defensive hypothesis of fruit pulp [29,44]. However, lack of pulp of sampled *C. humilis* fruits could relate to consumption by pulp feeders early after fruit fall [19] but also to pulp decay with fruit aging. In the later case, older defleshed seeds would have had longer time and thus higher chances of being attacked by seed predators. Though such a possibility cannot be ruled out, all evidence gathered during this and previous studies support the defensive role of *C. humilis* fruit pulp. For example, *C. dactyliperda* attacks on *C. humilis* (Authors *personal observation*) and date palms [33] fruits are sometimes unsuccessful because pulp secondary compounds deter (or even kill; [33]) these beetles. Also, *D. longicollis* attacks almost exclusively defleshed fruits or those with dry mesocarps [37]. Finally, Fedriani and Delibes [19] showed experimentally that fruits defleshed by seed dispersers experienced higher predation by borer insects (probably *D. longicollis* and other unidentified generalist seed predators) than whole intact fruits with pulp.

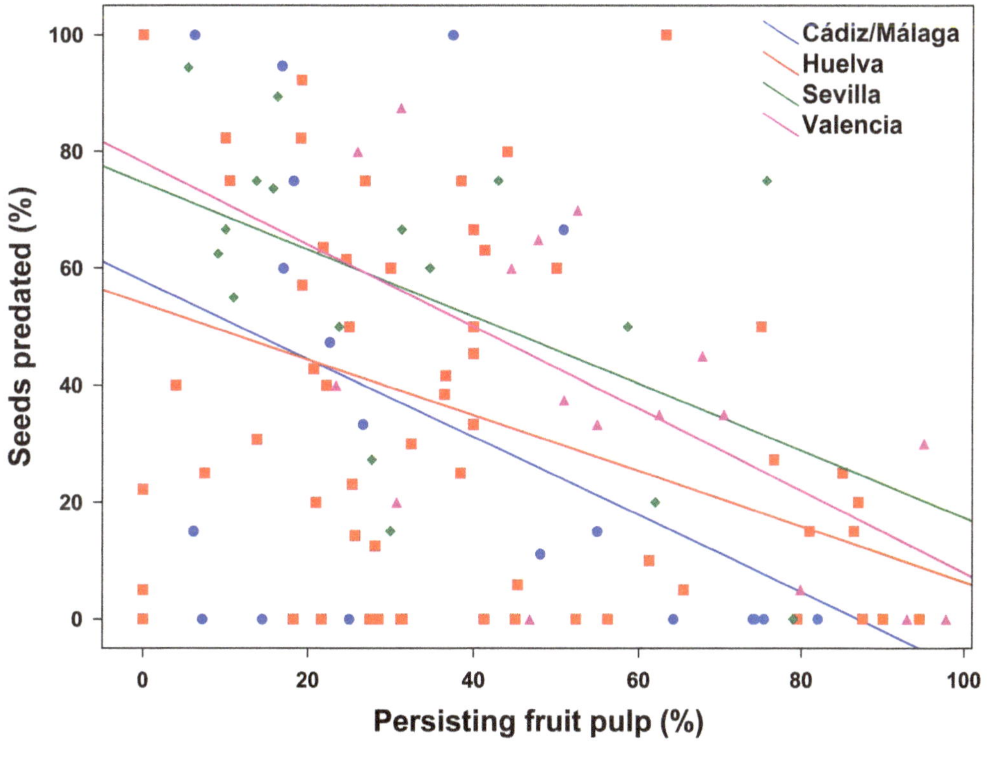

Figure 3. Negative linear relationships between percentage of seed predation and amount of persisting fruit pulp in each of the four target regions.

This is the first study on *C. humilis* seed predation across a geographical scale and reveals that introduced scolytines (and perhaps other beetles) are widespread seed predators, though their incidence was highly variable across the Iberian Peninsula. Seed predation strongly varied across a hierarchy of levels (among geographical regions, among populations, and within populations), with habitat degradation and the amount of defensive fleshy pulp influencing the level of seed predation. Interestingly, because human activity also can augment the population of some pulp consumers such as rabbits [45], it can lead to a number of synergic factors enhancing *C. humilis* proneness to seed predation. Further research concerning such possibilities as well as factors limiting *C. humilis* recruitment at different plant stages (seeds, seedlings, saplings) is clearly needed. This sort of study is particularly desirable given the charismatic nature of *C. humilis*, the high habitat degradation of the Mediterranean basin [46], and the current context of global change and on-going species invasions [2,26].

Supporting Information

Appendix S1 This file contains data on percentage of *C. humilis* seed predation by beetles for each palm in each population. Data on type of habitat (natural, humanized) and the mean percentage of persisting fruit pulp is also shown.

Appendix S2 This file contains data on the mean number of holes per seed for each *C. humilis* individual in each population.

Acknowledgments

We are indebted to Gemma Calvo for her assistance in several aspects of this study. Craig Benkman made numerous constructive comments to earlier drafts that significantly improved the manuscript. The laboratory of GIS and Remote Sensing (LAST, EBD) and, in particular, David Aragones, assisted us drawing Figure 1.

Author Contributions

Conceived and designed the experiments: JMF. Performed the experiments: MR JMF. Analyzed the data: JMF. Contributed reagents/materials/analysis tools: MR JMF. Wrote the paper: JMF MD MR.

References

1. Tylianakis JM, Didham RK, Bascompte J, Wardle DA (2008) Global change and species interactions in terrestrial ecosystems. Ecology Letters 11: 1351–1363.
2. Walther GR, Roques A, Hulme PE, Sykes MT, Pyšek P, et al. (2009) Alien species in a warmer world: risks and opportunities. Trends in Ecology & Evolution 24: 686–693.
3. Kiers ET, Palmer TM, Ives AR, Bruno JF, Bronstein JL (2010) Mutualisms in a changing world: An evolutionary perspective. Ecology Letters 13: 1459–1474.
4. Rose KE, Louda SM, Rees M (2005) Demographic and evolutionary impacts of native and invasive insect herbivores on *Cirsium canescens*. Ecology 86: 453–465.
5. Shiels AB, Drake DR (2011) Are introduced rats (*Rattus rattus*) both seed predators and seed dispersers in Hawaii? Biological Invasions 13: 883–894.
6. Wotton DM, Kelly D (2011) Frugivore loss limits recruitment of large-seeded trees. Proceedings of the Royal Society Series B: Biological Sciences 278: 3345–3354.
7. Hulme P, Benkman CW (2002) Granivory. Pages 132–154 in Herrera C and Pellmyr O, editors. Plant–animal interactions: an evolutionary approach. Blackwell Scientific Publications, New York, New York, USA.
8. Fedriani JM, Rey PJ, Garrido JL, Guitián J, Herrera CM, et al. (2004) Geographical variation in the potential of mice to constrain an ant-seed dispersal mutualism. Oikos 105: 181–191.
9. Fedriani JM, Boulay R (2006) Foraging by fearful frugivores: combined effect of fruit ripening and predation risk. Functional Ecology 20: 1070–1079.
10. Nuñez MA, Simberloff D, Relva MA (2008) Seed predation as a barrier to alien conifer invasions. Biological Invasions 10: 1389–1398.
11. Janzen DH (1982) Simulation of Andira fruit removal by bat reduces seed predation by *Cleogonus weevils*. Brenesia, 20: 165–170.
12. Herrera CM (1986) Vertebrate frugivores and their interaction with invertebrate fruit predators: supporting evidence from a Costa Rican dry forest. Oikos, 54: 185–188.
13. Forget PM, Muñoz E, Leigh EG (1994) Predation by rodents and bruchid beetles on seeds *Scheelea* palms on Barro Colorado Island, Panama. Biotropica 26: 420–426.
14. Janzen DH (1972) Escape in space by *Sterculia aoetala* seed from the bug Dysdercus fasciatus in a Costa Rican deciduous forest. Ecology 53: 350–361.
15. Andersen AN (1988) Insect seed predators may cause far greater losses than they appear to. Oikos 52: 337–340.
16. Silvius KM, Fragoso JMV (2002) Pulp handling by vertebrate seed dispersers increases palm seed predation by bruchid beetles in the northern Amazon. Journal of Ecology 90: 1024-1032.
17. Fragoso JMV, Silvius KM, Correa JA (2003) Long-distance seed dispersal by tapirs increases seed survival and aggregates tropical trees. Ecology 84:1998–2006.
18. Delobel A, Couturier G, Kahn F, Nilsson JA (1995)Trophic relationships between palms and bruchids (Coleoptera: Bruchidae: Pachymerini) in Peruvian Amazonia. Amazoniana 13: 209–219.
19. Fedriani JM, Delibes M (2011) Dangerous liaisons disperse the Mediterranean dwarf palm: fleshy-pulp defensive role against seed predators. Ecology92: 304–315.
20. Moore D (2001) Insects of palm flowers and fruits. in: Howard FW, Moore D, Giblin-Davis RM, Abad RG [Eds.] Insects on Palms. CABI Publishing, Wallingford, UK. pp. 233–266.
21. Gómez-Vives S (2004) Evolución e importancia del ataque y biología de *Coccotrypes dactyliperda* (Curculionidae: Scolytinae) perforador del fruto de la palmera datilera. Boletín De Sanidad Vegetal. Plagas 30: 497–506.
22. Delobel B, Delobel A (2006) Dietary specialization in European species groups of seed beetles (Coleoptera: Bruchidae: Bruchinae). Oecologia 149: 428–443.
23. Kirkendall L, Faccoli M (2010) Bark beetles and pinhole borers (Curculionidae, Scolytinae, Platypodinae) alien to Europe. Zookeys 56: 227–251.
24. Sauvard D, Branco M, Lakatos F, Faccoli M, Kirkendall L (2010) Weevils and Bark Beetles (Coleoptera, Curculionoidea). BioRisk 4: 219–266.
25. Médail F, Quézel P (1996) Climatic and phytoecological significance of the rediscovery in Mediterranean France of *Chamaerops humilis* L. (Palmae). Comptes Rendus de l'Academie des Sciences Serie III Sciences de la Vie 319: 139–145.
26. Alain R, Kenis M, Lees D ,Lopez-Vaamonde C, Rabitsch W, et al. (Eds), (2010) Alien terrestrial arthropods of Europe. BioRisk 4 (Special Issue). Pensoft Publishers, Sofia-Moscow.
27. Merlo ME, Alemán M, Cabello J, Peñas J (1993) On the Mediterrranean fan palm (*Chamaerops humilis*). Principes 37: 151–158.
28. Vilà M, Valladares F, Traveset A, Santamaría L, Castro P, editors (2008) Invasiones Biológicas, Colecciones divulgativas del CSIC. 216 pp.
29. Cipollini ML, Levey DJ (1997) Secondary metabolites of fleshy vertebrate-dispersed fruits: adaptive hypotheses and implications for seed dispersal. American Naturalist 150: 346–372.
30. Fedriani JM, Delibes M (2013) Pulp feeders alter plant interactions with subsequent animal associates. Journal of Ecology 101: 1581–1588.
31. Herrera J (1989) On the reproductive biology of the dwarf palm, *C. humilis* in southern Spain. Principes 33: 27–32.
32. Anstett MC (1999) An experimental study of the interaction between the dwarf palm (*C. humilis*) and its floral visitor *Derelomus chamaeropsis* throughout the life cycle of the weevil. Acta Oecologica 20: 551–558
33. Blumberg D (2008) Review: Date Palm Arthropod Pests and Their Management in Israel. *Phytoparasitica* 36: 411–448.
34. Palacios J (1973) *Dactylotrypes uyttenboogaarti* Egg. nuevo para la fauna ibérica (Col. Scolytidae). Miscellania Zoologica 3: 69–70.
35. Lombardero MJ, Novoa F (1994) Datos faunísticos sobre escolítidos ibéricos (Coleoptera: Scolytidae). Boletín de la Asociación Española de Entomología 18: 181–186.
36. Whitehead BP, Zach P, Kulfan J, Cicak A, Čunderlik I (2000) *Dactylotrypes longicollis* (Wollaston 1864)(Coleoptera, Scolytidae) introduced to the Slovak Republic. Anzeiger für Schädlingskunde 73: 17–18.
37. Bernabó E (1991) Sulla presenza a Genova di *Dactylotrypes longicollis* (Woll.)[= D. uyttenboogaarti Eggers]. Bollettino della Società entomologica italiana 122: 185–187.
38. Valladares F, Vilagrosa A, Peñuelas J, Ogaya R, Camarero JJ, et al. (2004) Estrés hídrico: ecofisiología y escalas de la sequía. En Valladares, F (ed.). Ecología del bosque mediterráneo en un mundo cambiante. M.M.A; EGRAF, S.A. Madrid. Págs.163–190.

39. Littell RC, Milliken GA, Stroup WW, Wolfinger RD, Schabenberger O (2006) SAS for Mixed Models, Second Edition, SAS Institute Inc. Cary, NC, USA.
40. Zorzenon FJ, Bergmann EC (1995) Occurrence of *Xyleborus ferrugineus* (Fabricius, 1801) (Coleoptera: Scolytidae) in fruits and seeds of two species of the genus Euterpe. Revista de Agricultura Piracicaba 7017–20.
41. Thompson JN (2005) The Geographic Mosaic of Coevolution. Chicago, University of Chicago Press.
42. Herrera CM (2009) Multiplicity in Unity. Plant Subindividual Variation and Interactions with Animals. The University Chicago Press, Chicago, IL.
43. Holzman JP, Bohonak AJ, Kirkendall LR, Gottlieb D, Hararis AR, et al. (2009) Inbreeding variability and population structure in the invasive haplodiploid palm-seed borer (*Coccotrypes dactyliperda*). Evol. Biol. 22: 1076–1087.
44. Rodríguez A, Alquézar B, Peña L (2013) Fruit aromas in mature fleshy fruits as signals of readiness for predation and seed dispersal. New Phytologist 197: 36–48.
45. Fedriani JM, Fuller TK, Sauvajot RM (2001) Does availability of anthropogenic food enhance densities of omnivorous mammals? An example with coyotes in southern California. Ecography 24: 325–331.
46. Thompson JD (2005) Plant evolution in the Mediterranean. Oxford: Oxford University Press. 293 pp.

21

Separating Macroecological Pattern and Process: Comparing Ecological, Economic, and Geological Systems

Benjamin Blonder[1]*, Lindsey Sloat[2], Brian J. Enquist[2,3], Brian McGill[4]

1 Sky School, University of Arizona, Tucson, Arizona, United States of America, **2** Department of Ecology and Evolutionary Biology, University of Arizona, Tucson, Arizona, United States of America, **3** Santa Fe Institute, Santa Fe, New Mexico, United States of America, **4** School of Biology and Ecology, University of Maine, Orono, Maine, United States of America

Abstract

Theories of biodiversity rest on several macroecological patterns describing the relationship between species abundance and diversity. A central problem is that all theories make similar predictions for these patterns despite disparate assumptions. A troubling implication is that these patterns may not reflect anything unique about organizational principles of biology or the functioning of ecological systems. To test this, we analyze five datasets from ecological, economic, and geological systems that describe the distribution of objects across categories in the United States. At the level of functional form ('first-order effects'), these patterns are not unique to ecological systems, indicating they may reveal little about biological process. However, we show that mechanism can be better revealed in the scale-dependency of first-order patterns ('second-order effects'). These results provide a roadmap for biodiversity theory to move beyond traditional patterns, and also suggest ways in which macroecological theory can constrain the dynamics of economic systems.

Editor: Rachata Muneepeerakul, Arizona State University, United States of America

Funding: BB was supported by a National Science Foundation pre-doctoral fellowship. LS was supported by the National Center for Ecological Analysis and Synthesis. BJE was supported by a National Science Foundation Advancing Theory in Biology grant. BJM was funded by a National Science Foundation Experimental Program to Stimulate Competitive Research grant EPS-0904155 at the University of Maine. The funders had no role in study design, data collection and analysis, decision to publish, or preparation of the manuscript.

Competing Interests: The authors have declared that no competing interests exist.

* Email: bblonder@gmail.com

Introduction

Decades of research have identified four central patterns that together describe the broad-scale organization of biological diversity [1]. These include the: abundance distribution of species [2]; the relationship between species richness and area [3,4]; the decrease in assemblage similarity with increasing distance [5]; and the spatial dispersion of individuals within species [6,7]. A central question is how local biotic and abiotic interactions and variation in rates of speciation and extinction influence these large-scale patterns of diversity [8]. Indeed, much of the ongoing debates within biodiversity science result from the fact that many different models have been proposed to explain these individual patterns [2,4,9–12]. More recently, several theories such as maximum entropy [13] and neutral theory [14–16] have claimed to be able to simultaneously predict these patterns [1].

A central problem for theories of biodiversity is that they all make similar predictions for these near-universal patterns despite beginning from disparate assumptions [17]. One potentially troubling implication for ecology is that these patterns may not reflect anything unique about organizational principles of biology or the functioning of ecological systems [11,18,19]. Instead, they may be a statistical inevitability for any complex system with a large number of variables influencing the system's dynamics [20–22]. If non-ecological systems show similar patterns, then the fundamental validity of theories of biodiversity that invoke ecological mechanisms as an explanation would be challenged. Stronger tests of theory require alternative approaches.

There is an opportunity to identify a different set of patterns that arise from only ecological processes, and which can therefore distinguish between ecological and non-ecological systems [23,24]. We hypothesize that distinguishing biodiversity theories using empirical patterns is possible with second-order effects but not with first-order effects. We define first-order effects as a set of functions that describe macroecological patterns across scales, and second-order effects as the scale-dependent parameters of these functions. We specifically hypothesize that:

(1) Any system where objects are partitioned in categories (species) across space and many variables interact multiplicatively will be described by a common set of functions, i.e. first-order effects. These first-order effects can be predicted based on common assumptions of multiple unified biodiversity theories [1] or are statistically inevitable consequences of the Central Limit Theorem [20,21]. Thus, any system should be characterized by an approximate log-normal species-abundance distribution (SAD) [25,26], an approximate [3,27] power law [28] species-area relationship (SAR), a monotonically declining Jaccard similarity-

distance function [5], and a positive intraspecific clustering function at different distances (i.e. clumped at all scales; see Methods) [6,29]. Many biodiversity theories predict some or all of these patterns [1].

(2) Ecological and non-ecological patterns, however, can be separated by changes in these patterns change with scale. Thus, quantifying the scale-dependent parameters, i.e. second-order effects [7,30–32] provide a novel way to assess mechanism in macroecology. For example, ecological processes (e.g. dispersal limitation, speciation) will have different scale dependences depending on the system of interest. Spatial scale may affect the slope of the species-area relationship [27] as well as the statistical moments of the species abundance distribution [33]. Some current unified theories of biodiversity are beginning to incorporate scale-dependence and these second-order effects into their predictions [13,14] while for others the role of scale remains unclear [19,34].

We hypothesize that ecological and non-ecological systems can be distinguished based on several patterns (Table 1). Our approach is to: 1) establish baseline expectations for first-order effects based on different biodiversity theories; 2) identify the potentially scale-dependent parameters of first-order effects; 3) plot these parameters as a function of spatial scale; and 4) detect changes in these functions from the baseline expectation. For example, the decay of similarity with distance pattern is predicted by several theories to be a negative exponential function (1). The slope of this function on logarithmic scale should be scale-invariant (2). However, a plot of the local slope of empirical data (3) might show a peak at large distance scales (4) indicative of a second-order effect that can only be explained by additional mechanisms not incorporated into the original biodiversity theories.

Here we use this approach to provide general insights into the degree to which non-ecological systems can be explained by ecological theory. We compared first- and second-order effects across a broad set of ecological and non-ecological systems. We compiled five large datasets that each describe the abundance, location, and identity of objects in multiple categories (species) throughout the continental United States, encompassing two ecological systems (North American birds and trees), two economic systems (US Census county business patterns, and a commercial database of franchise locations for several hundred major corporations) and one geological system (USGS mineral resources database) (Table 2). These datasets were chosen because they are either complete censuses or are known to be well sampled, have very large number of objects and categories, occur over the same large region, and have high spatial resolution (Figure 1).

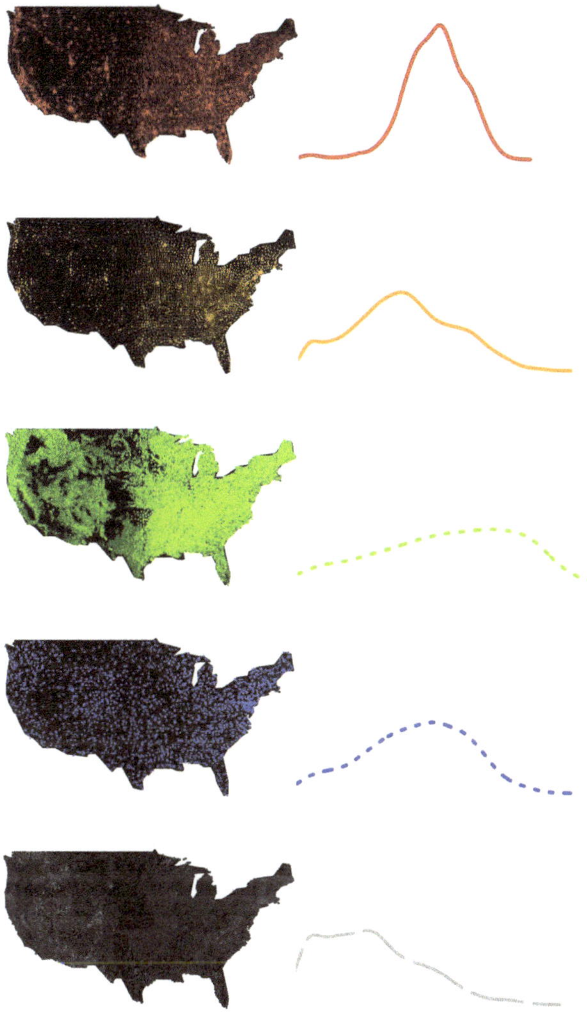

Figure 1. Distribution of objects across categories and space. Left column, site locations for each dataset (colored as described in Table 2). Site brightness is proportional to richness. Right column, relative abundance distribution for log-transformed abundance data at full scale (a first-order effect). All datasets are shown with the same axes.

Materials and Methods

Dataset assemblage

We generated community matrices for each of five datasets in which the *i*th row and *j*th column represented the abundance of

Table 1. First and second order effects.

Macroecological pattern	*First-order effect*	*Second-order effect*
Species abundance distribution	Log-normal (approximate)	Changes in mean, coefficient of variation, skewness, kurtosis at different area scales
Species area relationship	Power law (approximate)	Changes in local slope at different area scales
Similarity-distance relationship	Monotonic decreasing	Changes in local slope at different distance scales
Fraction of clumped species	Positive	Changes in local slope at different distance scales

First-order effects describe all datasets, while second-order effects may provide scale-dependent approaches for distinguishing datasets.

Table 2. Summary statistics for each economic, ecological, and geological dataset.

Dataset	Corporate locations	Industrial codes	Trees	Birds	Minerals
Type	Economy	Economy	Ecology	Ecology	Geology
Lines drawn as	solid red	solid orange	dotted green	dotted blue	dashed gray
# Species	455	3,777	384	584	746
# Individuals	660,935	7,628,863	11,887,262	1,640,449	587,571
# Sites	20,936	3,106	391,981	2,251	54,837
Most common five species	Subway, Shell, T-Mobile, McDonald's, BP	Offices of physicians (exc mental health), Independent artists, writers & performers, Offices of lawyers, Offices of dentists, Limited-service restaurants	Loblolly pine, Red maple, Sweetgum, Sugar maple, White oak	Red-winged blackbird, European starling, American robin, Mourning dove, American crow	Gold, Sand & gravel, Construction, Silver, Copper

Maps display higher species richness at each site in brighter colors. In all subsequent graphs we have used the line-coloring scheme shown here.

species *j* at site *i*. Each community matrix was augmented with the latitude and longitude of each site. Each dataset was clipped to sites contained within the continental United States. All datasets were transformed from their raw form to community matrices using R for data manipulation and MATLAB for GIS analyses. These datasets are available in Information S1.

Dataset 1: Corporate locations. We purchased a commercial dataset containing the street addresses and latitude/longitudes of all locations of several hundred major corporations (AggData, Inc.). The data represent a census obtained between 2008 and 2010. We used shapefiles for the United States Census zip code tabulation areas (ZCTAs) to assign these point occurrences into assemblages. These approximately 20,000 areas cover the entire United States. While ZCTAs have unique complex boundaries and variable areas, they each cover roughly equivalent population levels and are a good comparable assemblage unit for this study. We then determined the latitude and longitude of each assemblage as the centroid of each ZCTA.

Dataset 2: Industrial codes. We downloaded the United States Census County Business Patterns dataset, which counts the numbers of businesses of different size classes in each of the North American Industrial Classification System's (NAICS) nested categories, within each of the counties of every state of the United States. These data were valid for the 2007 census year. The data include some intentional inaccuracies (low-abundance data swapped between sites or abundances randomized) to comply with privacy laws, but these effects are small in magnitude and should not affect our analyses. We restricted our analysis to only the most specific (six-digit) level of NAICS classification in order to closely match between biological and business species. To further improve this correspondence, we also assumed that businesses that fell into different size classes (1–10, 10–100, 100–1000, 1000+ employees) within a given NAICS category represented different species. We also obtained shapefiles for county boundaries and determined the latitude and longitude of each assembly as the centroid of each county.

Dataset 3: Birds. We obtained data from the North American Breeding Bird Survey, which counts the abundance of the bird species observed along hundreds of multi-kilometer transect routes by multiple volunteer birders. We used data from the 2007 counts. We treated each route as an assemblage and determined its latitude and longitude as the midpoint of the route.

Dataset 4: Trees. We obtained data from the United States Forest Service's Forest Inventory of America, which counts the abundances of several hundred species of trees at hundreds of thousands of plots across the United States. At each plot, we used data from its most recent census, which ranged from 1985–2008. Plot data were pre-corrected for variable plot size and only included live trees. Because of privacy laws, these data contain intentional inaccuracies (plots on private land have their coordinates fuzzed and their abundances swapped) that are small in magnitude and do not affect our analyses. We treated each plot as an assemblage and used the plot center for the assemblage latitude and longitude.

Dataset 5: Minerals. We downloaded data from the United States Geological Survey's Mineral Resource Data System, which describes the locations of metallic and nonmetallic minerals throughout the world. We pooled the abundances of commodities, ores, and gangue at each site, because we were interested in geological processes and did not wish to stratify the data by economic value. Because each site contained a very low number of minerals (typically representing the useful output of a single mine) we chose to generate an equal-area grid (1000×1000) covering the bounding box of the continental United States, and pooled mineral abundances for all sites falling within each grid cell. We then defined the assemblage latitude and longitude as the center of the grid cell.

Data analysis

Species-abundance distribution. We sampled the abundance distribution at 100 spatial scales that logarithmically spanned a range from 0.1° to 40°. At each scale, we chose 500 random sites. We defined a small circle on the surface of the earth whose radius was determined by the current spatial scale and whose center was the location of the current site. We then intersected this circle with a polygon defining the boundary of the region of interest (here, the continental United States). We calculated the surface area of this new polygon (in km^2) using spherical geometry and the known radius of the earth to determine the effective area of the site. We then pooled abundances for all species at all sites enclosed within this polygon, applied a log transformation, and calculated the mean, coefficient of variation, skewness, and kurtosis of this distribution.

Species-area relationship. We determined the species-area relationship using an identical procedure as for the species-abundance distribution, but calculated species richness instead of abundance distribution moments within each polygon. We log-transformed both area and richness before analysis so that the local/global slope of the curve reflects the scale dependent/independent power-law scaling exponent.

Similarity-distance function. We sampled the similarity-distance function at 100 spatial scales that logarithmically spanned a range from 0.1° to 40°. At each scale, we chose 1000 random pairs of sites. We calculated the distance between assemblages (in km) as the minimum arc length along the surface of the earth joining the centers of these assemblages. Then, for each site within each pair of sites we generated small circles centered on the location of each site with a radius equal to the current spatial scale. We intersected each small circle with the boundary polygon of the region of interest (here, the continental United States). To obtain the assemblage area we used the sum of the areas of both polygons (in km^2) calculated by the same approach described for the abundance distribution. We also pooled all abundances within each polygon and calculated the Jaccard similarity (number of species in common divided by the number of species in either polygon). To simplify the display of information results were plotted only for assemblages whose summed area was in the 10–100 km^2 or the 1000 to 10000 km^2 bins.

Intraspecific clumping. We assessed the fraction of species that exhibited intraspecific clumping at distances ranging from 10 to 5000 km in 10 km intervals. An individual species was defined to be intraspecifically clumped at a given distance scale if its observed pairwise distance distribution exceeded the upper 95% quantile of 100 samples from a null pairwise distance distribution. We calculated the pairwise distance distribution as the vector of distances (accounting for the curvature of the earth, as defined for the similarity-distance function) between every pair of sites at which this species occurred. We determined the null pairwise distance distribution by counting the number of sites at which this species occurred, randomly assigning that many occurrences of this species to randomly chosen sites, and repeating the pairwise distance calculation. This method accounts for sites that are non-uniformly or non-randomly positioned, corrects distances for the curvature of the earth, and generates conclusions that are consistent with more established methods for detecting clumping (e.g. pair correlation/o-ring function [6]). However, this method is computationally much faster for large datasets, because distance and intersection calculations can be pre-computed a single time.

First-order effects. Statistics were based on the metrics calculated using the methods described in the previous section. For the species-abundance distribution, we used abundance data at the largest spatial scale. We fit several candidate distributions to the data (Pareto, power-bend [35], Poisson log-normal, log-series, Weibull) and identified the distribution with the lowest AIC. For the species-area relationship, we used the log-transformed area and richness values described in the previous section, then reported the slope of the model (i.e. the power law exponent). For the similarity-distance relationship, we constructed a linear model using the sampled similarities and log-transformed distances for 10^4 km^2 assemblages and reported the slope of the model. For these two analyses, all models were highly significant ($p<0.05$) but trivially so because of the very large degrees of freedom. For the clumping analysis, we reported the fraction of species that were significantly clumped at the 5000 km distance scale. Because of the very large number of degrees of freedom in all these analyses (up to 49,998), the standard error for every coefficient was much smaller than the coefficient value. We therefore did not present these uncertainty estimates or *p*-values because these statistical differences were unlikely to conclusively reflect biological differences.

Data visualization. We chose to show central trends in the data using LOESS (locally smoothed regression) and to quantify variation using error envelopes representing each middle quartile of a local subset of the data.

Results

We first quantified first-order effects in each ecological, economic, and geological dataset neglecting the effect of spatial scale. Our analyses show that, when data are aggregating at a continental spatial scale, each dataset is characterized by the expected first order effects (see Methods for details). All species abundance distributions were best fit by a log-normal distribution (ΔAIC to the next-best distribution >30), except for the tree dataset for which a log-normal or Weibull distribution were both appropriate ($\Delta AIC = 4$). All species-area relationships had log-log slopes ranging from 0.28 to 0.50. All distance decay relationships had log-linear slopes ranging from −0.17 to −0.72. Lastly, most datasets showed intraspecific clumping (7–52% of species significantly clumped). The only exception was for trees, but this dataset included many widely cultivated species. In contrast, we found that second-order effects can distinguish between these datasets, suggesting the operation of different processes structuring each system across a range of scales.

We found that for scales of up to $\sim 10^5$ km^2 the coefficient of variation for the species-abundance distribution (Figure 2) for skewness, and kurtosis was larger for the economic datasets than for the ecological or geological datasets. These differences indicate the existence of a process specific to these economic systems that generates higher fractions of rare species at local scales. For example, it could be the case that rapidly growing economies have both higher birth and death rates reflected in large numbers of new businesses. However, all datasets had lower positive skewness at large scales, consistent with dispersal limiting the spread of rare species across space.

Across all systems, species-area relationships displayed a range of slopes and curvature indicating scale-dependent processes of richness accumulation (Figure 3). Consistent with dispersal limitation and a transition to novel species pools, we found a small increase in slope at intermediate scales for the biological data, A more striking pattern is the decrease in slope at large scales of the economic data, consistent with convergence to similar species pools on both coasts.

The similarity-distance function showed a range of decay rates and minimum similarities when comparing datasets (Figure 4). Ecological datasets decayed faster than economic or geological datasets, and an ecological dataset (trees) had the lowest minimum similarity at large distances. Minimum bird similarity was higher than minimum tree similarity, presumably because of the high dispersal potential and large range size of many birds. Economic datasets showed an increase in similarity at very large distances, consistent with high similarity of species pools on both coasts. This change in similarity was weaker for the geological dataset. Changes in the similarity-distance function with assemblage area were also consistent with dispersal limitation being more important in biological systems. We found that in communities encompassing larger areas, ecological systems maintained their decay rates and reached comparable minimum similarities, but

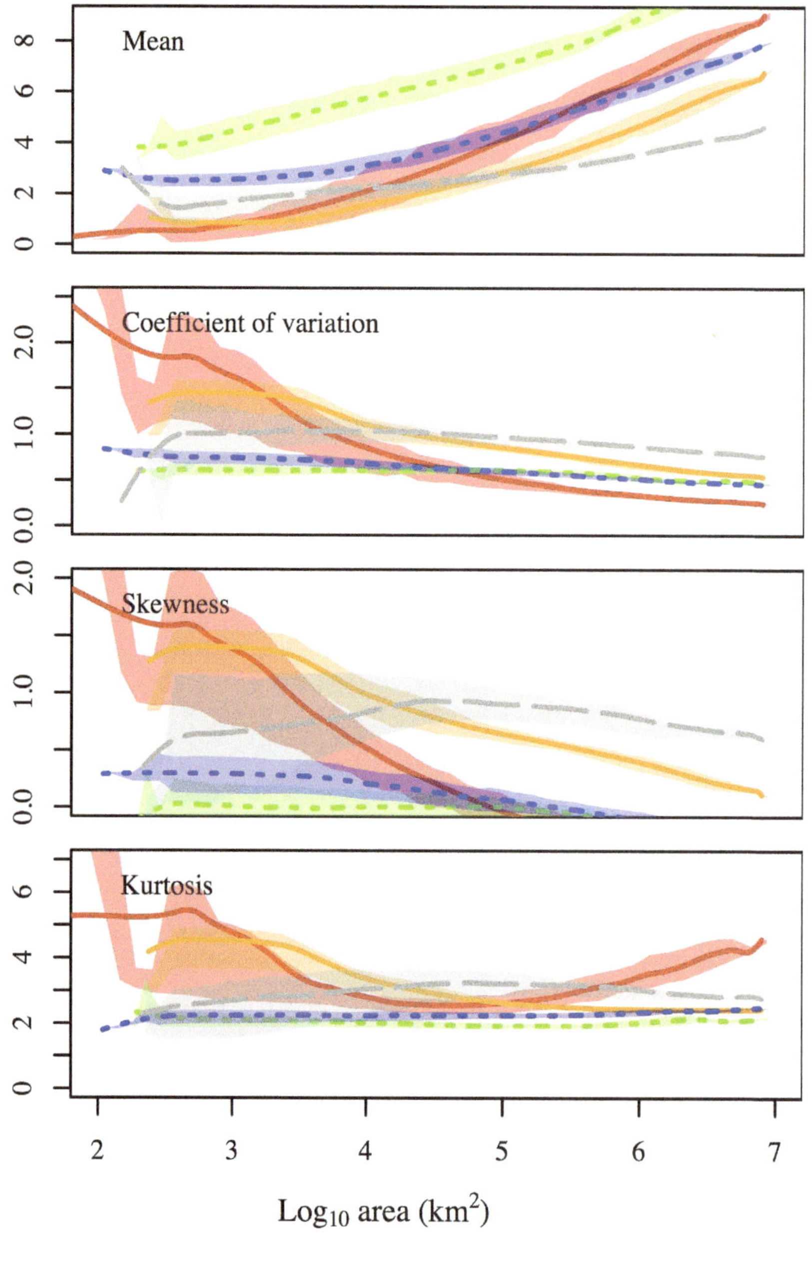

Figure 2. Central moments of the species-abundance distribution for log-transformed data. Line colors are described in Table 2.

that economic systems exhibited very limited decay and high minimum similarity.

The intraspecific clumping function showed more clumping at longer distances in ecological data sets compared to economic or geological datasets (Figure 5). The width of this leftmost part of the clumping function may provide insight into the average dispersal distance for species [6]. At intermediate distances, we found very low levels of clumping in all datasets, indicating that species distributions are spatially random at mid-continent scales. However, we also found increased clumping at whole-continent distances, especially in the economic datasets. This is broadly consistent with low dispersal limitation of businesses, and the high similarity of species pools on both coasts in economic systems.

Discussion

We have shown that spatial scaling can successfully separate four universal macroecological patterns. The consistency of first-order effects across ecological, economic, and geological systems [23,24] indicates that they provide little power to distinguish ecological mechanisms. We then identified second-order effects that were able to separate these five datasets – some of which are consistent with known ecological processes (e.g. dispersal limitation), and others not. Thus, future biodiversity theories should make simultaneous predictions for these scale-dependent second-order effects, in order to provide stronger tests and more separable predictions of theory. Our results indicate that theory especially should be able to make system-dependent predictions for the

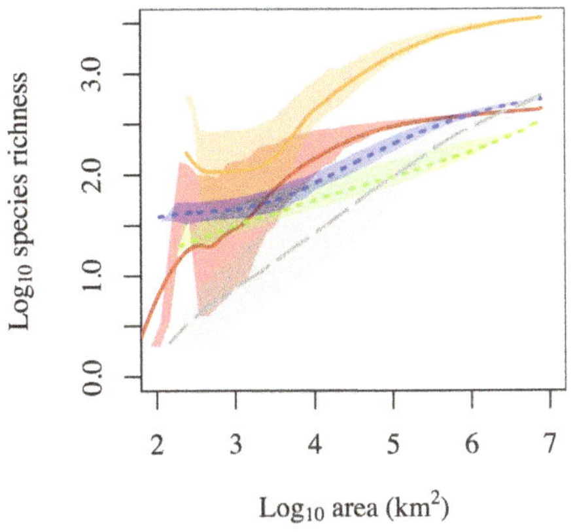

Figure 3. The species-area relationship distinguishes ecological datasets at large scales. Line colors are described in Table 2.

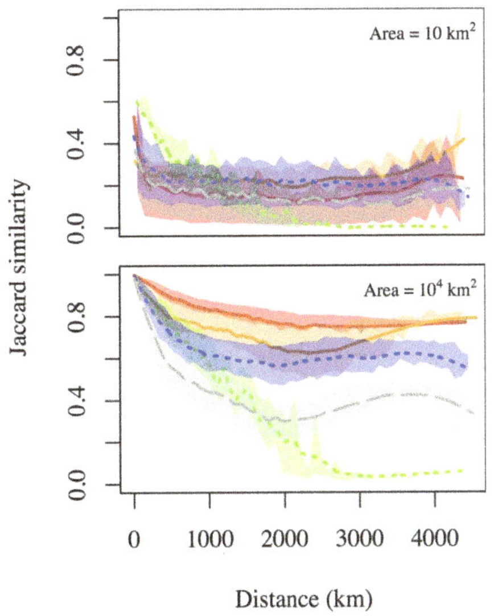

Figure 4. The decay in assemblage similarity with distance depends strongly on spatial scale. The rapidity of decrease and the minimum similarity are functions of dataset type and assemblage size. Line colors are described in Table 2.

spatial scaling of in skewness and kurtosis of the species abundance distribution, as well as of the intraspecific clumping function. Spatial scaling can become a powerful approach to distinguish mechanisms and guide the development and testing of more complex theories of biodiversity.

The general lack of dispersal limitation evident in economic systems is consistent with an 'everything is everywhere' perspective on economic diversity. For example, more than 80% of the major corporations remain the same in pairs of county-sized (10^4 km^2) communities at distances of up to 4500 km (Figure 4). Our results provide strong evidence for low beta diversity and high homogeneity of economic landscapes in the United States. Biodiversity theories will need to incorporate additional parameters to make scale-dependent predictions consistent with this finding.

Our results leave unresolved a potentially important zero-th order effect describing each system's state variables: the number of species and individuals found in each dataset. Although the value of these numbers set the scale of all first-order effects they may also ultimately constrain levels of variation in second-order effects. However, in all major theories of biodiversity, the number of individuals and the number of species are treated as free parameters [1]. Addressing the origin of the zeroth-order effect may provide as much insight as addressing the origin of second-order effects [36].

Our results question the importance of the species concept is to macroecological theory. All biodiversity theories and macroecological patterns are expressed in terms of species and individuals. For ecological systems these are natural and potentially preferred scales for understanding a system. However, there are many possible ways to partition objects in to categories for non-ecological systems, and it is unclear if any particular aggregation method should be preferred. For example, individuals businesses can be aggregated into NAICS codes, but the taxonomy and resolution of these codes is necessarily a human choice. Thus, macroecological theory may be applied best to biological species. However, many biodiversity theories are derived from very limited or no biological processes, suggesting that they should apply equally well to any partitioning of objects in to categories (e.g. taxon-invariance in the species-area relationship [37,38]). Therefore deviations from predictions, such as our second-order effects, should still reflect additional mechanisms.

We showed that our approach could be used for distinguishing different datasets or detecting situations where theory could be modified to better accommodate empirical data. We do not intend the approach to be used for null-hypothesis significance testing, i.e. statistically rejecting the null hypothesis of no second-order effects. There are two reasons: first, the specific form of first-order effects may depend on the exact mathematical formulation of a biodiversity theory which limits our ability to derive general equations; and second, the form of second-order effects is likely to be more interesting than simply rejecting the ecological null hypothesis. Nevertheless, it should be possible to develop a mathematical formalism to infer second-order effects, a goal that may be useful for the developers of next-generation biodiversity theories.

The similarities in economies and ecosystems may indicate a set of shared processes and constraints whose elucidation will have fundamental or practical implications [24]. Ecological principles and theories have been used to understand economic phenomena like competition, wealth distributions and the growth of cities [39–

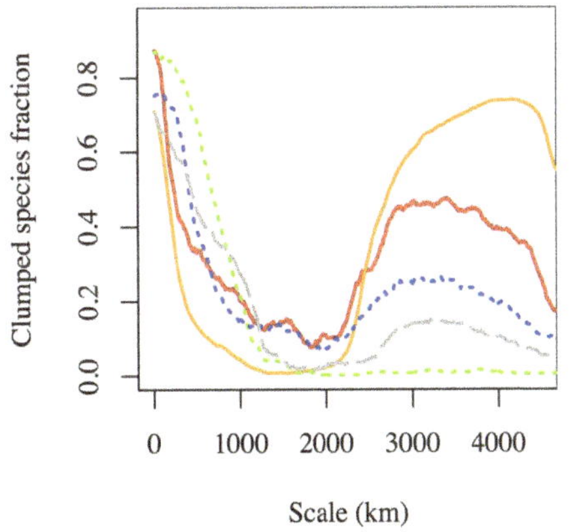

Figure 5. The fraction of species for which intra-specific clumping is consistently high at very small and very large scales. Among datasets, the clumping varies widely in magnitude with spatial scale. Line colors are described in Table 2.

41]. Because many first-order effects seem to occur regardless of system, extant macroecological theory may have practical consequences for other economic systems, e.g. financial networks [42]. For example, current work on up-scaling inexpensive local measurements of biodiversity for conservation purposes [27,43] may be relevant to economic reporting in developing regions, or in understanding the origin of other economic distributions [44]. Macroecological theory, because of its lack of intrinsic ecological mechanism, may also be applicable to many economic systems. In this way it may provide a more realistic understanding of limits to economic growth by identifying the first-order effects that provide universal and unavoidable constraints on economic systems, but also by identifying the zeroth- or second-order effects that may practically be modulated by policy shifts.

Supporting Information

Information S1 Community matrices for each dataset and MATLAB code to replicate all analyses.

Acknowledgments

We thank the many people whose efforts have made these continental-scale datasets publicly available. The National Center for Ecological Analysis and Synthesis (NCEAS) provided computational resources.

Author Contributions

Conceived and designed the experiments: BB. Performed the experiments: BB LS. Analyzed the data: BB. Contributed reagents/materials/analysis tools: BE BM. Contributed to the writing of the manuscript: BB LS BE BM.

References

1. McGill B (2010) Towards a unification of unified theories of biodiversity. Ecol Lett 13: 627–642.
2. MacArthur R (1965) Patterns of species diversity. Biol Rev 40: 510–533.
3. Storch D, Keil P, Jetz W (2012) Universal species-area and endemics-area relationships at continental scales. Nature 488: 78–81.
4. Connor E, McCoy E (1979) The statistics and biology of the species-area relationship. Am Nat 113: 791–833.
5. Morlon H, Chuyong G, Condit R, Hubbell S, Kenfack D, et al. (2008) A general framework for the distance-decay of similarity in ecological communities. Ecol Lett 11: 904–917.
6. Condit R, Ashton P, Baker P, Bunyavejchewin S, Gunatilleke S, et al. (2000) Spatial patterns in the distribution of tropical tree species. Science 288: 1414–1418.
7. Plotkin J, Potts M, Leslie N, Manokaran N, LaFrankie J, et al. (2000) Species-area curves, spatial aggregation, and habitat specialization in tropical forests. J Theor Biol 207: 81–99.
8. Weir JT, Schluter D (2007) The Latitudinal Gradient in Recent Speciation and Extinction Rates of Birds and Mammals. Science 315: 1574–1576.
9. Gauch H, Whittaker R (1972) Coenocline simulation. Ecology 53: 446–451.
10. Preston F (1960) Time and space and the variation of species. Ecology 41: 612–627.
11. May R (1975) Patterns of species abundance and diversity. In: Cody M, Diamond J, editors. Ecology and evolution of communities. Cambridge, MA: Harvard University Press.
12. Sugihara G (1980) Minimal community structure - an explanation of species abundance patterns. Am Nat 116: 770–787.
13. Harte J, Zillio T, Conlisk E, Smith AB (2008) Maximum entropy and the state-variable approach to macroecology. Ecology 89: 2700–2711.
14. Chave J, Leigh E (2002) A spatially explicit neutral model of beta-diversity in tropical forests. Theor Popul Biol 62: 153–168.
15. Hubbell S (2001) A unified theory of biodiversity and biogeography: Princeton University Press.
16. Rosindell J, Cornell SJ (2009) Species-area curves, neutral models, and long-distance dispersal. Ecology 90: 1743–1750.
17. Chave J, Muller-Landau H, Levin S (2002) Comparing classical community models: theoretical consequences for patterns of diversity. Am Nat 159: 1–23.
18. MacArthur R (1969) Patterns of communities in the tropics. Biol J Linn Soc 1: 19–30.
19. Tilman D (2004) Niche tradeoffs, neutrality, and community structure: A stochastic theory of resource competition, invasion, and community assembly. Proc Natl Acad Sci USA 101: 10854–10861.
20. McGill B (2003) Strong and weak tests of macroecological theory. Oikos 102: 679–685.
21. McGill B, Nekola JC (2010) Mechanisms in macroecology: AWOL or purloined letter? Towards a pragmatic view of mechanism. Oikos 119: 591–603.
22. Allen AP, Li BL, Charnov EL (2001) Population fluctuations, power laws and mixtures of lognormal distributions. Ecol Lett 4: 1–3.
23. Gaston K, Blackburn T, Lawton J (1993) Comparing animals and automobiles - a vehicle for understanding body size and abundance relationships in species assemblages. Oikos 66: 172–179.
24. Nekola JC, Brown JH (2007) The wealth of species: ecological communities, complex systems and the legacy of Frank Preston. Ecol Lett 10: 188–196.
25. McGill B, Etienne RS, Gray JS, Alonso D, Anderson MJ, et al. (2007) Species abundance distributions: moving beyond single prediction theories to integration within an ecological framework. Ecol Lett 10: 995–1015.
26. MacArthur R (1960) On the relative abundance of species. Am Nat 94: 25–36.
27. Harte J, Smith AB, Storch D (2009) Biodiversity scales from plots to biomes with a universal species-area curve. Ecol Lett 12: 789–797.
28. Preston F (1962) The canonical distribution of commonness and rarity: part I. Ecology 43: 185–215.
29. McGill BJ (2011) Linking biodiversity patterns by autocorrelated random sampling. Am J Bot 98: 481.
30. Holyoak M, Loreau M (2006) Reconciling empirical ecology with neutral community models. Ecology 87: 1370–1377.
31. Martiny JBH, Eisen JA, Penn K, Allison SD, Horner-Devine MC (2011) Drivers of bacterial β-diversity depend on spatial scale. Proc Natl Acad Sci USA 108: 7850.
32. McGill B (2010) Matters of scale. Science 328: 575–576.
33. Rosindell J, Cornell SJ (2013) Universal scaling of species-abundance distributions across multiple scales. Oikos 122: 1101–1111.
34. McKane A, Alonso D, Sole R (2000) Mean-field stochastic theory for species-rich assembled communities. Phys Rev E 62: 8466–8484.
35. Pueyo S (2006) Diversity: between neutrality and structure. Oikos 112: 392–405.
36. Wiens J (2011) The causes of species richness patterns across space, time, and clades and the role of "ecological limits". Q Rev Biol 86: 75–96.
37. Harte J, Kitzes J, Newman EA, Rominger AJ (2013) Taxon categories and the universal species-area relationship. The American Naturalist 181: 282–287.
38. Storch D, Šizling A (2008) The concept of taxon invariance in ecology: do diversity patterns vary with changes in taxonomic resolution? Folia Geobot 43: 329–344.
39. Bettencourt L, Lobo J, Helbing D, Kühnert C, West G (2007) Growth, innovation, scaling, and the pace of life in cities. Proc Natl Acad Sci USA 104: 7301–7306.
40. Brown J, Burnside W, Davidson A, Delong J, Dunn W, et al. (2011) Energetic limits to economic growth. BioScience 61: 19–26.
41. Decker E, Kerkhoff A, Moses M (2007) Global patterns of city size distributions and their fundamental drivers. PLoS ONE 2: e934.
42. Haldane AG, May RM (2011) Systemic risk in banking ecosystems. Nature 469: 351–355.
43. He F, Gaston K (2000) Estimating species abundance from occurrence. Am Nat 156: 553–559.
44. Stanley M, Amaral L, Buldyrev S, Havlin S, Leschhorn H, et al. (1996) Scaling behaviour in the growth of companies. Nature 379: 804–806.

22

Habitat and Forage Associations of a Naturally Colonising Insect Pollinator, the Tree Bumblebee *Bombus hypnorum*

Liam P. Crowther[1]*, Pierre-Louis Hein[1,2], Andrew F. G. Bourke[1]

1 School of Biological Sciences, University of East Anglia, Norwich Research Park, Norwich, Norfolk, United Kingdom, **2** IUT Nancy-Brabois, Rue du Doyen Urion, Villers-les-Nancy CS, France

Abstract

Bumblebees (*Bombus* species) are major pollinators of commercial crops and wildflowers but factors affecting their abundance, including causes of recent population declines, remain unclear. Investigating the ecology of species with expanding ranges provides a potentially powerful means of elucidating these factors. Such species may also bring novel pollination services to their new ranges. We therefore investigated landscape-scale habitat use and foraging preferences of the Tree Bumblebee, *B. hypnorum*, a recent natural colonist that has rapidly expanded its range in the UK over the past decade. Counts of *B. hypnorum* and six other *Bombus* species were made in March-June 2012 within a mixed landscape in south-eastern Norfolk, UK. The extent of different landscape elements around each transect was quantified at three scales (250 m, 500 m and 1500 m). We then identified the landscape elements that best predicted the density of *B. hypnorum* and other *Bombus* species. At the best fitting scale (250 m), *B. hypnorum* density was significantly positively associated with extent of both urban and woodland cover and significantly negatively associated with extent of oilseed rape cover. This combination of landscape predictors was unique to *B. hypnorum*. Urban and woodland cover were associated with *B. hypnorum* density at three and two, respectively, of the three scales studied. Relative to other *Bombus* species, *B. hypnorum* exhibited a significantly higher foraging preference for two flowering trees, *Crataegus monogyna* and *Prunus spinosa*, and significantly lower preferences for *Brassica napus*, *Glechoma hederacea* and *Lamium album*. Our study provides novel, quantitative support for an association of *B. hypnorum* with urban and woodland landscape elements. Range expansion in *B. hypnorum* appears to depend, on exploitation of widespread habitats underutilised by native *Bombus* species, suggesting *B. hypnorum* will readily co-exist with these species. These findings suggest that management could target bumblebee species with distinctive habitat requirements to help maintain pollination services.

Editor: Maura (Gee) Geraldine Chapman, University of Sydney, Australia

Funding: This work was supported in part by a studentship held by LPC funded by a Natural Environment Research Council Doctoral Training Award. The funders had no role in study design, data collection and analysis, decision to publish, or preparation of the manuscript.

Competing Interests: The authors have declared that no competing interests exist.

* Email: l.crowther@uea.ac.uk

Introduction

There is abundant evidence that bees and other insect pollinators are in global decline [1], [2], [3], [4]. Because 9.5% of global food production is attributable to wild insect pollination [5], pollinator losses threaten food security. Furthermore, such losses cannot be entirely mitigated by use of managed populations of honey bees (*Apis*), as these have been shown to complement but not replace the pollination services provided by wild pollinators across diverse agricultural systems [6].

In common with congeneric species elsewhere in the world, many bumblebee (*Bombus*) species in the UK have undergone declines over recent decades. Of the 20 recorded social species, three have gone nationally extinct and all but six are considered to have undergone range contractions [7]. Across species, the extent of these declines has been greater in species with later-starting colony cycles (i.e. with later nest foundation, build-up of worker numbers and production of sexuals) and smaller global ranges [7]. Because of the habitat associations of such species, this means that declines have occurred disproportionately in species that tend to be more dependent on specific habitats (e.g. unimproved grassland), which have been greatly reduced in extent [8]. In effect, the *Bombus* species that remain common and widespread tend to be generalists that can survive in a typical landscape mosaic of farmland, urbanised areas and isolated patches of semi-natural habitat, whereas those that have declined are restricted to fragmented patches of high-quality habitat. Since historical records indicate that many of these now-rare species were once widespread, structural changes have clearly altered the agricultural mosaic to reduce its suitability for many species [8]. In addition, it should be noted that, although widespread species are still extant in much of their historical ranges, there is insufficient evidence to conclude they have not also undergone declines in abundance.

The structural changes to agricultural habitats associated with bumblebee declines are reductions in floristic diversity, quantity of forage and abundance of nesting sites [9], [10]. Agriculturallymanaged grasslands have lower floristic diversity and density largely through heavy use of synthetic nitrogen and more frequent and earlier cutting regimes [11]. Therefore the bumblebee species that persist in the altered agricultural environment rely extensively on wild flowers in the margins of agricultural fields with briefly available peaks of forage from mass flowering crops [12]. Although mass flowering crops such as oilseed rape (*Brassica napus*) are being cultivated on larger scales, benefits to *Bombus* populations may be limited by the early timing and brevity of their flowering period relative to *Bombus* colony cycles [13].

Understanding the reasons behind range expansions of *Bombus* species may provide clues as to the factors that make the difference between population decline and population stability in the genus as a whole. In addition, range-expanding species potentially add to the suite of pollination services provided by native pollinators. This study therefore focuses on a notable outlier to the trend for population decline among *Bombus* species, the Tree Bumblebee *Bombus hypnorum* (Linnaeus), which is currently undergoing a rapid range expansion in the UK. First recorded in southern England in 2001 [14], *B. hypnorum* now occurs throughout much of England, Wales and southern Scotland, and so has undergone a northwards range expansion of almost 600 km in 12 years [15]. Although the possibility of accidental or deliberate introduction cannot be excluded, *B. hypnorum* is most likely to be a natural colonist of the UK, since it is not used or traded as a commercial pollinator. It presumably reached southern England by autonomous dispersion across the English Channel from the closest neighbouring area of the pre-2001 range, northern France. *B. hypnorum* has a large global range, encompassing most of Europe and Asia, and occupies a wide range of biotopes [14]. In pristine landscapes, *B. hypnorum* is associated primarily with boreal forests at higher latitudes and montane forests at lower latitudes. Consistent with these associations, the presence of *B. hypnorum* is predicted by length of boreal forest edges in Estonian populations [16]. However, there is no clear indication of its habitat use in anthropogenic landscapes such as those exemplified by the agricultural mosaic in the UK.

B. hypnorum appears to have an early-starting colony cycle [17], which, along with a large global range, would place it among species resilient to decline according to the analysis of Williams [7]. In addition, observations suggest that, in non-boreal, lowland areas, *B. hypnorum* favours human settlements [18]. It is also suspected of exhibiting a facultatively bivoltine colony cycle [19] and, unlike most UK *Bombus* species, nests in above-ground cavities [17], including holes in trees and walls. An early-starting colony cycle, association with human settlement, facultative bivoltinism and above-ground nesting may all be contributing to the range expansion of *B. hypnorum*. Nonetheless, it remains unclear what factors underlie this rapid expansion on a land mass in which other *Bombus* species are declining and subject to pressures of a changing agricultural landscape. Moreover, recent work suggests that *B. hypnorum* has spread within the UK despite relatively high levels of parasite prevalence and low levels of genetic diversity [20]. In addition, although climate change could be contributing to the UK expansion of *B. hypnorum*, an expansion mirrored by some native species in the far north of the UK [21], it is unclear why some species but not others should be affected by climate change. Furthermore, existing data suggest that the main pattern of change in the ranges of declining *Bombus* species in Europe is one of contraction towards the climatic range centre, modulated by the ecological suitability of available habitat [22]. This implies that even if climate change is contributing to the range expansion of *B. hypnorum* in the UK, it remains important to establish the ecological determinants of the success of this species.

Overall, no previous study has quantified the landscape-scale habitat use and foraging preferences of *B. hypnorum* in its new range in the UK. We therefore aimed, within this population, (i) to quantify the relationship between the frequency of occurrence of *B. hypnorum* and landscape elements potentially providing nesting and foraging habitats, and (ii) to quantify the foraging preferences of *B. hypnorum* relative to those of other *Bombus* species.

Materials and Methods

Observations of Bees

The study was conducted in south-eastern Norfolk, UK, an area with landscapes typical of the mixed agricultural, semi-natural and urban landscapes found in southern England as a whole. *B. hypnorum* was first recorded in Norfolk in 2008 [23]. Forty-two sampling sites were selected at random from a set of 120 possible locations identified using an Ordnance Survey map as accessible using public rights of way, within an area of approximately 26×23 km. Post-selection inspection showed that they comprised a broad mixture of urban and rural land cover types (Fig. S1; Table S1). At each sampling site, a single 200×2 m strip transect [24] (taken to include a vertical dimension 2 m high) was defined using existing physical landmarks. Each transect was placed along a linear feature, such as a field edge or other patch boundary. In order to increase temporal resolution, subset of transects ($N = 8$), selected at random across all sites, was visited at least every 8 days, and remaining transects ($N = 34$) were visited at least every 16 days. All visits took place between 26 March and 30 June 2012. Although this sampling period may have missed foraging activity by later-founded colonies or colonies founded (via facultative bivoltinism) by newly-produced queens of the year, it should have captured the bulk of *B. hypnorum* foraging activity, since, in the study year, the corresponding time period contained 82.3% of *B. hypnorum* records for England and Wales [23]. To avoid pseudoreplication of landscape metrics, transects were kept spatially separated (the mean distance between neighbouring transects was 3100 m, range 1600–6150 m).

On each transect visit, a single observer, walking at roughly 1.5 km h^{-1}, recorded all *Bombus* individuals observed on the transect (excluding socially parasitic species). The following data were collected for each bumblebee encountered: species, sex, caste (queen or worker), and forage plant visited (if any). Bumblebees were identified using [19], with *B. terrestris* and *B. lucorum* workers, which are difficult to distinguish in the field, being pooled (hereafter, '*Bombus terrestris* agg.'). When necessary, bees were temporarily caught to confirm species or sex. Transects were only surveyed in dry weather when the temperature was at least 10°C (before 1 May 2012) or 14°C (on or after 1 May 2012). The air temperature was recorded to the nearest 0.5°C with a digital thermometer; at each transect visit, values ranged from 10.0°C to 26.5°C. Since all sampling sites were on public rights of way and no endangered or protected organisms were sampled, no formal permissions were required for this work.

Sampling of Forage Composition

At each transect visit, a 2×2 m quadrat was placed at a randomised location within 5 m of each of the points 50 m, 100 m and 150 m along the transect's length. Abundances of all plant taxa with open flowers within the quadrat were recorded as presence/absence of open flowers in each of 25 equally-sized subdivisions of the quadrat. Species within the genera *Anchusa, Cirsium, Geranium, Malus, Papaver, Ranunculus, Rubus, Silene,*

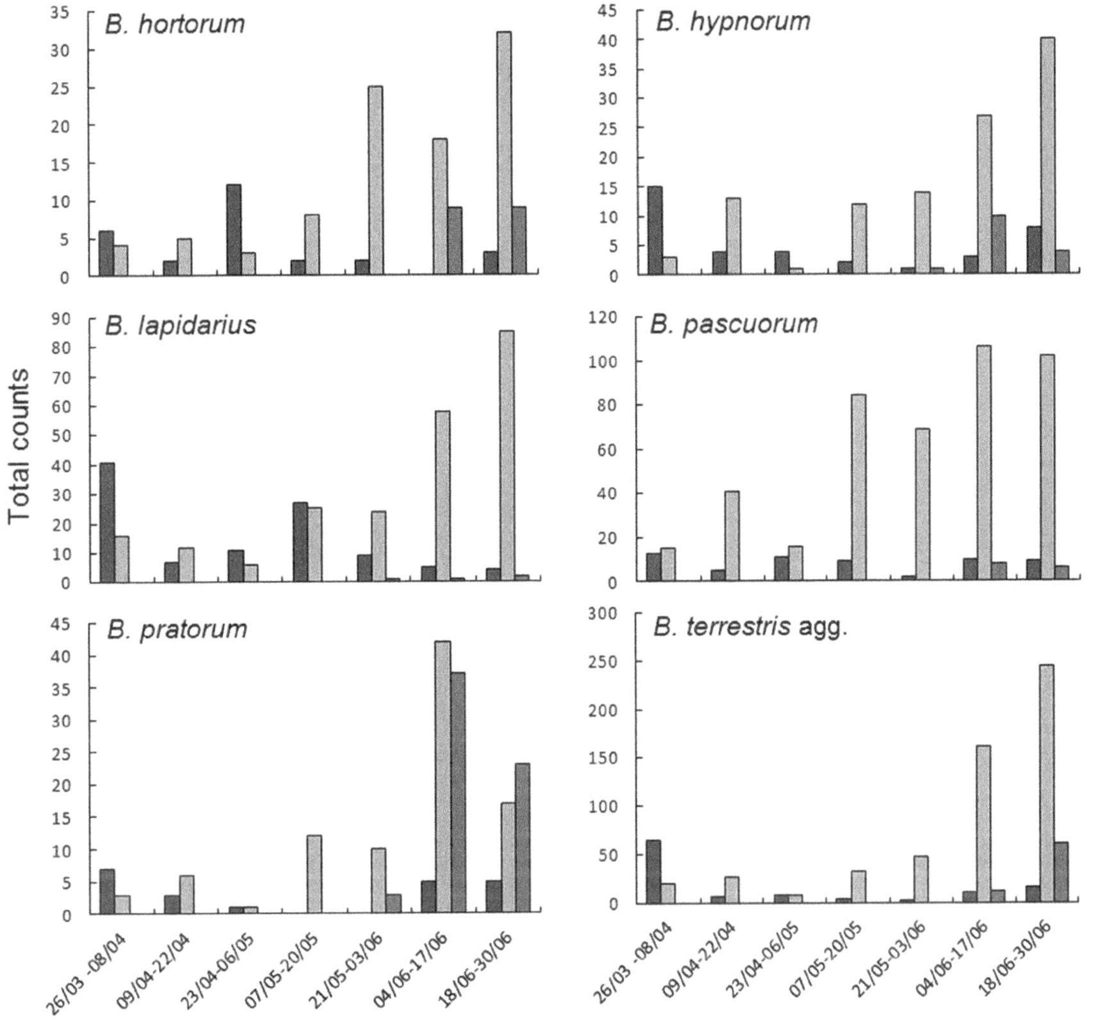

Figure 1. Phenologies of *Bombus* spp. across the 42 study sites (in south-eastern Norfolk, UK). Abundances shown as total counts on the transects per given two-week period (hence each period reflects approximately equal sampling effort; see 'Materials and methods'). Dates are expressed as dd/mm in the study year, 2012. Dark grey bars, queens; pale grey bars, workers; intermediate grey bars, males.

Table 1. Summary of landscape metrics for 42 *Bombus* transect sites in Norfolk, UK (means with range in parentheses).

Landscape metric	Information source	Scale		
		250 m	500 m	1500 m
% cover Cereals	Ground survey	28.0 (0–99.2)	29.9 (0–79.2)	35.4 (0–62.0)
% cover Intensive Grass	Ground survey	19.53 (0–93.4)	20.24 (0–90.3)	20.20 (6.2–54.6)
% cover Urban	OS MasterMap	18.8 (0–99.4)	20.1 (0.6–95.2)	17.4 (3.4–78.6)
% cover Other arable	Ground survey	9.7 (0–76. 4)	7.5 (0–60.1)	6.6 (0–22.1)
% cover Oilseed rape	Ground survey	9.3 (0–58.7)	8.2 (0–41.4)	8.8 (0–27.9)
% cover Woodland	OS MasterMap (checked via ground survey)	8.6 (0–45.9)	8.6 (0–30.6)	7.0 (0.8–18.9)
% cover Semi-natural	OS MasterMap (checked via ground survey)	3.2 (0–31.5)	2.7 (0–16.4)	2.5 (0–10.7)
% cover Species-rich grassland	Ground survey	1.9 (0–35.9)	1.2 (0–20.0)	0.5 (0–7.7)
% cover Field bean	Ground survey	0.2 (0–8.3)	0.5 (0–17.3)	0.3 (0–5.1)
Total edge (m)	OS MasterMap	3972 (0–7360)	13484 (0–25110)	32306 (2360–56820)

Scale refers to distance from the centre of the transect for which metrics were computed. Percentage cover is ranked in order of decreasing cover at the 250 m scale.

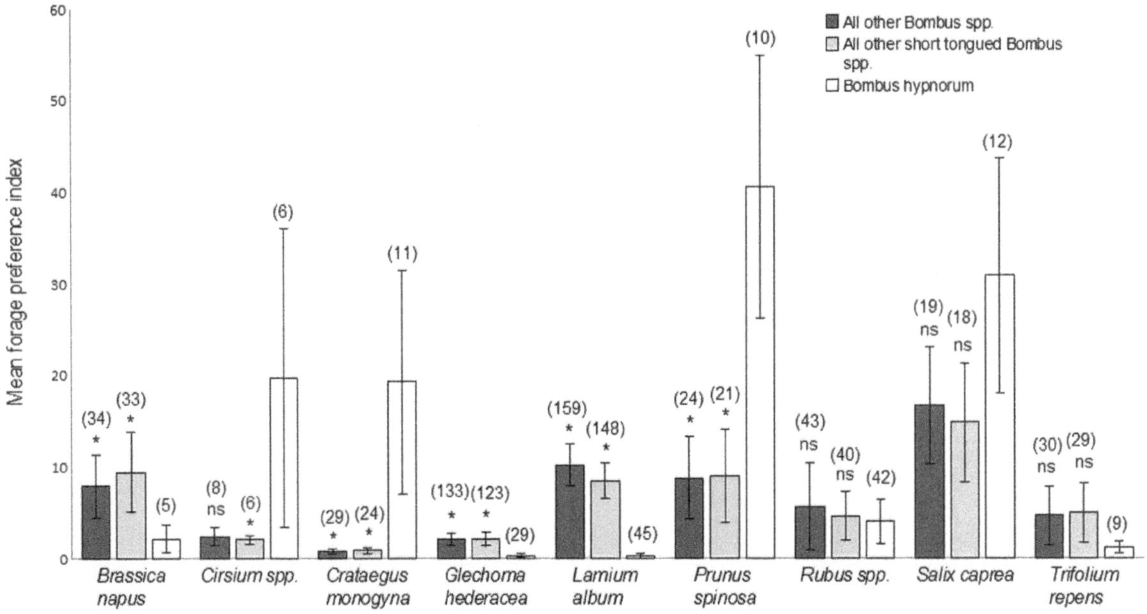

Figure 2. Foraging preference indices for all *Bombus* species excluding *B. hypnorum*, short-tongued *Bombus* species excluding *B. hypnorum* and *B. hypnorum*, for all plant taxa that co-occured with *B. hypnorum* on five or more transect visits. A foraging preference index >1 indicates a preference for visiting a given plant taxon (see 'Materials and methods'). Results of Mann-Whitney U-tests between median foraging preference indices of *Bombus* species groups and *B. hypnorum*: * $P<0.05$; ns, no significant difference. Error bar is ±1 S.E. Sample sizes (number of transect-visits at which plant taxon was present and focal bee species/group was foraging) in parentheses.

Solanum, Symphytum, Taraxacum, Ulex and *Vicia* were identified to genus level only. The quadrat was taken to be 2 m high, so that flowers on bushes within hedges and scrub vegetation were included.

Analysis of Foraging Preferences

We quantified the foraging preference of a given *Bombus* species for a focal forage plant taxon by comparing, within transects, the proportion of its visits to that taxon (out of its total visits, with sexes and castes being pooled) with the proportion of available flowers represented by the focal plant taxon [7]. If the given *Bombus* species had no preference for the focal plant, these proportions would be the same, whereas any deviation of the proportion of visits above or below the plant's relative abundance would indicate a foraging preference and non-preference, respectively. Preference was therefore calculated as $(\alpha - \beta)/\beta$ [7], where α = proportion of visits recorded by the given *Bombus* species to the focal plant taxon and β = proportion of flowering plants represented by that taxon. Where a *Bombus* species was recorded visiting a plant that was present on the transect but not recorded in the quadrats, the plant was assumed to have a relative abundance of 1%. In order to predict foraging patterns, we then calculated the quantity, $((\alpha - \beta)/\beta) + 1$ (hereafter, 'foraging preference index'). The relative abundance of a plant taxon, multiplied by the foraging preference index, predicted the expected proportion of foraging visits to the focal plant taxon. For example, if a plant taxon comprised 10% of the flowering plants and received 50% of visits from a given *Bombus* species, the foraging preference index was $((0.5-0.1)/0.1) + 1 = 5$. To avoid calculating foraging preference indices for plant taxa receiving very few visits, foraging preference indices were calculated for each plant taxon within the smallest set of plant taxa that collectively received more than 95% of observed visits.

Foraging preference indices for a given *Bombus* species were averaged across all transects for which the *Bombus* species co-occurred with the focal plant. We then compared the foraging preference indices of co-occurring bumblebee species groups [7]. For this purpose, given that tongue length in bumblebees affects their floral preferences, we grouped species by tongue length. Therefore, foraging preference indices were calculated for (i) all *Bombus* spp. excluding *B. hypnorum*, (ii) all short-tongued *Bombus* species excluding *B. hypnorum*, and (iii) *B. hypnorum*, and tested for significant differences between species groups using Mann-Whitney U tests. Short-tongued *Bombus* species were defined as all *Bombus* species in the data set with a proboscis length of less than 8 mm, i.e. *B. hypnorum*, *B. lapidarius*, *B. pratorum* and *B. terrestris* agg. *Bombus* species in the data set with proboscis lengths of more than 8 mm (*B. hortorum* and *B. pascuorum*) are hereafter referred to as long-tongued *Bombus* species. Proboscis lengths were taken from [25] and [26].

Classification of Land Cover

Ordnance Survey (OS) MasterMap data [27]were used to generate a vector map of the area surrounding each transect using ArcGIS 10.0 [28]. Each vector map consisted of polygons defining the extent of buildings, roads, fields and other land parcels, with each polygon possessing an attribute table containing its characteristics. All polygons that at their nearest point were within 1500 m of each transect centre were included.

The attribute tables associated with the polygons on these maps were used to reclassify the land cover of given areas as urban, woodland, semi-natural or farmland using the OS MasterMap fields. Urban areas were defined as any area covered by buildings, gardens or roads; woodland as any area covered by mature trees with canopy cover of greater than 40%; semi natural as areas covered by either scattered mature trees, scrub, osiers (*Salix* plantations), heathland, wetland or rough grassland; farmland as areas under arable cultivation or managed pasture (see Supporting Information for details). A ground survey was undertaken to

Table 2. Summary of final models of landscape predictors of *Bombus* density.

Bombus species/group	Optimum scale (m)	Fixed effects	Random effects	Dispersion estimate	AIC	ΔAIC			d.f.
						250 m	500 m	1500 m	
B. hortorum	250	1+ F_L+Date+SNA	Site	1.20	338.3	–	5.9	5.9	290
B. hypnorum	250	1+ Date+URB +WOO+OSR	Site	0.95	267.3	–	2.1	10.8	289
B. lapidarius	250	1+ F_S+SNA	Site+F_S	0.96	483.3	–	2	0.9	249
B. pascuorum	1500	1+ F_L+Date+SNA	Site+F_L	1.21	587.7	8.9	5.1	–	247
B. pratorum	250	1+ F_S+Date+URB	Site	1.48	414.8	–	1.4	2.4	290
B. terrestris agg.	250	1+ F_S+Date+SNA+TE	Site+F_S	2.87	911.0	–	4.9	2.4	247

All models are fitted to data from 338 visits to 42 transect sites. Optimum scale: landscape scale at which landscape metrics were calculated. Fixed effects: Date, date of transect-visit; F_L, visit-specific forage quality index for long-tongued *Bombus* species; F_S, visit-specific forage quality index for short-tongued *Bombus* species; OSR, % oilseed rape cover; SNA, % semi-natural cover; TE, total length of field edges; URB, % urban cover; WOO, % woodland cover. Random effects: Site, identity of site. Dispersion estimate: residual sum of squares divided by residual degrees of freedom. AIC: Akaike's Information Criterion. ΔAIC: comparison to final model at alternative landscape scales. d.f: degrees of freedom, calculated as: n -1 - number of parameters.

classify farmland further according to the crops being grown. This produced farmland classes of cereals, field bean (*Vicia faba*), intensive grass, oilseed rape (*Brassica napus*), other arable, and species-rich grass (see Supporting Information for details).

Computation of Landscape Metrics

The vector maps generated as described above were converted to 300×300 pixel grids (hereafter, 'rasters') with a 100 m^2 (i.e. 10 m×10 m) pixel size and with each pixel value representing the reclassified land cover classes at the centre of each grid cell. These rasters were clipped to three landscape scales corresponding to the circular areas lying within 250 m, 500 m and 1500 m, respectively, of the centre point of each transect. The radii of these areas were chosen to cover the range of *Bombus* worker foraging distances taken from the combined results of mass marking and genetic studies and to reflect that, while the number of estimated worker foraging distances remains relatively low, means tend to cluster between 250 m and 750 m and the highest maxima at around 1500 m [29], [30], [10], [31], [32]. Further rasters of 100 m^2 pixel size were created with a unique pixel value for each agricultural field and a null value for all other land covers, and these were clipped to the same three landscape scales as above. The rasters were then processed using FRAGSTATS 4.0 [33] to calculate the percentage cover of each land cover class, and the total length of agricultural field edges (hereafter, 'total edge'), within each landscape scale.

Calculation of Forage Quality Index

For both short-tongued and long-tongued *Bombus* species over each transect visit, we used the foraging preference indices weighted by the abundances of plants to calculate a forage quality index and so quantify the attractiveness of the transect as a foraging site on the visit date (see Text S1 for details).

Analysis of Landscape Predictors of *Bombus* Density

Poisson GLMMs using log link functions were constructed in R [34] using the lme4.0 package [35] to predict the density of each *Bombus* species (defined as counts per transect visit) from the landscape metrics, while controlling for phenological and local habitat variation. Initial data exploration indicated that species-rich grassland and field beans made up a particularly small proportion of the landscape and did not covary significantly with density of any *Bombus* species. These metrics were therefore omitted from further analyses.

Initial models included the following landscape metrics as fixed effects: total edge, percentage cover of urban, oilseed rape, semi-natural and woodland land-cover (hereafter, 'selected landscape metrics'), with all landscape metrics being measured at the intermediate, 500-metre scale. Initial models also included the following variables as fixed covariates; date, to control for the increase in numbers of *Bombus* workers over time as colonies grow in size; and the forage quality index appropriate to the proboscis length-class of the focal *Bombus* species, to control for the attractiveness of the given transect site. Akaike's Information Criteria (AIC) were used with the initial models to optimise the random component [36], which was allowed to be either just a random effect for site (grouping variable) or additionally to include effects for the forage quality indices (random slopes). The latter allowed for the attractiveness of a transect site to vary according to unmeasured landscape gradients, such as the quality of alternative forage sources. Alternative initial GLMMs were constructed using temperature instead of date as a fixed effect, but in every instance date produced a better fit and in no instance did this change the optimised random component.

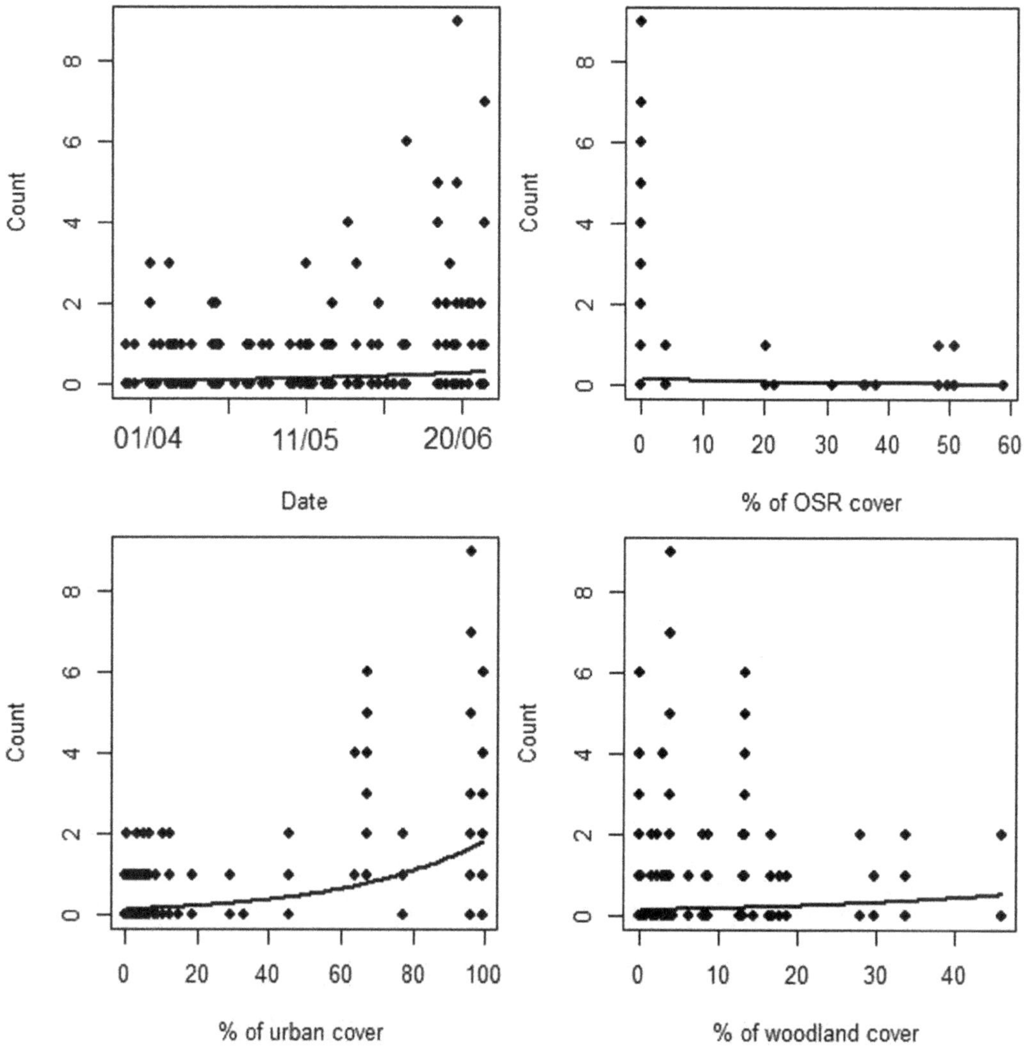

Figure 3. Relationship of *Bombus hypnorum* density with selected landscape metrics retained in the best-fitting final model (250 m scale: Tables 2, 3). Date expressed as dd/mm in study year (2012). Circles, counts of *B. hypnorum* observed at each transect-visit. Trend lines (plotted using LanguageR [44] from the final Poisson GLMM based on partial effects) are for illustrative purposes only; conclusions were based on results presented in Tables 2–4.

Using the optimised random component, we fitted three maximal models for each *Bombus* species, each with the selected landscape metrics at the three spatial scales (250 m, 500 m and 1500 m) and the appropriate forage quality index and date as fixed effects. In each case an alternative model with temperature instead of date was compared; however, using date always yielded a lower AIC value and hence a better-fitting model.

The fixed effects were then refined to include only significant effects (at $P<0.05$) using a backwards stepwise sequence of single term deletion likelihood ratio tests, the resulting models being

Table 3. Summary of final GLMM model of landscape predictors of *B. hypnorum* density at the optimal 250 m scale.

Fixed effect	Parameter Estimate	SE	Wald statistic	*P*
Intercept	-707.800	128.8000	-5.492	<0.001
Date	0.017	0.0031	5.477	<0.001
OSR	-0.040	0.0181	-2.190	<0.05
URB	0.026	0.0047	5.564	<0.001
WOO	0.029	0.0135	2.158	<0.05

The model is fitted to data from 338 visits to 42 transect sites. Date, date of transect-visit; OSR, % oilseed rape cover; URB,% urban cover; WOO, % woodland cover.

Table 4. Summary of likelihood ratios for landscape metrics at the three different scales retained as fixed effects in models predicting *Bombus* densities (data from 338 visits at 42 transect-sites).

Bombus species/group	Scale (m)	TE	OSR	SNA	URB	WOO
B. hortorum	250**	–	–	7.923	–	–
	500	–	–	–	–	–
	1500	–	–	–	–	–
B. hypnorum	250**	–	8.392	–	38.42	5.312
	500*	–	–	–	59.312	6.658
	1500	–	–	–	48.471	–
B. lapidarius	250**	–	–	5.1334	–	–
	500	–	–	–	3.1469	–
	1500*	–	4.2065	–	–	–
B. pascuorum	250	–	–	–	–	–
	500*	2.695	–	–	5.014	5.345
	1500**	–	–	9.618	–	–
B. pratorum	250**	–	–	–	3.435	–
	500*	–	–	–	–	–
	1500	–	–	1.22	–	2.797
B. terrestris agg.	250**	8.320	–	4.33	–	–
	500	2.932	–	–	–	–
	1500*	–	–	–	5.465	–

OSR, % oilseed rape cover; SNA, % semi-natural cover; TE, total length of field edges; URB, % urban cover; WOO, % woodland cover. ** model at landscape scale with lowest AIC; * model at landscape scale with intermediate AIC (Table 2); dash (–), fixed effect was not retained in the final model.

assessed with AIC values. The model with the lowest AIC compared to final models of the alternative scales is presented as the final model at the optimum scale for each *Bombus* species. Estimates of dispersion were also calculated as the Pearson residual sum of squares divided by residual degrees of freedom.

Finally, we compared the relative strength of the influence of different landscape elements on bee density across *Bombus* species and landscape scales. For this, we compared the fixed effects of the retained landscape metrics across models using the results of the likelihood ratio tests, which test the likelihood of a model including the focal landscape metric relative to the same model lacking the focal landscape metric.

Results

Phenology and Landscape Metrics

In total, 2048 *Bombus* records were accumulated across 338 person-visits to 42 transects. Individual species contributions to the total records were as follows: *B. hortorum*, 140; *B. hypnorum*, 162; *B. lapidarius*, 334; *B. pascuorum*, 506; *B. pratorum*, 175; and *B. terrestris* agg., 731. *B. hypnorum* was recorded at 26 of the 42 transect sites and records comprised 37 queens, 110 workers and 15 males. *B. hypnorum* queens and workers were observed throughout the study period, whereas males were only observed after 28 May 2012 (Fig. 1). Our estimates of landscape metrics for the 42 sites showed that cereals represented the most frequent cover class at all scales, followed by intensive grass and urban (Table 1).

Foraging Preferences

B. hypnorum was observed foraging in 140 of the total of 162 *B. hypnorum* records, distributed over 24 transect sites, during the study period. Eighteen plant taxa received 95% of all observed *Bombus* foraging visits (Table S2) and hence were used to calculate the foraging preference indices. Of these plant taxa, only nine co-occurred with *B. hypnorum* on sufficient transects to allow meaningful comparisons (taken as five or more transects).

Among these nine plant taxa, *B. hypnorum* exhibited the strongest foraging preferences for *Cirsium* spp., *Crataegus monogyna, Prunus spinosa* and *Salix caprea* (Fig. 2). Relative to other *Bombus* species and other short-tongued *Bombus* species, *B. hypnorum* showed a significantly higher preference for *Crataegus monogyna* and *Prunus spinosa* (Fig. 2). In addition, relative to other short-tongued *Bombus* species but not other *Bombus* species, *B. hypnorum* showed a significantly higher preference for *Cirsium* spp. (Fig. 2). *B. hypnorum* also exhibited significantly lower preferences for *Brassica napus, Glechoma hederacea* and *Lamium album* relative to other *Bombus* species and other short-tongued *Bombus* species (Fig. 2).

Landscape Predictors of *Bombus* Density

Final models showed that in all cases except that of *B. pascuorum*, the density of each *Bombus* species was best predicted by the selected landscape metrics at the 250 m scale (Table 2; Tables S5–S9). The density of *B. pascuorum* was best predicted by the selected landscape metrics at the 1500 m scale (Table 2).

B. hypnorum density at the best-fitting scale (250 metres) showed a significant positive association with percentage urban cover (likelihood ratio = 38.419, $P<0.001$) and percentage woodland cover (likelihood ratio = 5.312, $P<0.05$) and a significant negative association with percentage oilseed rape cover (likelihood ratio = 8.392, $P<0.01$). The following fixed effects had no significant effect on *B. hypnorum* density: forage quality index (likelihood ratio = 2.866, $P = 0.09$), total edge (likelihood ratio = 0.044, $P = 0.83$) and percentage semi-natural cover (likelihood ratio = 1.663, $P = 0.20$) (Fig. 3; Tables 2–4). No other *Bombus* species or species group showed the combination of significant predictors of density exhibited by *B. hypnorum* (Tables 2, 4).

For most *Bombus* species, density was predicted optimally by different landscape metrics across the different scales (Table 4). For *B. hypnorum* and *B. terrestris* agg. alone, common fixed effects exhibited a significant influence across the different scales. For *B. hypnorum*, common effects were urban cover (all scales) and woodland cover (250 m and 500 m scales). For *B. terrestris* agg., the common effect was total edge (250 and 500 m scales) (Table 4). *B. pascuorum* density was predicted by a different set of predictors at the 500 m scale (urban cover and woodland cover) than at its optimal 1500 m scale (semi-natural cover) (Table 4).

Finally, the comparison of final models using likelihood tests showed that the effect of urban cover at all scales on *B. hypnorum* density was much stronger and/or more consistent (i.e. was associated with higher likelihoods) than that of any other landscape element retained in the final models for any other *Bombus* species (Table 4).

Discussion

To elucidate the ecological factors underpinning range expansion in a naturally colonising bumblebee, we investigated the landscape predictors and foraging preferences of the Tree Bumblebee, *Bombus hypnorum*, in a representative area within its recently-colonised UK range. We found that, at the best fitting scale (250 m), *B. hypnorum* density was significantly positively associated with extent of both urban and woodland cover and significantly negatively associated with extent of oilseed rape cover. This combination of significant landscape predictors was unique to *B. hypnorum*. In addition, urban cover was associated with *B. hypnorum* density at all three scales studied and woodland cover was associated with *B. hypnorum* density at two of the three scales. Moreover, urban cover had a stronger and/or more consistent effect on *B. hypnorum* density relative to that of any other landscape elements retained in final models for other *Bombus* species. *B. hypnorum* exhibited a significantly higher foraging preference for two flowering trees, *Crataegus monogyna* and *Prunus spinosa*, relative to all other *Bombus* species studied (and hence independently of proboscis length). It also showed significantly lower preferences for the mass-flowering crop *Brassica napus* and two common, early-flowering herbaceous plants, *Glechoma hederacea* and *Lamium album*, relative to all other *Bombus* species studied (and so, again, independently of proboscis length).

The strong association shown by this study between *B. hypnorum* density and urban landscape elements (Table 4) provides quantitative support for an association previously suspected from natural-history observations alone [18], [19], [17]. The marked difference between the predictive power of urban landscape elements in explaining *B. hypnorum* density and that of any other combination of landscape element or *Bombus* species suggests the existence of quantitative ecological differences between *B. hypnorum* and other *Bombus* species. While other widespread *Bombus* species have been shown to utilise urban landscapes at high densities [37], [38], and conversely another *Bombus* species (*B. vosnesenskii*) has been shown to nest at lower densities in urbanised areas [39], this result indicates that not only does *B. hypnorum* use such landscapes in addition to the agricultural mosaic but also that urban elements represent the major component of its habitat use. This conclusion is strengthened by the fact that our final models for other *Bombus* species included significant effects for the elements that these species are known to favour in agricultural environments such as semi-natural areas and total edge [12], [38], suggesting that, if *B. hypnorum* used them widely, this would have been detected.

Our findings show that *B. hypnorum* makes use of resources offered by urban areas that are not utilised to the same extent by other *Bombus* species. Urban environments offer varied and often abundant forage in gardens, parks and waste-ground, but other species have equal access to this. Therefore a likely possibility is that *B. hypnorum* can nest at higher densities in the urban environment, relative to other environments, because of the greater availability of above-ground cavities suitable for nesting. This explanation would also be consistent with our finding that *B. hypnorum* density was significantly associated with extent of woodland cover. We note, however, that worker density is not necessarily an accurate predictor of nest density (and hence population size) in the social Hymenoptera. Hence landscape elements associated with higher *B. hypnorum* counts in our study could be facilitating either higher nest densities, or greater worker numbers per nest or a combination of these.

B. hypnorum may become an important crop pollinator in the UK. While it has a lower preference for *B. napus* (by far the most extensive commercial mass flowering crop in the study area) than other *Bombus* species, its early phenology and preference for *C. monogyna* and *P. spinosa* suggest it could act as a frequent pollinating visitor to spring-flowering tree-fruit crops. *P. spinosa* has several cultivated congeners, for example *Prunus avium* (sweet cherry) and *P. domestica* (plum), which typically have a flowering phenology similar to that of *P. spinosa* [40], [41]. While the economic importance of fruit crops is not as high as that of some arable mass flowering crops (e.g. *B. napus, Vicia faba*), they have been shown to be at greater risk from pollinator declines, both globally and in the European Union [5]. Overall, because *Cirsium* species, *C. monogyna* and *P. spinosa* are abundant in hedges, woodland edges and urban waste-ground, it appears that *B. hypnorum* is likely to encounter these preferred forage plants in most landscapes within the UK. *B. hypnorum* is also likely to benefit from the presence of numerous species of cultivated plants in urban environments, although such species (Table S2) could not have been identified as preferred forage species in the present study because they did not occur frequently enough across the study sites. However, it needs noting that many sources of variation were not controlled for in our calculation of foraging preference indices, including forage plant phenology and conspecific density variation.

Significant landscape predictors of bee density for *Bombus* species other than *B. hypnorum* (Table 2; Tables S5–S9) included extent of semi-natural cover (negative effect in *B. hortorum*, *B. lapidarius* and *B. terrestris* agg., positive effect in *B. pascuorum*), extent of urban cover (positive effect in *B. pratorum*) and total edge (negative effect in *B. terrestris* agg.). From previous studies identifying landscape features utilised by *Bombus* species [10], [12], [42], a positive effect of extent of semi-natural cover on *Bombus* density, as found in *B. pascuorum*, and a positive effect of total edge, might be expected. The negative effects of extent of semi-natural cover found in *B. hortorum*, *B. lapidarius* and *B. terrestris* agg., and the negative effect of total edge found in *B. terrestris* agg., are therefore unexpected. A possible explanation as regards the effects of semi-natural cover and total edge is that that a greater extent of such features caused a 'dilution effect' whereby densities per transect were reduced, in contrast to higher densities that can occur in resource-rich patches in a matrix of poorer-quality habitat [43]. Supporting this interpretation, in *B.*

pascuorum (Table 2), in which the optimum scale was larger (1500 m), extent of semi-natural cover had a positive effect on bee density. The positive effect of extent of urban cover in *B. pratorum* accords with this species, like *B. hypnorum* but probably to a lesser degree, nesting in above-ground cavities [17].

In conclusion, this study demonstrates that, at both a landscape and within-patch scale, *B. hypnorum* and native *Bombus* species exhibit different patterns of resource utilisation. This finding could help explain the otherwise puzzlingly rapid range expansion of *B. hypnorum* within the UK. It also implies that *B. hypnorum* is unlikely to represent a competitive threat to native *Bombus* species. *B. hypnorum* may become an important pollinator of tree-fruit crops, although further research would be required to determine the efficiency with which it pollinates such crops. Because it has habitat associations differing from those of other *Bombus* species, *B. hypnorum* will also potentially provide pollination services in landscape contexts in which other species are scarce. However, in addition to the variation explained by the landscape predictors of *B. hypnorum* occurrence detected in the current study, there is clearly much unexplained variation underlying the incidence of this species (Fig. 3), which requires investigation in future work. Overall, from our current results, it appears that the quantitatively unusual combination of landscape elements favoured by *B. hypnorum* extends the range of resource utilisation patterns previously described within bumblebees. This suggests that, even in the presence of land use changes that bring about pollinator declines, management could target species with different requirements to help maintain pollination services.

Supporting Information

Figure S1 Sampling locations.

Table S1 sampling locations.

Table S2 Summary of forage plant abundances.

Table S3 Correlation of landscape variables.

Table S4 Summary of forage plant relative abundances across sites.

Tables S5 Summaries of final models for other *Bombus* species densities.

Table S6 Summaries of final models for other *Bombus* species densities.

Table S7 Summaries of final models for other *Bombus* species densities.

Table S8 Summaries of final models for other *Bombus* species densities.

Table S9 Summaries of final models for other *Bombus* species densities.

Text S1 Supporting methods detailing land cover classification and calculation of forage quality index.

Acknowledgments

We thank Katy Appleton, Claire Carvell, Richard Davies, Mike Edwards and Jenny Gill for help and comments and Stuart Roberts for use of BWARS information. This work was supported in part by a studentship held by LPC funded by a Natural Environment Research Council Doctoral Training Award.

Author Contributions

Conceived and designed the experiments: LC AB. Performed the experiments: LC PLH. Analyzed the data: LC. Contributed to the writing of the manuscript: LC AB.

References

1. Brown MJF, Paxton RJ (2009) The conservation of bees: a global perspective. Apidologie 40: 410–416.
2. Potts SG, Biesmeijer JC, Kremen C, Neumann P, Schweiger O, et al. (2010) Global pollinator declines: trends, impacts and drivers. Trends in Ecology & Evolution 25: 345–353.
3. Bommarco R, Lundin O, Smith HG, Rundlöf M (2012) Drastic historic shifts in bumble-bee community composition in Sweden. Proceedings of the Royal Society B: Biological Sciences 279: 309–315.
4. Vanbergen AJ, Insect Pollinators Initiative (2013) Threats to an ecosystem service: pressures on pollinators. Frontiers in Ecology and the Environment 11: 251–259.
5. Gallai N, Salles J-M, Settele J, Vaissiere BE (2009) Economic valuation of the vulnerability of world agriculture confronted with pollinator decline. Ecological Economics 68: 810–821.
6. Garibaldi LA, Steffan-Dewenter I, Winfree R, Aizen MA, Bommarco R, et al. (2013) Wild Pollinators Enhance Fruit Set of Crops Regardless of Honey Bee Abundance. Science 339: 1608–1611.
7. Williams P (2005) Does specialization explain rarity and decline British bumblebees? - A response to Goulson et al. Biological Conservation 122: 33–43.
8. Goulson D, Hanley ME, Darvill B, Ellis JS (2006) Biotope associations and the decline of bumblebees (Bombus spp.). Journal of Insect Conservation 10: 95–103.
9. Carvell C, Roy DB, Smart SM, Pywell RF, Preston CD, et al. (2006) Declines in forage availability for bumblebees at a national scale. Biological Conservation 132: 481–489.
10. Knight ME, Osborne JL, Sanderson RA, Hale RJ, Martin AP, et al. (2009) Bumblebee nest density and the scale of available forage in arable landscapes. Insect Conservation and Diversity 2: 116–124.
11. Pywell RF, Meek WR, Hulmes L, Hulmes S, James KL, et al. (2011) Management to enhance pollen and nectar resources for bumblebees and butterflies within intensively farmed landscapes. Journal of Insect Conservation 15: 853–864.
12. Diekoetter T, Kadoya T, Peter F, Wolters V, Jauker F (2010) Oilseed rape crops distort plant-pollinator interactions. Journal of Applied Ecology 47: 209–214.
13. Westphal C, Steffan-Dewenter I, Tscharntke T (2009) Mass flowering oilseed rape improves early colony growth but not sexual reproduction of bumblebees. Journal of Applied Ecology 46: 187–193.
14. Goulson D, Williams PH (2001) Bombus hypnorum (Hymenoptera: Apidae), a new British bumblebee? British Journal of Entomology and Natural History 14: 129–131.
15. Bees Wasps and Ants Recording Society (2013) Bombus hypnorum mapping project. Roberts SPM, editor. http://www.bwars.com/index.php?q=content/bombus-hypnorum-2013, accessed 2014 April 15.
16. Sepp K, Mikk M, Mand M, Truu J (2004) Bumblebee communities as an indicator for landscape monitoring in the agri-environmental programme. Landscape and Urban Planning 67: 173–183.
17. Benton T (2006) Bumblebees. London: Collins.
18. Loken A (1973) Studies on Scandinavian Bumblebees. Norsk Entomologisk Tidsskrift 20.
19. Edwards M, Jenner M (2005) Field Guide to the Bumblebees of Great Britain and Ireland. Willingdon: Ocelli Limited.
20. Jones CM, Brown MJF (2014) Parasites and genetic diversity in an invasive bumblebee. Journal of Animal Ecology: doi: 10.1111/1365-2656.12235.
21. Plowright CMS, Plowright RC, Williams PH (1997) Replacement of *Bombus muscorum* by *Bombus pascuorum* in Northern Britain? Canadian Entomologist 129: 985–990.
22. Williams PH, Araújo MB, Rasmont P (2007) Can vulnerability among British bumblebee (*Bombus*) species be explained by niche position and breadth? Biological Conservation 138: 493–505.

23. Bees Wasps and Ants Recording Society (2013) *Bombus hypnorum* UK records. Roberts SPM, editor, available at: https://data.nbn.org.uk/Taxa/NHMSYS0000875501, accessed 2013-12-03.
24. Heard MS, Carvell C, Carreck NL, Rothery P, Osborne JL, et al. (2007) Landscape context not patch size determines bumble-bee density on flower mixtures sown for agri-environment schemes. Biology Letters 3: 638–641.
25. Pekkarinen A (1979) Morphometric, colour and enzyme variation in bumblebees (Hymenoptera, Apidae, Bombus) in Fennoscandia and Denmark. Acta Zoologica Fennica 158: 1–60.
26. Williams PH (1989) Why are there so many species of bumble bees at Dungeness? Botanical Journal of the Linnean Society 101: 31–44.
27. Ordnance Survey (2012) OS MasterMap Topography Layer [GML geospatial data]. Using: EDINA Digimap Ordnance Survey Service, <http://edina.ac.uk/digimap>: Ordnance Survey, GB.
28. Esri Inc (2011) ArcGIS Desktop: Release 10.0.
29. Darvill B, Knight ME, Goulson D (2004) Use of genetic markers to quantify bumblebee foraging range and nest density. Oikos 107: 471–478.
30. Osborne JL, Martin AP, Carreck NL, Swain JL, Knight ME, et al. (2008) Bumblebee flight distances in relation to the forage landscape. Journal of Animal Ecology 77: 406–415.
31. Carvell C, Jordan WC, Bourke AFG, Pickles R, Redhead JW, et al. (2012) Molecular and spatial analyses reveal links between colony-specific foraging distance and landscape-level resource availability in two bumblebee species. Oikos 121: 734–742.
32. Charman TG, Sears J, Green RE, Bourke AFG (2010) Conservation genetics, foraging distance and nest density of the scarce Great Yellow Bumblebee (Bombus distinguendus). Molecular Ecology 19: 2661–2674.
33. McGarigal K, Cushman SA, Ene E (2012) FRAGSTATS v4: Spatial pattern analysis program for categorical and continuous Maps. Computer software program produced by the authors at the University of Massachusetts, Amherst. Available at the following web site: http://www.umass.edu/landeco/research/fragstats/fragstats.html.
34. R Development Core Team (2011) R: A Language and Environment for Statistical Computing. Vienna: R Foundation for Statistical Computing.
35. Bates D, Maechler M, Bolker B (2012) lme4: Linear mixed-effects models using S4 classes, Version R package version 0.999999-0.
36. Zuur AF, Ieno EN, Walker N, Saveliev AA, Smith GM (2009) Mixed Effects Models and Extensions in Ecology with R. New York: Springer Science & Business Media.
37. Chapman RE, Wang J, Bourke AFG (2003) Genetic analysis of spatial foraging patterns and resource sharing in bumble bee pollinators. Molecular Ecology 12: 2801–2808.
38. Osborne JL, Martin AP, Shortall CR, Todd AD, Goulson D, et al. (2008) Quantifying and comparing bumblebee nest densities in gardens and countryside habitats. Journal of Applied Ecology 45: 784–792.
39. Jha S, Kremen C (2013) Resource diversity and landscape-level homogeneity drive native bee foraging. Proceedings of the National Academy of Sciences of the United States of America 110: 555–558.
40. Yeboah Gyan K, Woodell SRJ (1987) Flowering Phenology, Flower Colour and Mode of Reproduction of Prunus spinosa L. (Blackthorn); Crataegus monogyna Jacq. (Hawthorn); Rosa canina L. (Dog Rose); and Rubus fruticosus L. (Bramble) in Oxfordshire, England. Functional Ecology 1: 261–268.
41. Usenik V, Stampar F (2011) The Effect of Environmental Temperature on Sweet Cherry Phenology. European Journal of Horticultural Science 76: 1–5.
42. Kennedy CM, Lonsdorf E, Neel MC, Williams NM, Ricketts TH, et al. (2013) A global quantitative synthesis of local and landscape effects on wild bee pollinators in agroecosystems. Ecology Letters 16: 584–599.
43. Carvell C, Osborne JL, Bourke AFG, Freeman SN, Pywell RF, et al. (2011) Bumble bee species' responses to a targeted conservation measure depend on landscape context and habitat quality. Ecological Applications 21: 1760–1771.
44. Baayen RH (2011) languageR: Data sets and functions with "Analyzing Linguistic Data: A practical introduction to statistics". R package version 1.4 ed.

23

Accumulation and Connectivity of Coarse Woody Debris in Partial Harvest and Unmanaged Relict Forests

Robert C. Morrissey[1,2]*, Michael A. Jenkins[2], Michael R. Saunders[2]

1 Department of Forest Ecology, Faculty of Forestry and Wood Sciences, Czech University of Life Sciences, Prague, Czech Republic, **2** Hardwood Tree Improvement and Regeneration Center, Department of Forestry and Natural Resources, Purdue University, West Lafayette, Indiana, United States of America

Abstract

When a tree dies, it continues to play an important ecological role within forests. Coarse woody debris (CWD), including standing deadwood (SDW) and downed deadwood (DDW), is an important functional component of forest ecosystems, particularly for many dispersal-limited saproxylic taxa and for metapopulation dynamics across landscapes. Processes, such as natural disturbance or management, modify forest composition and structure, thereby influencing CWD abundance and distribution. Many studies have compared older forests to forests managed with even-aged silvicultural systems and observed a prolonged period of low CWD occurrence after harvesting. With fine-scale spatial data, our study compares the long-term impacts of light partial harvesting on the CWD structure of eastern deciduous hardwood forests. We mapped and inventoried DDW and SDW using variable radius plots based on a 10 m×10 m grid throughout an unmanaged, structurally-complex relict forest and two nearby forests that were partially harvested over 46 years ago. The relict stand had significantly larger individual pieces and higher accumulations of DDW and SDW than both of the partially harvested stands. Connectivity of CWD was much higher in the relict stand, which had fewer, larger patches. Larger pieces and higher proportion of decay-resistant species (e.g. *Quercus* spp.) in the relict forest resulted in slower decomposition, greater accumulation and increased connectivity of CWD. Partial harvests, such that occur with selection forestry, are generally considered less disruptive of ecosystem services, but this study highlights the long-term impacts of even light partial harvests on CWD stocks and distribution. When planning harvesting events, forest managers should also consider alternative methods to ensure the sustainability of deadwood resources and function.

Editor: Christopher Carcaillet, Ecole Pratique des Hautes Etudes, France

Funding: This work was funded by the Hardwood Tree Improvement and Regeneration Center, and the Department of Forestry and Natural Resources at Purdue University. The funders had no role in study design, data collection and analysis, decision to publish, or preparation of the manuscript.

Competing Interests: The authors have declared that no competing interests exist.

* Email: robcmorrissey@gmail.com

Introduction

Even after death, trees continue to influence ecological processes and biodiversity of forest ecosystems [1]. Coarse woody debris (CWD), both standing dead trees (SDW) and downed deadwood (DDW), provides regeneration substrate for some plants, serves as habitat for many insect and wildlife taxa, mitigates runoff and erosion from slopes, contributes to soil development, and provides long-term sequestration of carbon and other nutrients [2]. Because CWD decays slowly, it can potentially influence these ecological processes at local, stand, and landscape levels for many decades. As forest managers have attempted to integrate a broader array of ecological objectives into their planning, provision of CWD to preserve structural attributes and ecological processes has become increasingly emphasized.

The importance of CWD as habitat has been well documented for many organisms including birds [3], arthropods [4], fungi [5], and small mammals and amphibians [6]. Given the abundance of CWD in mature forests, it is conceivable that many species evolved to rely on a suite of habitats based on abundance, species, size, decay stages, and position (standing vs. fallen) of CWD [1], [2]. In these forests, CWD-dependent species with limited dispersal should predominate, and the connectivity and spatial arrangement of CWD should be critical to the abundance of species and individuals [7]. Availability, connectivity, patch size and abundance of CWD are critical to supporting and maintaining diverse natural communities [8]. Connectivity, an emergent property across multiple ecological scales, strongly influences the population and metapopulation dynamics of multiple taxa.

Not surprisingly, the literature is replete with differences among species regarding the importance of connectivity of CWD at the stand scale [5], [7], [9]. Organisms that are highly mobile are likely unaffected at stand-level scales [10], but there may be population fitness benefits associated with high connectivity for these species, e.g. prey avoidance, perch sites [11]. For saproxylic taxa that may be dispersal-limited, including many insects, bryophytes, lichens, amphibians, and reptiles, the scale of connectivity or, conversely, fragmentation, may influence an individual's ability to disperse and occupy new habitat [12], [13], [14]. Thus, changes to spatiotemporal patterns of habitat continuity in managed and fragmented forests likely influence species dispersal, success, and fitness.

Understanding the spatiotemporal dynamics of CWD is essential to framing strategies for forest and wildlife management

Table 1. Coarse woody debris piece summaries.

a) Downed deadwood				
Tract	**n**	**LED (cm)**	**Length (m)**	**Volume (m^3)**
Relict	924	34.3±0.7 A	11.2±0.2 A	1.19±0.08 A
Single-harvest	288	26.0±0.9 B	8.7±0.2 B	0.48±0.06 B
Double-harvest	153	31.3±2.2 B	7.6±0.3 C	0.84±0.21 B
b) Standing deadwood				
Tract	**n**	**DBH (cm)**	**Height (m)**	**Volume (m^3)**
Relict	182	46.8±2.5 A	13.2±0.7 A	2.27±0.26 A
Single-harvest	96	36.5±2.4 A	10.9±0.8 A	1.24±0.28 A
Double-harvest	57	28.7±3.7 B	7.0±0.5 B	0.56±0.16 B

Total number of pieces (n) and mean (± standard error) of individual piece sizes of sampled a) downed deadwood (minimum diameter of 10 cm) and b) standing deadwood in relict, single-harvest, and double-harvest stands. Large end diameter (LED), length, and volume of downed deadwood, and diameter at 1.37 m (DBH), total height, and volume of standing deadwood are presented. We tested for differences ($p<0.05$) of log-transformed values using a generalized linear model and post-hoc multiple comparisons with Tukey's honestly significant difference. Different letters indicate significant differences among stands.

[15], and is thus an essential element of biological conservation in forest ecosystems. Spatiotemporal CWD dynamics are largely influenced by tree mortality and decay rates. Tree mortality is not evenly distributed over time or space [1] and is related to disturbance history, stand structure, and site factors [2], all of which influence patterns of CWD dispersion. Several studies of old-growth forests have observed high abundance and aggregated patterns of DDW [5], [7], [10], but patterns of SDW varied across many studies [16], [17]. Light partial harvests, which include thinnings, selection cut, and some timber stand improvement treatments, generally disproportionally remove low vigor and individual or small groups of trees, in order to favor more vigorous, higher quality trees throughout the stand. Therefore, these harvests may impact future spatiotemporal inputs of CWD by influencing CWD size, species, volume, input rate, and spatial distribution, which thereby can both reduce the abundance of CWD over time and either homogenize or concentrate CWD distribution, depending on the management regime. Light partial harvests have typically been considered to have a short-term and minor ecological impact, but studies of CWD impacts related to known management events is critical to understand long-term dynamics of ecological processes.

We examined the quantity, quality, and spatial arrangement of deadwood in a relict forest and two managed forests that were partially harvested more than 46 years ago. All stands are isolated forests within a fragmented agricultural matrix; thus, connectivity at a stand level may be especially important to sustain dispersal-limited taxa under the assumption that highly connected populations are less likely to succumb to local random extinction events. We chose not to consider individual species or taxa in this study, but rather focus on structural connectivity of the CWD within stands. Based on differences in disturbance histories of these stands, we hypothesized: 1) the overall volume of CWD would be less in managed stands; 2) average piece size and abundance of CWD would be less in managed stands than in the relict, unmanaged stand; 3) the distribution of CWD volume among decay stages would shift towards less decayed material in managed stands relative to the relict stand; and 4) distribution of CWD would exhibit higher levels of connectivity in less disturbed stands.

Methods

Study sites

Study sites were located on the Davis Purdue Agricultural Center property; permissions were granted by the Woodlands Management Committee through the Department of Forestry and Natural Resources of Purdue University. They included three *Quercus*-dominated deciduous forest stands within a fragmented, agricultural landscape in east-central Indiana, U.S.A. (40°15′26″, 85°09′16″). These stands have gently rolling, low relief topography with elevation changes <5 m across the entire area. Since the middle of the 20th century, all stands have been impacted by Dutch elm disease and elm phloem necrosis, both of which resulted in the death of most elm trees before they recruit into the overstory canopy. The relict forest is the largest stand (20.6 ha), of which we sampled the 8.5 ha interior. Records indicate that this stand has not been managed, except for the removal of some dead and dying trees in 1941, 1948, and 1955 (volume removed of 5.3 m^3 ha^{-1}). The smallest stand (2.8 ha) was sampled in its entirety; this stand, referred to as the double-harvest stand henceforth, was selectively harvested in 1951 (volume removed of 9.8 m^3 ha^{-1}) and again in 1964 (volume removed of 7.8 m^3 ha^{-1}). The third stand (2.9 ha) was also selectively harvested in 1951 (volume removed of 8.1 m^3 ha^{-1}); this stand will henceforth be referred to as the single-harvest stand. According to historical records, all trees removed from these stands were overstory individuals ≥40 cm at 1.37 m (DBH). For a more thorough description of these sites, see [18].

Data collection

We sampled CWD on a 10 m by 10 m grid using probability proportional to size (PPS) sampling methods. At each grid point, DDW was sampled using point relascope sampling [19] with an angle of 60°, and SDW was sampled using a 10-factor prism. For each individual DDW piece with a small end diameter (SED) ≥ 10 cm diameter, we measured the large end diameter (LED), SED, and the length of the piece. When pieces were long enough to still have intact multiple branches within a partially intact crown, we chose the single most linear piece to measure the SED and length. In rare cases, branches of intact crowns had sufficiently large diameters and were sampled as independent pieces. For

SDW, pieces had to be ≥1.3 m in height, but no minimum diameter limit was used; height was recorded to either top of crown, if still intact, or top of the standing stem.

To estimate individual DDW piece volumes, we assumed a conic-paraboloid shape to minimize bias of estimates and field measures required [20]. We calculated volume (V; m^3) using:

$$V = \frac{L}{12} \times \left(5A_l + 5A_s + 2\sqrt{A_l A_s}\right) \quad (5.1)$$

where L is DDW length (m), A_l is cross-sectional area (cm) at large end, and A_s is cross-sectional area (cm) at small end. Because we were only interested in historical inputs of DDW as a function of forest dynamics, we did not account for decreased volumes of collapsed pieces in advanced states of decay, as recommended by [20]. Volume for SDW with intact or partially intact crowns was based on unpublished local volume tables using a 10 cm top diameter (Table S1). For sheared stems, volume was estimated with equation 5.1 using height for length, breast height diameter to calculate A_l, and a form class of 78 was used to estimate SED and calculate A_s; form class mathematically describes the shape of the tree bole and provides an estimate of the bole diameter at any given length (or height in this case) based on the taper of the bole and DBH [21].

We classified each piece according to its decay stage; when a piece is sufficiently large, it often had different sections of the piece in differing decay stages, so we assigned decay stage based on the section most representative of the piece based on volume. For DDW, we classified decay stage as: 'sound' if recently dead with wood still hard and bark largely present, either loose or tight; 'intermediate-decay' as bark mostly gone, bole periphery softened whereby a blade can penetrate the outer layer; and 'high-decay' as little to no bark remains, bole periphery is very soft and extends in to the core, often partially incorporated into forest floor or vegetation has begun to colonize. For SDW, we had only two categories: 'sound' and 'intermediate-decay', as described above.

Data analyses

We compared mean length, LED, and volume of individual DDW among all three stands using a generalized linear model (GLM) to account for the unbalanced number of pieces sampled in each stand. We used a log transformation of the response variable and a Gaussian distribution of errors. The same procedure was used for individual SDW pieces to examine DBH, height, and volume. To test for differences of DDW and SDW abundance among stands, we used a subset of the sampling grid (30 m by 30 m) distribution to limit correlation among plots related to sampling of individual pieces from multiple different plots. We used a GLM procedure on the square root transformed response variable and a Gaussian distribution of errors. Post-hoc tests of comparison were conducted using Tukey's honest significant differences method.

We tested for differences between stands based on volume ha^{-1} by size classes (≤30 cm, >30 cm and ≤60 cm, >60 cm) and decay stages using a chi-square test. For density, we could only test for differences in the number of pieces of DDW by decay stages; DDW by size classes and SDW for decay and size classes had small bin values that could not be tested formally, although we do present them graphically.

Estimates of connectivity were developed to characterize the potential for dispersal-limited organisms to traverse the stand based on their affinity for CWD of a certain form or size. We assumed organisms could disperse 10 m, or the distance between two grid cells, if CWD was present in adjacent cells; thus, data included SDW and DDW pieces that were tallied from at least two gridpoints. Connectivity was defined by the number of gridpoints that tallied a CWD piece divided by the total number of gridpoints, and reported as a percentage by diameter size classes (10 cm width). We standardized patch area and patch density using per hectare values; patches of CWD were defined either by the edge of the sample area, adjacent plots that tallied no CWD, or some combination of the two. Connectivity interactions between DDW and SDW were not substantiated because the PPS methods differ both in angle used to select sample pieces (i.e., DDW is sampled with 60° and SDW is sampled with 1° 48′) and also the size criterion used for each sample procedure (i.e., DDW uses length of piece and SDW uses diameter).

Given the time since harvest in these stands and the lack of stumps, we assumed that all CWD was related to natural processes rather than the harvest events. All tests were conducted with an alpha level of 0.05, and all analyses were done using R [22].

Table 2. Coarse woody debris stand-level summaries.

a) Downed deadwood			
Tract	**n**	**Pieces ha^{-1}**	**Volume ha^{-1} ($m^3\ ha^{-1}$)**
Relict	817	242.8±17.0 A	130.8±4.2 A
Single-harvest	392	154.8±14.8 B	39.9±2.5 B
Double-harvest	234	199.3±28.1 AB	59.8±6.4 B
b) Standing deadwood			
Tract	**n**	**Pieces ha^{-1}**	**Volume ha^{-1} ($m^3\ ha^{-1}$)**
Relict	817	21.7±1.9 A	24.3±1.0 A
Single-harvest	392	18.4±2.1 A	11.2±0.7 B
Double-harvest	234	41.2±9.1 A	7.6±0.7 B

Total plots (n), and mean (± standard error) number of pieces ha^{-1} and volume ha^{-1} of a) downed deadwood (minimum diameter of 10 cm) and standing deadwood in relict, single-harvest, and double-harvest stands. To minimize correlation between plots, we used a subset of plots based on 30 m×30 m spacing in each stand. We tested for differences ($p<0.05$) of square root-transformed values using a generalized linear model and post-hoc multiple comparisons with Tukey's honestly significant difference. Different letters indicate significant differences among stands.

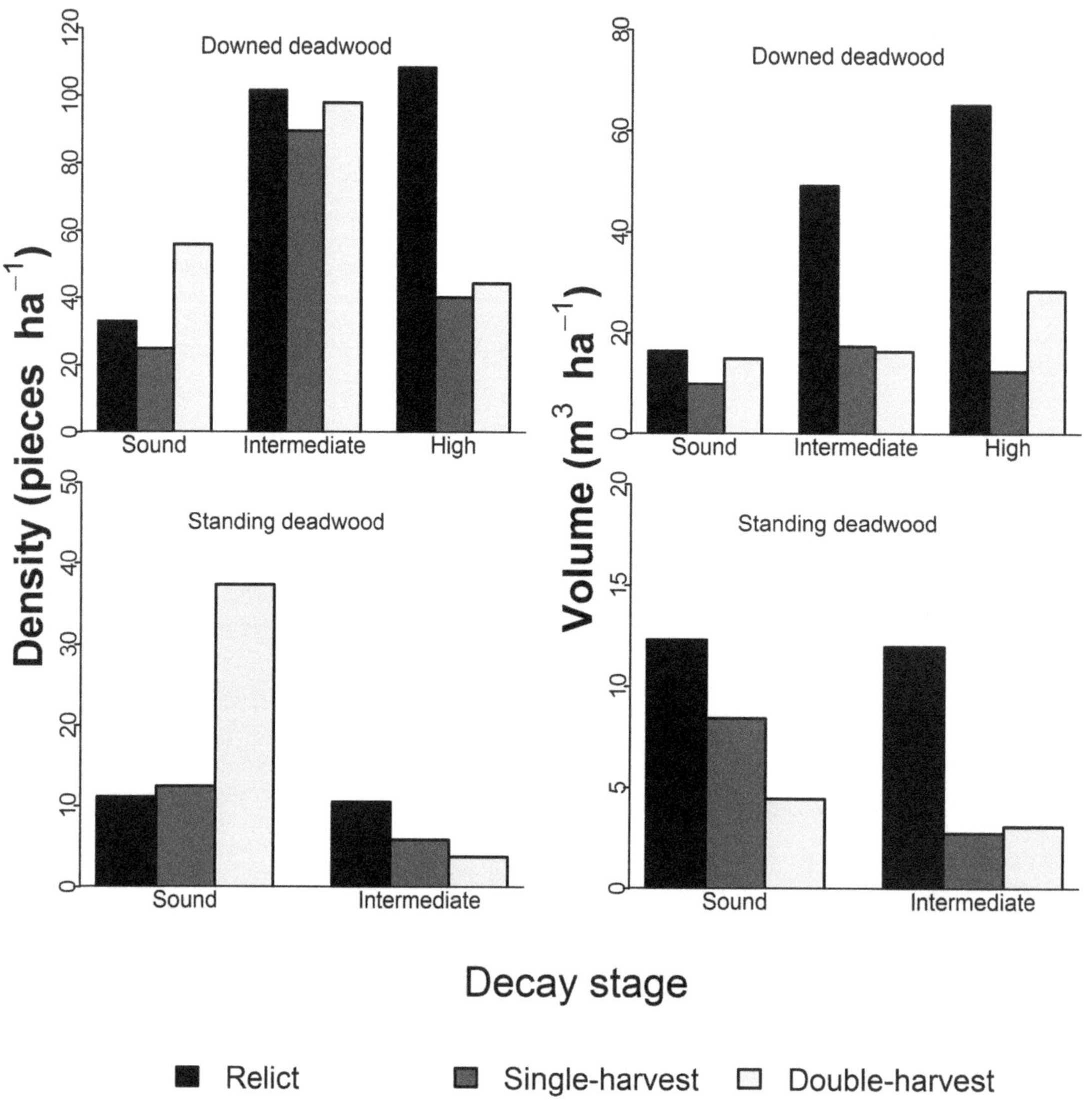

Figure 1. Coarse woody debris decay stage summaries. Downed deadwood (DDW; top row) and standing deadwood (SDW; bottom row) number of pieces ha^{-1} and volume ha^{-1} by decay stage for relict, single-harvest, and double-harvests stands. Decay stages were assigned as: sound, intermediate, or high; for SDW, sound and intermediate categories were used. For a more thorough description, see 2.2 Data Collection.

Results

Individual pieces of DDW in the relict stand were significantly larger in LED, length, and volume compared to the partial-harvest stands (Table 1; all $p \leq 0.003$). Mean volume of individual DDW pieces within the relict stand was 142% larger compared to the double-harvest stand and 248% larger than those of the single-harvest stand. Individual SDW pieces of the relict and single-harvest stands were not significantly different (all $p \geq 0.202$); however, individual piece means of DBH, height, and volume of both stands differed from those of the double-harvest stand (all $p \leq 0.02$). Within the double-harvest stand, the population of SDW was dominated by smaller pieces with correspondingly low volumes.

Mean number of DDW pieces ha^{-1} within the relict stand was significantly higher compared to the single-harvest stand ($p = 0.033$), but did not differ from the double-harvest stand ($p = 0.134$; Table 2); the partial-harvest stands were not significantly different from one another ($p = 0.975$). Significant differences between the relict and partial-harvest stands were evident for volume ha^{-1} of DDW (all $p < 0.001$). The relict stand had 328% and 219% more volume compared to the single-harvest and double-harvest stands, respectively; the partial-harvest stands were not significantly different from one another ($p = 0.918$). Mean density of SDW pieces ha^{-1} did not differ among all three stands (all $p \geq 0.764$), but SDW volume ha^{-1} of the relict stand was significantly higher (both $p \leq 0.000$) than that of both partial-harvest stands, which were not different from one another ($p = 0.773$). Although density of SDW pieces ha^{-1} did not significantly differ among stands (all $p < 0.245$), the double-harvest stand had 41.2 ± 9.1 (mean ± SD) pieces ha^{-1} on average, almost double the relict stand. The SDW volume ha^{-1} of the relict stand was 217% greater than that of the single-harvest stand and 320% greater than that of the double-harvest stand.

We observed a significant difference in the number of DDW pieces ha^{-1} ($\chi^2 = 15.2$, $p = 0.004$) and volume ha^{-1} ($\chi^2 = 131.0$, $p < 0.001$) by decay stages among stands (Figure 1). In the relict stand, density of high-decay DDW pieces in the relict stand (108.4

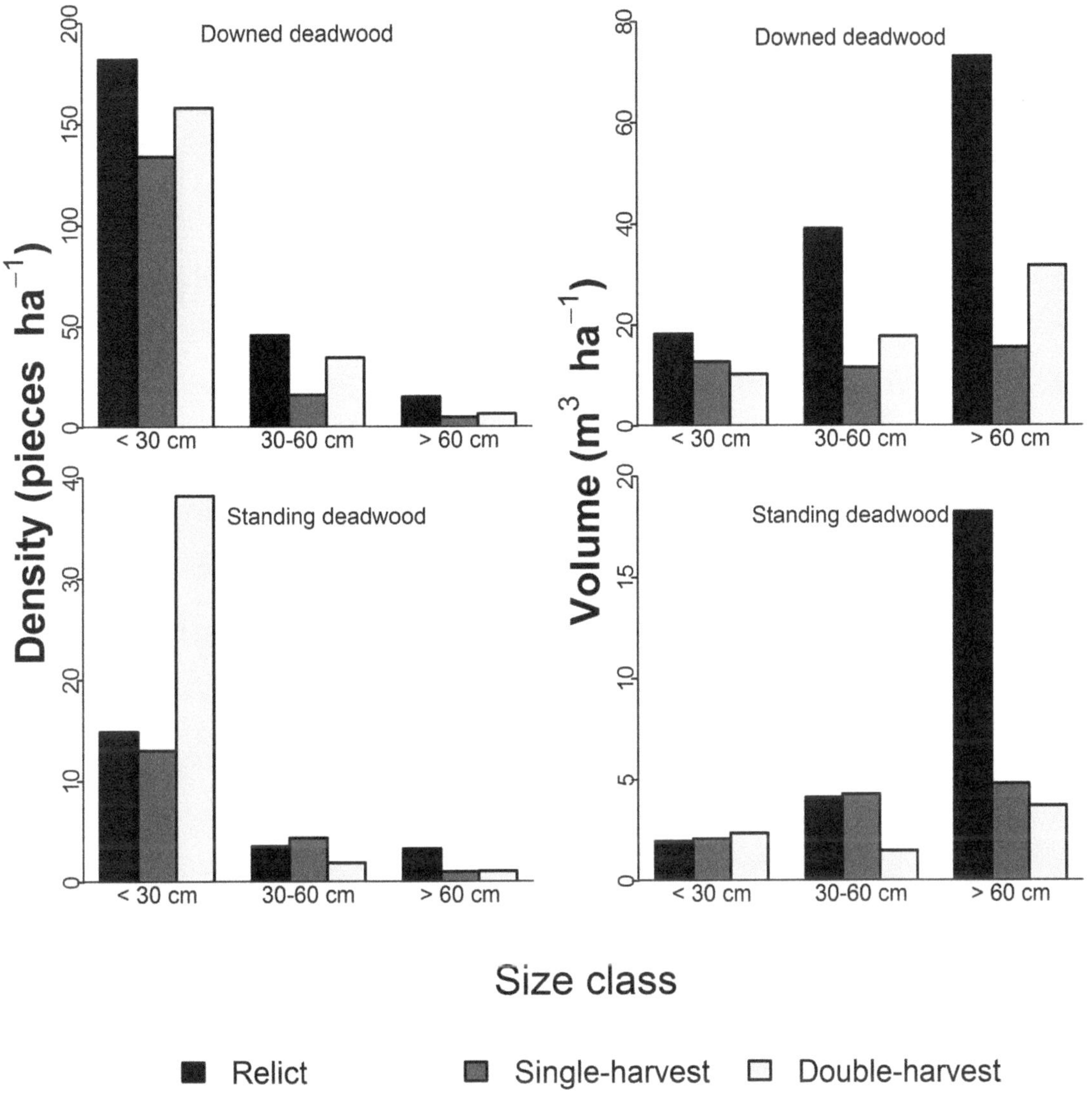

Figure 2. Coarse woody debris size class summaries. Downed deadwood (DDW; top row) and standing deadwood (SDW; bottom row) number of pieces ha^{-1} and volume ha^{-1} by size class for relict, single-harvest, and double-harvests stands. Size classes were defined as diameters of ≤30 cm, >30 cm and ≤60 cm, and >60 cm; DDW diameter was based on large end diameter, and we used diameter at 1.37 m (DBH) for SDW.

pieces ha^{-1}) was much higher than either of the partial-harvest stands (≤44.4 pieces ha^{-1}), and volumes in intermediate- and high-decay, 49.4 m^3 ha^{-1} and 65.1 m^3 ha^{-1}, respectively, were more than twice the amount in either of the partial-harvest stands. Density of SDW pieces by decay stage could not be tested because there were too few pieces in the intermediate-decay stage, but there was more than three times the number of sound SDW pieces (37.4 pieces ha^{-1}) in the double-harvest stand compared to the other stands (Figure 1). Volume ha^{-1} of SDW by decay stage was significantly different among stands ($\chi^2 = 177.6$, $p<0.001$); the relict stand had 12 m^3 ha^{-1} in the intermediate decay stage compared to <5 m^3 ha^{-1} in each of the partial-harvest stands.

For size class distributions, we could not test for differences in the number of DDW or SDW pieces ha^{-1} among stands because the largest size classes had very low values (Figure 2). All stands exhibited high densities of DDW <30 cm, similar to the living tree distribution [18]. The double-harvest stand had more than double the density of SDW pieces, 38.2 pieces ha^{-1}, in the <30 cm size class compared to the other two stands. Volume of DDW ha^{-1} across size classes was different across stands ($\chi^2 = 103.8$, $p< 0.001$), with the relict stand having larger volumes in all size classes, most notably in classes >30 cm. Volume of SDW ha^{-1} across size classes differed across stands ($\chi^2 = 76.3$, $p<0.001$), attributable to the large volume of large diameter (≥60 cm) SDW, 18.2 m^3 ha^{-1}, compared to 4.8 m^3 ha^{-1} and 3.7 m^3 ha^{-1} for the single- and double-harvest stands, respectively.

Connectivity of DDW and SDW declined as minimum diameter increased in a similar pattern for all stands, but the relict stand always displayed much higher connectivity (Figures 3, 4, and 5). Mean patch area ha^{-1} of DDW generally decreased as minimum diameter increased, but the partial-harvest stands area did not vary as much by minimum diameter class compared to the relict stand (Figure 3). Mean patch area ha^{-1} of the relict stand DDW exhibited two abrupt decreases at minimum LEDs of 40 cm and 60 cm; beyond 60 cm, all three stands had similar mean patch area ha^{-1}. For SDW, patch sizes show a similar flat profile for partial-harvest stands, and mean patch area ha^{-1} of the relict stand abruptly decreases beyond a diameter of 40 cm. Number of

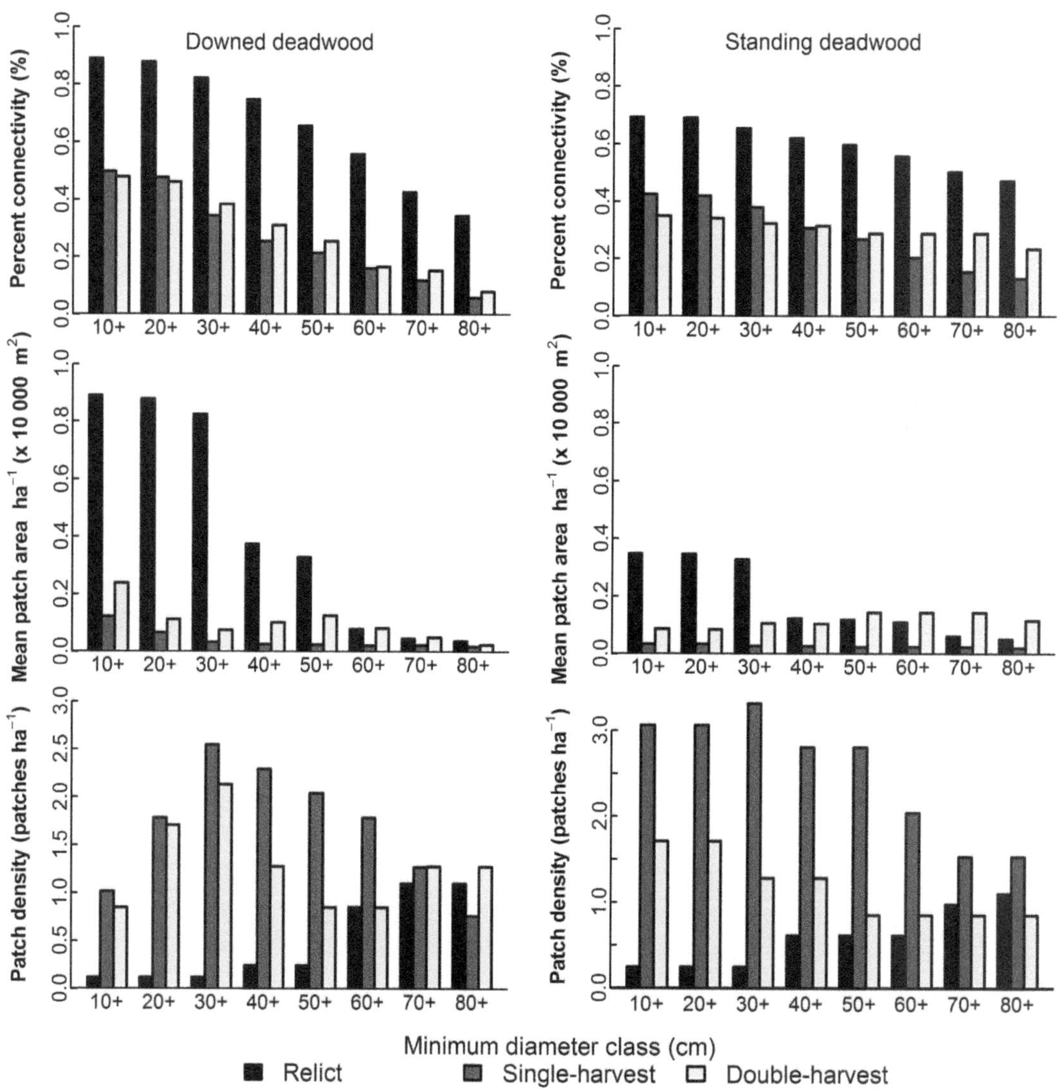

Figure 3. Coarse woody debris connectivity and patch-level summaries. Percent of all plots connected based upon presence of CWD (top row), mean patch area ha^{-1} (middle row), and number of patches ha^{-1} (bottom row) by 10 cm diameter classes for relict, single-harvest, and double-harvests stands. Downed deadwood (left column) diameter was based on large end diameter, and we used diameter at 1.37 m (DBH) for standing deadwood (right column); each diameter class indicates pieces of indicated size and larger were included in the analysis.

patches ha^{-1} generally shows an inverse relationship to mean patch area ha^{-1}, and the partial-harvest stands tended to have a greater number of smaller patches (Figures 4 and 5). Beyond a diameter of 60 cm for DDW and SDW, number of patches was more similar across all stands, although the single-harvest stand generally exhibited a greater number of patches than the other two stands. Although we used no metrics to compare overlay of DDW and SDW sample volume $plot^{-1}$, differences between stands were apparent (Figure 6). In the relict stand, plots with large SDW volumes tend to occur in areas with little or no DDW; we observed a similar pattern in the partial-harvest stands, but volumes of DDW and SDW throughout the stands were much lower. There were very few plots with no DDW or SDW sampled in the relict stand, but there are numerous patches throughout the partial-harvest stands with little or no DDW or SDW sampled.

Discussion

Almost half a century later, the effects of light partial harvesting on the type, amount, and spatial distribution of CWD remains evident. The relict forest had significantly larger pieces, greater volume, and higher connectivity of DDW and SDW compared to both partially harvested stands. Partial harvesting has altered species composition and structure relative to the relict stand [18], and the abundance and spatial distribution of CWD are a function of the resulting mortality and decomposition rates [23], [24].

Density-independent mortality related to windthrow, senescence, disease, insects, or some combination of these agents occurred in all stands, but they may affect species and/or different size trees in differing capacities [25]. For example, Dutch elm disease has been present in all stands since about the 1950s [26]. *Ulmus* trees, most of which were <20 cm DBH, were more abundant in the double-harvest stand (86 trees ha^{-1}), but were also common in the relict (41 trees ha^{-1}) and single-harvest stands (25 trees ha^{-1}). Given the relatively small size of *Ulmus* trees in the

Figure 4. Downed deadwood connectivity examples. Display of connected plots for downed deadwood (DDW) in relict, single-harvest, and double-harvests stands for pieces ≥30 cm (top row) and ≥60 cm (bottom row) based on large end diameter. Large dots indicate plots that sampled at least one DDW piece of the stated size, and small dots indicate plots with no DDW pieces of designated size sampled.

understory, they are a relatively ephemeral source of CWD throughout all stands with limited impact on long-term volumes and connectivity. In contrast, windthrow and senescence were likely the primary disturbance types in the relict stand given the higher density of larger trees compared to the partially harvested stands [18]. Relict stand CWD volume ha^{-1} was significantly higher than both partially harvested stands similar to several other studies [27], [28], [29]. This observation is presumably related to the changes in the size distribution of living trees, and the associated change in potential exogenous natural disturbances, particularly wind, related to tree size.

Removal of individual or small groups of large canopy trees allows for the release, growth and development of advance regeneration, or the establishment of new regeneration. These areas tend to have higher stem densities of smaller stems that will experience endogenous disturbances, particularly density-dependent mortality, as the stand develops rather than exogenous natural disturbances. Although these areas may contribute numerous CWD pieces, their cumulative low volume dictates a more ephemeral source of CWD because of faster rates of decomposition [2], [29]. Using a chronosequence approach, one study [30] indicated that CWD dynamics of managed stands were more unidirectional related to succession in contrast to virgin forests that exhibited a more cyclical pattern. In boreal mesic forests of southern Finland, researchers observed larger average diameter pieces of DDW and SDW in old-growth stands versus managed stands, with considerably more large-diameter (≥40 cm) SDW in old-growth stands and twice as many CWD stems < 10 cm in the managed stands compared to old-growth stands [27]. In boreal *Picea* forests of Sweden, significant differences in the availability of large diameter, highly-decayed CWD logs between selectively managed and semi-natural forests were observed, even after about 100 years [31]. In contrast, no differences were evident in DDW piece sizes in old-growth and selection cut stands of *A. saccharum*-dominated stands in Quebec, Canada, although SDW average diameter was lower in diameter-limit cuts [32]. Intermediate harvests can also influence species composition.

The overstory composition of the relict and partially harvested stands differed, which likely influenced decay rates and stand-level abundance of CWD. *Quercus* species dominated the overstory of all stands, but the relict stand *Quercus* basal area was 17.1 m^2 ha^{-1} compared to only 7.9 m^2 ha^{-1} and 8.6 m^2 ha^{-1} in the single-harvest and double-harvest stands, respectively [18]. The partially harvested stands had a higher abundance of *Acer*, *Carya*, and *Fraxinus* individuals, which replaced the harvested overstory *Quercus* trees. At an old-growth deciduous forest in Indiana,

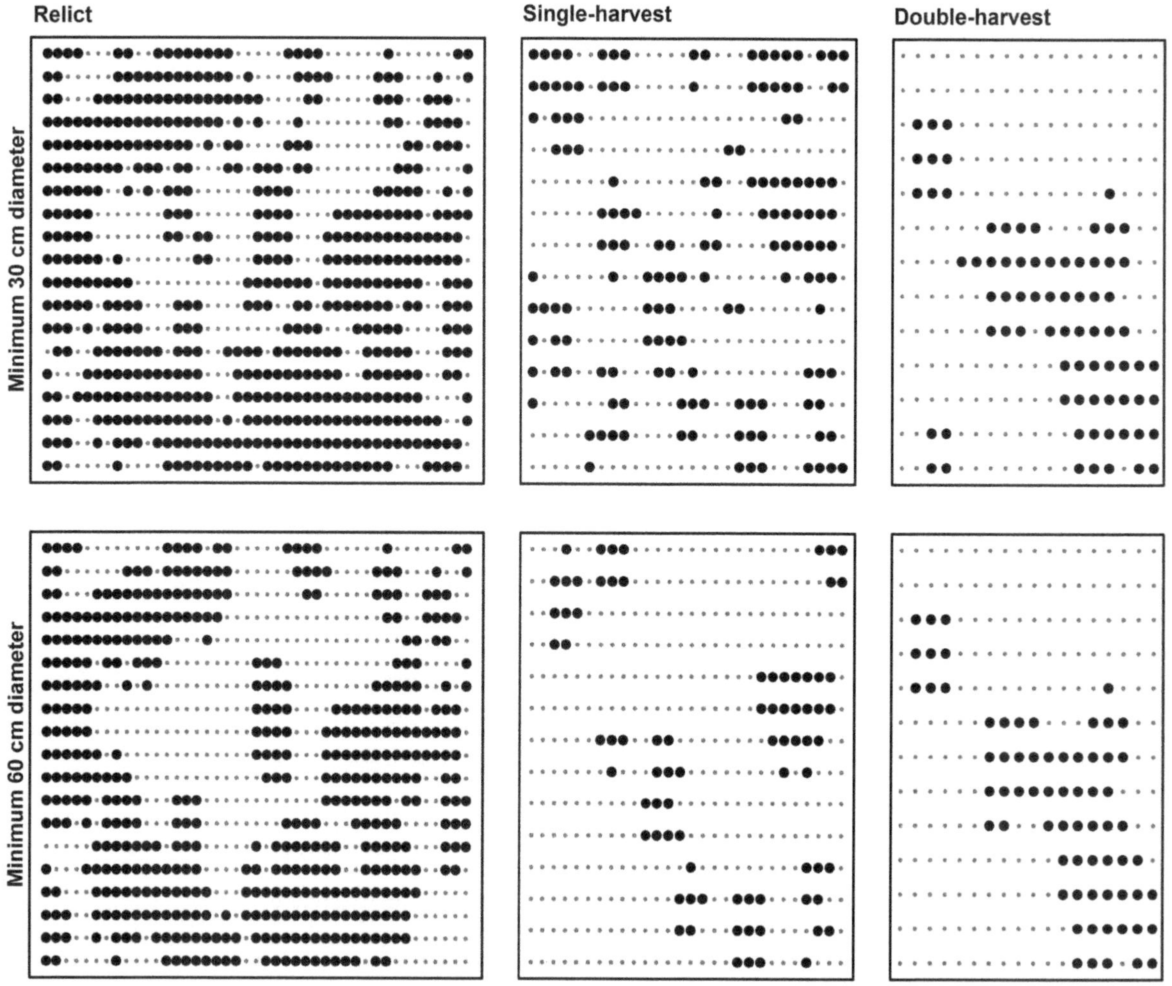

Figure 5. Standing deadwood connectivity examples. Display of connected plots for standing deadwood (SDW) in relict, single-harvest, and double-harvests stands for pieces ≥30 cm (top row) and ≥60 cm (bottom row) based on diameter at 1.37 m height. Large dots indicate plots that sampled at least one SDW piece of the stated size, and small dots indicate plots with no SDW pieces of designated size sampled.

U.S.A., it was predicted that 95% loss of CWD density of *Quercus*, *Carya*, and *Acer* species would occur at 171, 86, and 66 years, respectively [29]. Although we were unable to consistently identify CWD pieces to genus, the higher volumes of CWD within intermediate and high-decay stages likely reflects of the interaction of piece size and species composition within the relict stand; larger, more decay-resistant pieces will remain in situ compared to smaller pieces that decay quicker.

Rates of inputs (mortality events) and outputs (decomposition) are modified by management decisions, thus, CWD stores within a stand will also be modified. Assuming random mortality events with a 1% coefficient of variation over time [33], combined with slow decomposition rates related to large piece sizes and species composition, CWD can accumulate over time. This phenomenon may explain the aggregated patterns of CWD observed in one study [34] despite randomly dispersed mortality, as well as aggregated patterns of CWD reported for other old-growth forests [7], [10], [35]. Harvesting often alters rates of inputs and outputs, thus influencing potential CWD stores. One study observed that accumulation of CWD was slower after selective harvests, and that CWD levels had not returned to primeval forest levels even after 100 years [28]. Similarly, it has been proposed that selective logging operations interrupted the continuity of CWD inputs over time, thus influencing proportions of decaying logs and large CWD pieces [31]. Because there are fewer slow-decaying large pieces and a greater number of smaller CWD pieces derived from aggregated mortality events due to density-dependent competition within newly-formed harvest gaps, CWD is more ephemeral in selection harvests due to faster rates of decomposition of smaller pieces. Faster decomposition of CWD stores will result in lower volumes and decreased connectivity.

Decreased connectivity of DDW and SDW resulted in a larger number of smaller patches in the partially harvested stands, notably in diameter classes <60 cm. As minimum diameter class size increases, differences between all stands become minimized around 60 cm. Although this is partly an artifact of the methodology used whereby larger pieces are sampled more frequently across larger areas, connectivity, patch size, and number of patches were regulated by large diameter CWD. For taxa that can utilize smaller pieces, connectivity was still maintained in the in the partially harvested stands, being more than half of that observed in the relict stand, although the patch size and density were not comparable. Abundance of DDW and SDW sampled at each plot (Figure 6) indicated an inverse spatial relationship between the two pools of CWD; areas with abundant DDW tended to have low volumes of SDW and vice-versa. This

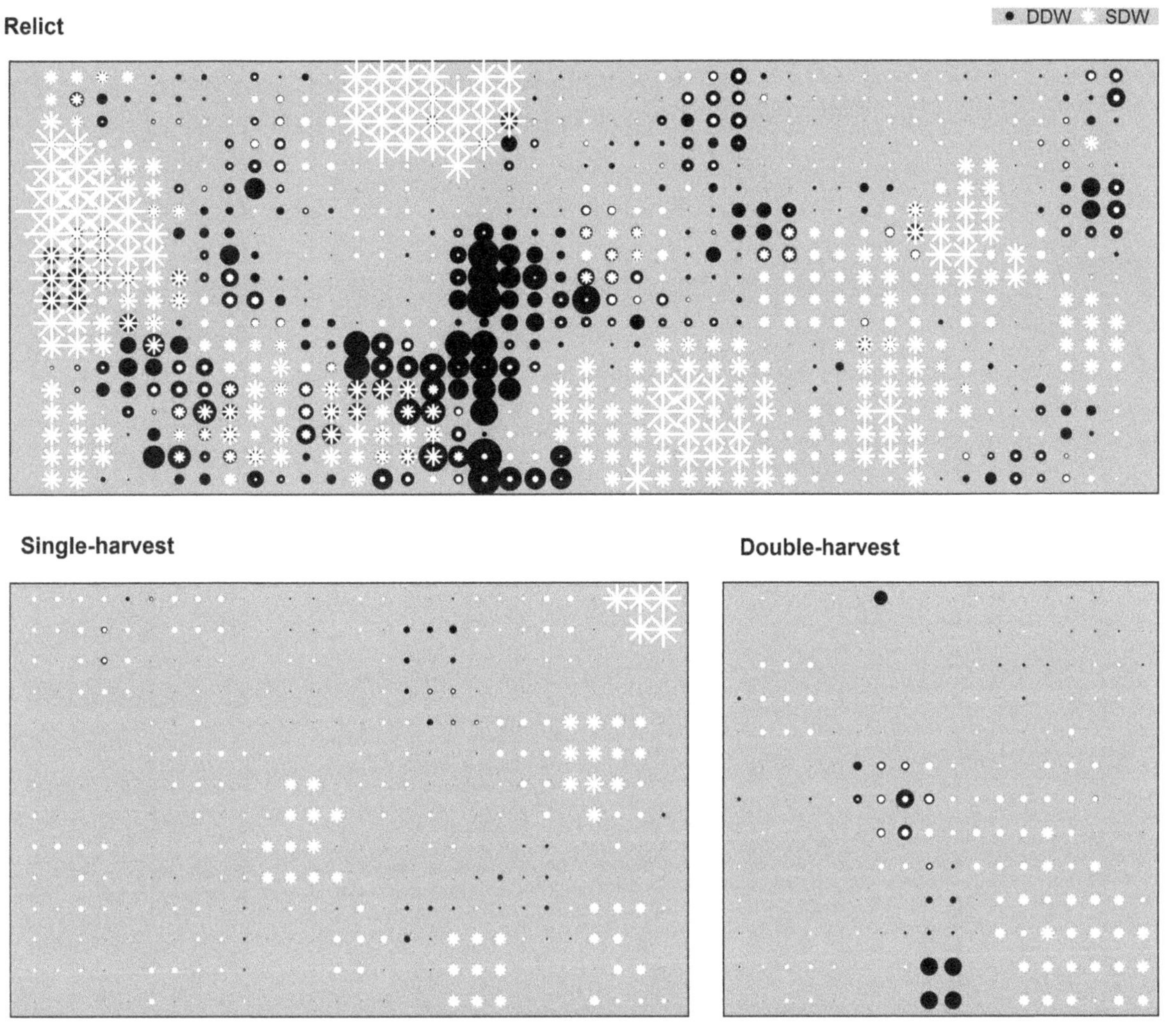

Figure 6. Coarse woody debris spatial distributions by volume. Each panel displays the volume (m^3) of downed deadwood (DDW; black circles) and standing deadwood (SDW; white stars) sampled at each plot location for relict, single-harvest, and double-harvests stands; size of symbol within each pane is standardized.

may suggest a dynamic relationship between the two pools whereby SDW eventually contribute to DDW pools and enhances connectivity. In the partially harvested stands, large patches with little or no CWD are indicative of areas that were previously harvested but now contain vigorous trees; many of these areas will be devoid of large CWD for the foreseeable future.

Traditional forest management practices can reduce CWD stores in forest stands for extended periods of time; differences in CWD connectivity and abundance among the three stands remained evident 46+ years after harvesting. Because the partially harvested stands have received no management since the initial harvest, this scenario differs from conventional uneven-aged, selection management in the region where most trees larger than 60 cm are harvested, stands are entered at short intervals, often 10–15 years, and removals tend to include declining individuals. This management approach reduces CWD inputs and drives reduced accumulation and connectivity across the stand. Extended rotation and entry times are often cited as ecologically viable harvest practices that promote greater heterogeneity in forest structure. However, these options may not be economically feasible or sufficient to preserve CWD pools in many management regimes. Several other studies have observed low levels of CWD long after active management operations ceased [28], [30], [31], and it may take a century or more for CWD abundance to return to the range of natural levels without adequate planning. Retaining individual and groups of trees within harvested stands can help preserve current and future CWD associated dispersal-limited taxa, and enhance biodiversity [36]. Using an "island" approach (*sensu* [37]), within stands, especially in fragmented forest landscapes, may be more appropriate to preserve dispersal-limited, smaller populations and metapopulation dynamics across the landscape akin to the "lifeboat hypothesis". We propose that differences in abundance and connectivity are attributable to decreased CWD stores in the managed stands. Although most studies of CWD focus on abundance within forests, understanding of CWD spatiotemporal dynamics may prove more critical to understand sustainability of CWD and its associated functions within forests.

Acknowledgments

We appreciate co-operation and logistical support from the staff of the Davis Purdue Agricultural Center. J. Dyer, K. Leffel, D. Luchik, A. Meier, M. Schuessler, J. Stark and C. Sullivan assisted with field data collection. Also thank you to B. Graham, J. Shields, and C. Zellers for critique and discussion in project development. We would also like to thank two anonymous reviewers for their comments and critiques to improve the manuscript.

Author Contributions

Conceived and designed the experiments: RCM MAJ MRS. Performed the experiments: RCM. Analyzed the data: RCM. Contributed reagents/materials/analysis tools: RCM MAJ MRS. Contributed to the writing of the manuscript: RCM MAJ MRS.

References

1. Franklin JF, Shugart HH, Harmon ME (1987) Tree death as an ecological process. Bioscience 37: 550–556.
2. Harmon ME, Franklin JF, Swanson FJ, Sollins P, Gregory SV, et al. (1986). Ecology of coarse woody debris in temperate ecosystems. Adv Ecol Res 15: 133–302.
3. Linden DW, Roloff GJ, Kroll AJ (2012) Conserving avian richness through structure retention in managed forests of the Pacific Northwest, USA. For Ecol Manage 284: 174–184.
4. Grove SJ (2002) Saproxylic insect ecology and the sustainable management of forests. Annu Rev Ecol Syst 33: 1–23.
5. Ylisirniö AL, Berglund H, Aakala T, Kuuluvainen T, Kuparinen AM, et al. (2009) Spatial distribution of deadwood and the occurrence of five saproxylic fungi in old-growth timberline spruce forests in northern Finland. Scand J For Res 24: 527–540.
6. Butts SR, McComb WC (2000) Associations of forest-floor vertebrates with coarse woody debris in managed forests of western Oregon. J Wildl Manage 64: 95–104.
7. Edman M, Jonsson BG (2001) Spatial Pattern of Downed Logs and Wood-Decaying Fungi in an Old-growth *Picea abies* Forest. J Veg Sci 12: 609–620.
8. Turner MG, Gardner RH, O'Neill RV (2001) Landscape ecology in theory and practice. New York: Springer-Verlag. 406 p.
9. Jönsson MT, Edman M, Jonsson BG (2008) Colonization and extinction patterns of wood-decaying fungi in a boreal old-growth *Picea abies* forest. J Ecol 96: 1065–1075.
10. Komonen A (2005) Local spatial pattern in the occurrence of two congeneric wood-decaying fungi in an old-growth boreal forest. Scand J For Res 20: 393–399.
11. McComb WC (2003) Ecology of coarse woody debris and its role as habitat for mammals. In: Zabel CJ, Anthony RG, editors. Mammal Community Dynamics: Management and Conservation in the Coniferous Forests of Western North America. New York: Cambridge University Press. 374–404.
12. Schiegg K (2000) Effects of deadwood volume and connectivity on saproxylic insect species diversity. Ecoscience 7: 290–298.
13. Laaka-Lindberg S, Korpelainen H, Pohjamo M (2006) Spatial distribution of epixylic hepatics in relation to substrate in a boreal old-growth forest. J Hattori Bot Lab 100: 311–323.
14. Manning AD, Cunningham RB, Lindenmayer DB (2013) Bringing forward the benefits of coarse woody debris in ecosystem recovery under different levels of grazing and vegetation density. Biol Conserv 157: 204–214.
15. Riffell S, Verschuyl J, Miller D, Wigley TB (2011) Biofuel harvests, coarse woody debris, and biodiversity – A meta-analysis. For Ecol Manage 261: 878–887.
16. Szwagrzyk J, Czerwczak M (1993) Spatial patterns of trees in natural forests of East-Central Europe. J Veg Sci 4: 469–476.
17. Aakala T, Kuuluvainen T, Grandpré LD, Gauthier S (2007) Trees dying standing in the northeastern boreal old-growth forests of Quebec: spatial patterns, rates, and temporal variation. Can J For Res 37: 50–61.
18. Morrissey RC (2013) Long-term disturbance histories of managed and natural mixed-hardwood forests of Central Indiana. Ph.D. Dissertation, Purdue University.
19. Gove JH, Ducey MJ, Ståhl G, Ringvall A (2001) Point relascope sampling: a new way to assess downed coarse woody debris. J For 99: 4–11.
20. Fraver S, Ringvall A, Jonsson BG (2007) Refining volume estimates of down woody debris. Can J For Res 37: 627–633.
21. Mesavage C, Girard JW (1946) Tables for estimating board foot volume of timber. USDA Forest Service, Washington, DC. 94.
22. R Core Team (2014) R: A language and environment for statistical computing. R Foundation for Statistical Computing, Vienna: Austria. ISBN 3-900051-07-0, URL: http://www.R-project.org/.
23. Siitonen J (2001) Forest Management, Coarse Woody Debris and Saproxylic Organisms: Fennoscandian Boreal Forests as an Example. Ecol Bull 49: 11–41.
24. Harmon ME (2009) Woody detritus mass and its contribution to carbon dynamics of old-growth forests: the temporal context. IN: Wirth C, Gleixner G, Heimann M, editors. Old-growth forests: function, fate and value – an overview old-growth forests (Ecological Studies). Berlin- Heidelberg: Springer. 159–190.
25. Turcotte RM, Elliott TR, Fajvan MA, Park YL, Snider DA, et al. (2012) Effects of ice storm damage on hardwood survival and growth in Ohio. North J Appl For 29: 53–59.
26. Parker GR, Leopold DJ (1983) Replacement of *Ulmus americana* L. in a mature east-central Indiana woods. Bull Torrey Bot Club 110: 482–488.
27. Siitonen J, Martikainen P, Punttila P, Rauh J (2000) Coarse woody debris and stand characteristics in mature managed and old-growth boreal mesic forests in southern Finland. For Ecol Manage 128: 211–225.
28. Sippola AL, Lehesvirta T, Renvall P (2001) Effects of Selective Logging on Coarse Woody Debris and Diversity of Wood-Decaying Polypores in Eastern Finland. Ecol Bull 49: 243–254.
29. MacMillan PC (1988) Decomposition of coarse woody debris in an old-growth Indiana forest. Can J For Res 18: 1353–1362.
30. Vandekerkhove K, De Keersmaeker L, Menke N, Meyer P, Verschelde P (2009) When nature takes over from man: Dead wood accumulation in previously managed oak and beech woodlands in North-western and Central Europe. For Ecol Manage 258: 425–435.
31. Bader P, Jansson S, Jonsson BG (1995) Wood-inhabiting fungi and substratum decline in selectively logged boreal spruce forests. Biol Conserv 72: 355–362.
32. Angers VA, Messier C, Beaudet M, Leduc A (2005) Comparing composition and structure in old-growth and harvested (selection and diameter-limit cuts) northern hardwood stands in Quebec. For Ecol Manage 217: 275–293.
33. Runkle JR (2000) Canopy tree turnover in old-growth mesic forests of eastern North America. Ecology 81: 554–567.
34. Aakala T, Fraver S, Palik BJ, D'Amato AW (2012) Spatially random mortality in old-growth red pine forests of northern Minnesota Can J For Res 42: 899–907.
35. Ylisirniö AL, Berglund H, Aakala T, Kuuluvainen T, Kuparinen AM, et al. (2009) Spatial distribution of deadwood and the occurrence of five saproxylic fungi in old-growth timberline spruce forests in northern Finland. Scand J For Res 24: 527–540.
36. Gustafsson L, Baker SC, Bauhus J, Beese WJ, Brodie A, et al. (2012). Retention Forestry to Maintain Multifunctional Forests: A World Perspective. Bioscience 62: 633–645.
37. MacArthur RH, Wilson EO (1967) The Theory of Island Biogeography. Princeton: Princeton University Press. 203.

Permissions

All chapters in this book were first published in PLOS ONE, by The Public Library of Science; hereby published with permission under the Creative Commons Attribution License or equivalent. Every chapter published in this book has been scrutinized by our experts. Their significance has been extensively debated. The topics covered herein carry significant findings which will fuel the growth of the discipline. They may even be implemented as practical applications or may be referred to as a beginning point for another development.

The contributors of this book come from diverse backgrounds, making this book a truly international effort. This book will bring forth new frontiers with its revolutionizing research information and detailed analysis of the nascent developments around the world.

We would like to thank all the contributing authors for lending their expertise to make the book truly unique. They have played a crucial role in the development of this book. Without their invaluable contributions this book wouldn't have been possible. They have made vital efforts to compile up to date information on the varied aspects of this subject to make this book a valuable addition to the collection of many professionals and students.

This book was conceptualized with the vision of imparting up-to-date information and advanced data in this field. To ensure the same, a matchless editorial board was set up. Every individual on the board went through rigorous rounds of assessment to prove their worth. After which they invested a large part of their time researching and compiling the most relevant data for our readers.

The editorial board has been involved in producing this book since its inception. They have spent rigorous hours researching and exploring the diverse topics which have resulted in the successful publishing of this book. They have passed on their knowledge of decades through this book. To expedite this challenging task, the publisher supported the team at every step. A small team of assistant editors was also appointed to further simplify the editing procedure and attain best results for the readers.

Apart from the editorial board, the designing team has also invested a significant amount of their time in understanding the subject and creating the most relevant covers. They scrutinized every image to scout for the most suitable representation of the subject and create an appropriate cover for the book.

The publishing team has been an ardent support to the editorial, designing and production team. Their endless efforts to recruit the best for this project, has resulted in the accomplishment of this book. They are a veteran in the field of academics and their pool of knowledge is as vast as their experience in printing. Their expertise and guidance has proved useful at every step. Their uncompromising quality standards have made this book an exceptional effort. Their encouragement from time to time has been an inspiration for everyone.

The publisher and the editorial board hope that this book will prove to be a valuable piece of knowledge for researchers, students, practitioners and scholars across the globe.

List of Contributors

Jeffrey A. Falke and Kristina M. McNyset
Department of Fisheries and Wildlife, Oregon State University, Corvallis, Oregon, United States of America

Jason B. Dunham
Forest and Rangeland Ecosystems Science Center, U.S. Geological Survey, Corvallis, Oregon, United States of America

Christopher E. Jordan
Northwest Fisheries Science Center, National Oceanographic and Atmospheric Agency, Seattle, Washington, United States of America

Gordon H. Reeves
Pacific Northwest Research Station, United States Forest Service, Corvallis, Oregon, United States of America

Kenneth F. Kellner and Robert K. Swihart
Department of Forestry and Natural Resources, Purdue University, West Lafayette, Indiana, United States of America

Robert C. Szava-Kovats and Meelis Pärtel
Institute of Ecology and Earth Sciences, University of Tartu, Tartu, Estonia

Shirley Saenz and Tomas Walschburger
Northern Andes Program, The Nature Conservancy, Bogota, Colombia

Juan Carlos González
Northern Andes Program, The Nature Conservancy, Quito, Ecuador

Jorge León
Latin America Program, The Nature Conservancy, Cartagena, Colombia

Bruce McKenney
Global Conservation Lands Program, The Nature Conservancy, Charlottesville, Virginia, United States of America

Joseph Kiesecker
Global Conservation Lands Program, The Nature Conservancy, Fort Collins, Colorado, United States of America

Darren Norris
Programa de Pós-Graduação em Biodiversidade Tropical, Universidade Federal do Amapá, Macapá, Amapá, Brazil
Coordenação de Biodiversidade, Instituto Nacional de Pesquisas da Amazônia, Manaus, Amazonas, Brazil

Marie-Josée Fortin
Department of Ecology & Evolutionary Biology, University of Toronto, Ontario, Canada

William E. Magnusson
Coordenação de Biodiversidade, Instituto Nacional de Pesquisas da Amazônia, Manaus, Amazonas, Brazil

Sara M. Santos and António Mira
CIBIO - Research Center in Biodiversity and Genetic Resources, University of Évora, Évora, Portugal

Rui Lourenço
LabOr - Laboratory of Ornithology, Instituto de Ciências Agrárias e Ambientais Mediterrânicas, University of Évora, Évora, Portugal

Pedro Beja
EDP Biodiversity Chair, CIBIO - Research Center in Biodiversity and Genetic Resources, University of Porto, Vairão, Portugal

Pedro Giovâni da Silva and Malva Isabel Medina Hernández
Programa de Pós-Graduação em Ecologia, Departamento de Ecologia e Zoologia, Universidade Federal de Santa Catarina, Florianó polis, Santa Catarina, Brazil

Adrián Leiva
Departamento de Ciencias Naturales, Universidad Técnica Particular de Loja (UTPL), Sección de Ecología, Loja, Ecuador

Carlos Iñiguez-Armijos
Departamento de Ciencias Naturales, Universidad Técnica Particular de Loja (UTPL), Sección de Ecología, Loja, Ecuador
Institute for Landscape Ecology and Resources Management (ILR), Research Centre for Bio Systems, Land Use and Nutrition (IFZ), Justus-Liebig-University Giessen, Giessen, Germany

Hans-Georg Frede and Lutz Breuer
Institute for Landscape Ecology and Resources Management (ILR), Research Centre for Bio Systems, Land Use and Nutrition (IFZ), Justus-Liebig-University Giessen, Giessen, Germany

Henrietta Hampel
Universidad de Cuenca, Quinta Balzaín, Cuenca, Ecuador

Lluís Brotons
CEMFOR-CTFC, Forest Sciences Center of Catalonia, Solsona, Catalonia, Spain
CREAF, Centre for Ecological Research and Forestry Applications, Cerdanyola del Vallés, Catalonia, Spain

María José Madeira and Benjamin J. Gómez-Moliner
Department of Zoology and Animal Cell Biology, University of the Basque Country, UPV/EHU, Vitoria-Gasteiz, Spain
Systematics, Biogeography and Population Dynamics Research Group, Lascaray Research Center, University of the Basque Country, UPV/EHU, Vitoria-Gasteiz, Spain

Aritz Ruiz-Gonzàlez
Department of Zoology and Animal Cell Biology, University of the Basque Country, UPV/EHU, Vitoria-Gasteiz, Spain
Systematics, Biogeography and Population Dynamics Research Group, Lascaray Research Center, University of the Basque Country, UPV/EHU, Vitoria-Gasteiz, Spain
Conservation Genetics Laboratory, National Institute for Environmental Protection and Research, ISPRA, Ozzano dell'Emilia, Bologna, Italy

Mikel Gurrutxaga
Systematics, Biogeography and Population Dynamics Research Group, Lascaray Research Center, University of the Basque Country, UPV/EHU, Vitoria-Gasteiz, Spain
Department of Geography, University of the Basque Country, UPV/EHU, Vitoria-Gasteiz, Spain

Samuel A. Cushman
U.S. Forest Service, Rocky Mountain Research Station, Flagstaff, AZ, United States of America

Ettore Randi
Conservation Genetics Laboratory, National Institute for Environmental Protection and Research, ISPRA, Ozzano dell'Emilia, Bologna, Italy
Department 18/Section of Environmental Engineering, Aalborg University, Aalborg, Denmark

Jeremy W. Fox, Geoff Legault and Jodie A. Einarson
Department of Biological Sciences, University of Calgary, Calgary, Alberta, Canada

David A. Vasseur
Department of Ecology and Evolutionary Biology, Yale University, New Haven, Connecticut, United States of America

Claire D. Stevenson-Holt
Centre for Wildlife Conservation, University of Cumbria, Ambleside, Cumbria, United Kingdom

Kevin Watts
Centre for Ecosystems, Society and Biosecurity, Forest Research, Farnham, Surrey, United Kingdom

Chloe C. Bellamy
Centre for Ecosystems, Society and Biosecurity, Forest Research, Roslin, Midlothian, United Kingdom

Owen T. Nevin
School of Medical and Applied Sciences, Central Queensland University, Gladstone, Queensland, Australia

Andrew D. Ramsey
School of Biological and Forensic Sciences, University of Derby, Derby, Derbyshire, United Kingdom

Samantha Low-Choy and Kerrie Mengersen
School of Mathematical Sciences, Queensland University of Technology, Brisbane, Queensland, Australia

Julie Vercelloni
School of Mathematical Sciences, Queensland University of Technology, Brisbane, Queensland, Australia
Australian Institute of Marine Science, Townsville, Queensland, Australia

M. Julian Caley
Australian Institute of Marine Science, Townsville, Queensland, Australia

Mohsen Kayal
Bren School of Environmental Science and Management, University of California Santa Barbara, Santa Barbara, California, United States of America

James R. Welch and Carlos E. A. Coimbra Jr
Escola Nacional de Saúde Pública, Fundação Oswaldo Cruz, Rio de Janeiro, Rio de Janeiro, Brazil

Eduardo S. Brondízio
Department of Anthropology, Indiana University, Bloomington,Indiana, United States of America
Anthropological Center for Training and Research on Global Environmental Change, Indiana University, Bloomington, Indiana, United States of America

Scott S. Hetrick
Anthropological Center for Training and Research on Global Environmental Change, Indiana University, Bloomington, Indiana, United States of America

Dennis Rödder
Zoological Research Museum Alexander Koenig, Bonn, Germany

Jan O. Engler
Zoological Research Museum Alexander Koenig, Bonn, Germany
Department of Wildlife Sciences, University of Göttingen, Göttingen, Germany

Niko Balkenhol
Department of Wildlife Sciences, University of Göttingen, Göttingen, Germany

Katharina J. Filz
Department of Biogeography, Trier University, Trier, Germany
Museum of Natural History Dortmund, Dortmund, Germany

Jan C. Habel
Department of Ecology and Ecosystemmanagement, Technical University Munich, Freising-Weihenstephan, Germany

Juan Carlos Castilla, Alejandro Delgado, Verónica Ortiz and Manuel Varas
Centro de Conservación Marina, Estación Costera de Investigaciones Marinas, Las Cruces, Facultad de Ciencias Biológicas, Pontificia Universidad Católica de Chile, Santiago, Chile

Patricio H. Manríquez and María Elisa Jara
Laboratorio de Ecología y Conducta de la Ontogenia Temprana (LECOT), Centro de Estudios Avanzados en Zonas Àridas (CEAZA), Coquimbo, Chile

William T. Bean
Humboldt State University, Arcata, California, United States of America

Robert Stafford
California Department of Fish and Game, Los Osos, California, United States of America

H. Scott Butterfield
The Nature Conservancy, San Francisco, California, United States of America

Justin S. Brashares
Department of Environmental Science, Policy and Management, University of California, Berkeley, California, United States of America

Stuart C. Brown, Rebecca E. Lester and Laurie Laurenson
School of Life and Environmental Sciences, Deakin University, Warrnambool, Victoria, Australia

Vincent L. Versace
Greater Green Triangle University, Department of Rural Health, Flinders University, Warrnambool, Victoria, Australia
Centre for Environmental Management, Faculty of Science and Technology, Federation University Australia, Ballarat, Victoria, Australia

Jonathon Fawcett
Sinclair-Knight-Merz, Bendigo, Victoria, Australia

Diego Alvarez-Berastegui
Balearic Islands Coastal Observing and Forecasting System (SOCIB), Palma de Mallorca, Balearic Islands, Spain

Lorenzo Ciannelli
College of Earth, Ocean, and Atmospheric Sciences (CEOAS), Oregon State University, Corvallis, Oregon, United States of America

Alberto Aparicio-Gonzalez, Patricia Reglero, Manuel Hidalgo, Jose Luis López-Jurado and Francisco Alemany
Instituto Español de Oceanografía - Centre Oceanográfic de les Balears (COB-IEO), Palma de Mallorca, Balearic Islands, Spain

Joaquín Tintoré
Balearic Islands Coastal Observing and Forecasting System (SOCIB), Palma de Mallorca, Balearic Islands, Spain
Institute for Advanced Studies (IMEDEA), Consejo Superior de Investigaciones Científicas (CSIC) and the University of the Balearic Islands (UIB). Esporles, Spain

Marta Rodríguez and Miguel Delibes
Departament of Conservation Biology. Estación Biológica de Doñana (CSIC), Sevilla, Spain

José Ma. Fedriani
Departament of Conservation Biology. Estación Biológica de Doñana (CSIC), Sevilla, Spain
Department of Ecological Modelling. Helmholtz Centre for Environmental Research GmbH-UFZ, Leipzig, Germany
Centre for Applied Ecology "Prof. Baeta Neves", Institute Superior of Agronomy (University of Lisbon), Lisbon, Portugal

Benjamin Blonder
Sky School, University of Arizona, Tucson, Arizona, United States of America

Lindsey Sloat
Department of Ecology and Evolutionary Biology, University of Arizona, Tucson, Arizona, United States of America

Brian J. Enquist
Department of Ecology and Evolutionary Biology, University of Arizona, Tucson, Arizona, United States of America
Santa Fe Institute, Santa Fe, New Mexico, United States of America

Brian McGill
School of Biology and Ecology, University of Maine, Orono, Maine, United States of America

Liam P. Crowther and Andrew F. G. Bourke
School of Biological Sciences, University of East Anglia, Norwich Research Park, Norwich, Norfolk, United Kingdom

Pierre-Louis Hein
School of Biological Sciences, University of East Anglia, Norwich Research Park, Norwich, Norfolk, United Kingdom
IUT Nancy-Brabois, Rue du Doyen Urion, Villers-les- Nancy CS, France

Robert C. Morrissey
Department of Forest Ecology, Faculty of Forestry and Wood Sciences, Czech University of Life Sciences, Prague, Czech Republic
Hardwood Tree Improvement and Regeneration Center, Department of Forestry and Natural Resources, Purdue University, West Lafayette, Indiana, United States of America

Michael A. Jenkins and Michael R. Saunders
Hardwood Tree Improvement and Regeneration Center, Department of Forestry and Natural Resources, Purdue University, West Lafayette, Indiana, United States of America

Index